Microbial Diversity, Interventions and Scope

Shiwani Guleria Sharma •
Neeta Raj Sharma • Mohit Sharma
Editors

Microbial Diversity, Interventions and Scope

Springer

Editors
Shiwani Guleria Sharma
School of Bioengineering and Biosciences
Lovely Professional Univeristy
Phagwara, Punjab, India

Neeta Raj Sharma
School of Bioengineering and Biosciences
Lovely Professional Univeristy
Phagwara, Punjab, India

Mohit Sharma
Molecular Genetics Laboratory
Dayanand Medical College and Hospital
Ludhiana, Punjab, India

ISBN 978-981-15-4098-1 ISBN 978-981-15-4099-8 (eBook)
https://doi.org/10.1007/978-981-15-4099-8

This Springer imprint is published by the registered company Springer Nature Singapore Pte Ltd.
The registered company address is: 152 Beach Road, #21-01/04 Gateway East, Singapore 189721, Singapore

Dedicated to budding microbiologists

Preface

The book is a comprehensive reference book containing information on relationships of microbes with different fields of science. Main concern of scientific society is to understand how microorganisms effect environment. Recycling of natural resources and development of recombinant DNA technology, nanotechnology, bioremediation, and industries bring microbiology in limelight. The chapters of this book focus on microbial ecosystem and exploitation of microbes and their activity for human benefit. Each portion of this book emphasizes on new fields with essays highlighting introduction and important discoveries and developments in microbiology.

The book contains contributions from many experts and will be a valuable resource for undergraduate and graduate students, doctoral scholars, scientists, and researchers associated with microbiology. This book will be beneficial for academicians as many universities throughout world have microbiology as a subject that cannot be completed without understanding diversity, interventions, and scope of microbes. The book will be beneficial for graduate students to understand the relationship of microbes in various fields, that is, agriculture, nanotechnology, genetic engineering, medical science, forensic science, and space. The book will also be beneficial to researchers and scientists as they can understand the impact of and explore new opportunities in this field.

Phagwara, Punjab, India — Shiwani Guleria Sharma
Phagwara, Punjab, India — Neeta Raj Sharma
Ludhiana, Punjab, India — Mohit Sharma

Contents

About the Editors

Shiwani Guleria Sharma is an Assistant Professor at the School of Bioengineering and Biosciences at Lovely Professional University, Punjab, India. She holds a PhD in Microbiology from DR Yashwant Singh Parmar University of Horticulture and Forestry, Himachal Pradesh. Her research interests include agricultural microbiology, microbial enzymology and molecular biology. She has published articles in national and international, peer-reviewed journals. She has received various awards, including INSPIRE fellowship by DST-GOI, Gold Medal for doctorate from DR YSP UHF, H.P., state merit scholarships and a best research award from the Society for Advancement of Human and Nature, H.P.

Neeta Raj Sharma holds a Ph.D. in Biochemistry from Jiwaji University Gwalior, and heads the School of Bioengineering and Biosciences as Additional Dean at Lovely Professional University, Punjab. She has experience in multidisciplinary areas such as microbial biochemistry and biotechnology, and waste management. She has been instrumental in conducting various international funded research projects. Dr. Sharma is a member of a number of societies, including the Association of Biotechnology and Pharmacy, Indian Science Congress, Association of Microbiologists of India, Indo-US Collaboration of Engineering Education, International Science Congress Association, and the Indian Society of Agricultural Biochemists. She has published over 55 articles in respected journals and book chapters, and holds 20 patents.

Mohit Sharma is an Associate Biotechnologist at the Molecular Genetics Laboratory at Dayanand Medical College and Hospital, Ludhiana, Punjab. His interests include human genetics and cytogenetics, molecular plant microbial interactions, vector cloning, bioinformatics, molecular biology, and cell biology. He holds a Ph.D. in Biotechnology from Shoolini University of Biotechnology and Management Sciences, Himachal Pradesh, India. He has also worked on the isolation and characterization of hypovirulence-associated mycovirus from Fusarium species from apple orchards in Himachal Pradesh. He has published research articles in various national and international, peer-reviewed journals.

Abbreviations

AMF	Arbuscular mycorrhizal fungi
BOD	Biological oxygen demand
CAED	Computer Aided Enzyme Design
CDC	Centre for Disease Control
COD	Chemical oxygen demand
DGGE	Denaturing gradient gel electrophoresis
FBI	Federal Bureau of Investigation
FMT	Fecal microbial transplantation
GEMS	Genetically Engineered Microorganisms
HMP	Human Microbiome Project
ISR	Induced systemic resistance
ISS	International Space Station
ITS	Internal transcribed spacer
MPS	Massive parallel sequencing
MS	Mass spectrometry
NAATs	Nucleic acid amplification tests
NSG	Next generation sequencing
PMA	Propidium monoazide
PNAS	Proceedings of National Academies of Science
SCFAs	Short chain fatty acids
SDM	Site-directed mutagenesis
SOM	Soil organic matter
SRB	Sulfates reducing bacteria
STDs	Sexually transmitted diseases

Microbial Ecosystem and Anthropogenic Impacts

1

Lalita Vithal Baragi, Dhiraj Dhondiram Narale, Sangeeta Mahableshwar Naik, and K. M. Rajaneesh

Abstract

Oceans are the most vulnerable sites for anthropogenic waste from domestic as well as industrial origin. Usually, marine ecosystems are exposed to most anthropogenic stressors ranging from sewage disposal to nuclear waste contaminants. Most recent threats to marine ecosystems are ocean warming and ocean acidification (related to anthropogenic emission of CO_2), oil (tarball), and (micro) plastic contamination, which is proved to have a devastating impact on the marine ecosystem. Microbes are abundantly present in marine ecosystems playing essential roles in ecosystem productivity and biogeochemistry. Generally, microbial communities are the initial responders of these stressors. Altered microbial communities in response to these stressors can, in turn, have adverse impact on the marine ecosystem and later on humans. In this review, we highlight the effect of oil pollution, microplastics, and increased CO_2 on the marine microbial ecosystem. The information on the impacts of such stressors on microbial communities will be valuable to formulate appropriate remediation approaches for future use.

L. V. Baragi (✉)
Applied and Environmental Biotechnology Laboratory, Department of Biological Sciences, BITS, Pilani, KK Birla Goa Campus, Sancoale, Goa, India
e-mail: 2017proj21@goa.bits-pilani.ac.in

D. D. Narale
CSIR-National Institute of Oceanography, Dona Paula, Goa, India

S. M. Naik
National Center for Coastal Research, Government of India, Ministry of Earth Science, Chennai, India

K. M. Rajaneesh
King Fahd University of Petroleum and Minerals, Dhahran, Saudi Arabia

S. G. Sharma et al. (eds.), *Microbial Diversity, Interventions and Scope*,
https://doi.org/10.1007/978-981-15-4099-8_1

Keywords

Microbial ecosystem · Anthropogenic stressors · Oil pollution · Microplastics · Ocean warming · Ocean acidification

1.1 Introduction

It is well known that two-thirds of planet Earth is covered by marine waters (Charette and Smith 2010). These waters have a significant role in the global biogeochemical cycles, which sustains the life in the ocean (Schlesinger 1997; Sarmiento and Gruber 2006). Even though the standing crop of marine ecosystems represents 1% of the terrestrial biomass, it contributes to approximately half of the biomass produced on Earth (Gruber et al. 2009; Hader et al. 2011). Among the marine ecosystems, one-half of global primary production occurs in the oceans (Falkowski et al. 1998; Field et al. 1998). Oceans also control the climate and weather pattern. Therefore, the ocean has a significant effect on the biosphere and much of life on Earth.

In recent years, exploitation of the ocean by human-derived activities such as fishing, tourism, oil exploration, maritime transport, and industrial activities has a substantial impact on the marine ecosystem (Nogales et al. 2011; Halpern et al. 2007, 2008). These activities affect different trophic levels of marine food web, which comprises microorganisms to animal predators. According to a recent study, the vast region of the world ocean is forecasted to have medium to high impact from these stressors (Halpern et al. 2008). Most recent stressors to these ecosystems are ocean warming and ocean acidification and oil (tarball) and microplastic contamination, which are proved to have a devastating impact on the marine ecosystem, mainly biology of the ecosystem.

Ocean warming and ocean acidification are the result of increasing concentration of atmospheric CO_2 (Caldeira and Wickett 2003; Orr et al. 2005) and are now being recognized as a major responsible factor for change in the biological system of the oceans (Lovejoy and Hannah 2005). Presently, the concentration of atmospheric CO_2 has reached 400 ppm from 280 ppm from preindustrial revolution (NOAA/ESRL; Stocker et al. 2013) with ~0.5% year^{-1} rising rate (Forster et al. 2007). Approximately one-third of the anthropogenic CO_2 generated is absorbed by the oceans and will help moderate future climate change (Sabine et al. 2004). This resulted in a decrease in pH by 0.1 unit (referred to as ocean acidification) and rise in temperature by 0.85 °C (referred to as ocean warming) (Raven et al. 2005). By the end of this century, the concentration of atmospheric CO_2 is predicted to reach 800–1000 μatm by the "business as usual" CO_2 emission scenario climate models, which will further decrease the pH (0.3–0.4 units) and increase the temperature (1–4 °C) (Stocker et al. 2013; Caldeira and Wickett 2003). Elevated CO_2 concentration will result in an increase in H^+ concentration (100–150%), which will negatively affect the marine organisms, especially calcifying organisms (Haugan and Drange 1996; Brewer 1997).

Oil pollution is another threat to the marine environment presently. Oil spills mainly arise from either accidents or oily discharges from ships (Solberg 2012). Operational discharges from tankers cause the majority of the oil pollution cases. The effect of the spills on the marine ecosystem depends on several factors such as the quantity and quality of spilled oil, the sensitivity of the organisms exposed to the oil, location, depth, season, and meteorological and oceanic conditions (Fukuyama et al. 1998). These oil spills mainly have negative consequences on the ecology of the marine ecosystem (Fukuyama et al. 1998).

To some extent, oceans are also used as dumping sites for debris from human activities. Marine debris comprised of manufactured solid material, of which 60–80% consists of plastic (Gregory and Ryan 1997). According to a study, an estimated 1.3 plastic items can be found for every m^2 of shoreline worldwide, which is believed to be a significant threat to the marine ecosystem (Bravo et al. 2009). More than 267 species worldwide are impacted by this debris either by ingestion or entanglement (Gall and Thompson 2015). Plastics having size range between 333 μm and 5 mm are called microplastics. Smaller particles (<1 μm) also exist in marine waters but are less often detected (Arthur et al. 2009). The most commonly found micro-debris particles include polyethylene, polypropylene, and polystyrene (Andrady 2011). These are ubiquitous in marine environments. However, the harmful effects of microplastic on the marine food web are less known (Derraik 2002). Possible threats to the marine food web may include physical harm from ingestion, leaching of toxic additives, and desorption of persistent, bioaccumulative, and toxic chemicals (Nobre et al. 2015). Current literature revealed that some planktons and many classes of invertebrates and vertebrates are known to ingest and accumulate microplastics, as the size of microplastic falls in the same size range as their natural food (Wright et al. 2013).

Ocean warming, ocean acidification, oil (tarball) pollution and microplastic contamination have instant and long-lasting effects on marine organisms, from basic to complex life forms and from the cellular to the community levels. Keeping in mind the extent and the tenacity of the impact from these threats, workable and effective methods for remediation are essential. This chapter delivers a synopsis of the cause of the three key recent threats to marine environments, ecological processes, and transformation in the marine food web. Prospects for possible solutions are also discussed.

1.2 Anthropogenic Impact on Marine Organisms

1.2.1 Rising Atmospheric Carbon Dioxide: Ocean Acidification and Warming

Since the industrial revolution, due to rapid industrialization, the carbon dioxide concentration (pCO_2) has increased from 280 ppm to the present level of 400 ppm (NOAA/ESRL; Stocker et al. 2013). This has resulted in 0.1 unit reduction in pH (referred to as ocean acidification) and 0.85 °C rise in seawater temperature (referred to as ocean warming) (Raven et al. 2005). By the end of this century, the concentration of atmospheric CO_2 is predicted to reach 800–1000 μatm by the "business as usual" CO_2 emission scenario climate models, which will further decrease the pH (0.3–0.4 units) and increase the temperature (1–4 °C) (Stocker et al. 2013; Caldeira and Wickett 2003). Significant warming and acidification incidents have occurred in the earth's history, resulting in considerable changes in marine communities (Pelejero et al. 2010; Hönisch et al. 2012) and numerous mass extinctions (Pelejero et al. 2010; Clarkson et al. 2015). Though these acidification and warming incidents are not the same as present as their pace was much slower compared to the present (Pelejero et al. 2010). Thus, based on past histories, we cannot forecast the impacts of acidification and warming on marine organisms, and hence in recent years, there is increasing research in this area.

In marine organisms, ocean acidification affects their metabolism, acid-base balance, and calcification (Chan et al. 2012; Pörtner et al. 2004; Pörtner 2008; Ries et al. 2009; Nilsson et al. 2012; Andersson and Gledhill 2013; Lane et al. 2013; Mostofa 2016). The marine organisms show species-specific (Hendriks et al. 2010; Kroeker et al. 2010, 2013; Harvey et al. 2013) and habitat-specific response (Andersson et al. 2008; Clark et al. 2009) to ocean acidification. Bacterial community structure and diversity is also known to change in acidified conditions (Allgaier et al. 2008; Kerfahi et al. 2014; Witt et al. 2011; Lidbury et al. 2012). Phytoplankton may either benefit from rising pCO_2 or be affected by the related reduction in pH depending on species (Gao and Campbell 2014; Torstensson et al. 2015). In phytoplankton, acidification influences their growth, energy allocation, photosynthesis, calcification, carbon acquisition, cellular fluxes, particulate carbon production, elemental composition, and biochemical composition (Rost et al. 2006; Rickaby et al. 2010; Sett et al. 2014; Bautista-Chamizo et al. 2016; Jin and Gao 2016; Kottmeier et al. 2016). Phytoplankton shows species-specific response to acidification such as positive, neutral, and negative (Gao et al. 2012; Johnson et al. 2013; Baragi and Anil 2016; Jin and Gao 2016). Such variation in response might be due to variation in carbon-concentrating mechanism (CCM), which uses high metabolic energy to transport and convert HCO_3^- into CO_2 around RuBisCO in the cell under CO_2-limited condition (Raven et al. 2008; Reinfelder 2011). Under the acidified condition, the elevated CO_2 downregulates CCM, thus reducing the energy demand of the cell (Burkhardt et al. 2001; Beardall and Raven 2004; Spijkerman 2008; Holtz et al. 2015; Wu et al. 2015).

A meta-analysis reported the effect of moderate acidified condition (936 μatm) on all animal taxa (corals, echinoderms, mollusks, crustaceans, fishes) with more significant impact on those having weak acid-base regulation abilities and calcified structures (corals, echinoderms, and mollusks); however, crustaceans are relatively unaffected (Wittmann and Pörtner 2013). Compared to non-calcifying organisms, calcifying organisms are highly susceptible to acidification (Hendriks et al. 2010; Kroeker et al. 2010; Byrne and Przeslawski 2013); however, few calcifying organisms are robust to acidification either due to their adaptive capacity to naturally acidic habitats (Talmage and Gobler 2011) or due to the higher buffering ability of local waters (Range et al. 2012). Recently, it has been observed that ocean acidification affects some non-calcifying organisms (Wage et al. 2016; Borges et al. 2018). Invertebrates are capable of tolerating acidification by investing more energy in compensatory mechanisms like maintenance of acid-base homeostasis rather than basic mechanisms like growth and reproduction (Pörtner 2008; Melzner et al. 2009; Kroeker et al. 2013; Xu et al. 2016).

On the other hand, warming significantly affects growth and metabolism of marine organisms (Eppley 1972; Brown et al. 2004), consequently changing their abundance and distribution (Thomas et al. 2004; Harley et al. 2006). Organisms show species-specific response to temperature and rely on the organism's thermal tolerance window and its capability to adapt to varying temperature. Beyond these limits, the rising temperature can impose physiological stress and a decline in biochemical and metabolic processes such as development and growth in the organisms (Pörtner and Farrell 2008; Poloczanska et al. 2014). Under warming situations, organisms might display poleward migrations as organisms generally adapt to warming by shifting to regions with optimal temperature (Nguyen et al. 2012; Kamya et al. 2014; Poloczanska et al. 2014). Moreover, warming is also predicted to have severe negative impacts on tropical species than temperate as the former species have already comparatively small thermal windows and are usually living near their maximum thermal limit (Stillman and Paganini 2015). Phytoplankton shows enhanced photosynthesis, growth, calcification, and reduction in size in response to warming (De Bodt et al. 2010; Müller et al. 2014; Sett et al. 2014). In natural plankton community, warming increased phytoplankton abundance (Lewandowska et al. 2014) and changed the distribution of phytoplankton groups (Thomas et al. 2012). Warming is the significant reason for the speedy reduction (~1% yearly) of phytoplankton biomass globally (Boyce et al. 2010), with diatoms being the significantly affected group (Toseland et al. 2013). Further, warming is forecasted to cause a reduction in tropical phytoplankton diversity and a poleward shift in the species' thermal niches (Thomas et al. 2012).

In the future climatic scenario, both acidification and warming are known to co-occur. Some studies reported positive synergistic effect of these stressors on some species of phytoplankton wherein it caused an increase in growth and repair rate of UV-damaged PSII machinery of microalgae (Connell and Russell 2010; Fiorini et al. 2011; Li et al. 2012). However, other studies reported the insignificant synergistic effect of these stressors on cyanobacteria (Fu et al. 2007; Hutchins et al. 2007) and coccolithophores (De Bodt et al. 2010). Some microalgal species showed temperature-dependent response to acidification, wherein the optimum temperature for the growth, carbon fixation, and calcification increased under elevated pCO_2 in contrast to ambient pCO_2 concentration (Sett et al. 2014). Our understanding of the effect of acidification and warming is limited only to single species of microalgae. However, few studies have been carried out to know the response of the natural microalgal community to these stressors (Kim et al. 2006; Calbet et al. 2014; Sommer et al. 2015). Moreover, very limited snapshots are available for prolonged adaptation of microalgae to acidification (Lohbeck et al. 2012; Jin et al. 2013; Low-Décarie et al. 2013; Jin and Gao 2016). Marine invertebrates show species-specific response to the synergistic effect of acidification and warming. Acidification is known to shorten the thermal window of some species, thus making them highly susceptible to warming (Schalkhausser et al. 2013). However, in other species, there was no such effect observed (Zittier et al. 2015).

In invertebrates, compared to adults, the larval and initial life phases of invertebrates are severely susceptible to acidification and warming (Dupont et al. 2010a; Kroeker et al. 2013). Thus, it is crucial to emphasize on the effects of altering environmental conditions on larval phases as they signify a bottleneck for populations that try to tolerate the altering environmental conditions (Havenhand et al. 2008; Dupont et al. 2010b). Any influence on larval development and growth will show a high impact on population. The larval sensitivity to these stressors differs with individuals and taxa because of the variation in the maternal nutritional and energetic investment and history (Byrne et al. 2009; Przeslawski and Webb 2009; Donelson et al. 2012). Some studies have observed significant "carryover" effects from one life stage to another (transgenerational effect) (Kurihara 2008; Putnam and Gates 2015; Manno et al. 2016; Borges et al. 2018). For example, Parker et al. (2015) reported that larvae and juveniles of barnacle were able to survive better under acidified conditions due to positive carryover effects from an adult. Thus, it is essential to investigate the impact of these stressors on different life stages through parental and transgenerational effects.

Acidification and warming may indirectly drive ecological change through biotic interactions, which are the crucial "pressure point" (Gaylord et al. 2015). Studies also discovered that most remarkable impacts of acidification and warming would arise through changed species interactions (Rossoll et al. 2012; De Kluijver et al. 2013; Poore et al. 2013; Kroeker et al. 2014). Recently, it is observed that food supply alleviates the negative effects of these stressors on marine invertebrates (Melzner et al. 2011; Thomsen et al. 2013; Asnaghi et al. 2014; Pansch et al. 2014; Uthicke et al. 2015, 2016; Ramajo et al. 2016).

1.2.2 Oil (Petroleum/Tarball) Pollution

Petroleum is one of the common contaminants in the aquatic environment. As a consequence of rising global demand for energy, there is increased crude oil exploration and transportation in the marine environment, thus making them vulnerable to crude oil pollution (National Research Council 2003). Oil spill as a result of accidents or discharge of ballast waters is a common occurrence nowadays. On the contrary to the common belief, even the small oil spills and their repetitive nature can have an immediate adverse biological impact on marine biota, thereby affecting the marine ecosystem functioning (Brussaard et al. 2016). In addition to the tanker-derived oil pollution, land-derived inputs due to urbanization and industrialization coupled with domestic petroleum production also contribute to coastal oil pollution (Zakaria et al. 2000). The compounds such as isoprenoid alkanes, steranes, hopanes, and polycyclic aromatic hydrocarbons (PAH) have been proposed as the molecular compounds or biomarker to identify the sources of oil pollution (Zakaria et al. 2000).

Of all the marine life forms, planktonic organisms are prone to oil spill contamination. Planktons are on the mercy of currents and hence cannot avoid the crude oil areas, compelling them into the polluted waters and causing unplanned encounters with the polluted regions. The crude oil effect on marine species has gathered attention from ages; however, mostly higher organisms have been in the limelight compared to the marine microbes. Phytoplankton plays a vital role in biogeochemistry of the marine ecosystem, and hence any alterations in the ecosystem can have a profound effect on the food web dynamics. PAHs, a significant fraction of crude oil, are primarily responsible for crude oil toxicity in phytoplankton (Ozhan and Bargu 2014). They even accumulate in the sediment, posing a severe threat to the benthic community (e.g., Ozhan and Bargu 2014). Oil pollution can result in acute and chronic effects of phytoplankton productivity and community composition which can alter the whole planktonic ecosystem. Therefore, any alterations due to contaminants can result in ecosystem alterations (Othman et al. 2018), while some work has suggested a positive impact of crude oil on water chemistry which has enhanced the phytoplankton biomass. Most studies have shown that it has negatively altered the growth of phytoplankton. This impact is also determined by the concentration of crude oil exposed to the phytoplankton (Huang et al. 2011). Several factors influence the harmfulness of crude oil to phytoplankton, and it is not clear. Echeveste et al. (2011) found that the phytoplankton cellular size was key factor in determining the susceptibility to PAHs, with the pico-sized group showing the synergistic relationship between PAHs and UV radiation. Temperature is also another critical factor influencing the toxicity of crude oil in phytoplankton as demonstrated by Huang et al. (2011). They showed that *Skeletonema costatum* was highly tolerant to water-accommodated fraction (WAF) during winter but in summer even low concentration of WAF restricted their growth. They attributed this to the increased metabolic rate due to increased temperature resulting in more excellent absorption on toxicants.

Zooplankton is also the key player in marine food web dynamics, biogeochemical processes, and fish population dynamics (Banse 1995; Castonguay et al. 2008; Alcaraz et al. 2010). The impact of crude oil on zooplankton depends on several factors such as type of species, life stages, size, oil concentration, chemical dispersant, exposure time, temperature, salinity, UV radiation, etc. (Almeda et al. 2013a, b). The impact of oil on the zooplankton can also be reduced or counteracted by the presence of another organism which is important in the fate or degradation of crude oil in the marine environment (Almeda et al. 2013b). The effect of hydrocarbons on marine includes changes in feeding behavior, growth, and reproduction (Almeda et al. 2013a, b).

1.2.3 Plastic Pollution

In the ocean, plastic debris is ubiquitous and abundantly reported in natural habits. Plastic waste in the oceans was firstly reported in the 1970s (Carpenter et al. 1972; Carpenter and Smith 1972; Colton and Knapp 1974); afterward, they had got little public attention. Over a period, the ocean contains over 150 million metric tons of plastic (MacArthur et al. 2016). About 8 million metric tons of plastics waste enters marine environment, annually (Jambeck et al. 2015; Science Daily, 12 February 2015). By 2050, there will be excess plastic (based on weight) in the oceans compared to fish (MacArthur et al. 2016). Depending upon polymers used, plastic material can persevere up to several years (2–450 years) (Kibria 2018). Thus, plastic pollution is persistent in the oceans and has adverse ecological effects. Accumulation of marine plastic litter over a period has openly threatened marine biota.

In nature, plastic litter gets fragmented by UV radiation, hydrolysis, oxidation physical abrasion, and/or biodegradation into micro- or nanoscopic particles. Based on the size, they are referred to as nano- (<1 μm) and microplastics (1–5 mm) (Germanov et al. 2018). However, in most scientific articles, particles smaller than 5 mm are referred to as microplastics (Arthur et al. 2009; Andrady 2011; Hidalgo-Ruz et al. 2012). Depending on the source of origin, these fragmented plastics are called as secondary microplastics (Andrady 2011; Mrowiec 2017; Germanov et al. 2018). The primary microplastics arise from makeup products, dyes, fabrics, and waste from plastic industry (Mrowiec 2017; Germanov et al. 2018). A particular concern over microplastic debris over large size plastic litter is that they remained under-investigated due to their non-visibility to the naked eye. Incidences on the entanglement or ingestion of plastic material by marine species are extensively documented worldwide (Laist 1997; Clapham et al. 1999; Mascarenhas et al. 2004). Overall, the accumulation of large plastics remains stable or disappears, whereas that of microplastic rises (Eriksen et al. 2014). Scientific investigations reported the devastating effect of microplastics at higher trophic organisms. However, the possible impact of microplastics is under-evaluated on marine microorganisms, which are the foundation of marine food web. Considering this fact, the present section is restricted toward only the influencing mechanisms of microplastics on diverse marine microorganisms.

Plastics are incredibly resistant to biodegradation because of their high molecular weight and hydrophobicity. However, some microbial species have the ability to biodegrade the plastic material (Sivan et al. 2006; Shah et al. 2008; Mor and Sivan 2008; Harshvardhan and Jha 2013). The biodegradation process is generally initiated by surface assimilation of organic molecules on the microplastic, which further supports bacterial colonization (biofilm formation) (Gilan et al. 2004; Mor and Sivan 2008; Balasubramanian et al. 2010). Scanty information is available on the exact mechanisms involved in the biodegradation. Possibly biodegradation could be due to the interplay of various oxidative mechanisms caused by the microorganisms alone or in combination with the atmospheric oxygen, and the mechanisms would be complicated (Glass and Swift 1989). However, studies have proved that biodegradation process may reduce the weight of plastic molecules (Harshvardhan and Jha 2013) and surface hydrophobicity (Gilan et al. 2004).

The microplastic-associated microbial community is remarkably different from those present in the adjacent water (Ogonowski et al. 2018; Parrish and Fahrenfeld 2019). In water, the microbial population and metabolic processes depend on the quality of dissolved organic matter (Ruiz-González et al. 2015; Pernthaler 2017).

The microplastic-derived dissolved organic carbon attracts suitable microbial population, which could differ from the natural one. This difference in the utilization of carbon substances could partially explain the profile difference between the microplastic microbes and microbes present in adjacent water (Arias-Andres et al. 2018). Microplastics are known to release carbon thus, can affect the other free-living bacterial population by adding to dissolved organic carbon pool (Arias-Andres et al. 2018) and particulate organic matter bioavailability (Zhang et al. 2016). Considering the additional volume of anthropogenic dissolved organic carbon escaping from the tons of plastic per year (up to 23,600 metric tons from 4.8 to 12.7 million tons of plastics in 2010) entering marine environment (Romera-Castillo et al. 2018) could influence the natural carbon cycle in the ocean.

The genetic diversity within the microplastic biofilms increases gene transfer, which can affect the metabolic diversity of different microorganisms (Fazey and Ryan 2016; Rummel et al. 2017; Arias-Andres et al. 2018). This could further modulate various biochemical functions (Flemming et al. 2016) and bioadsorption capabilities of different toxic chemical and metal molecules on microorganisms. It is, therefore, the microbial biofilm formation on the consistently accumulating microplastics in the environment that can not only influence the fate of microplastic itself (Rummel et al. 2017) but also their impacts on the working of whole microbiomes (Arias-Andres et al. 2018). Ecologically also, plastic debris has an essential role in the spread of invasive, harmful microorganisms and algae species (Maso et al. 2003; Bryant et al. 2016; NOAA Marine Debris Program 2017) and thus changes in microbial biogeography. Biofilms established on microplastic attract other fouling organisms (biofouling), which do enhance microplastic sinking (Kaiser et al. 2017). The developed biofilm increases the sinking velocity of negatively buoyant microplastics, whereas macrofouling causes positively buoyant microplastics to sink (Kaiser et al. 2017). Further sinking in marine aggregates could further impact other planktonic and benthic feeding organisms (Long et al. 2015).

Plastic can concentrate contaminants such as persistent organic pollutants and heavy metals and increase their concentration up to 10^6 order (Mato et al. 2001). Moreover, plastic can act as vectors of contaminants which are already present in the waters, thus increasing their harmfulness to marine species (Bejgarn et al. 2015). Enormous literature is available on the effect of nanoparticles together with contaminants on higher-trophic-level organisms. However, the datasets on the toxicological effect of nano- or microplastic particles on common marine microbes (bacterial, phytoplankton) are limited. These studies are mostly based on synthetic microplastic particles under laboratory conditions. Hydrophobic contaminants such as persistent organic pollutants (POPs) get readily accumulated on the surface of plastics (Avio et al. 2017; Bhattacharya et al. 2010). The degraded or aged microplastics have uneven outer layers, which further promote POP adsorption (Cole et al. 2011). However, contradictorily, recent laboratory studies revealed microplastic surfaces modulate the toxicity of pollutants and made them less available to the microorganisms (Garrido et al. 2019; Yi et al. 2019). A recent study observed that polyethylene microplastics alone do not have any inhibitory effect on

Isochrysis galbana, whereas the negative effect was observed upon exposure to chlorpyrifos (CPF) in lethal concentration (2–3 mg L^{-1}) (Garrido et al. 2019). However, upon following incubation, chlorpyrifos gets adsorbed on the microplastic surface and not much available for microalgae. Similarly, Yi et al. (2019) observed adsorption of triphenyltin chloride (TPTCl) on polystyrene surface control bioavailability and toxicity of TPTCl to green algae.

Under laboratory conditions, microplastic toxicity exhibits conflicting results mainly due to confounding factors of microplastic size and dosages (Long et al. 2017). In *Chlorella pyrenoidosa*, upon exposure to polystyrene microplastics, dose-dependent adverse effect was observed at initial growth phases. Initially, in *C. pyrenoidosa*, reduced photosynthetic activity, indistinct pyrenoids, and impaired cell membranes were detected. Later, cellular wall thickening together with homo- and hetero-aggregation triggers cell growth and algal photosynthesis (Mao et al. 2018). Likewise, a study by Sjollema et al. (2016) observed no effect of polystyrene microplastics on photosynthetic activity but had effect on the growth rate of marine flagellate *Dunaliella tertiolecta*, only at high exposure level (250 mg L^{-1}) with smaller particle size (0.05 mm). However, in marine diatom *Phaeodactylum tricornutum*, the polystyrene nanoparticles (50 and 100 nm) possibly have effects at different physiological and cellular levels (Sendra et al. 2019). At higher concentrations (50 mg L^{-1}), smallest (50 nm) nanoparticles significantly damage the photosynthetic apparatus; damage DNA, depolarization of mitochondria, and cell membrane; and later inhibit chlorophyll content and population growth. An adverse effect of micro-PVC (~1 mm) was observed on the growth and chlorophyll fluorescence of *Skeletonema costatum* at high level (50 mg L^{-1}); this might have resulted from the obstruction of alveoli and impairment of cell surface (Zhang et al. 2017). Compared to microplastics, nanoplastics might more actively interact with microalgal membrane by covering or obstructing the pores or gas exchange (Bhattacharya et al. 2010). The physiological adaptive strategies of test species also greatly influence the effect of microplastics. Seoane et al. (2019) observed that the marine diatom *Chaetoceros neogracil* modulates the oil body level to overcome the stress produced upon polystyrene microbead exposure.

In conclusion, the marine microbial population could have a diverse impact on increasing anthropogenic plastic pollution. Overall, the "plastisphere" could control the microgeography and microbial diversity in the oceans.

1.3 Concluding Remarks

Human activities have caused marine ecosystems and, in turn, microbial communities to suffer a lot. Anthropogenic stressors like ocean warming and ocean acidification (related to anthropogenic emission of CO_2), oil (tarball) pollution, and (micro) plastic contamination have been proved to have a devastating impact on the marine ecosystem, which can have severe consequences on socioeconomic levels. Available pieces of literature have revealed significant alterations in marine species community dynamics. Based on the available literature, we are yet not confident enough to predict the functioning of the marine ecosystem in the future, mainly if the stressor is either still present or increases. Therefore, to know the effect of these stressors, we need to focus on multidisciplinary approaches/holistic frameworks linking modeling, observations, and experiments including new technologies. Further, there is a need to promote awareness within stakeholders and governments. Thus, with this approach, we may able to take necessary steps to slow or minimize the impacts of these stressors on the marine microbial ecosystem and in turn on human populations.

Acknowledgments The first author expresses her sincere thanks to the Department of Science and Technology-SERB, New Delhi, for providing the National Postdoctoral Fellowship.

References

Alcaraz M, Almeda R, Calbet A, Saiz E, Duarte CM, Lasternas S, Agustí S, Santiago R, Movilla J, Alonso A (2010) The role of arctic zooplankton in biogeochemical cycles: respiration and excretion of ammonia and phosphate during summer. Polar Biol 33:1719–1731. https://doi.org/10.1007/s00300-010-0789-9

Allgaier M, Riebesell U, Vogt M, Thyrhaug R, Grossart H-P (2008) Coupling of heterotrophic bacteria to phytoplankton bloom development at different pCO2 levels: a mesocosm study. Biogeosciences 5:1007–1022

Almeda R, Wambaugh Z, Chai C, Wang Z, Liu Z, Buskey EJ (2013a) Effects of crude oil exposure on bioaccumulation of polycyclic aromatic hydrocarbons and survival of adult and larval stages of gelatinous zooplankton. PLoS One 8:20–21. https://doi.org/10.1371/journal.pone.0074476

Almeda R, Wambaugh Z, Wang Z, Hyatt C, Liu Z, Buskey EJ (2013b) Interactions between zooplankton and crude oil: toxic effects and bioaccumulation of polycyclic aromatic hydrocarbons. PLoS One 8:1–21. https://doi.org/10.1371/journal.pone.0067212

Andersson AJ, Gledhill D (2013) Ocean acidification and coral reefs: effects on breakdown, dissolution, and net ecosystem calcification. Annu Rev Mar Sci 5:321–348

Andersson AJ, Mackenzie FT, Bates NR (2008) Life on the margin: implications of ocean acidification on Mg-calcite, high latitude and cold-water marine calcifiers. Mar Ecol Prog Ser 373:265–273

Andrady AL (2011) Microplastics in the marine environment. Mar Pollut Bull 62(8):1596–1605

Arias-Andres M, Klümper U, Rojas-Jimenez K, Grossart H-P (2018) Microplastic pollution increases gene exchange in aquatic ecosystems. Environ Pollut 237:253–261

Arthur C, Baker J, Bamford H (eds) (2009) Proceedings of the international research workshop on the occurrence, effects and fate of micro-plastic marine debris, September 9–11, 2008. NOAA Technical Memorandum NOS-OR&R-30

Asnaghi V, Mangialajo L, Gattuso J-P, Francour P, Privitera D, Chiantore M (2014) Effects of ocean acidification and diet on thickness and carbonate elemental composition of the test of juvenile sea urchins. Mar Environ Res 93:78–84

Avio CG, Gorbi S, Regoli F (2017) Plastics and microplastics in the oceans: from emerging pollutants to emerged threat. Mar Environ Res 128:2–11

Balasubramanian V, Natarajan K, Hemambika B, Ramesh N, Sumathi CS, Kottaimuthu R, Kannan VR (2010) High-density polyethylene (HDPE) degrading potential bacteria from marine ecosystem of Gulf of Mannar, India. Lett Appl Microbiol 51:205–211

Banse K (1995) Zooplankton: pivotal role in the control of ocean production. ICES J Mar Sci 52:265–277. https://doi.org/10.1016/1054-3139(95)80043-3

Baragi LV, Anil AC (2016) Synergistic effect of elevated temperature, pCO_2 and nutrients on marine biofilm. Mar Pollut Bull 105:102–109

Bautista-Chamizo E, De Orte MR, DelValls TA, Riba I (2016) Simulating CO_2 leakages from CCS to determine Zn toxicity using the marine microalgae Pleurochrysis roscoffensis. Chemosphere 144:955–965

Beardall J, Raven JA (2004) The potential effects of global climate change on microalgal photosynthesis, growth and ecology. Phycologia 43:26–40

Bejgarn S, MacLeod M, Bogdal C, Breitholtz M (2015) Toxicity of leachate from weathering plastics: an exploratory screening study with *Nitocra spinipes*. Chemosphere 132:114–119

Bhattacharya P, Lin S, Turner JP, Ke PC (2010) Physical adsorption of charged plastic nanoparticles affects algal photosynthesis. J Phys Chem C 114:16556–16561

Borges FO, Figueiredo C, Sampaio E, Rosa R, Grilo TF (2018) Transgenerational deleterious effects of ocean acidification on the reproductive success of a keystone crustacean (*Gammarus locusta*). Mar Environ Res 138:55–64

Boyce DG, Lewis MR, Worm B (2010) Global phytoplankton decline over the past century. Nature 466:591–596

Bravo M, de los Ángeles Gallardo M, Luna-Jorquera G, Núñez P, Vásquez N, Thiel M (2009) Anthropogenic debris on beaches in the SE Pacific (Chile): results from a national survey supported by volunteers. Mar Pollut Bull 58(11):1718–1726

Brewer PG (1997) Ocean chemistry of the fossil fuel CO_2 signal: the haline signal of "business as usual". Geophys Res Lett 24:1367–1369

Brown JH, Gillooly JF, Allen AP, Savage VM, West GB (2004) Toward a metabolic theory of ecology. Ecology 85:1771–1789

Brussaard CPD, Peperzak L, Beggah S, Wick LY, Wuerz B, Weber J, Samuel Arey J, Van Der Burg B, Jonas A, Huisman J, Van Der Meer JR (2016) Immediate ecotoxicological effects of short-lived oil spills on marine biota. Nat Commun 7:1–11. https://doi.org/10.1038/ncomms11206

Bryant JA, Clemente TM, Viviani DA, Fong AA, Thomas KA, Kemp P, Karl DM, White AE, DeLong EF (2016) Diversity and activity of communities inhabiting plastic debris in the North Pacific Gyre. mSystems 1(3):e00024–e00016. https://doi.org/10.1128/mSystems.00024-16

Burkhardt S, Amoroso G, Riebesell U, Sültemeyer D (2001) CO_2 and HCO_3- uptake in marine diatoms acclimated to different CO_2 concentrations. Limnol Oceanogr 46:1378–1391

Byrne M, Przeslawski R (2013) Multistressor impacts of warming and acidification of the ocean on marine invertebrates' life histories. Integr Comp Biol 53:582–596

Byrne M, Ho M, Selvakumaraswamy P, Nguyen HD, Dworjanyn SA, Davis AR (2009) Temperature, but not pH, compromises sea urchin fertilization and early development under near-future climate change scenarios. Proc R Soc Lond B Biol Sci 276:1883–1888

Calbet A, Sazhin AF, Nejstgaard JC, Berger SA, Tait ZS, Olmos L, Sousoni D, Isari S, Martínez RA, Bouquet J-M (2014) Future climate scenarios for a coastal productive planktonic food web resulting in microplankton phenology changes and decreased trophic transfer efficiency. PLoS One 9:e94388

Caldeira K, Wickett ME (2003) Oceanography: anthropogenic carbon and ocean pH. Nature 425:365–365

Carpenter EJ, Smith KL Jr (1972) Plastics on the Sargasso Sea surface. Science 175:1240–1241

Carpenter EJ, Anderson SJ, Harvey GR, Miklas HP, Peck BB (1972) Polystyrene spherules in coastal water. Science 178:749–750

Castonguay M, Plourde S, Robert D, Runge JA, Fortier L (2008) Copepod production drives recruitment in a marine fish. Can J Fish Aquat Sci 65:1528–1531. https://doi.org/10.1139/F08-126

Chan VB, Li C, Lane AC, Wang Y, Lu X, Shih K, Zhang T, Thiyagarajan V (2012) CO2-driven ocean acidification alters and weakens integrity of the calcareous tubes produced by the serpulid tubeworm, Hydroides elegans. PLoS One 7:e42718

Charette MA, Smith WHF (2010) The volume of Earth's ocean. Oceanography 23(2):112–114

Clapham PJ, Young SB, Brownell JR (1999) Baleen whales: conservation issues and the status of the most endangered populations. Mammal Rev 29:35–60

Clark D, Lamare M, Barker M (2009) Response of sea urchin pluteus larvae (Echinodermata: Echinoidea) to reduced seawater pH: a comparison among a tropical, temperate, and a polar species. Mar Biol 156:1125–1137

Clarkson MO, Kasemann SA, Wood RA, Lenton TM, Daines SJ, Richoz S, Ohnemueller F, Meixner A, Poulton SW, Tipper ET (2015) Ocean acidification and the Permo-Triassic mass extinction. Science 348:229–232

Cole M, Lindeque P, Halsband C, Galloway TS (2011) Microplastics as contaminants in the marine environment: a review. Mar Pollut Bull 62:2588–2597

Colton JB, Knapp FD (1974) Plastic particles in surface waters of the northwestern Atlantic. Science 185:491–497

Connell SD, Russell BD (2010) The direct effects of increasing CO_2 and temperature on non-calcifying organisms: increasing the potential for phase shifts in kelp forests. Proc R Soc Lond B Biol Sci 277(1686):1409–1415. https://doi.org/10.1098/rspb.2009

De Bodt C, Oostende NV, Harlay J, Sabbe K, Chou L (2010) Individual and interacting effects of pCO_2 and temperature on Emiliania huxleyi calcification: study of the calcite production, the coccolith morphology and the coccosphere size. Biogeosci Discuss 6:1401–1412

De Kluijver A, Soetaert K, Czerny J, Schulz KG, Boxhammer T, Riebesell U, Middelburg J (2013) A 13C labelling study on carbon fluxes in Arctic plankton communities under elevated CO_2 levels. Biogeosciences 10:1425–1440

Derraik JG (2002) The pollution of the marine environment by plastic debris: a review. Mar Pollut Bull 44(9):842–852

Donelson J, Munday P, McCormick M, Pitcher C (2012) Rapid transgenerational acclimation of a tropical reef fish to climate change. Nat Clim Chang 2:3032

Dupont S, Dorey N, Thorndyke M (2010a) What meta-analysis can tell us about vulnerability of marine biodiversity to ocean acidification? Estuar Coast Shelf Sci 89:182–185

Dupont S, Ortega-Martínez O, Thorndyke M (2010b) Impact of near-future ocean acidification on echinoderms. Ecotoxicology 19:449–462

Echeveste P, Agustí S, Dachs J (2011) Cell size dependence of additive versus synergetic effects of UV radiation and PAHs on oceanic phytoplankton. Environ Pollut 159:1307–1316. https://doi.org/10.1016/j.envpol.2011.01.023

Eppley RW (1972) Temperature and phytoplankton growth in the sea. Fish Bull 70:1063–1085

Eriksen M, Lebreton LCM, Carson HS, Thiel M, Moore CJ, Borerro JC, Galgani F, Ryan PG, Reisser J (2014) Plastic pollution in the world's oceans: more than 5 trillion plastic pieces weighing over 250,000 tons afloat at sea. PLoS One 9(12):e111913

Falkowski PG, Barber RT, Smetacek V (1998) Biogeochemical controls and feedbacks on ocean primary production. Science 281:200–206

Fazey FMC, Ryan PG (2016) Biofouling on buoyant marine plastics: an experimental study into the effect of size on surface longevity. Environ Pollut 210:354–360

Field CB, Behrenfeld MJ, Randerson JT, Falkowski P (1998) Primary production of the biosphere: integrating terrestrial and oceanic components. Science 281:237–240

Fiorini S, Middelburg JJ, Gattuso J-P (2011) Effects of elevated CO_2 partial pressure and temperature on the coccolithophore Syracosphaera pulchra. Aquat Microb Ecol 64:221–232

Flemming H-C, Wingender J, Szewzyk U, Steinberg P, Rice SA, Kjelleberg S (2016) Biofilms: an emergent form of bacterial life. Nat Rev Microbiol 14:563–575

Forster P, Ramaswamy V, Artaxo P, Berntsen T, Betts R, Fahey DW, Haywood J et al (2007) Changes in atmospheric constituents and in radiative forcing. In: Climate change 2007: the physical science basis. Contribution of Working Group I to the Fourth

Fu FX, Warner ME, Zhang Y, Feng Y, Hutchins DA (2007) Effects of increased temperature and CO2 on photosynthesis, growth, and elemental ratios in marine Synechococcus and Prochlorococcus (cyanobacteria). J Phycol 43:485–496

Fukuyama K Shigenaka G, Blaricom GRV (1998) Oil spill impacts and the biological basis for response guidance: an applied synthesis of research on three subarctic intertidal communities. National Oceanic and Atmospheric Administration (NOAA) Tech Memo. NOS ORCA 125

Gall SC, Thompson RC (2015) The impact of debris on marine life. Mar Pollut Bull 92 (1–2):170–179

Gao K, Campbell DA (2014) Photophysiological responses of marine diatoms to elevated CO2 and decreased pH: a review. Funct Plant Biol 41:449–459

Gao K, Xu J, Gao G, Li Y, Hutchins DA, Huang B, Wang L, Zheng Y, Jin P, Cai X (2012) Rising CO_2 and increased light exposure synergistically reduce marine primary productivity. Nat Clim Chang 2:519–523

Garrido S, Linares M, Campillo JA, Albentosa M (2019) Effect of microplastics on the toxicity of chlorpyrifos to the microalgae *Isochrysis galbana*, clone t-ISO. Ecotoxicol Environ Saf 173:103–109

Gaylord B, Kroeker KJ, Sunday JM, Anderson KM, Barry JP, Brown NE, Connell SD, Dupont S, Fabricius KE, Hall-Spencer JM (2015) Ocean acidification through the lens of ecological theory. Ecology 96:3–15

Germanov ES, Marshall AD, Bejder L, Fossi MC, Loneragan NR, N.R. (2018) Microplastics: no small problem for filter-feeding megafauna. Trends Ecol Evol 33(4):227–232

Gilan I, Hadar Y, Sivan A (2004) Colonization, biofilm formation and biodegradation of polyethylene by a strain of Rhodococcus ruber. Appl Microbiol Biotechnol 65:97–104

Glass JE, Swift G (eds) (1989) Agricultural and synthetic polymers, biodegradation and utilization, ACS symposium series 433. American Chemical Society, Washington, DC, pp 9–64

Gregory MR, Ryan PG (1997) Pelagic plastics and other seaborne persistent synthetic debris: a review of Southern Hemisphere perspectives. In: Coe JM, Rogers DB (eds) Marine debris—sources, impacts and solutions. Springer, New York, pp 49–66

Gruber N, Gloor M, Mikaloff Fletcher SE, Doney SC, Dutkiewicz S, Follows MJ, Gerber M, Jacobson AR, Joos F, Lindsay K, Menemenlis D (2009) Oceanic sources, sinks, and transport of atmospheric CO_2. Glob Biogeochem Cycles 23(1):GB1005

Häder D-P, Helbling EW, Williamson CE, Worrest RC (2011) Effects of UV radiation on aquatic ecosystems and interactions with climate change. Photochem Photobiol Sci 10(2):242–260

Halpern BS, Selkoe KA, Micheli F, Kappel CV (2007) Evaluating and ranking the vulnerability of global marine ecosystems to anthropogenic threats. Conserv Biol 21:1301–1315

Halpern BS, Walbridge S, Selkoe KA et al (2008) A global map of human impact on marine ecosystems. Science 319:948–952

Harley CD, Randall Hughes A, Hultgren KM, Miner BG, Sorte CJ, Thornber CS, Rodriguez LF, Tomanek L, Williams SL (2006) The impacts of climate change in coastal marine systems. Ecol Lett 9:228–241

Harshvardhan K, Jha B (2013) Biodegradation of low-density polyethylene by marine bacteria from pelagic waters, Arabian Sea, India. Mar Pollut Bull 77:100–106

Harvey BP, Gwynn-Jones D, Moore PJ (2013) Meta-analysis reveals complex marine biological responses to the interactive effects of ocean acidification and warming. Ecol Evol 3:1016–1030

Haugan PM, Drange H (1996) Effects of CO_2 on the ocean environment. Energy Convers Manag 37:1019–1022

Havenhand JN, Buttler F-R, Thorndyke MC, Williamson JE (2008) Near-future levels of ocean acidification reduce fertilization success in a sea urchin. Curr Biol 18:R651–R652

Hendriks IE, Duarte CM, Álvarez M (2010) Vulnerability of marine biodiversity to ocean acidification: a meta-analysis. Estuar Coast Shelf Sci 86:157164

Hidalgo-Ruz V, Gutow L, Thompson RC, Thiel M (2012) Microplastics in the marine environment: a review of the methods used for identification and quantification. Environ Sci Technol 46 (6):3060–3075

Holtz L-M, Wolf-Gladrow D, Thoms S (2015) Simulating the effects of light intensity and carbonate system composition on particulate organic and inorganic carbon production in Emiliania huxleyi. J Theor Biol 372:192–204

Hönisch B, Ridgwell A, Schmidt DN, Thomas E, Gibbs SJ, Sluijs A, Zeebe R, Kump L, Martindale RC, Greene SE (2012) The geological record of ocean acidification. Science 335:1058–1063

Huang Y-J, Jiang Z-B, Zeng J-N, Chen Q-Z, Zhao Y, Liao Y, Shou L, Xu X (2011) The chronic effects of oil pollution on marine phytoplankton in a subtropical bay, China. Environ Monit Assess 176:517–530. https://doi.org/10.1007/s10661-010-1601-6

Hutchins D, Fu F-X, Zhang Y, Warner M, Feng Y, Portune K, Bernhardt P, Mulholland M (2007) CO2 control of Trichodesmium N2 fixation, photosynthesis, growth rates, and elemental ratios: implications for past, present, and future ocean biogeochemistry. Limnol Oceanogr 52:1293–1304

Jambeck JR, Geyer R, Wilcox C, Siegler TR, Perryman M, Andrady A, Narayan R, Law KL (2015) Plastic waste inputs from land into the ocean. Science 347(6223):768–771

Jin P, Gao K (2016) Reduced resilience of a globally distributed coccolithophore to ocean acidification: confirmed up to 2000 generations. Mar Pollut Bull 103:101–108

Jin P, Gao K, Villafañe VE, Campbell DA, Helbling EW (2013) Ocean acidification alters the photosynthetic responses of a coccolithophorid to fluctuating ultraviolet and visible radiation. Plant Physiol 162:2084–2094

Johnson V, Brownlee C, Rickaby R, Graziano M, Milazzo M, Hall-Spencer J (2013) Responses of marine benthic microalgae to elevated CO_2. Mar Biol 160:1813–1824

Kaiser D, Kowalski N, Waniek JJ (2017) Effects of biofouling on the sinking behavior of microplastics. Environ Res Lett 12:124003

Kamya PZ, Dworjanyn SA, Hardy N, Mos B, Uthicke S, Byrne M (2014) Larvae of the coral eating crown-of-thorns starfish, Acanthaster planci in a warmer-high CO_2 ocean. Glob Chang Biol 20:3365–3376

Kerfahi D, Hall-Spencer JM, Tripathi BM, Milazzo M, Lee J, Adams JM (2014) Shallow water marine sediment bacterial community shifts along a natural CO_2 gradient in the Mediterranean Sea off Vulcano, Italy. Microb Ecol 67:819828

Kibria G (2018) Plastic pollution sources, global production, global hot spots, impacts on biodiversity & seafood, adsorption of organic & inorganic chemicals, and mitigation. https://doi.org/10.13140/RG.2.2.35028.24967

Kim JM, Lee K, Shin K, Kang JH, Lee HW, Kim M, Jang PG, Jang MC (2006) The effect of seawater CO2 concentration on growth of a natural phytoplankton assemblage in a controlled mesocosm experiment. Limnol Oceanogr 51(4):1629–1636

Kottmeier DM, Rokitta SD, Rost B (2016) Acidification, not carbonation, is the major regulator of carbon fluxes in the coccolithophore Emiliania huxleyi. New Phytol 211:126–137

Kroeker KJ, Kordas RL, Crim RN, Singh GG (2010) Meta-analysis reveals negative yet variable effects of ocean acidification on marine organisms. Ecol Lett 13:1419–1434

Kroeker KJ, Kordas RL, Crim R, Hendriks IE, Ramajo L, Singh GS, Duarte CM, Gattuso JP (2013) Impacts of ocean acidification on marine organisms: quantifying sensitivities and interaction with warming. Glob Chang Biol 19:1884–1896

Kroeker KJ, Sanford E, Jellison BM, Gaylord B (2014) Predicting the effects of ocean acidification on predator-prey interactions: a conceptual framework based on coastal molluscs. Biol Bull 226:211–222

Kurihara H (2008) Effects of CO2-driven ocean acidification on the early developmental stages of invertebrates. Mar Ecol Prog Ser 373:275–284

Laist DW (1997) Impacts of marine debris: entanglement of marine life in marine debris including a comprehensive list of species with entanglement and ingestion records. In: Coe JM, Rogers DB (eds) Marine debris—sources, impacts and solutions. Springer, New York, pp 99–139

Lane AC, Mukherjee J, Chan VB, Thiyagarajan V (2013) Decreased pH does not alter metamorphosis but compromises juvenile calcification of the tube worm Hydroides elegans. Mar Biol 160:1983–1993

Lewandowska AM, Boyce DG, Hofmann M, Matthiessen B, Sommer U, Worm B (2014) Effects of sea surface warming on marine plankton. Ecol Lett 17:614623

Li Y, Gao K, Villafañe V, Helbling E (2012) Ocean acidification mediates photosynthetic response to UV radiation and temperature increase in the diatom *Phaeodactylum tricornutum*. Biogeosciences 9:3931–3942

Lidbury I, Johnson V, Hall-Spencer JM, Munn CB, Cunliffe M (2012) Community-level response of coastal microbial biofilms to ocean acidification in a natural carbon dioxide vent ecosystem. Mar Pollut Bull 64:1063–1066

Lohbeck KT, Riebesell U, Reusch TB (2012) Adaptive evolution of a key phytoplankton species to ocean acidification. Nat Geosci 5:346–351

Long M, Moriceau B, Gallinari M, Lambert C, Huvet A, Raffray J, Soudant P (2015) Interactions between microplastics and phytoplankton aggregates: impact on their respective fates. Mar Chem 175:39–46

Long M, Lambert C, Paul-Pont I, Hegaret H, Moriceau B, Huvet A, Soudant P (2017) Interactions between polystyrene microplastics and marine phytoplankton lead to species-specific hetero-aggregation. Environ Pollut 228:454–463

Lovejoy TE, Hannah L (2005) Climate change and biodiversity. Yale University Press, New Haven, 418 pp

Low-Décarie E, Jewell MD, Fussmann GF, Bell G (2013) Long-term culture at elevated atmospheric CO2 fails to evoke specific adaptation in seven freshwater phytoplankton species. Proc R Soc Lond B Biol Sci 280:20122598

MacArthur E, Waughray D, Stuchtey MR (2016) The new plastics economy: rethinking the future of plastics. World Economic Forum, Geneva

Manno C, Peck VL, Tarling GA (2016) Pteropod eggs released at high pCO_2 lack resilience to ocean acidification. Sci Rep 6:25752

Mao Y, Ai H, Chen Y, Zhang Z, Zeng P, Kang L, Li W, Gu W, He Q, Li H (2018) Phytoplankton response to polystyrene microplastics: perspective from an entire growth period. Chemosphere 208:59–68

Mascarenhas R, Santos R, Zeppelini D (2004) Plastic debris ingestion by sea turtle in Paraiba, Brazil. Mar Pollut Bull 49(4):354–355

Maso M, Garces E, Pages F, Camp J (2003) Drifting plastic debris as a potential vector for dispersing harmful algal bloom (HAB) species. Sci Mar 67(1):107–111

Mato Y, Isobe T, Takada H, Kanehiro H, Ohtake C, Kaminuma T (2001) Plastic resin pellets as a transport medium for toxic chemicals in the marine environment. Environ Sci Technol 35:318–324

Melzner F, Gutowska M, Langenbuch M, Dupont S, Lucassen M, Thorndyke MC, Bleich M, Pörtner H-O (2009) Physiological basis for high CO2 tolerance in marine ectothermic animals: pre-adaptation through lifestyle and ontogeny? Biogeosciences 6:2313–2331

Melzner F, Stange P, Trübenbach K, Thomsen J, Casties I, Panknin U, Gorb SN, Gutowska MA (2011) Food supply and seawater pCO_2 impact calcification and internal shell dissolution in the blue mussel Mytilus edulis. PLoS One 6:e24223

Mor R, Sivan A (2008) Biofilm formation and partial biodegradation of polystyrene by the actinomycete *Rhodococcus ruber*. Biodegradation 19(6):851–858

Mostofa KM (2016) Reviews and syntheses: ocean acidification and its potential impacts on marine ecosystems. Biogeosciences 13:1767

Mrowiec B (2017) Plastic pollutants in water environment. Environ Protect Nat Resour 28 4 (74):51–55

Müller MN, Lebrato M, Riebesell U, Barcelos e Ramos J, Schulz KG, Blanco Ameijeiras S, Sett S, Eisenhauer A, Stoll HM (2014) Influence of temperature and CO_2 on the strontium and magnesium composition of coccolithophore calcite. Biogeosciences 11:1065–1075

National Oceanic and Atmospheric Administration Marine Debris Program (2017) Report on marine debris as a potential pathway for invasive species. National Oceanic and Atmospheric Administration Marine Debris Program, Silver Spring

National Research Council, N (2003) Oil in the sea: inputs, fates, and effects. National Academy Press, Washington, DC. https://doi.org/10.17226/10388

Nguyen HD, Doo SS, Soars NA, Byrne M (2012) Noncalcifying larvae in a changing ocean: warming, not acidification/hypercapnia, is the dominant stressor on development of the sea star Meridiastra calcar. Glob Chang Biol 18:2466–2476

Nilsson GE, Dixson DL, Domenici P, McCormick MI, Sørensen C, Watson S-A, Munday PL (2012) Near-future carbon dioxide levels alter fish behaviour by interfering with neurotransmitter function. Nat Clim Chang 2:201–204

NOAA/ESRL. www.esrl.noaa.gov/gmd/ccgg/trends

Nobre CR, Santana MFM, Maluf A, Cortez FS, Cesar A, Pereira CDS, Turra A (2015) Assessment of microplastic toxicity to embryonic development of the sea urchin Lytechinus variegatus (Echinodermata: Echinoidea). Mar Pollut Bull 92(1–2):99–104

Nogales B, Lanfranconi MP, Piña-Villalonga JM, Bosch R (2011) Anthropogenic perturbations in marine microbial communities. FEMS Microbiol Rev 35(2):275–298

Ogonowski M, Motiei A, Ininbergs K, Hell E, Gerdes Z, Udekwu KI, Bacsik Z, Gorokhova E (2018) Evidence for selective bacterial community structuring on microplastics. Environ Microbiol 20(8):2796–2808

Orr JC, Fabry VJ, Aumont O, Bopp L, Doney SC, Feely RA, Key RM (2005) Anthropogenic ocean acidification over the twenty-first century and its impact on calcifying organisms. Nature 437 (7059):681–686

Othman HB, Lanouguere E, Got P, Hlaili AS, Leboulanger C (2018) Structural and functional responses of coastal marine phytoplankton communities to PAH mixtures. Chemosphere 209:908–919. https://doi.org/10.1016/j.chemosphere.2018.06.153

Ozhan K, Bargu S (2014) Can crude oil toxicity on phytoplankton be predicted based on toxicity data on benzo(a)pyrene and naphthalene? Bull Environ Contam Toxicol 92:225–230. https://doi.org/10.1007/s00128-013-1181-6

Pansch C, Schaub I, Havenhand J, Wahl M (2014) Habitat traits and food availability determine the response of marine invertebrates to ocean acidification. Glob Chang Biol 20:765–777

Parker LM, O'Connor WA, Raftos DA, Portner HO, Ross PM (2015) Persistence of positive carryover effects in the oyster, Saccostrea glomerata, following transgenerational exposure to ocean acidification. PLoS One 10:e0132276

Parrisha K, Fahrenfeld NL (2019) Microplastic biofilm in fresh- and wastewater as a function of microparticle type and size class. Environ Sci Water Res Technol 5(3):495–505

Pelejero C, Calvo E, Hoegh-Guldberg O (2010) Paleo-perspectives on ocean acidification. Trends Ecol Evol 25:332–344

Pernthaler J (2017) Competition and niche separation of pelagic bacteria in freshwater habitats. Environ Microbiol 19:2133–2150

Poloczanska E, Hoegh-Guldberg O, Cheung W, Pörtner HO, Burrows MT, (2014) Cross-chapter box on observed global responses of marine biogeography, abundance, and phenology to climate change. In: Climate change 2014: impacts, adaptation, and vulnerability. Part A: global and sectoral aspects. Contribution of Working Group II to the Fifth Assessment Report of the Intergovernmental Panel on Climate Change, pp 123–127

Poore AG, Graba-Landry A, Favret M, Brennand HS, Byrne M, Dworjanyn SA (2013) Direct and indirect effects of ocean acidification and warming on a marine plant–herbivore interaction. Oecologia 173:1113–1124

Pörtner H-O (2008) Ecosystem effects of ocean acidification in times of ocean warming: a physiologist's view. Mar Ecol Prog Ser 373:203–217

Pörtner HO, Farrell AP (2008) Physiology and climate change. Science 322:690–692

Pörtner HO, Langenbuch M, Reipschläger A (2004) Biological impact of elevated ocean CO_2 concentrations: lessons from animal physiology and earth history. J Oceanogr 60:705–718

Przeslawski R, Webb A (2009) Natural variation in larval size and developmental rate of the northern quahog Mercenaria mercenaria and associated effects on larval and juvenile fitness. J Shellfish Res 28:505–510

Putnam HM, Gates RD (2015) Preconditioning in the reef-building coral Pocillopora damicornis and the potential for trans-generational acclimatization in coral larvae under future climate change conditions. J Exp Biol 218:2365–2372

Ramajo L, Marba N, Prado L, Peron S, Lardies MA, Rodriguez-Navarro AB, Vargas CA, Lagos NA, Duarte CM (2016) Biomineralization changes with food supply confer juvenile scallops (Argopecten purpuratus) resistance to ocean acidification. Glob Chang Biol 22:2025–2037

Range P, Piló D, Ben-Hamadou R, Chícharo MA, Matias D, Joaquim S, Oliveira AP, Chícharo L (2012) Seawater acidification by CO_2 in a coastal lagoon environment: effects on life history traits of juvenile mussels Mytilus galloprovincialis. J Exp Mar Biol Ecol 424–425:89–98

Raven J, Caldeira K, Elderfield H, Hoegh-Guldberg O, Liss P, Riebesell U, Shepherd J, Turley C, Watson A (2005) Ocean acidification due to increasing atmospheric carbon dioxide. The Royal Society, London

Raven JA, Cockell CS, De La Rocha CL (2008) The evolution of inorganic carbon concentrating mechanisms in photosynthesis. Philos Trans R Soc Lond Ser B Biol Sci 363:2641–2650

Reinfelder JR (2011) Carbon concentrating mechanisms in eukaryotic marine phytoplankton. Mar Sci 3:291–315

Rickaby RE, Henderiks J, Young JN (2010) Perturbing phytoplankton: response and isotopic fractionation with changing carbonate chemistry in two coccolithophore species. Clim Past 6:771–785

Ries JB, Cohen AL, McCorkle DC (2009) Marine calcifiers exhibit mixed responses to CO2-induced ocean acidification. Geology 37:1131–1134

Romera-Castillo C, Pinto M, Langer TM, Álvarez-Salgado XA, Herndl GJ (2018) Dissolved organic carbon leaching from plastics stimulates microbial activity in the ocean. Nat Commun 9:1430

Rossoll D, Bermúdez R, Hauss H, Schulz KG, Riebesell U, Sommer U, Winder M (2012) Ocean acidification-induced food quality deterioration constrains trophic transfer. PLoS One 7:e34737

Rost B, Richter KU, Riebesell U, Hansen PJ (2006) Inorganic carbon acquisition in red tide dinoflagellates. Plant Cell Environ 29:810–822

Ruiz-González C, Niño-García JP, Lapierre J-F, del Giorgio PA (2015) The quality of organic matter shapes the functional biogeography of bacterioplankton across boreal freshwater ecosystems. Glob Ecol Biogeogr 24:1487–1498

Rummel CD, Jahnke A, Gorokhova E, Kühnel D, Schmitt-Jansen M (2017) Impacts of biofilm formation on the fate and potential effects of microplastic in the aquatic environment. Environ Sci Technol Lett 4:258–267

Sabine CL, Feely RA, Gruber N, Key RM, Lee K, Bullister JL et al (2004) The oceanic sink for CO2. Science 305:367–371

Sarmiento JL, Gruber N (2006) Ocean biogeochemical dynamics. Princeton University Press, Princeton

Schalkhausser B, Bock C, Stemmer K, Brey T, Pörtner H-O, Lannig G (2013) Impact of ocean acidification on escape performance of the king scallop, Pecten maximus, from Norway. Mar Biol 160:1995–2006

Schlesinger WH (1997) Biogeochemistry: an analysis of global change. Academic, San Diego

Sendra M, Staffieri E, Yeste MP, Moreno-Garrido I, Gatica JM, Corsi I, Blasco J (2019) Are the primary characteristics of polystyrene nanoplastics responsible for toxicity and ad/absorption in the marine diatom *Phaeodactylum tricornutum*? Environ Pollut 249:610–619

Seoane M, González-Fernández C, Soudant P, Huvet A, Esperanza M, Cid Á, Paul-Pont I (2019) Polystyrene microbeads modulate the energy metabolism of the marine diatom *Chaetoceros neogracile*. Environ Pollut 251:363–371

Sett S, Bach LT, Schulz KG, Koch-Klavsen S, Lebrato M, Riebesell U (2014) Temperature modulates coccolithophorid sensitivity of growth, photosynthesis and calcification to increasing seawater pCO2. PLoS One 9:e88308

Shah AA, Fariha H, Abdul H, Safia A (2008) Biological degradation of plastics: a comprehensive review. Biotechnol Adv 26:2467–2650

Sivan A, Szanto M, Pavlov V (2006) Biofilm development of the polyethylenedegrading bacterium Rhodococcus ruber. Appl Microbiol Biotechnol 72:346–352

Sjollema SB, Redondo-Hasselerharm P, Leslie HA, Kraak MHS, Vethaak AD (2016) Do plastic particles affect microalgal photosynthesis and growth? Aquat Toxicol 170:259–261

Solberg AHS (2012) Remote sensing of ocean oil-spill pollution. Proc IEEE 100(10):2931–2945

Sommer U, Paul C, Moustaka-Gouni M (2015) Warming and ocean acidification effects on phytoplankton--from species shifts to size shifts within species in a mesocosm experiment. PLoS One 10:e0125239

Spijkerman E (2008) What physiological acclimation supports increased growth at high CO_2 conditions? Physiol Plant 133:41–48

Stillman JH, Paganini AW (2015) Biochemical adaptation to ocean acidification. J Exp Biol 218:1946–1955

Stocker TF, Dahe Q, Plattner G-K (2013) Climate change 2013: the physical science basis. Cambridge University Press, Cambridge

Talmage SC, Gobler CJ (2011) Effects of elevated temperature and carbon dioxide on the growth and survival of larvae and juveniles of three species of Northwest Atlantic bivalves. PLoS One 6:e26941

Thomas CD, Cameron A, Green RE, Bakkenes M, Beaumont LJ, Collingham YC, Erasmus BF, De Siqueira MF, Grainger A, Hannah L (2004) Extinction risk from climate change. Nature 427:145–148

Thomas MK, Kremer CT, Klausmeier CA, Litchman E (2012) A global pattern of thermal adaptation in marine phytoplankton. Science 338:1085–1088

Thomsen J, Casties I, Pansch C, Körtzinger A, Melzner F (2013) Food availability outweighs ocean acidification effects in juvenile Mytilus edulis: laboratory and field experiments. Glob Chang Biol 19:1017–1027

Torstensson A, Hedblom M, Mattsdotter Bjork M, Chierici M, Wulff A (2015) Long-term acclimation to elevated pCO2 alters carbon metabolism and reduces growth in the Antarctic diatom Nitzschia lecointei. Proc R Soc B Biol Sci 282(1815):20151513

Toseland A, Daines SJ, Clark JR, Kirkham A, Strauss J, Uhlig C, Lenton TM, Valentin K, Pearson GA, Moulton V (2013) The impact of temperature on marine phytoplankton resource allocation and metabolism. Nat Clim Chang 3:979–984

University of Georgia (2015) Magnitude of plastic waste going into the ocean calculated: 8 million metric tons of plastic enter the oceans per year. Science Daily, 12 February

Uthicke S, Logan M, Liddy M, Francis D, Hardy N, Lamare M (2015) Climate change as an unexpected co-factor promoting coral eating seastar (Acanthaster planci) outbreaks. Sci Rep 5:8402

Uthicke S, Ebert T, Liddy M, Johansson C, Fabricius KE, Lamare M (2016) Echinometra sea urchins acclimatised to elevated pCO_2 at volcanic vents outperform those under present-day pCO2 conditions. Glob Chang Biol 22(7):2451–2461

Wage J, Lerebours A, Hardege JD, Rotchell JM (2016) Exposure to low pH induces molecular level changes in the marine worm, Platynereis dumerilii. Ecotoxicol Environ Saf 124:105–110

Witt V, Wild C, Anthony K, Diaz-Pulido G, Uthicke S (2011) Effects of ocean acidification on microbial community composition of, and oxygen fluxes through, biofilms from the Great Barrier Reef. Environ Microbiol 13:2976–2989

Wittmann AC, Pörtner H-O (2013) Sensitivities of extant animal taxa to ocean acidification. Nat Clim Chang 3:995–1001

Wright SL, Thompson RC, Galloway TS (2013) The physical impacts of microplastics on marine organisms: a review. Environ Pollut 178:483–492

Wu Y, Beardall J, Gao K (2015) Physiological responses of a model marine diatom to fast pH changes: special implications of coastal water acidification. PLoS One 10:e0141163

Xu X, Yang F, Zhao L, Yan X (2016) Seawater acidification affects the physiological energetics and spawning capacity of the Manila clam *Ruditapes philippinarum* during gonadal maturation. Comp Biochem Phys A 196:20–29

Yi X, Chi T, Li Z, Wang J, Yu M, Wu M, Zhou H (2019) Combined effect of polystyrene plastics and *triphenyltin chloride* on the green algae *Chlorella pyrenoidosa*. Environ Sci Pollut Res 26 (15):15011–15018

Zakaria MP, Horinouchi A, Tsutsumi S, Takada H, Tanabe S, Ismail A (2000) Oil pollution in the Straits of Malacca, Malaysia: application of molecular markers for source identification. Environ Sci Technol 34:1189–1196

Zhang Y, Xiao W, Jiao N (2016) Linking biochemical properties of particles to particle attached and free-living bacterial community structure along the particle density gradient from freshwater to open ocean. J Geophys Res Biogeosci 121:2261–2274

Zhang C, Chen X, Wang J, Tan L (2017) Toxic effects of microplastic on marine microalgae Skeletonema costatum: interactions between microplastic and algae. Environ Pollut 220:1282–1288

Zittier ZMC, Bock C, Lannig G, Pörtner HO (2015) Impact of ocean acidification on thermal tolerance and acid–base regulation of Mytilus edulis (L.) from the North Sea. J Exp Mar Biol Ecol 473:16–25

Part I

Microbes in Agriculture

Recent Advances in Plant-Microbe Interaction

2

Jayakumar Pathma, Gurusamy Raman, Rajendiran Kamaraj Kennedy, and Laxman Sonawane Bhushan

Abstract

The association of plants and microbes has begun since their evolution. Microbes and plants have coevolved and interacted with each other to meet their demands. Their relationship might be cordial symbiotic as in case of interaction between plants and beneficial microbes or detrimental as in case of interaction between plants and phytopathogens. Numerous genera of microbes are known to be associated with the plants and their rhizosphere. The interaction among these diverse microbial communities and their ability to excel the competition decides the overall plant health. In the past decades, agricultural microbiologists had given more emphasis to plant growth-promoting rhizosphere microbes and soil-borne phytopathogens and their interactions, which has resulted in the identification and use of promising microbial strains with biocontrol and biofertilizing properties. With recent advancement in molecular diagnostics, it is evidenced that in addition to rhizosphere microbes, the interactions between plant microbiomes, *viz.* epiphytes and endophytes, colonizing the entire plant and the plant genome (holobiont) significantly affect the fitness of the plant. Scientific studies evidence that the plant genotype, biostage, soil biogeochemistry and microbe-microbe interaction decide the nature of associated microbiomes. Recent research shows that artificial inoculation of beneficial microbiomes instead of a single or a

J. Pathma (✉)
Department of Entomology, School of Agriculture, Division of Research and Development, Lovely Professional University, Phagwara, Punjab, India

G. Raman
Department of Life Sciences, Yeungnam University, Gyeongsan, Gyeongsangbuk-do, South Korea

R. K. Kennedy
Department of Biotechnology, St. Joseph's College, Trichy, Tamil Nadu, India

L. S. Bhushan
Department of Entomology, School of Agriculture, Lovely Professional University, Phagwara, Punjab, India

S. G. Sharma et al. (eds.), *Microbial Diversity, Interventions and Scope*,
https://doi.org/10.1007/978-981-15-4099-8_2

consortium of microbial strains would improve the success rate of establishment and functioning of the introduced microbial community. This chapter highlights the recent advancements in plant-microbe interaction and ways it could be explored and exploited to enhance plant health, thereby improving crop production qualitatively and quantitatively supporting sustainable agriculture.

Keywords

Plant microbiome · Holobiont · Microbial diversity · Beneficial microbes · Phytopathogens · Molecular tools · Sustainable agriculture

2.1 Introduction

Majority of the terrestrial plants are harboured in soil which is a huge and richest reservoir of diverse microbes (Tringe et al. 2005). It is estimated that a gram of soil contains 10^7 microbial species, of which bacterial diversity alone ranges up to 5×10^4 with a population of 10^{10} bacterial cells gm^{-1} of soil (Gans et al. 2005; Roesch et al. 2007; Raynaud and Nunan 2014). Microbes and plants have coevolved and are interdependent. Microbial association with host plants could be ectophytic or endophytic, and their intimacy may be beneficial to both the host plant and the microbe or may be favourable to the associated microbes alone posing a health risk to the host plant. Vogl (1898) cultured the first symptomless endophyte from *Lolium temulentum* seeds. Hartmann et al. (2007) documented that as early as 1901, Hiltner was able to predict the role of plant root exudates in shaping different microbial communities associated with plants and emphasized that the 'plant microflora' composition decides the resistance of plants towards pathogens. Plant root acts as a bridge paving the entry of selected soil bacteria into plants which multiply within the plants either as benign endophytes enhancing plant growth or as phytopathogens hampering plant growth. Beneficial benign endophytes offer a variety of services to the associated host plants such as plant growth promotion, yield enhancement and plant protection against various biotic (phytopathogens, invertebrate herbivores) and abiotic (temperature, drought, salinity, heavy metals) stress by influencing the host plants' metabolism. The host plants recognize microbe-associated molecular patterns (MAMPs) and initiate immune responses which modulates the association and multiplication of the microbe (Rosenblueth and Martinez-Romero 2006). Majority of the microbes inhabiting the plant rhizosphere belonging to the genera *Pseudomonas, Bacillus, Rhizobium, Azospirillum, Acetobacter, Streptomyces, Trichoderma, Glomus, Acaulospora, Scutellospora, Enterphospora*, etc. have evidenced to play a beneficial role in plant health and ecological fitness. Most of them colonize the rhizosphere region, while few others possess intracellular and intercellular endosymbiosis with host plant and act as an interface between the plant and the soil medium channelling the effective transmission of nutrients, minerals and water from soil. The microbes rely on plants for dwelling space and derive their nutrition from root exudates and in turn supply plants with nutrients, vitamins, growth-promoting hormones and disease-evading biomolecules as well as trigger plant immune system, thereby protecting the plant from various biotic and abiotic stresses. On the other hand, soil also contains various phytopathogenic species belonging to the genera *Rhizoctonia, Fusarium, Phomphosis, Phytophthora, Agrobacterium*, etc., which causes dreaded disease in plants. The ratio and competitive ability between the good and bad microbes decides the overall plant health and fitness. Until the last decade, more emphasis was given to explore the potential rhizosphere microbes to utilize them as biofertilizers and biopesticides. There are growing evidence that the composition of plant microbiomes (those living as endophytes and epiphytes in all plant parts), their networking and signalling decide the plant health. Recent advances in molecular diagnostics like metagenomics, metabolomics, proteomics, high-throughput sequencing, etc. have opened new insights and a better understanding of uncultivable microbes and their role in maintaining plant health. This chapter briefs about plant-associated microbiomes, factors affecting their establishment and their role in preserving the health of the plants, techniques used in studying the holobiont and their potential application in agriculture to boost the yield in the most sustainable manner.

2.2 Plant Microbiomes

Plant microbiome includes all microbial partners associated with plants underground and aboveground. Underground microbes include epiphytic and endophytic microbes colonizing the roots (rhizosphere and rhizoplane), while aboveground microbes are those inhabiting the phyllosphere as endophytes and epiphytes and include microbes dwelling in caulosphere (stem), phylloplane (leaves), anthosphere (flowers) and carposphere (fruits). Plants offer space, protection, nutrients and supports the dissemination of associated microbes. The microbes, in turn, provide substances that help in seed germination, plant growth and development and resistance to salinity, drought and water stress and activate plants' defence against herbivore pest and phytopathogens (Stanley and Fagan 2002). Though a microbe belonging to a specific taxon provides a specific functional advantage, other members in the microbiome are essential to support or synergize the effect of the key candidate. Thus, knowledge on holobiont (plants and associated microbes) will help us understand their evolution, biodiversity, interdependence and functionality of the ecosystem which has ample applications in food security and safety. Attempts made by evolutionary biologists to study the evolution of plant microbiomes showed that arbuscular mycorrhizal (AM) symbioses were primogenital which might have helped in the establishment of terrestrial plants (Brundrett 2002; Parniske 2008). Evolutionary theory also suggests that endophytes that are systemic and vertically transmitted in grasses pose greater resistance to host plants against invertebrate herbivores and phytopathogens as compared to horizontally transmitted endophytes inhabiting woody plants (Stanley and Fagan 2002). Studies on genetic linkages between fungal and bacterial symbiosis opens up the possibility of the evolvement of rhizobial root nodule symbiosis from functional aspects of mycorrhiza (Parniske 2008; Oldroyd et al. 2009).

The rhizosphere is the biologically active interface and per gram of root contains nearly 10^{11} microbial cells with 30,000 diverse species which determines plant health (Berendsen et al. 2012; Pathma et al. 2019). The total surface area of phyllosphere has been estimated to be approximately 10^9 km^2 globally which acts as a house for various beneficial and pathogenic microbes with a microbial density of 10^7 cells/cm^2 of leaf surface (Lindow and Brandl 2003; Farre-Armengol et al. 2016). Early studies focussed on plant-microbe interaction at tissue level, but the advancement of biochemical and molecular diagnostics has enabled us to understand the interaction of microbe at the cellular level of plants which acts as the interface for molecular conversation between plants and microbes. The microbial effectors delivered into the host plant cell shapes the dialogue between them. This acts as the beginning for a plethora of changes at the cellular level which could be compatible leading to beneficial symbiotic interactions (rhizobacteria, mycorrhiza) or incompatible leading to detrimental pathogenic infections by phytopathogenic fungi, bacteria and viruses (Panstruga and Kuhn 2015). In both cases, genetic signatures of the host plant and the associated microbes are the drivers that shape their associations and the outcome. The host plant produces specific cues which are recognized by the microbes enabling their orientation with the host plants. Also microbes produce specific chemical cues which attract or repel another microbe and decide the taxonomical diversity of the microbial community associated with the host plant. In case of pathogenic association, virulence which decides the success of infection is the combination product of plant-pathogen interaction and not the individual trait of plant or the pathogen. Both host plant and pathogen genotype are key factors deciding the success of infection (Ebert and Hamilton 1996).

Plants' immune system initially recognizes all microbial infection as harmful invasions after which they discriminate between pathogenic and beneficial microbes (Pel and Pieterse 2013). Microbes have specific molecular signatures, and microbial infections induce plants' systemic resistance. In case of infection of plants with microbes, the pattern recognition receptors (PRRs) present in the plasma membrane of plants recognize microbe- or pathogen-associated molecular patterns (MAMPs or PAMPs) and activate the MAMP- or PAMP-triggered immunity (MTI/PTI) that inhibits infection in case of phytopathogen, while the MTI does not evade beneficial infections. There occurs continuous molecular signalling between the host plant and associated microbe and phytohormones, viz. jasmonic acid, salicylic acid and ethylene which plays an important role in the defence responses triggered by both beneficial and pathogenic microbes. Successful phytopathogenic microbes secrete effector proteins which alter resistance signalling, and plants, in turn, had evolved a more specific immune response called effector-triggered immunity (ETI) where the microbial effector proteins are recognized and handled by plant resistance (R) proteins. PTI provides immunity before the pathogen gains entry into the plant, while R proteins come to rescue once the infection occurs (Glazebrook 2005; Chisholm et al. 2006; Van Wees et al. 2008; Trda et al. 2015).

2.3 Key Players Shaping and De-shaping Plant Microbiomes

Soil and host plants were considered to be the key factors responsible for shaping the plant-associated microbiomes. However, ranking their degree of influence in deciding the associated microbiomes are debatable as the results of scientific experiments were contradictory. Factors influencing plant microbiomes are illustrated (Fig. 2.1). Few investigations emphasized that diversity of plant-associated microbial community was greatly influenced by host plants (Grayston et al. 1998; Costa et al. 2006), while few studies highlighted the role of soil factors in shaping the plant microbiome (Buyer et al. 1999; Girvan et al. 2003; Horner-Devine et al. 2004; Fierer and Jackson 2006). Other abiotic factors include temperature, humidity, precipitation, wind patterns and light (quality and quantity), etc., which vary with season and have a profound effect on plant physiology and biochemistry, which in turn has a cascading impact on the native microflora of the soil and thereby the host plant. Marschner et al. (2001) reported that plant genotype and soil type should be considered as dependent variables to analyse their complex interactions in shaping associated microflora for better understanding and accurate results.

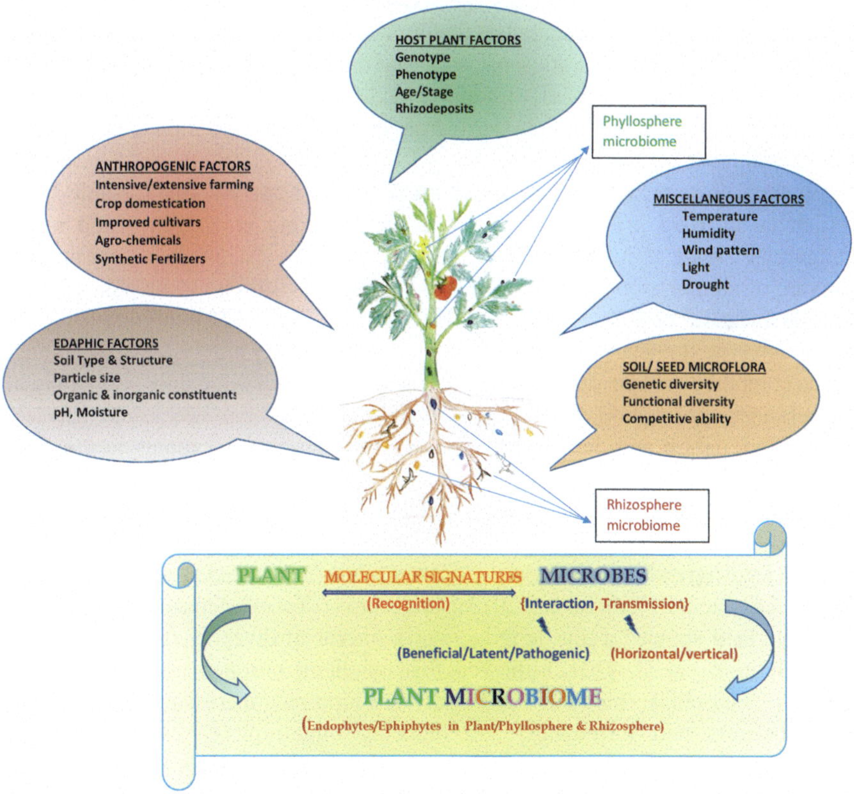

Fig. 2.1 Factors influencing plant-microbe interaction

2.3.1 Crop Domestication

Plants and their microbial partners have coevolved, and this statement holds good for all organisms on Earth and their associated symbiotic microbes. Anthropogenic activities and crop domestication have significantly affected the plant genetic makeup due to continuous selection for a preferred trait which was mostly focussed on yield and quality enhancement. This selection not only increased the desired alleles in the subsequent progenies but had also swept away genomic sections close to target regions leading to the loss in diversity of other desired traits related to plant morphology, biochemistry, etc., which may offer protection against insect herbivores, phytopathogens or influence the patterns of nutrition acquisition as well as recruitment of beneficial microbes from the soil, etc. Loss of genetic diversity due to domestication has been reported in crops like paddy (Ram et al. 2007), wheat (Haudry et al. 2007), barley (Bulgarelli et al. 2015), common bean (Bitocchi et al. 2013), sugar beet (Zachow et al. 2014) and lettuce (Cardinale et al. 2015). Germida and Siciliano (2001) reported that ancient plant races possessed diverse rhizobacterial community structure with pseudomonads being predominant followed by *Aureobacter*, while the modern cultivars exhibited less rhizobacterial diversity. Wild ancestral cultivars of legumes showed ability to attract and colonize diverse rhizobacterial species as compared to the domesticated pea (*Pisum sativum*), broad bean (*Vicia faba*), soya bean (*Glycine max*) and chick pea (*Cicer arietinum*) (Mutch and Young 2004; Kim et al. 2014). Similar studies on mycorrhizal association with wild primitive ancestors and modern crop cultivars showed that the primitive cultivars showed higher preference for mycorrhizal colonization and dependence as compared to improved modern cultivars of wheat (Kapulnik and Kushnir 1991; Hetrick et al. 1993; Zhu et al. 2001), breadfruit (Xing et al. 2012) and maize (Sangabriel-Conde et al. 2015). However, there were few exceptions where modern cultivars of annual crops were more responsive to mycorrhizal symbiosis as compared to ancestral ones in a meta-analysis (Lehmann et al. 2012).

Though crop domestication aimed to improve the quality and quantity of production as compared to their wild relatives, domesticated crops were poor supporters of self-sustained production systems and were unable to tap ecosystem services provided by nature and demanded assistance through external inputs in the form of synthetic fertilizers and plant protection chemicals which in turn polluted the ecosystem and hampered numerous beneficial interactions between host plant and microbes due to the loss of soil microbial diversity. The bacterial communities in domesticated agricultural fields were different from the adjacent native tall grass prairie ecosystem (Fierer et al. 2013), and conversion of Amazon rainforest to cultivable land showed a drastic reduction in microbial diversity (Rodrigues et al. 2013). Ramirez et al. (2012) showed that synthetic nitrogen amendments will suppress microbial biomass and soil respiration increasing copiotrophs (Actinobacteria and Firmicutes) and reducing oligotrophs (Acidobacteria and Verrucomicrobia). Weese et al. (2015) reported that continuous use of nitrogen fertilizers had resulted in reduced evolution of mutualistic rhizobia. Anthropogenic interventions and crop domestication affected the soil physiochemical properties and thereby reduced its microbial density and diversity (Garcia-Palacios et al. 2013). In fact, human interference has significantly disrupted the coevolutionary pattern between host plants and their beneficial and pathogenic microbial counterparts and had posed a serious threat to healthy sustainable crop production.

2.3.2 Plant Genotype

Plant genotype is a key modulator of its microbiome composition. Host plant DNA fingerprints have proved its significance in drafting the diversity of root-associated microbes (Ofek et al. 2014; Matthews et al. 2019). Increase in evolutionary distance between plant species shows a proportional increase in the diversity of the assembled microbial community (Bouffaud et al. 2014). Apart from different plant species, genotypic variation within the same plant species also shows a profound difference among the associated microbiomes (Inceoglu et al. 2011; Peiffer et al. 2013). Plant genotype decides its phenotype including leaf morphological features like hairs, stomata, veins, etc., which influences microbial colonization (Lindow and Brandl 2003). Similarly, root architecture as influenced by the host plant genotype also affects microbial colonization. Fitzpatrick et al. (2018) showed that the diversity of endophytes increased with an increase in root hair density, while the diversity of rhizosphere microbes decreased with an increase in root length in case of angiosperms. Variation among microbial communities associated with different cultivars of potato (Weinert et al. 2011), maize (Peiffer et al. 2013), sweet potato (Marques et al. 2014) and barley (Bulgarelli et al. 2015) has been documented. Plant genotype determines the chemistry of root exudates and their blend. Root exudates contain sugars, organic acids, amino acids, flavonoids, nucleotides, enzymes and antimicrobial compounds and supply the rhizosphere with carbon-rich compounds which chemotactically attract or deter the soil microbes. Root exudates' chemical composition and proportion determines the quality and quantity of associated rhizosphere microbes (Micallef et al. 2009). Thus, rhizodeposits which are specific to plant genotype influence microbial community assemblage. It reduces the diversity of associated microbes but enriches the abundance of microbes belonging to specific taxa (Bulgarelli et al. 2012; Lundberg et al. 2012). The citric acid found in root exudates of cucumber was evidenced to attract *B. amyloliquefaciens* SQR9, while fumaric acid from root exudates of banana attracted *B. subtilis* N11 and stimulated biofilm formation (Zhang et al. 2014). Rhizodeposits of paddy primarily contained amino acids, viz. alanine, histidine, glycine, proline and valine, and carbohydrates, viz. glucose, mannose, arabinose, galactose and glucuronic acid, which facilitates the orientation of endophytic bacteria *Bacillus pumilus* and *Corynebacterium flavescens* (Bacilio-Jimenez et al. 2003). Plant roots also secrete compounds such as phenols and terpenoids which play a defensive role and suppress infection by phytopathogens. Cinnamic acid derivatives, namely, phenylpropanoids, were secreted by the roots of barley plant infected by *Fusarium graminearum* (Lanoue et al. 2010). Badri et al. (2013) documented that phytochemicals, especially phenolic compounds, played a major role in recruiting microbes in *Arabidopsis* rhizosphere.

For instance, canavanine, the amino acid present in the root exudates, attracted a particular group of microbes and deterred few other taxa, thereby shaping soil microbial community. Lebeis et al. (2015) reported the role of salicylic acid in sculpturing the root-associated microbiomes in *Arabidopsis*.

The composition of sugars and organic acid in root exudates varies in quantity and quality with plant species and developmental stage which in turn modulates antibiotic biosynthesis and offers protection against soilborne phytopathogens (Kravchenko et al. 2003). Acetosyringone and hydroxyacetosyringone (phenylpropanoids) secreted by damaged plants serve as an attractant for *Agrobacterium tumefaciens* (Dixon 1995). Rhizodeposits contain flavonoids which were evidenced to regulate quorum-sensing (QS) signals for *Pseudomonas aeruginosa* PAO1 (Vandeputte et al. 2011) and nodulation genes in rhizobia (Hassan and Mathesius 2012). Chen et al. (2015) evidenced that the vacuolar sugar transporter gene in *Arabidopsis* (*SWEET2*) controlled the glucose efflux from *Arabidopsis* roots, thereby inhibiting infection by *Pythium*. Szoboszlay et al. (2016) reported that the bulk soil treated with a flavonoid 7,4′-dihydroxyflavone commonly found in *Medicago sativa* root exudates showed enhanced species richness of bacteria belonging to the taxa *Acidobacteria*, *Nocardioidaceae*, *Thermomonosporaceae* and *Gaiella*. Candidate gene approach by mutation studies revealed the effect of plant genotype on the microbial community of *Arabidopsis thaliana* phyllosphere microbes. Cuticle formation was affected in *pec1* and *lac*s mutants, and this condition increased the diversity of microbial community composition as well as bacterial abundance. Additionally, ethylene signalling gene (*ein2*) of the host also influenced the composition of the microbial community (Bodenhausen et al. 2014). Vellend and Agrawal (2010) recorded that four main processes, namely, dispersal, drift, speciation and selection, influence the microbial community composition and its diversity.

2.3.3 Plant Developmental Stage

Structural and functional diversity of the microbial community associated with the plants is dynamic and changes throughout the plant phenology. Plant age and stage is another driver that shapes the associated microbial community. Microbial association and interaction starts from seed material. Seeds acquire their microbiome from the parent plant and transport them to the new environment by seed dispersal. In turn, the microbiome protects the seeds from pathogenic infections. Seed-borne microbes gain a competitive advantage and close association with the host plants on seed germination as compared to the opportunistic microbes from the surrounding soil. Though plant genotype determines the chemistry of root exudates, their titre is influenced by the age of the plant. Thus, the age of the plant impacts the rhizodeposits which concurrently influences the associated rhizosphere microbial community (Bulgarelli et al. 2013). Quantity and quality of rhizodeposits of the same plant species vary with age. In most cases, especially in annuals, the rhizodeposits decrease with increase in plant age. Rhizodeposits of young plants were rich in low molecular weight compounds as compared to older plants (Vestergard et al. 2008). Plant age had a significant correlation with molecular and functional diversity of rhizosphere microbes as evidenced in case of many other crops including maize (Baudoin et al. 2002), *Medicago* (Mougel et al. 2006), wheat, pea and sugar beet (Houlden et al. 2008). Epiphytic bacteria are found to exceed in number as compared to the endophytes. Younger plants were found to have a higher population of endophytes as compared to the mature ones as evidenced by the concentration of endophytic bacterium *Herbaspirillum* in paddy, and this might be attributed to the fact that the non-pathogenic endophytes could not withstand the plant defence mechanisms which increases with the age of the plant (James et al. 2002). Denaturing gradient gel electrophoresis (DGGE) analysis of the *Arabidopsis* rhizosphere bacterial communities showed that exudates during seed germination attract more diverse rhizosphere microbes and the titre of exudates slowly declines with plants aging resulting in negligible differences between microbes in rhizosphere and the bulk soils (Micallef et al. 2009). Metatranscriptomics analysis of *Arabidopsis* rhizosphere microbiome showed that different stages of development of plants, viz. seedling, vegetative, bolting and flowering, expressed unique transcripts as the plants select the subset of associated microbes and shape their assemblage to tap their services (Chaparro et al. 2014). Sugiyama et al. (2014) observed that in soyabean the population of *Bacillus, Bradyrhizobium* and *Stenotrophomonas* was higher in the flowering stage as compared to other vegetative and pod-setting phase; however, no such differences were traced with the fungal communities.

2.3.4 Microbe-Microbe Interaction

Microbial species present in a community also influence the survival and performance of their microbial counterparts belonging to different taxa by the process of niche construction or modification. Competition for space and nutrition, production of secondary metabolites, effector proteins, polysaccharides, induction of plant defences, etc. by the microbial partners decide the species richness of the microbial community occupying the particular niche in the host plant. Primary microbes produce effector proteins and secondary metabolites that modify the host metabolism and establishment of other secondary microbes, thereby shaping the plant microbiomes. Certain microbes produce exopolysaccharides (EPS), phytoalexins, etc., which protect themselves and the other bacterial immigrants. Arbuscular mycorrhizal fungi (AMF) infection in plants was evidenced to alter the other microbial members in the community including bacteria which might be due to the effect of antibiotics or stimulatory compounds they produce (Marschner et al. 2001; Vestergard et al. 2008). Poza-Poza-Carrion et al. (2013) evidenced that the primary colonizers, viz. *Pseudomonas fluorescens, P. syringae* and *Erwinia herbicola*, determine the colonization of lettuce leaf by a human pathogenic strain of *Salmonella enterica.* The resident epiphytes assist the colonization by the immigrant by providing resistance to desiccation. Investigations on *Arabidopsis* leaf microbiome sampled in different seasons documented six microbial hubs including fungi (*Udeniomyces*, *Dioszegia*), oomycete (*Albugo*) and bacteria (*Caulobacter* and two species of order Burkholderiales). An artificial infestation of *Arabidopsis* with *Albugo laibachii* has a negligible effect on the phyllosphere microbiome structure, while *Dioszegia* sp. infestation showed 100-fold reduction in *Caulobacter* sp. evidencing the disproportionate role played by microbial hubs in structuring the microbiomes (Agler et al. 2016). Hyperparasitism of primary colonizers is another mechanism used by microbes in shaping their community structure. *Pythium oligandrum*, an oomycete, effectively parasitizes another oomycete, *Phytophthora infestans*, causing late blight of potato. This opens up an avenue for use of the mycoparasitic *Pythium oligandrum* as an effective biocontrol agent of *Phytophthora* (Horner et al. 2012). Fungal endophytes *Neotyphodium* sp. and *Epichloe* sp. harboured in fescue species produced secondary metabolites loline which not only protected the plant from herbivory but also shaped the establishment of epiphytic microbes such as *Burkholderia ambifaria* that could utilize lolines as a source of carbon and nitrogen (Roberts and Lindow 2014).

2.3.5 Soil Factors

Soil type and physio-chemistry, viz. texture, structure, water retention potential, nutrient availability, pH, organic matter content, etc., decide the native microbial community structure and functioning evidencing their role in nutrient cycling in bulk soils. The same is true with rhizosphere soils as the soil physiochemical properties influence the availability of root exudates and in turn its role in microbial recruitment (Ho et al. 2017). Root exudates of seedlings of *Pinus radiata* that are grown in phosphate-deficient soils have been documented to produce double the amount of amino acids and amides than under normal conditions (Bowen 1969). Soil type plays a major role in determining the rhizobial community in soya bean, as compared to the plant genotype which has also been reported to influence rhizosphere microbiology. DGGE and sequence analysis showed that members belonging to Acidobacteria, Actinobacteria, Bacteroidetes, Proteobacteria, Firmicutes, Nitrospirae and Verrucomicrobia (Xu et al. 2009) and fungi belonging to Ascomycetes and Basidiomycetes (Wang et al. 2009) were predominant inhabitants of soyabean rhizosphere. Both the above experiments were carried out by the same research group in which they documented that black soils (Mollisol) supported the diversity of rhizobacteria, while dark brown soil (Alfisol) supported the diversity of fungal communities. Comparison of the bacterial community in differently sized soil particles from field subjected to long-term fertilization by 16S rRNA genes and TRFLP analyses showed that fine particles harboured diverse microflora and included members of *Holophaga* and *Acidobacterium*, while coarse particles supported lesser diversity and were enriched with α-Proteobacteria. Additionally, this study evidenced that soil particle size initially determined the specificity of the associated microbial taxon as compared to fertilizer amendments (Sessitsch et al. 2001). Studies on soil microbial diversity at a continental scale by ribosomal DNA fingerprinting showed that microbial diversity was affected by ecosystem type and the major factor being soil pH with neutral soils supporting rich diversity and acidic soils have a poor diversity (Fierer and Jackson 2006). Long-term fertilization significantly impacted the soil pH, soil carbon content and community diversity of bacteria and mycorrhiza in maize rhizosphere (Toljander et al. 2008). Soil deficient in nitrogen enhances the plant grown in it to secrete more of flavonols and flavones which initiates rhizobia-legume symbiosis evidencing the effect of soil chemistry in orchestrating plant rhizosphere microbiomes (Davidson and Robson 1986; Zhang et al. 2009). Similarly, plants grown in iron-deficient soils are evidenced to excrete more phenolic compounds via their roots, which greatly impacts the microbial community colonizing the rhizosphere region (Jin et al. 2014). In addition to the use of synthetic fertilizers and agrochemicals that deprive the soil microbial diversity and efficacy, cropping systems (Xiong et al. 2015a, b), and other agriculture practices including logging (Hartmann et al. 2014), soil tillage (Souza et al. 2016), etc. disturbs the soil integrity, its aggregation patterns, infiltration capacity and organic carbon content indirectly impacting structural and functional diversity of associated microbiomes. Conservation agriculture, which supports zero tillage and practises organic manuring positively influences, strengthens and stabilizes microbial community diversity and biomass (Wang et al. 2017).

2.3.6 Miscellaneous Environmental Factors

Several hypotheses including 'niche theory' and 'neutral theory' attempt to explain the species assemblage in microbial community, and both the theories emphasize the selective role of environmental variables on species assemblage (Mendes et al. 2014). Factors like soil temperature and water availability also influence the plant microbiome. Soil moisture has been evidenced to influence the composition of root exudates. Plants grown in conditions with limited soil moisture showed an increase in amino acid production which in turn affected the microbiology of rhizosphere (Katznelson et al. 1955). The temperature has a profound impact on the plant metabolism, on biochemistry and in turn on the root exudates. Studies revealed that root exudates of strawberry plants grown in low soil temperatures ranging between 5 and 10 °C produced more amino acids that affected the pathogenicity of *Rhizoctonia fragariae* and its infection in strawberry as compared to plants grown in 20–30 °C (Husain and McKeen 1963). Warmer and humid conditions of the tropics favour microbial richness especially those in the phyllosphere as compared to temperate ones (Vorholt 2012). Copeland et al. (2015) reported the effect of seasonal variation in phyllosphere microbiome succession. Toljander et al. (2008) from their scientific investigations hypothesized the impact of temporal variation and crop harvest in the bacterial community composition. Studies based on RNA operon copy numbers indicate that in environments subjected to disturbance as in case of agro-ecosystems, the microbial community contains organisms highly responsive to nutrient inputs but metabolically less active (Nemergut et al. 2016). Drought conditions are known to impact microbes associated with Poaceae plant roots causing a shift in the community structure (Santos-Medellín et al. 2017).

2.4 Role of Plant Microbiome in Preserving Plant Health

All living organisms have core microbiomes which act as secondary genomes which are tenfold larger than the host genome, and they decide the overall health and fitness of the host plant in a given ecosystem (Berg 2009; Bulgarelli et al. 2013; Gopal et al. 2013). The advent of molecular tools such as functional genomics and system biology approach depicts that the plant microbiomes are highly structured and forms complex networks which play a key role in plant health as well as the functioning of the ecosystem. Key stone species shapes the microbial hub taxa which in turn impacts the plant performance (van der Heijden and Hartmann 2016). Plant microbiomes, apart from determining the overall plant health and its ecological fitness by promoting plant growth (by enhancing nutrient and mineral availability and secretion of plant growth regulators) and evading abiotic stress (temperature, salinity, drought and heavy metals) and biotic stress (by reducing pests and disease incidence), also play a key role in biogeochemical cycle by the way of nitrogen fixation, denitrification, carbon fixation and release, methanogenesis, mineral fixation, solubilization, etc. Microbes interacting with plants in the ago-ecosystem are responsible for the release of a significant amount of methane and nitrous oxide from the system leading to greenhouse effects. Zolla et al. (2013) documented that core microbiome of soil under study contained members of *Aminobacter, Acidiphilum, Bacillus, Burkholderia* and *Phormidium* which were involved in alleviating abiotic stress in plants grown in them. Beneficial bacteria belonging to genera *Azospirillum, Achromobacter, Azotobacter, Bacillus, Burkholderia, Enterobacter, Microbacterium, Methylobacterium, Pseudomonas, Rhizobium, Pantoea, Paenibacillus* and *Variovorax* have been reported to promote plant growth (Pathma and Sakthivel 2013) and provide tolerance to abiotic (Grover et al. 2011) and biotic stress (Pathma et al. 2019).

Beneficial bacteria support plant growth directly by nitrogen fixation; phosphorous, potassium, calcium, zinc and silica solubilization and mobilization (Edwards and Burrows 1988); production of plant growth regulators such as indole-3-acetic acid (IAA), cytokinins, gibberellins and aminocyclopropane-1-carboxylate (ACC) deaminase; etc. (Glick 1995; Penrose and Glick 2002). They indirectly promote plant growth by evading biotic and abiotic stress. Beneficial microbes produce antimicrobial compounds, including antibiotics, siderophores, hydrogen cyanide and hydrolytic enzymes such as pectinase, chitinase, DNAse, lipase etc., which protect the host plant from the invading phytopathogens and herbivores. Among the plant-associated microbes members of genus *Pseudomonas, Bacillus* and *Streptomyces* are known to be prolific producers of antibiotics that protect the host plants from phytopathogenic invasions. The compounds produced include phenazines, phloroglucinols, phenolics, pyrrole-type compounds, polyketides, peptides, bacteriocins, lantibiotics, cyclic lipoheptapeptide, macrolactones, phospholipids, coumarins, aminopolyols, adenine nucleotide analogues, polyacetylene derivatives, aminoglycoside, quinones, etc. (Pathma et al. 2011). Members of *Pseudomonas* and *Bacillus* with inhibitory effect against many phytopathogens including *Xanthomonas* spp., *Agrobacterium tumefaciens, Erwinia amylovora,*

Colletotrichum spp., *Fusarium* spp., *Rhizoctonia solani, Helminthosporium* sp., *Pestalotia theae*, *Macrophomina phaseolina* and *Sarocladium oryzae* have been reported (Ho et al. 2017; Pathma et al. 2019). *P. fluorescens* WCS417r and *P. fluorescens* CHA0 induced systemic resistance in carnation and tomato, respectively, and protected them from infection by *F. oxysporum* (Van Peer et al. 1991; Ardebili et al. 2011). Endophytes *P. fluorescens* 89B-61, *Achromobacter* sp. F2feb.44, *B. licheniformis* AE6 and *Streptomyces* sp. Zapt10 were used to induce systemic resistance in cucumber and protect it from downy mildew caused by *Pseudoperonospora cubensis* (Kloepper and Ryu 2006; Sen et al. 2014). Similarly, plant-associated microbiomes also evidenced protection against herbivore pests. This includes control of lepidopterans, coleopterans and nematodes by *Brevibacillus laterosporus* (Ho et al. 2017); cotton aphids by *Bacillus pumilus* INR-7 (Stout et al. 2002); and wheat aphids by a mixture of *Pseudomonas* sp. strain 6 K and *Bacillus* sp. strain 6 (Naeem et al. 2018). PGPR strains were also lethal to blue-green aphids (Kempster et al. 2002), green peach aphids (Boughton et al. 2006) and termites (Sindhu et al. 2011).

Plant growth-promoting rhizobacterium, *Burkholderia phytofirmans* PsJN, was reported to impart cold tolerance in grapevine plants inoculated with it (Barka et al. 2006). Verma et al. (2015) reported plant growth-promoting properties of a psychrotolerant epiphytic strain *Methylobacterium phyllosphaerae* IARI-HHS2-67, isolated from wheat phyllosphere. *Pseudomonas* sp. DSMZ 13134, *Bacillus simplex* and *B. amyloliquefaciens* subsp. *plantarum* were reported to protect maize against cold stress (Bradacova et al. 2016), and members of *Pseudomonas, Arthrobacter, Flavobacterium, Pedobacter* and *Flavimonas* protected tomato seedlings from chilling injury (Subramanian et al. 2016). *Pseudomonas* sp. strain AKM-P6 offered protection to sorghum against increased temperature (Ali et al. 2009), while *P. putida* AKMP7 provided thermotolerance and growth promotion in wheat under heat stress (Ali et al. 2011). The mechanisms involved included increased production of cellular metabolites, proteins, amino acids such as proline, chlorophyll and sugars and reduced activity of antioxidant enzymes and reduced membrane damage. *Pseudomonas stutzeri, P. aeruginosa* and *P. fluorescens* provided halo tolerance in tomato plants (Tank and Saraf 2010), endophytic *P. pseudoalcaligenes* offered salinity tolerance in paddy (Jha et al. 2011), rhizobacteria *Dietzia natronolimnaea* protected wheat from salt stress (Bharti et al. 2016) and *Achromobacter piechaudii* ARV8 and *P. fluorescens* Pf1 protected tomato (Mayak et al. 2004) and green gram (Saravanakumar et al. 2011), respectively, from water stress. *Kluyvera ascorbata* protected canola from nickel toxicity (Burd et al. 1998), while *Photobacterium halotolerans* strain MELD1 offered protection to *Vigna unguiculata* ssp. *sesquipedalis* against mercury toxicity (Mathew et al. 2015). An endophyte *Achromobacter xylosoxidans* F3B detoxified aromatic pollutants from *Chrysopogon zizanioides* and *A. thaliana* (Ho et al. 2013).

Thus, plant microbiomes are potential reservoirs that could be tuned and recruited systematically so as to offer maximum beneficial services for agriculture and thereby to mankind. The continuous evolution of plants and associated microbes both good and bad needs constant research to update knowledge and find prominent solutions to the selection pressure caused by phytopathogens. Microbial diversity can be employed as efficient biomarkers to identify healthy microbiomes and utilize them systematically in breeding and biological control programmes which will help us conserve the biodiversity preserving ecosystem health and ensure self-sustainable agricultural production systems (Berg et al. 2017).

2.5 Molecular Tools for Analysing Microbial Community Diversity and Their Interaction with Host Plant

Though microbial culturing techniques and biochemical analysis had appreciably contributed for studies on microbial taxonomy and functional diversity as well as their interaction with host plants until the last century, the advent of molecular techniques has added new insights. Some molecular diagnostic tools used for studying plant-microbe interaction are depicted (Table 2.1). Culture-dependent methods enabled us to study only a small portion ($< 1\%$) of microbes especially confining it to aerobic bacteria or particular taxa, viz. *Pseudomonas, Bacillus*, etc., (Staley and Konopka 1985). The limitations were addressed by microscopic analysis of environmental samples that could enable visualization of the live or fixed microbe by using high-resolution techniques, viz. confocal microscopy, electron microscopy and fluorescence microscopy including fluorescence in situ hybridization (*FISH*) and photoswitchable fluorophores for single-molecule localization microscopy (SMLM) (Coltharp and Xiao 2012). Though these techniques provide high-resolution images, identification and classification of the microbes in a community becomes challenging even for an experienced taxonomist and at times misguiding where biochemical profiling and molecular fingerprinting had come to the rescue (Hugerth and Andersson 2017). An array of techniques including the use of small subunit (SSU) of the ribosomal RNA (rRNA) gene, ribosomal internal transcribed spacer (ITS) (Woese and Fox 1977; Pace et al. 1985), denaturing gradient gel electrophoresis (DGGE) (Muyzer et al. 1993), analysis of phospholipid-derived fatty acids (*PLFA*) (Tunlid et al. 1985; Buyer et al. 1999; Willers et al. 2015), in vivo expression technology (IVET) (Osbourn et al. 1987; Rainey et al. 1997), terminal restriction fragment length polymorphism (T-RFLP) analysis (Liu et al. 1997; Lukow et al. 2000), automated ribosomal intergenic spacer analysis (ARISA) (Fisher and Triplett 1999), fluorescence induction promoter traps (Rediers et al. 2005), microarray based on 16SrRNA (Ehrenreich 2006; Sanguin et al. 2006), 454 pyrosequencing, analysis of total nucleic acids from the environment, metagenomics (Handelsman 2004; Erkel et al. 2006; Leveau 2007), ultradeep sequencing (Velicer et al. 2006), transcriptome analysis (Mark et al. 2005; Yuan et al. 2008), flow cytometry for in situ antifungal gene expression (De Werra et al. 2008), use of isotope probes (Haichar et al. 2008), real-time PCR (RT-PCR), chromatography techniques, Fourier transform infrared spectroscopy (FTIR), nuclear magnetic resonance (NMR) (Wu et al. 2009), differential fluorescence induction (DFI), signature tagged mutagenesis (STM), single-molecule real-time (SMRT) sequencing (Walder et al. 2017), etc. provides clear, detailed insights on plant microbiomes. Multiphasic approaches,

Table 2.1 Molecular diagnosis of plant-microbe interaction

S. no	Host plant specimen	Associated microbial domain studied	Diagnostic tool	Reveals	References
1	Soybean	Bacteria	Shotgun metagenomics	Identification of beneficial functional traits	Mendes et al. (2014)
2	*Jacobaea vulgaris*	Bacteria	16S rRNA sequencing and shotgun metagenomics	Assembly of bacterial rhizosphere communities	Yan et al. (2017)
3	*Lilium davidii* var. *unicolor*	Bacterial and fungal communities	Illumina MiSeq sequencing of 16S rRNA and ITS gene amplicons	Identification of biological control agents against soilborne pathogens	Shang et al. (2016)
4	*Panax notoginseng*	Fungi	Illumina MiSeq of fungal ITS gene markers	Rhizospheric soil and root endogenous fungal diversity for plant health and soil fertility	Tan et al. (2017)
5	Boreal pine forest (*Calluna vulgaris, Vaccinium myrtillus* and *Vaccinium vitis-idaea*)	Bacteria	qPCR and next-generation sequencing (NGS) of 16S rRNA	Root- and hyphae-associated bacteria may have a functionally more active role in soil microbiome	Timonen et al. (2017)
6	*Saccharum* sp.	Bacteria	16S rRNA gene multiplex amplicon sequencing	Identification of community-based culture collection (CBC) having novel plant growth-promoting traits	Armanhi et al. (2018)
7	*Vitis vinifera*	Bacterial and fungi genera	16S rRNA gene sequencing and of the internal transcribed spacer (ITS)	Identification of dominated bacterial genera in microbial community of *Vitis vinifera* phyllosphere present in French Mediterranean region	Singh et al. (2018)
8	*Populus deltoides* and *P. trichocarpa*	Bacteria	16S rDNA PCR followed by Sanger sequencing	Genomes of plant-associated bacteria are surprisingly consistent across phylogenetically diverse bacterial taxa and that some functions are even shared with PA eukaryotes	Levy et al. (2018)

(continued)

Table 2.1 (continued)

S. no	Host plant specimen	Associated microbial domain studied	Diagnostic tool	Reveals	References
9	*Capsicum annuum*	Bacteria	16S rRNA PCR-DGGE analysis	95% of in vitro microbiome promotes plant growth and stress resistance	Marasco et al. (2012)
10	*Vitis vinifera* cv. barbera	Bacteria	16S rRNA PCR-DGGE analysis	Root-associated bacterial microbiome improves plant adaptation to drought through a water stress-induced promotion ability	Rolli et al. (2015)
11	*Arabidopsis thaliana* Col-0	Bacteria	Transcriptome analysis	A subset of commensals trigger expression of defence-related genes and thereby may contribute to plant health upon pathogen encounter	Vogel et al. (2016)
12	*Arabidopsis thaliana*	Bacteria	16S rRNA gene sequencing	Salicylic acid controls colonization of specific bacterial families in roots	Lebeis et al. (2015)
13	*Arabidopsis thaliana*	Fungus (*Piriformospora indica*)	Gene expression analysis by qRT-PCR	Suppression of innate immune system in roots confers the colonization of fungus (*Piriformospora indica*) in *Arabidopsis* roots	Jacobs et al. (2011)
14	*Arabidopsis thaliana*	Bacteria	16S rRNA and whole-genome sequencing	Identification of taxonomic overlap between the leaf and root microbiota	Bai et al. (2015)
15	*Crotalaria pumila*	Bacteria (*Methylobacterium* sp. Cp3)	16S rRNA gene sequencing	Seed endophyte bacteria (*Methylobacterium* sp. Cp3) colonize in the root cortex cells and xylem vessels of the stem under metal stress	Sanchez-Lopez et al. (2018)

metagenomics, metaproteomics, metatranscriptomics including next-generation sequencing (NGS) technologies and bioinformatics, had provided a novel, deeper and comprehensive insights on plant-microbe interaction (Turner et al. 2013a, b; Mendes et al. 2014; Hacquard et al. 2015). Ramirez-Flandes et al. (2019) showed that genes controlling the redox potential of the microbes could possibly be used to characterize the microbial assemblies in the corresponding microbiomes which are interlinked to the energetics of the ecosystem, thereby enabling differentiation among microbes in highly dynamic complex associations.

2.6 Engineering Plant Microbiomes for Sustainable Production Systems

Extensive research that would unlock the complexity of plant-associated microbiomes and their ecology will provide us with clues to understand the process of microbial assembly as well as their links and importance in plant performance. The advent of molecular tools had evidenced the advantage of transferring the core soil microbiome over the previous practice of inoculating the crops with single strain or consortium of beneficial microbes since plant growth promotion or biotic and abiotic stress evasion was evidenced to be the combined function of the rhizosphere microbiome instead of a single taxon. Thus, the practice of inoculating the core microbiome as such by transferring the disease suppressive soils as rhizosphere substitutes can improve the success rate of the use of microbes for improved crop production and protection (Berendsen et al. 2012). Construction of synthetic microbiomes with beneficial microbes and inoculating the plant with it artificially is one of the techniques in microbiome engineering. Mueller and Sachs (2015) emphasized the use of host phenotype as a probe for the selection of members of the synthetic microbiome. This technique is termed as host-mediated microbiome engineering. One simple cost-effective means of engineering root microbiomes is mixing up of disease suppressive soils with disease conducive soils which had proved its potential in controlling black root rot in tobacco (Kyselkova et al. 2009), *Rhizoctonia* infection in sugar beet (Mendes et al. 2011) and common scab in potato (Rosenzweig et al. 2012). Analyses of soil metagenome evidenced that core microbiome of the soil contained 17 bacterial communities belonging to Actinobacteria, Firmicutes and Proteobacteria with biocontrol properties that were responsible for the disease suppressive nature of the soil. Among all the studies, bacteria belonging to Pseudomonadaceae was identified as key players responsible for disease suppression (Gopal et al. 2013). Cutting-edge molecular biology techniques, viz. next-generation sequencing, transcriptome profiling of multispecies (Schenk et al. 2012), bioinformatics tools (Lee et al. 2012) and advanced spectroscopy that helps in identification of microbial bioactive molecules (Watrous et al. 2012; Badri et al. 2013), provided deeper insights on the microbiome and the success rate of their use in sustainable agriculture. Apart from data on operational taxonomic units (OUTs), comprehensive data collection on OTU α and β diversity, spatial and temporal persistence, metabolic networking and their studies in crop model will assist in proficient assembling of robust microbiomes especially those associated with the rhizosphere as well as successful establishment and functioning of the introduced microbiomes in the new ecosystem (Shade and Handelsman 2012; Lozupone et al. 2012; Scheuring and Yu 2013). Identification and use of members of core microbiomes of plants and genetically engineering them with genes encoding essential proteins or compounds that help in crop protection improves the efficacy of the technique as the engineered microbe infects the host plant efficiently and transfers the required trait. For instance, *Pantoea agglomerans* (33.1), an endophyte of sugarcane with growth-promoting activity, was engineered with cry1Ac7 gene, and it provided excellent control of lepidopteran borer of sugarcane *Diatraea saccharalis*. Similar genetic modification with *Bacillus thuringiensis* δ-endotoxin has been attempted in endophytes *Clavibacter xyli* and *Herbaspirillum seropedicae*. Since the endophyte colonizes the sugarcane tissue internally and hence the larval stage of the borer pest occupies the same niche, it could not escape the cry toxins. Thus, this mode of delivery enhances the success rate of the biopesticides than being applied as a foliar spray (Downing et al. 2000; Quecine et al. 2014).

2.7 Future Perspectives and Conclusion

Projects funded by the National Institutes of Health and European Union intensified research on human microbiomes, which opened up new avenues in the field of human medicine. Similar co-ordinated and focussed research on plant microbiome will help us identify and appreciate beneficial microbiomes and integrate them in crop pest and nutrient management programmes and reap benefits of the services they provide. Understanding the plant microbiome and their interaction, which has a huge impact on the holobiont and the associated ecosystem, is an important research prospect and is highly challenging. Sampling techniques and data processing protocols need to be standardized as per research priorities in order to fill the knowledge gaps. In-depth understanding of the microbiome community and their functional diversity by advance molecular approaches like metagenomics and metabolomics can reveal the uncultivable hidden microbial partners and their role in plant health and other ecosystem services and help us utilize this versatile bioresource for sustained eco-friendly agricultural production systems which is the need of the hour to feed the increasing population with depleting resources without posing pollution pressure on the production system. Though plant-microbe interaction studies have been done since the beginning of the twentieth century, it is still a brooding area of research that could benefit the human community by their indispensable role not only in agriculture sector but also in medicine and environmental protection. Genetically modified crops can greatly impact the native microbiomes of the plant leading to unpredicted changes in the diversity of associated microbes which may be fruitful or detrimental. Hence, it is equally important to consider the plant-associated microbiomes while designing better performing hybrids or GM crops. Hence, the use of biotechnological tools in crop breeding and biocontrol programmes should be designed in a way to recruit beneficial communities as well as to minimize the loss of microbial biodiversity so as to provide a self-sustained production system that enhances plant health, ecological fitness and performance.

Acknowledgement The authors wish to thank Dr. R. Nagarajaprakash, group leader, Chemical Sciences Research Group, Lovely Professional University, Punjab, India, for his constructive criticism and support in preparing this book chapter.

References

Agler MT, Ruhe J, Kroll S et al (2016) Microbial hub taxa link host and abiotic factors to plant microbiome variation. PLoS Biol 14:e1002352

Ali SZ, Sandhya V, Grover M et al (2009) *Pseudomonas* sp. strain AKM-P6 enhances tolerance of sorghum seedlings to elevated temperatures. Biol Fertil Soils 46:45–55

Ali SZ, Sandhya V, Grover M et al (2011) Effect of inoculation with a thermotolerant plant growth promoting *Pseudomonas putida* strain AKMP7 on growth of wheat (*Triticum* spp.) under heat stress. J Plant Interact 6:239–246

Ardebili ZO, Ardebili NO, Hamdi SMM (2011) Physiological effects of *Pseudomonas fluorescens* CHA0 on tomato (*Lycopersicon esculentum* Mill.) plants and its possible impact *on Fusarium oxysporum* f. sp. *lycopersici*. Aust J Crop Sci 5:1631–1638

Armanhi JSL, de Souza RSC, Damasceno NDB et al (2018) A community-based culture collection for targeting novel plant growth-promoting bacteria from the sugarcane microbiome. Front Plant Sci 8:2191

Bacilio-Jimenez M, Aguilar-Flores S, Ventura-Zapata E et al (2003) Chemical characterization of root exudates from rice (Oryza sativa) and their effects on the chemotactic response of endophytic bacteria. Plant Soil 249:271–277

Badri DV, Chaparro JM, Zhang R et al (2013) Application of natural blends of phytochemicals derived from the root exudates of arabidopsis to the soil reveal that phenolic related compounds predominantly modulate the soil microbiome. J Biol Chem 288:4502–4512

Bai Y, Muller DB, Srinivas G et al (2015) Functional overlap of the *Arabidopsis* leaf and root microbiota. Nature 528:364–369

Barka EA, Nowak J, Clement C (2006) Enhancement of chilling resistance of inoculated grapevine plantlets with a plant growth-promoting rhizobacterium, *Burkholderia phytofirmans* strain PsJN. Appl Environ Microbiol 72:7246–7252

Baudoin E, Benizri E, Guckert A (2002) Impact of growth stage on the bacterial community structure along maize roots, as determined by metabolic and genetic fingerprinting. Appl Soil Ecol 19:135–145

Berendsen RL, Pieterse CM, Bakker PA (2012) The rhizosphere microbiome and plant health. Trends Plant Sci 17:478–486

Berg G (2009) Plant–microbe interactions promoting plant growth and health: perspectives for controlled use of microorganisms in agriculture. Appl Microbiol Biotechnol 84:11–18

Berg G, Martina K, Daria R et al (2017) Plant microbial diversity is suggested as the key to future biocontrol and health trends. FEMS Microbiol Ecol 93:1–9

Bharti N, Pandey SS, Barnawal D et al (2016) Plant growth promoting rhizobacteria Dietzia natronolimnaea modulates the expression of stress responsive genes providing protection of wheat from salinity stress. Sci Rep 6:34768

Bitocchi E, Bellucci E, Giardini A et al (2013) Molecular analysis of the parallel domestication of the common bean (*Phaseolus vulgaris*) in Mesoamerica and the Andes. New Phytol 197:300–313

Bodenhausen N, Bortfeld-Miller M, Ackermann M et al (2014) A synthetic community approach reveals plant genotypes affecting the phyllosphere microbiota. PLoS Genet 10:e1004283

Bouffaud ML, Poirier MA, Mulle D et al (2014) Root microbiome relates to plant host evolution in maize and other Poaceae. Environ Microbiol 16:2804–2814

Boughton AJ, Hoover K, Felton GW (2006) Impact of chemical elicitor applications on greenhouse tomato plants and population growth of the green peach aphid, *Myzus persicae*. Entomol Exp Appl 120:175–188

Bowen G (1969) Nutrient status effects on loss of amides and amino acids from pine roots. Plant Soil 30:139–142

Bradacova K, Weber NF, Morad-Talab N et al (2016) Micronutrients (Zn/Mn), seaweed extracts, and plant growth-promoting bacteria as cold-stress protectants in maize. Chem Biol Technol Agric 3:19

Brundrett MC (2002) Coevolution of roots and mycorrhizas of land plants. New Phytol 154:275–304

Bulgarelli D, Rott M, Schlaeppi K et al (2012) Revealing structure and assembly cues for Arabidopsis root-inhabiting bacterial microbiota. Nature 488:91–95

Bulgarelli D, Schlaeppi K, Spaepen S et al (2013) Structure and functions of the bacterial microbiota of plants. Annu Rev Plant Biol 64:807–838

Bulgarelli D, Garrido-Oter R, Munch PC et al (2015) Structure and function of the bacterial root microbiota in wild and domesticated barley. Cell Host Microbe 17:392–403

Burd GI, Dixon DG, Glick BR (1998) A plant growth-promoting bacterium that decreases nickel toxicity in seedlings. Appl Environ Microbiol 64:3663–3668

Buyer JS, Roberts DP, Cohen RE (1999) Microbial community structure and function in the spermosphere as affected by soil and seed type. Can J Microbiol 45:138–144

Cardinale M, Grube M, Erlacher A et al (2015) Bacterial networks and co-occurrence relationships in the lettuce root microbiota. Environ Microbiol 17:239–252

Chaparro JM, Badri DV, Vivanco JM et al (2014) Rhizosphere microbiome assemblage is affected by plant development. ISME J 8:790–803

Chen HY, Huh JH, Yu YC et al (2015) The Arabidopsis vacuolar sugar transporter SWEET2 limits carbon sequestration from roots and restricts Pythium infection. Plant J83:1046–1058

Chisholm ST, Coaker G, Day B et al (2006) Host-microbe interactions: shaping the evolution of the plant immune response. Cell 124:803–814

Coltharp C, Xiao J (2012) Super resolution microscopy for microbiology. Cell Microbiol 14:1808–1818

Copeland JK, Yuan L, Layeghifard M et al (2015) Seasonal community succession of the phyllosphere microbiome. Mol Plant-Microbe Interact 28:274–285

Costa R, Gotz M, Mrotzek N et al (2006) Effects of site and plant species on rhizosphere community structure as revealed by molecular analysis of microbial guilds. FEMS Microbiol Ecol 56:236–249

Davidson IA, Robson MJ (1986) Effect of contrasting patterns of nitrate application on the nitrate uptake, N_2-fixation, nodulation and growth of white clover. Ann Bot 57:331–338

De Werra P, Baehler E, Huser A et al (2008) Detection of plant-modulated alterations in antifungal gene expression in Pseudomonas fluorescens CHA0 on roots by flow cytometry. Appl Environ Microbiol 74:1339–1349

Dixon RA (1995) Stress-induced phenylpropanoid metabolism. Plant Cell 7:1085–1097

Downing KJ, Leslie G, Thompson JA (2000) Biocontrol of the sugar-cane borer *Eldana saccharina* by expression of the *Bacillus thuringiensis* cry1Ac7 and *Serratia marcescens* chiA gene in sugar-cane associated bacteria. Appl Environ Microbiol 66:2804–2810

Ebert D, Hamilton WD (1996) Sex against virulence: the coevolution of parasitic diseases. Trends Ecol Evol 11(2):79–82

Edwards CA, Burrows I (1988) The potential of earthworm composts as plant growth media. In: Edwards CA, Neuhauser E (eds) Earthworms in waste and environmental management. SPB Academic Press, The Hague, pp 21–32

Ehrenreich A (2006) DNA microarray technology for the microbiologist: an overview. Appl Microbiol Biotechnol 73:255–273

Erkel C, Kube M, Reinhardt R, Liesack W (2006) Genome of rice cluster I archaea – the key methane producers in the rice rhizosphere. Science 313:370–372

Farre-Armengol G, Filella I, Llusia J et al (2016) Bidirectional interaction between phyllospheric microbiotas and plant volatile emissions. Trends Plant Sci 21:854–860

Fierer N, Jackson RB (2006) The diversity and biogeography of soil bacterial communities. Proc Natl Acad Sci 103:626–631

Fierer N, Ladau J, Clemente JC et al (2013) Reconstructing the microbial diversity and function of pre-agricultural tall grass prairie soils in the United States. Science 342:621–624

Fisher MM, Triplett EW (1999) Automated approach for ribosomal intergenic spacer analysis of microbial diversity and its application to freshwater bacterial communities. Appl Environ Microbiol 65:4630–4636

Fitzpatrick CR, Copeland J, Wang PW et al (2018) Assembly and ecological function of the root microbiome across angiosperm plant species. Proc Natl Acad Sci 115:1157–1165

Gans J, Woilinsky M, Dunbar J (2005) Computational improvements reveal great bacterial diversity and high metal toxicity in soil. Science 309:1387–1390

Garcia-Palacios P, Milla R, Delgado-Baquerizo M et al (2013) Side-effects of plant domestication: ecosystem impacts of changes in litter quality. New Phytol 198:504–513

Germida JJ, Siciliano SD (2001) Taxonomic diversity of bacteria associated with the roots of modern, recent and ancient wheat cultivars. Biol Fertil Soils 33:410–415

Girvan MS, Bullimore J, Pretty JN et al (2003) Soil type is the primary determinant of the composition of the total and active bacterial communities in arable soils. Appl Environ Microbiol 69:1800–1809

Glazebrook J (2005) Contrasting mechanisms of defense against biotrophic and necrotrophic pathogens. Annu Rev Phytopathol 43:205–227

Glick BR (1995) The enhancement of plant growth by free-living bacteria. Can J Microbiol 41:109–117

Gopal M, Gupta A, Thomas GV (2013) Bespoke microbiome therapy to manage plant diseases. Front Microbiol 4:355

Grayston SJ, Wang SQ, Campbell CD et al (1998) Selective influence of plant species on microbial diversity in the rhizosphere. Soil Biol Biochem 30:369–378

Grover M, Ali SZ, Sandhya V et al (2011) Role of microorganisms in adaptation of agriculture crops to abiotic stresses. World J Microbiol Biotechnol 27:1231–1240

Hacquard S, Garrido-Oter R, Gonzalez A et al (2015) Microbiota and host nutrition across plant and animal kingdoms. Cell Host Microbe 17:603–616

Haichar FE, Marol C, Berge O et al (2008) Plant host habitat and root exudates shape soil bacterial community structure. ISME J 2:1221–1230

Handelsman J (2004) Metagenomics: application of genomics to uncultured microorganisms. Microbiol Mol Biol Rev 68:669–685

Hartmann A, Rothballer M, Schmid M (2007) Lorenz Hiltner, a pioneer in rhizosphere microbial ecology and soil bacteriology research. Plant Soil 312:7–14

Hartmann M, Niklaus PA, Zimmermann S et al (2014) Resistance and resilience of the forest soil microbiome to logging-associated compaction. ISME J 8:226–244

Hassan S, Mathesius U (2012) The role of flavonoids in root-rhizosphere signalling: opportunities and challenges for improving plant-microbe interactions. J Exp Bot 63:3429–3444

Haudry A, Cenci A, Ravel C et al (2007) Grinding up wheat: a massive loss of nucleotide diversity since domestication. Mol Biol Evol 24:1506–1517

Hetrick BAD, Wilson GWT, Cox TS (1993) Mycorrhizal dependence of modern wheat cultivars and ancestors: a synthesis. Can J Bot 71(3):512–518

Ho YN, Hsieh JL, Huang CC (2013) Construction of a plant–microbe phytoremediation system: combination of vetiver grass with a functional endophytic bacterium, *Achromobacter xylosoxidans* F3B, for aromatic pollutants removal. Bioresour Technol 145:43–47

Ho YN, Mathew DC, Huang C (2017) Plant-microbe ecology: interactions of plants and symbiotic microbial communities. In: Yousaf Z (ed) Plant ecology – traditional approaches to recent trends. https://doi.org/10.5772/intechopen.69088

Horner NR, Grenville-Briggs LJ, van West P (2012) The oomycete *Pythium oligandrum* expresses putative effectors during mycoparasitism of *Phytophthora infestans* and is amenable to transformation. Fungal Biol 116:24–41

Horner-Devine M, Lage M, Hughes J et al (2004) A taxa-area relationship for bacteria. Nature 432:750–753

Houlden A, Timms-Wilson TM, Day MJ et al (2008) Influence of plant developmental stage on microbial community structure and activity in the rhizosphere of three field crops. FEMS Microbiol Ecol 65:193–201

Hugerth LW, Andersson AF (2017) Analysing microbial community composition through amplicon sequencing: from sampling to hypothesis testing. Front Microbiol 8:1561

Husain S, McKeen W (1963) Interactions between strawberry roots and Rhizoctonia fragariae. Phytopathology 53:541–545

Inceoglu O, Al-Soud WA, Salles JF et al (2011) Comparative analysis of bacterial communities in a potato field as determined by pyrosequencing. PLoS One 6:e23321

Jacobs S, Zechmann B, Molitor A et al (2011) Broad-spectrum suppression of innate immunity is required for colonization of *Arabidopsis* roots by the fungus *Piriformospora indica*. Plant Physiol 156:726–740

James EK, Gyaneshwar P, Mathan N et al (2002) Infection and colonization of rice seedlings by the plant growth-promoting bacterium *Herbaspirillum seropedicae* Z67. Mol Plant-Microbe Interact 15:894–906

Jha Y, Subramanian R, Patel S (2011) Combination of endophytic and rhizospheric plant growth promoting rhizobacteria in *Oryza sativa* shows higher accumulation of osmoprotectant against saline stress. Acta Physiol Plant 33:797–802

Jin CW, Ye YQ, Zheng SJ (2014) An underground tale: contribution of microbial activity to plant iron acquisition via ecological processes. Ann Bot 113:7–18

Kapulnik Y, Kushnir U (1991) Growth dependency of wild, primitive and modern cultivated wheat lines on vesicular–arbuscular mycorrhiza fungi. Euphytica 56:27–36

Katznelson H, Rouatt J, Payne T (1955) The liberation of amino acids and reducing compounds by plant roots. Plant Soil 7:35–48

Kempster VN, Scott ES, Davies KA (2002) Evidence for systemic, cross-resistance in white clover (*Trifolium repens*) and annual medic (*Medicago truncatula* var. truncatula) induced by biological and chemical agents. Biocontrol Sci Tech 12:615–623

Kim DH, Kaashyap M, Rathore A et al (2014) Phylogenetic diversity of Mesorhizobium in chickpea. J Biosci 39:513–517

Kloepper JW, Ryu CM (2006) Bacterial endophytes as elicitors of induced systemic resistance. In: Schulz BJE, Boyle CJC, Sieber TN (eds) Microbial root endophytes. Springer, Berlin/Heidelberg, pp 33–52

Kravchenko LV, Azarova TS, Shaposhnikov AI et al (2003) Root exudates of tomato plants and their effect on the growth and antifungal activity of *Pseudomonas* strains. Microbiology 72:37–41

Kyselkova M, Kopecky J, Frapolli M et al (2009) Comparison of rhizobacterial community composition in soil suppressive or conducive to tobacco black root rot disease. ISME J 3:1127–1138

Lanoue A, Burlat V, Henkes GJ et al (2010) De novo biosynthesis of defense root exudates in response to Fusarium attack in barley. New Phytol 185:577–588

Lebeis SL, Paredes SH, Lundberg DS et al (2015) Salicylic acid modulates colonization of the root microbiome by specific bacterial taxa. Science 349:860–864

Lee HC, Lai K, Lorenc MT et al (2012) Bioinformatics tools and databases for analysis of next-generation sequence data. Brief Funct Genomics 11:12–24

Lehmann A, Barto EK, Powell JR et al (2012) Mycorrhizal responsiveness trends in annual crop plants and their wild relatives – a meta-analysis on studies from 1981 to 2010. Plant Soil 355:231–250

Leveau JHJ (2007) The magic and menace of metagenomics: prospects for the study of plant growth-promoting rhizobacteria. Eur J Plant Pathol 119:279–300

Levy A, Salas Gonzalez I, Mittelviefhaus M et al (2018) Genomic features of bacterial adaptation to plants. Nat Genet 50:138

Lindow SE, Brandl MT (2003) Microbiology of the phyllosphere. Appl Environ Microbiol 69:1875–1883

Liu WT, Marsh TL, Cheng H et al (1997) Characterization of microbial diversity by determining terminal restriction fragment length polymorphisms of genes encoding 16S rRNA. Appl Environ Microbiol 63:4516–4522

Lozupone CA, Stombaugh JI, Gordon JI et al (2012) Diversity, stability and resilience of the human gut microbiota. Nature 489:220–230

Lukow T, Dunfield PF, Liesack W (2000) Use of the T-RFLP technique to assess spatial and temporal changes in the bacterial community structure within an agricultural soil planted with transgenic and non-transgenic potato plants. FEMS Microbiol Ecol 32:241–247

Lundberg DS, Lebeis SL, Paredes SH et al (2012) Defining the core Arabidopsis thaliana root microbiome. Nature 488:86–90

Marasco R, Rolli E, Ettoumi B et al (2012) A drought resistance-promoting microbiome is selected by root system under desert farming. PLoS One 7:e48479

Mark GL, Dow JM, Kiely PD et al (2005) Transcriptome profiling of bacterial responses to root exudates identifies genes involved in microbe–plant interactions. Proc Natl Acad Sci 102:17454–17459

Marques JM, da Silva TF, Vollu RE et al (2014) Plant age and genotype affect the bacterial community composition in the tuber rhizosphere of field-grown sweet potato plants. FEMS Microbiol Ecol 88:424–435

Marschner P, Crowley D, Lieberei R (2001) Arbuscular mycorrhizal infection changes the bacterial 16S rDNA community composition in the rhizosphere of maize. Mycorrhiza 11:297–302

Mathew DC, Ho Y-N, Gicana RG et al (2015) A rhizosphere-associated symbiont, *Photobacterium* spp. strain MELD1, and its targeted synergistic activity for phytoprotection against mercury. PLoS One 10:e0121178

Matthews A, Pierce S, Hipperson H et al (2019) Rhizobacterial community assembly patterns vary between crop species. Front Microbiol 10:581

Mayak S, Tirosh T, Glick BR (2004) Plant growth-promoting bacteria that confer resistance to water stress in tomatoes and peppers. Plant Sci 166:525–530

Mendes R, Kruijt K, de Bruijn I et al (2011) Deciphering the rhizosphere microbiome for disease-suppressive bacteria. Science 332:1097–1100

Mendes LW, Kuramae EE, Navarrete AA et al (2014) Taxonomical and functional microbial community selection in soybean rhizosphere. ISME J 8:1577–1587

Micallef SA, Channer S, Shiaris MP et al (2009) Plant age and genotype impact the progression of bacterial community succession in the Arabidopsis rhizosphere. Plant Signal Behav 4:777–780

Mougel C, Offre P, Ranjard L et al (2006) Dynamic of the genetic structure of bacterial and fungal communities at different developmental stages of *Medicago truncatula* Gaertn. cv. *Jemalong* line J5. New Phytol 170:165–175

Mueller UG, Sachs JL (2015) Engineering microbiomes to improve plant and animal health. Trends Microbiol 10:606–617

Mutch LA, Young JP (2004) Diversity and specificity of *Rhizobium leguminosarum* biovar *viciae* on wild and cultivated legumes. Mol Ecol 13:2435–2444

Muyzer G, de Waal EC, Uitterlinden AG et al (1993) Profiling of complex microbial populations by denaturing gradient gel electrophoresis analysis of polymerase chain reaction-amplified genes coding for 16S rRNA. Appl Environ Microbiol 59:695–700

Naeem M, Aslam Z, Khaliq A et al (2018) Plant growth promoting rhizobacteria reduce aphid population and enhance the productivity of bread wheat. Braz J Microbiol 49:9–14
Nemergut DR, Knelman JE, Ferrenberg S et al (2016) Decreases in average bacterial community rRNA operon copy number during succession. ISME J 10:1147–1156
Ofek M, Voronov-Goldman M, Hadar Y et al (2014) Host signature effect on plant root-associated microbiomes revealed through analyses of resident vs. active communities. Environ Microbiol 16:2157–2167
Oldroyd GED, Harrison MJ, Paszkowski U (2009) Reprogramming plant cells for endosymbiosis. Science 324:753–754
Osbourn AE, Barber CE, Daniels MJ (1987) Identification of plant-induced genes of the bacterial pathogen *Xanthomonas campestris* pathovar *campestris* using a promoter-probe plasmid. EMBO J 6:23–28
Pace NR, Stahl DA, Lane DJ et al (1985) Analyzing natural microbial populations by rRNA sequences. ASM News 51:4–12
Panstruga R, Kuhn H (2015) Introduction to a virtual special issue on cell biology at the plant–microbe interface. New Phytol 207:931–938
Parniske M (2008) Arbuscular mycorrhiza: the mother of plant root endosymbioses. Nat Rev Microbiol 6:763–775
Pathma J, Sakthivel N (2013) Molecular and functional characterization of bacteria isolated from straw and goat manure based vermicompost. Appl Soil Ecol 70:33–47
Pathma J, Rahul GR, Kamaraj Kennedy R, Subashri R, Sakthivel N (2011) Secondary metabolite production by bacterial antagonists. J Biol Control 25:165–181
Pathma J, Raman G, Sakthivel N (2019) Microbiome of rhizospheric soil and vermicompost and their applications in soil fertility, pest and pathogen management for sustainable agriculture. In: Panpatte DG, Jhala YK (eds) Soil fertility management for sustainable development. Springer Nature Singapore, Singapore, pp 189–210
Peiffer JA, Spor A, Koren O et al (2013) Diversity and heritability of the maize rhizosphere microbiome under field conditions. Proc Natl Acad Sci 110:6548–6553
Pel MJ, Pieterse CM (2013) Microbial recognition and evasion of host immunity. J Exp Bot 64:1237–1248
Penrose DM, Glick BR (2002) Methods for isolating and characterizing ACC deaminase containing plant growth promoting rhizobacteria. Physiol Plant 118:10–15
Poza-Carrion C, Suslow T, Lindow S (2013) Resident bacteria on leaves enhance survival of immigrant cells of *Salmonella enterica*. Phytopathology 103:341–351
Quecine MC, Araujo WL, Tsui S et al (2014) Control of *Diatraea saccharalis* by the endophytic *Pantoea agglomerans* 33.1 expressing cry1Ac7. Arch Microbiol 196:227–234
Rainey PB, Heithoff DM, Mahan MJ (1997) Single-step conjugative cloning of bacterial gene fusions involved in microbe-host interactions. Mol Gen Genet 256:84–87
Ram SG, Thiruvengadam V, Vinod KK (2007) Genetic diversity among cultivars, landraces and wild relatives of rice as revealed by microsatellite markers. J Appl Genet 48:337–345
Ramirez KS, Craine JM, Fierer N (2012) Consistent effects of nitrogen amendments on soil microbial communities and processes across biomes. Glob Chang Biol 18:1918–1927
Ramirez-Flandes S, Gonzalez B, Ulloa O (2019) Redox traits characterize the organization of global microbial communities. Proc Natl Acad Sci 116:3630–3635
Raynaud X, Nunan N (2014) Spatial ecology of bacteria at the microscale in soil. PLoS One 9: e87217
Rediers H, Rainey PB, Vanderleyden J et al (2005) Unravelling the secret lives of bacteria: use of in vivo expression technology and differential fluorescence induction promoter traps as tools for exploring niche-specific gene expression. Microbiol Mol Rev 69:217–261
Roberts E, Lindow S (2014) Loline alkaloid production by fungal endophytes of fescue species select for particular epiphytic bacterial microflora. ISME J 8:359–368

Rodrigues JL, Pellizari VH, Mueller R et al (2013) Conversion of the Amazon rainforest to agriculture results in biotic homogenization of soil bacterial communities. Proc Natl Acad Sci 110:988–993

Roesch LFW, Fulthorpe RR, Riva A et al (2007) Pyrosequencing enumerates and contrasts soil microbial diversity. ISME J 1:283–290

Rolli E, Marasco R, Vigani et al (2015) Improved plant resistance to drought is promoted by the root-associated microbiome as a water stress-dependent trait. Environ Microbial 17:316–331

Rosenblueth M, Martinez-Romero E (2006) Bacterial endophytes and their interactions with hosts. Mol Plant-Microbe Interact 19:827–837

Rosenzweig N, Tiedje JM, Quensen JF et al (2012) Microbial communities associated with potato common scab-suppressive soil determined by pyrosequencing analyses. Plant Dis 96:718–725

Sanchez-Lopez A, Pintelon I, Stevens V et al (2018) Seed endophyte microbiome of *Crotalaria pumila* unpeeled: identification of plant-beneficial methylobacteria. Int J Mol Sci 19:291

Sangabriel-Conde W, Maldonado-Mendoza IE, Mancera-Lopez ME et al (2015) Glomeromycota associated with Mexican native maize landraces in Los Tuxtlas, Mexico. Appl Soil Ecol 87:63–71

Sanguin H, Remenant B, Dechesne A et al (2006) Potential of a 16S rRNA-based taxonomic microarray for analyzing the rhizosphere effects of maize on Agrobacterium spp. and bacterial communities. Appl Environ Microbiol 72:4302–4312

Santos-Medellin C, Edwards J, Liechty Z et al (2017) Drought stress results in a compartment-specific restructuring of the rice root-associated microbiomes. MBio 8:e00764-17

Saravanakumar D, Kavino M, Raguchander T et al (2011) Plant growth promoting bacteria enhance water stress resistance in green gram plants. Acta Physiol Plant 33:203–209

Schenk PM, Carvalhais LC, Kazan K (2012) Unravelling plant–microbe interactions: can multi-species transcriptomics help? Trends Biotechnol 30:177–184

Scheuring I, Yu DW (2013) How to assemble a beneficial microbiome in three easy steps. Ecol Lett 15:1300–1307

Sen K, Sengupta C, Saha J (2014) PGPR consortium alleviating downy mildew of cucumber. Int J Plant Anim Environ Sci 4:150–159

Sessitsch A, Weilharter A, Gerzabek MH et al (2001) Microbial population structures in soil particle size fractions of a long-term fertilizer field experiment. Appl Environ Microbiol 67:4215–4224

Shade A, Handelsman J (2012) Beyond the venn diagram: the hunt for the core microbiome. Environ Microbiol 14:4–12

Shang Q, Yang G, Wang Y et al (2016) Illumina-based analysis of the rhizosphere microbial communities associated with healthy and wilted Lanzhou lily (*Lilium davidii* var. unicolor) plants grown in the field. WJMB 32:95

Sindhu SS, Rakshiya YS, Verma MK (2011) Biological control of termites by antagonistic soil microorganisms. In: Singh A, Parmar N, Kuhad RC et al (eds) Bioaugmentation, biostimulation and biocontrol, soil biology. Springer, Berlin/Heidelberg, pp 261–310

Singh P, Gobbi A, Santoni S et al (2018) Assessing the impact of plant genetic diversity in shaping the microbial community structure of *Vitis vinifera* phyllosphere in the Mediterranean. Front Life Sci 11:35–46

Souza RC, Mendes IC, Reis-Junior FB et al (2016) Shifts in taxonomic and functional microbial diversity with agriculture: how fragile is the Brazilian Cerrado? BMC Microbiol 16:42

Staley JT, Konopka A (1985) Measurements of in situ activities of non-photosynthetic microorganisms in aquatic and terrestrial habitats. Annu Rev Microbiol 39:321–346

Stanley HF, Fagan WF (2002) Fungal endophytes: common host plant symbionts but uncommon mutualists. Integr Comp Biol 42:360–368

Stout MJ, Zehnder GW, Baur ME (2002) Potential for the use of elicitors of plant defense in arthropod management programs. Arch Insect Biochem 51:222–235

Subramanian P, Kim K, Krishnamoorthy R et al (2016) Cold stress tolerance in psychrotolerant soil bacteria and their conferred chilling resistance in tomato (*Solanum lycopersicum* Mill.) under low temperatures. PLoS One 11:e0161592

Sugiyama A, Ueda Y, Zushi T et al (2014) Changes in the bacterial community of soybean rhizospheres during growth in the field. PLoS One 9:1–9

Szoboszlay M, White-Monsant A, Moe LA et al (2016) The effect of root exudate 7,4-′-dihydroxyflavone and naringenin on soil bacterial community structure. PLoS One 11: e0146555

Tan Y, Cui Y, Li H et al (2017) Rhizospheric soil and root endogenous fungal diversity and composition in response to continuous *Panax notoginseng* cropping practices. Microbiol Res 194:10–19

Tank N, Saraf M (2010) Salinity-resistant plant growth promoting rhizobacteria ameliorates sodium chloride stress on tomato plants. J Plant Interact 5:51–58

Timonen S, Sinkko H, Sun H et al (2017) Ericoid roots and mycospheres govern plant-specific bacterial communities in boreal forest humus. Microb Ecol 73:939–953

Toljander JF, Santos-Gonzalez JC, Tehler A et al (2008) Community analysis of arbuscular mycorrhizal fungi and bacteria in the maize mycorrhizosphere in a long-term fertilization trial. FEMS Microbiol Ecol 65:323–338

Trda L, Boutrot F, Claverie J et al (2015) Perception of pathogenic or beneficial bacteria and their evasion of host immunity: pattern recognition receptors in the frontline. Front Plant Sci 6:219

Tringe SG, Mering CV, Kobayashi A (2005) Comparative metagenomics of microbial communities. Science 308:554–557

Tunlid A, Baird BH, Trexler MB et al (1985) Determination of phospholipid ester-linked fatty acids and poly-b-hydroxybutyrate for the estimation of bacterial biomass and activity in the rhizosphere of the rapa plant *Brassica napus*. Can J Microbiol 31:1113–1119

Turner TR, James EK, Poole PS (2013a) The plant microbiome. Genome Biol 14:209

Turner T, Ramakrishnan K, Walshaw J et al (2013b) Comparative metatranscriptomics reveals kingdom level changes in the rhizosphere microbiome of plants. ISME J 7:2248–2258

van der Heijden MG, Hartmann M (2016) Networking in the plant microbiome. PLoS Biol 14: e1002378

Van Peer R, Niemann G, Schippers B (1991) Induced resistance and phytoalexin accumulation in biological control of Fusarium wilt of carnation by Pseudomonas sp. strain WCS 417r. Phytopathology 81:728–734

Van Wees SCM, Ent SV, Pieterse CMJ (2008) Plant immune responses triggered by beneficial microbes. Curr Opin Plant Biol 11:443–448

Vandeputte OM, Kiendrebeogo M, Rasamiravaka T et al (2011) The flavanone naringenin reduces the production of quorum sensing-controlled virulence factors in Pseudomonas aeruginosa PAO1. Microbiology 157:2120–2132

Velicer GJ, Raddatz G, Keller H, Deiss S, Lanz C, Dinkelacker I, Schuster SC (2006) Comprehensive mutation identification in an evolved bacterial cooperator and its cheating ancestor. Proc Natl Acad Sci 103:8107–8112

Vellend M, Agrawal A (2010) Conceptual synthesis in community ecology. Q Rev Biol 85:183–206

Verma P, Yadav AN, Khannam KS et al (2015) Assessment of genetic diversity and plant growth promoting attributes of psychrotolerant bacteria allied with wheat (*Triticum aestivum*) from the northern hills zone of India. Ann Microbiol 65:1885–1899

Vestergard M, Hendry F, Castro JIR et al (2008) Rhizosphere bacterial community composition responds to arbuscular mycorrhiza, but not to reductions in microbial activity induced by foliar cutting. FEMS Microbiol Ecol 64:78–89

Vogel C, Bodenhausen N, Gruissem W et al (2016) The *Arabidopsis* leaf transcriptome reveals distinct but also overlapping responses to colonization by phyllosphere commensals and pathogen infection with impact on plant health. New Phytol 212:192–207

Vogl A (1898) Mehl und die anderen mehl produkte der cerealien und leguminosen. Zeitschrift Nahrungsmittle untersuchung, Hgg Warlenkunde 21:25–29

Vorholt JA (2012) Microbial life in the phyllosphere. Nat Rev Microbiol 10:828–840

Walder F, Schlaeppi K, Wittwer R et al (2017) Community profiling of *Fusarium* in combination with other plant-associated fungi in different crop species using SMRT sequencing. Front Plant Sci 28:8

Wang G, Xu Y, Jin J et al (2009) Effect of soil type and soybean genotype on fungal community in soybean rhizosphere during reproductive growth stages. Plant Soil 317:135–144

Wang Y, Li C, Tu C et al (2017) Long-term no-tillage and organic input management enhanced the diversity and stability of soil microbial community. Sci Total Environ 609:341–347

Watrous J, Roach P, Alexandrov T et al (2012) Mass spectral molecular networking of living microbial colonies. Proc Natl Acad Sci 109:1743–1752

Weese DJ, Heath KD, Dentinger BT et al (2015) Long-term nitrogen addition causes the evolution of less-cooperative mutualists. Evolution 69:631–642

Weinert N, Piceno Y, Ding GC et al (2011) PhyloChip hybridization uncovered an enormous bacterial diversity in the rhizosphere of different potato cultivars: many common and few cultivar-dependent taxa. FEMS Microbiol Ecol 75:497–506

Willers C, Jansen van Rensburg PJ, Claassens S (2015) Microbial signature lipid biomarker analysis–an approach that is still preferred, even amid various method modifications. J Appl Microbiol 118(6):1251–1263

Woese CR, Fox GE (1977) Phylogenetic structure of the prokaryotic domain: the primary kingdoms. Proc Natl Acad Sci 74:5088–5090

Wu CH, Bernard SM, Andersen GL et al (2009) Developing microbe–plant interactions for applications in plant-growth promotion and disease control, production of useful compounds, remediation and carbon sequestration. Microb Biotechnol 2:428–440

Xing X, Koch AM, Jones AM et al (2012) Mutualism breakdown in breadfruit domestication. Proc Biol Sci 22:1122–1130

Xiong W, Li Z, Liu H et al (2015a) The effect of long-term continuous cropping of black pepper on soil bacterial communities as determined by 454 pyrosequencing. PLoS One 10:e0136946

Xiong W, Zhao Q, Zhao J, Xun W et al (2015b) Different continuous cropping spans significantly affect microbial community membership and structure in a vanilla-grown soil as revealed by deep pyrosequencing. Microb Ecol 70:209–218

Xu Y, Wang G, Jin J et al (2009) Bacterial communities in soybean rhizosphere in response to soil type, soybean genotype, and their growth stage. Soil Biol Biochem 41:919–925

Yan Y, Kuramae EE, de Hollander M, Klinkhamer PG et al (2017) Functional traits dominate the diversity-related selection of bacterial communities in the rhizosphere. ISME J 11:56–66

Yuan Z, Liu P, Saenkham P et al (2008) Transcriptome profiling and functional analysis of *Agrobacterium tumefaciens* reveals a general conserved response to acidic conditions (pH 5.5) and a complex acid-mediated signalling involved in agrobacterium–plant interactions. J Bacteriol 190:494–507

Zachow C, Muller H, Tilcher R et al (2014) Differences between the rhizosphere microbiome of *Beta vulgaris* ssp. *maritima*-ancestor of all beet crops-and modern sugar beets. Front Microbiol 5:415

Zhang J, Subramanian S, Stacey G et al (2009) Flavones and flavonols play distinct critical roles during nodulation of *Medicago truncatula* by *Sinorhizobium meliloti*. Plant J 57:171–183

Zhang N, Wang D, Liu Y (2014) Effects of different plant root exudates and their organic acid components on chemotaxis, biofilm formation and colonization by beneficial rhizosphere-associated bacterial strains. Plant Soil 374:689–700

Zhu YG, Smith SE, Barritt AR et al (2001) Phosphorus (P) efficiencies and mycorrhizal responsiveness of old and modern wheat cultivars. Plant Soil 237:249–255

Zolla G, Badri DV, Bakker MG et al (2013) Soil microbiomes vary in their ability to confer drought tolerance to *Arabidopsis*. Appl Soil Ecol 68:1–9

Microbes in Crop Production: Formulation and Application

3

Pankaj Prakash Verma, Rahul Mahadev Shelake, Parul Sharma, Jae-Yean Kim, Suvendu Das, and Mohinder Kaur

Abstract

Agriculture depends upon expensive inputs of pesticides and chemical fertilizers to increase crop yields. This dependence on agrochemicals poses risks to human and environmental health such as disruption of nutrient cycling and demolition of beneficial microbial communities for higher crop production. Over the last decade, soil microbes have been widely exploited to enhance the crop production and plant and soil health management. The higher crop yields are reported after inoculation with plant growth-promoting microbes (PGPM). The PGPM signify as an effective and promising way to improve quality food production without environmental or human health hazard. This chapter will explore the current research and trends in microbial exploitation in growth promotion of different agricultural crops. We further discuss the key mechanisms underlying growth promotion and technological advances in bioformulation development to increase shelf life. Recent uses, development, and application of microbial formulation for managing a sustainable environmental system are also discussed.

P. P. Verma (✉)
Institute of Agriculture and Life Science, Gyeongsang National University, Jinju, Republic of Korea

Department of Basic Sciences, Dr. Y.S. Parmar University of Horticulture and Forestry, Solan, Himachal Pradesh, India

R. M. Shelake · J.-Y. Kim
Division of Applied Life Science (BK21 Plus), Plant Molecular Biology and Biotechnology Research Center, Gyeongsang National University, Jinju, Republic of Korea

P. Sharma · M. Kaur
Department of Basic Sciences, Dr. Y.S. Parmar University of Horticulture and Forestry, Solan, Himachal Pradesh, India

S. Das
Institute of Agriculture and Life Science, Gyeongsang National University, Jinju, Republic of Korea

S. G. Sharma et al. (eds.), *Microbial Diversity, Interventions and Scope*,
https://doi.org/10.1007/978-981-15-4099-8_3

Keywords

Bioformulation · Holobiont · Microbial diversity · Plant-soil-microbe interaction · Rhizo-microbiome · Sustainable agriculture

3.1 Introduction

The global human population is expected to increase approx. 9 billion from its current population of 7.3 billion by 2050 (Rodriguez and Sanders 2015). The increased population and global climate change have posed a serious threat to crop production and food security. The widespread use of mineral fertilizers and agrochemicals (like fungicides, insecticides, herbicides, etc.) in crop production for higher crop yields remains a common practice. The growing food and fiber demand has led to the expansion of conventional agricultural practices, which is neither economic nor environment friendly (Trivedi et al. 2017). These trends pose a series of unprecedented challenge to worldwide food and agriculture production leading to sustainably intensify food and agricultural crop production and find solutions to combat phytopathogens and abiotic stress.

The application of plant growth-promoting microbes (PGPM) in agriculture represents an economically attractive and environment friendly alternative to extensive chemical fertilization. The collective set of rhizospheric microbes is known as rhizosphere microbiome or rhizo-microbiome (Bulgarelli et al. 2013). A continued exploration and manipulation of rhizo-microbiome and their interactions with plant is a prerequisite for development of efficient microbial formulations (bioformulation). The application of bioformulations can enhance crop growth, vigor, and nutrient use efficiency and provide protection from phytopathogens and biotic and abiotic stress tolerance (Ahmad et al. 2018).

The widespread commercial use of PGPM requires a good screening and mass multiplication procedures that can promote quality, quantity, and product formulation with enhanced shelf life and bioactivity (Gopalakrishnan et al. 2016). In addition, new sustainable approaches will ensure competitive crop yields, crop protection, and soil health improvement. In this chapter, we discuss about soil microbes, role of PGPM in plant health management, and selection criteria of bioformulations.

3.2 Soil Microbes

Soil comprises a living and dynamic ecosystem containing approximately 90–100 million bacteria along with around 0.2 million fungi (per gram soil). Most of the beneficial PGPM inhabit around the plant roots. The rhizo-microbiome depends on the plant root exudates like organic acids, amino acids, sugars, etc. that provide carbon as a food source (Glick 2018). The plant roots exude chemicals including signaling molecules and metabolites accessible to microbes. The plant-microbe interaction is considered beneficial, neutral, or detrimental for the plant growth. This interaction depends on the plant and specific microbes inhabiting the rhizosphere.

Soil microbial community consists of mixed populations that include bacteria, actinomycetes, fungi, algae, protozoa, and viruses. Nearly all soils contain a mixture of microbial populations. Among them, bacterial community is generally much higher than other groups. All microbial groups are important in bringing about numerous transformations and making up the soil environment. The microbial communities also contribute to various soil ecosystem functions including global biogeochemical cycling (C, N, P, Fe, etc.), organic matter cycling, soil aggregation, etc. Soil organisms influence the soil structure and aggregate formation, which are hotspots of microbial activity and diversity. Soil structure is thus both the cause and the product of soil biodiversity (Havlicek and Mitchell 2014).

The soil organic matter (SOM) decomposition is carried out by the activity of hydrolytic enzymes secreted by bacteria and fungi (primary decomposers). These primary decomposers determine both the magnitude of carbon (C) stored in soils and the rate at which nutrients become available to plants (Shelake et al. 2019). The high soil organic carbon (SOC) content improves the soil biological (microbial biomass), chemical, and physical properties, such as enhanced biological activity, improved soil structure, higher water-holding capacity, soil fertility, and sorption of organic and inorganic pollutants (Bhogal et al. 2018; Shelake et al. 2019). The growth and development of crops/plants is mainly affected by the soil microbial diversity, mineral nutrients, and physical properties of the soil.

3.3 Plant Growth-Promoting Microbes in Sustainable Agriculture

In recent times, agriculture faces numerous challenges like limited nutrient resources, extensive losses by phytopathogens, environmental deterioration through depletion of resources (air, water, and soil), and food security (Kroll et al. 2017). Sustainable agriculture involves a wide range of approaches to meet the growing food demand and fiber requirements without harming the environment (Barea 2015). This integrates three key objectives: healthy environment, economic profitability, and socioeconomic equity. The agricultural crop productivity is sturdily influenced by the activities of soil microbial communities. The microbial communities vary with soil type, soil pH and EC (electrical conductivity), availability of nutrients, and vegetation type (Wang et al. 2018). The exploitation of these beneficial microbial communities is of vital importance to agriculture for sustainable crop production and food safety. Soil microbes derive their energy and nutrients from decomposing organic substrate in the soil. They are involved in SOM transformation and nutrient immobilization and various soil processes ultimately improving soil fertility and productivity (Sharma et al. 2017a, b).

The PGPM are defined as the root-/rhizosphere-inhabiting microbes capable of colonizing root surface and can promote plant growth. The PGPM are divided into two distinctive groups: plant growth-promoting rhizobacteria (PGPR) and plant

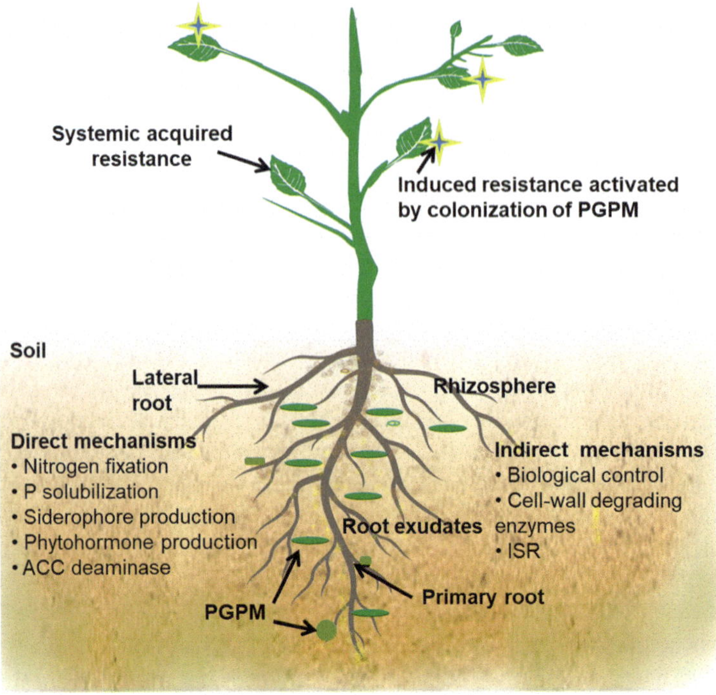

Fig. 3.1 Mechanisms used by PGPM for enhancing plant growth

growth-promoting fungi (PGPF) (Mishra et al. 2017). The term PGPR was coined by Kloepper and Schroth (1978) to beneficial soil bacteria inhabiting rhizosphere and able to colonize and promote plant growth. They are involved directly or indirectly in the growth and development of plant (Fig. 3.1). The mode of action by PGPR includes the nitrogen fixation, nutrient solubilization/mobilization, siderophore production, phytohormone production, and ACC-deaminase activity. Indirect effects include biological control through antibiotic production, cell-wall degrading enzyme activity, and induced systemic resistance (ISR) (Verma et al. 2016; Ahmad et al. 2018). The PGPF are nonpathogenic soilborne saprophytic filamentous fungi that facilitate plant growth. Several reported PGPF belong to fungal genera *Trichoderma*, *Aspergillus*, *Piriformospora*, *Fusarium*, *Penicillium*, *Phoma*, and arbuscular mycorrhizal (AM) fungi (Hossain et al. 2017). The PGPF colonize plant roots, stimulate growth, and suppress phytopathogens. They produce plant hormones, hydrolytic enzymes, antifungal metabolites, nutrient solubilization, organic matter degradation, and ISR in plants (Mishra et al. 2017).

The microbial use and application in crop production and soil health management is important for achieving sustainable agriculture. The use of PGPM largely excludes the use of chemically synthesized fertilizers, pesticides, and growth regulators and can increase crop productivity with environmental restoration. Understanding the rhizosphere structure and function will allow to harness plant-microbial interactions and improved crop productivity (Ahkami et al. 2017).

3.4 Plant-Soil-Microbe Interactions in the Rhizosphere

The term "rhizosphere" was coined by Lorenz Hiltner (1904), to describe the area around plant roots, inhabited and influenced by diverse microbial species and plant root exudates. This influence results from the release of organic compounds, also referred as rhizodeposition. The rhizodeposits include root exudates (sugars, amino acids, organic acids, etc.), insoluble materials (sloughed cells and root mucilage), dead fine roots, lysates, and gases, such as CO_2 (by root and microbial respiration) and ethylene (Cheng and Gershenson 2007). As a result, the rhizosphere soils are regarded as *mesotrophic*, favoring the microbial growth (bacteria, fungi, archaea, and viruses), and the bare soils are described to have *oligotrophic* environments (Dessaux et al. 2016). This chemically unique and complex environment supports the growth of remarkably diverse and unique microbial populations.

The rhizo-microbiome composition is complex and dynamic, controlled by several biotic and abiotic factors. The abiotic factors include the physicochemical properties of soil and environmental parameters, whereas biotic factors include the chemicals secreted by bacteria and plant together with their biological activities (Haldar and Sengupta 2015). The root exudate chemistry dictates the rhizosphere microbial communities (Ahmad et al. 2018). The rhizo-microbiome mediates interactions via the production and secretion of signaling molecules by both plants and microbes.

The signaling in the rhizosphere can be divided into three groups:

Microbe-microbe (via quorum-sensing molecules like N-acyl homoserine lactones (AHLs), diketopiperazines (DKPs), and diffusible signal factor (DSF).

The second group includes *plants to microbe* (via plant-secreted molecules, e.g., root exudates).

The third group contains *microbes to plants* (via microbially produced compounds like lipopolysaccharides, peptidoglycans, flagellin, and chitin).

This signaling between plants and rhizosphere microbes resulted in shaping the rhizo-microbiome, inducing systemic resistance (by priming) sustaining plant health, growth, nutrition, and stress tolerance (Venturi and Keel 2016).

3.5 PGPM Affect Root Growth and Development

Soil microbial communities are recognized to play crucial roles in agricultural and natural ecosystems. Their activities have a positive impact on chemical, biological, and physical soil properties (Levy et al. 2018). The rhizo-microbiome also depends upon the soil type and the composition of root exudates (like organic acids, sugars, amino acids, enzymes, fatty acids, phenolics, coumarins, anthocyanins, and flavonoids) secreted by the host plant (Chaparro et al. 2013; Badri and Vivanco 2009). The ability of rhizobacteria to colonize rhizosphere depends on their chemotactic response toward root exudates. This chemical communication among plants and rhizo-microbes results in altered microbial community structure, plant health, and growth. For example, plant roots exude rosmarinic acid which stimulates *quorum-sensing* response, influencing bacterial population in the rhizosphere (Corral-Lugo et al. 2016). Additionally, *salicylic acid* influences the colonization of specific bacterial families within the roots, thereby altering the microbial community structure (Lebeis et al. 2015). The beneficial rhizosphere microbes include PGPR, PGPF, and protozoa that have been reported for their positive effects on plant growth and development (Mendes et al. 2013, Weidner et al. 2017). The PGPR affect root system architecture (temporal and spatial distribution of roots in soil) by altering the cell division and differentiation (in primary root), thereby affecting root hair formation and lateral root development (Verbon and Liberman 2016).

Several PGPR species have been identified to increase lateral root formation and shoot growth and inhibit primary root growth (by decreasing the cell elongation) of plants. Some PGPR species are shown to induce cell division and differentiation at both the root apical meristem and lateral root emergence sites. The cell division is positively or negatively affected depending upon the type of species within the meristem. For example, *Pseudomonas simiae* WCS417 increases cell division, whereas *Bacillus megaterium* decreases cell division and growth conditions. The differentiation is induced close to the root tip in PGPR-inoculated plants, due to which root hairs emerge close to the root tip. As a result, root hair density and length increases upon colonization. Thus, rhizo-microbiome affects root growth and development by manipulating the host endogenous mechanisms by regulating postembryonic root development (Verbon and Liberman 2016).

3.6 Microbes in Crop Production

The plant health depends upon the interactions between living organisms and their environment. Both plants and microbes, the components of rhizosphere can be engineered, and the soil can also be amended to promote growth and development (Dessaux et al. 2016). Genetic engineering of crop plants has resulted in pathogen resistance, high metal concentration resistance, etc. In contrast, there are few reports of PGPR engineering to render it more effective, for example, a chitinase gene (isolated from *Bacillus subtilis*) was inserted into *Burkholderia vietnamiensis*, a PGPR, to suppress Fusarium wilt (cotton), sheath blight (wheat), and gray mold (tomato) (Zhang et al. 2012). A recent method involves engineering of set of microbial population rather than single strain. Alternate way consists of ecological engineering (plant-microbe interaction). In general, plants and their associated microbes are considered as a holobiont or superorganism rather than as "individual" (Dessaux et al. 2016). The microbes play a crucial role in plant adaptation to changing environments. The holobiont paradigm in plant world is transforming our understanding (Vandenkoornhuyse et al. 2015).

Plant and microbial engineering by modern techniques such as transgenic production involves several environmental and ethical issues. Emerging trend is the application of microbial formulations as an excellent alternative to agrochemicals. These microbial inoculants can substantially lessen the use of inorganic fertilizers and pesticides in agricultural crops, thereby enhancing the nutrient uptake and stimulating growth and protection against phytopathogens (Ahmad et al. 2018). The PGPM play a vital role in agricultural systems (Table 3.1). They increase the uptake of primary nutrients (*biofertilizers*), produce phytohormones (*phytostimulators*), and suppress diseases or phytopathogens (*biopesticide*) enhancing plant growth and development (Trabelsi and Mhamdi 2013). Different microbial inoculants are already commercialized and used for several crops (Table 3.2).

3.7 Microbial Formulations and Application

Many pot and field studies have shown that plants inoculated with PGPM stimulate growth and yield. The microbial formulations are defined as the preparations of single or consortia strains of known microbes in a user-friendly and organic or inorganic carrier material. The specific number of cells (differs among species, e.g., 10^6–10^7 cells/plant of *Azospirillum brasilense*) is needed to reach the threshold to obtain the anticipated response in plants (Bashan et al. 2014). Various kinds of bioformulations being used in agriculture include nitrogen fixers, potassium (K) and phosphorus (P) solubilizers and mobilizers, growth-promoting AM fungi and cyanobacteria, and other useful microbes (Table 3.3). The bioformulation thus includes the desired microbe, suitable carrier material, sticking agents, and osmoprotectant (Sahu and Brahmaprakash 2016). The development of PGPM-based formulations with multifarious PGP and biocontrol activity with improved shelf life could pave the way for its commercialization. They provide a suitable microenvironment, physical protection, and structure to the introduced microbes. The development of techniques for mass multiplication of pure inoculants would offer a potential solution for allowing extensive use of biofertilizers. The main advantage of PGPM-based formulations is the choice of desired microbial formulation, the carrier material selection, and delivery methods (Zayed 2016).

3.7.1 Selection of Appropriate Microbes

The development of successful PGPM formulation is a multistep process, which starts with the isolation of beneficial microbes from plants, in vitro screening, characterization of PGP, and antagonistic activities, followed by its testing in greenhouse and field. The development process varies depending on the microbial group (bacteria, fungi, yeast, viruses, and nematodes) used for bioformulation. For example, bacteria and yeast are produced by liquid fermentation, whereas fungi are

Table 3.1 Different microbes being reported as biocontrol agents, biofertilizers, and phytostimulators

Genus	Species
Bacteria	
Acinetobacter sp.	*A. lwoffii, A. baumannii, A. calcoaceticus*
Aneurinibacillus sp.	*A. aneurinilyticus, A. terranovensis, A. migulanus, A. danicus*
Arthrobacter sp.	*A. protophormiae, A. pokkalii, A. agilis*
Azospirillum sp.	*A. brasilense, A. lipoferum, A. amazonense*
Azotobacter sp.	*A. salinestris, A. chroococcum, A. beijerinckii, A. paspali, A. armeniacus, A. nigricans, A. salinestri*
Bacillus sp.	*B. subtilis, B. amyloliquefaciens, B. pumilus, B. mojavensis, B. velezensis, B. thuringiensis, B. licheniformis, B. cereus, B. safensis, B. methylotrophicus, B. megaterium, B weihenstephanensis, B. edaphicus, B. pantothenticus, B. subtilisformis, B. circulans, B. altitudinis, B. simplex, B. firmus, B. pasteurii, B. mycoides, B. sphaericus, B. brevis, B. coagulans, B. mucilaginosus*
Brevibacterium sp.	*B. halotolerans, B. iodinum, B. linens, B. frigoritolerans*
Burkholderia sp.	*B. pyrrocinia, B. cepacia, B. ambifaria, B. phytofirmans, B. phymatum*
Cellulosimicrobium sp.	*C. funkei, C. cellulans, C. terreum*
Chryseobacterium sp.	*C. indologenes, C. hispalense, C. cucumeris, C. elymi*
Enterobacter sp.	*E. aerogenes, E. cloacae, E. radicincitans, E. sakazakii, E. agglomerans*
Klebsiella sp.	*K. pneumonia, K. oxytoca*
Lysobacter sp.	*L. antibioticus, L. enzymogenes*
Novosphingobium sp.	*N. oryzae, N. pentaromativorans*
Ochrobactrum sp.	*O. anthropi, O. cytisi, O. intermedium*
Paenibacillus sp.	*P. polymyxa, P. mucilaginosus, P. illinoisensis, P. brasilensis, P. oenotherae, P. hemerocallicola, P. graminis, P. odorifer, P. expansum, P. azotofixans, P. macerans, P. peoriae*
Pantoea sp.	*P. agglomerans, P. dispersa, P. ananatis*
Paraburkholderia sp.	*P. phytofirmans, P. kururiensis, P. fungorum, P. tropica*
Pseudomonas sp.	*P. putida, P. fluorescens, P. aeruginosa, P. stutzeri, P. protegens, P. chlororaphis, P. brassicacearum, P. nitroreducens, P. geniculate, P. jesenii, P. migulae, P. tolaasii, P. picketti, P. savastanoi, P. cepacia, P. corrugate, P. striata, P. marginalis, P. oryzihabitans, P. gessardii, P. synxantha*
Sinorhizobium sp.	*S. meliloti, S. fredii, S. kostiense*
Serratia sp.	*S. marcescens, S. proteamaculans, S. nematodiphila, S. liquefaciens, S. plymuthica*
Sphingomonas sp.	*S. paucimobilis*
Stenotrophomonas sp.	*S. maltophilia, S. acidaminiphila*
Rhizobium sp.	*R. pusense, R. leguminosarum, R. tropici, R. etli, R. phaseoli, R. trifolii, R. japonicum, R. lupine, R. meliloti*

(continued)

Table 3.1 (continued)

Genus	Species
Fungus	
Acremonium sp.	*A. strictum, A. zeae*
Arbuscular mycorrhizal (AM) fungi	*Rhizophagus intraradices, Funneliformis mosseae, Glomus intraradices, Glomus mosseae, Gigaspora* sp., *Acaulospora* sp., *Scutellospora* sp., *Sclerocystis* sp.
Beauveria sp.	*B. bassiana, B. brongniartii, B. araneola*
Aspergillus sp.	*A. terreus, A. niger, A. aculeatus, A. oryzae, A. nidulans, A. fumigatus, A. flavus, A. versicolor, A. awamori*
Chaetomium sp.	*C. globosum*
Clonostachys sp.	*C. rosea, C. solani, C. rhizophaga*
Isaria sp.	*I. fumosorosea, I. javanica, I. poprawskii, I. farinosa*
Metarhizium sp.	*M. brunneum, M. robertsii, M. anisopliae*
Purpureocillium sp.	*P. lilacinum*
Penicillium sp.	*P. simplicissimum, P. bilaii, P. vermiculatum, P. expansum, P. citrinum*
Syncephalastrum sp.	*S. racemosum*
Talaromyces sp.	*T. flavus, T. wortmannii, T. pinophilus*
Trichoderma sp.	*T. viride, T. asperellum, T. harzianum, T. atroviride, T. polysporum, T. koningiopsis, T. gamsii, T. virens, T. longibrachiatum, T. hamatum, T. reesei, T. citrinoviride, T. brevicompactum, T. koningii, T. arundinaceum, T. ovalisporum*

produced by solid-state fermentation technology. The viruses and nematodes (possessing PGP traits) are scaled up by means of their alternate host or tissue culture method (Gopalakrishnan et al. 2016). It is important to select multiple compatible consortia forming beneficial associations with rhizo-microbiome, thus having a better chance to survive and provide multiple benefits to the host plant/crop, as compared to the single-strain bioformulations (Singh and Trivedi 2017; Wallenstein 2017). The PGPM formulation should possess:

1. High rhizosphere competency
2. Ability to enhance the plant growth
3. Highly competitive saprophytic ability and be more efficient
4. The ease of mass production or multiplication
5. The broad spectrum of action
6. Reliable control
7. Environmentally friendly and compatibility with other rhizobacteria
8. The ability to tolerate heat, desiccation, oxidizing agents, and UV radiations (Nakkeeran et al. 2005)

Table 3.2 Plant growth-promoting microorganisms, their application, and effect on different crops

Crops/plants	Microbe	Growth condition	Effect on plant	Application mode	References
Rice (*Oryza sativa*)	*Enterobacter* sp.	Growth room (under salt stress)	Seedling growth	Seed treatment (10^8 cfu/ml)	Sarkar et al. (2018)
	Bacillus spp.	Field	Enhanced yield and control blast diseases	Bacterial culture suspension (8×10^9 cfu/ml)	Rais et al. (2018)
	Bacillus amyloliquefaciens RWL-1	Growth chamber (under salinity stress)	Increased growth attributes, increased essential amino acids	Bacterial culture suspension (20 ml to root zone)	Shahzad et al. (2017)
	Bacillus amyloliquefaciens RWL-1	Growth chamber	Higher endogenous salicylic acid production	Bacterial culture suspension (2 ml to root zone)	Shahzad et al. (2016)
	Pseudomonas fluorescens	Field	Increase in seed germination, seedling vigor characters, yield	500 g/fed and 1000 g/fed (powder inoculation) for nursery and soil, respectively, and 1 liter fed^{-1} of liquid for foliar spray	Elekhtyar (2015)
Wheat (*Triticum aestivum*)	*Pseudomonas* sp. P34	In vitro	Promote root growth and dry matter accumulation	Bacterial culture suspension (10^9 cfu/ml)	Liu et al. (2018a, b)
	Aneurinibacillus aneurinilyticus, *Aeromonas* sp., and *Pseudomonas* sp.	Pot	Increased root and shoot length, increased germination and weight (fresh and dry) of plant	Seed treatment	Kumar et al. (2018)
	Bacillus sp. and *Brevibacterium halotolerans*	Pot	Increase plant growth-promoting traits and tolerate salt stress	Seed treatment (2.5×10^7 cells/seed)	Ansari and Ahmad (2018)
	Arthrobacter protophormiae, Dietzia natronolimnaea and *Bacillus subtilis*	Hydroponic (under salinity and draught)	Confer abiotic stress tolerance (salt and drought), improved growth	Seedling treatment	Barnawal et al. (2017)

	Bacillus megaterium, *Bacillus safensis*, *Enterobacter aerogenes*	Pot and field	Increased dry weight of root and shoot, increased length and seed weight	Seed treatment (10^7 cfu/ml, coated in carrier material)	Mukhtar et al. (2017)
	Bacillus megaterium, *Arthrobacter chlorophenolicus*, *Enterobacter* sp.	Pot and field	Increased grain and straw yield, plant height, nutrient acquisition, and micronutrient (Fe, Cu, Mn, and Zn) content in grain	Seed treatment (10^7 cfu/ml)	Kumar et al. (2014)
Maize	*Pseudomonas* sp., strain DSMZ 113134	Field	Increased Mg, K, and S contents	Soil treatment (pdx 22.7 kg/ha)	Holečková et al. (2018)
	Azospirillum spp.	Pot (under drought condition)	Enhanced drought tolerance and improved root and shoot gr	Seedling treatment (5×10^7 cfu/ml)	García et al. (2017)
Barley (*Hordeum vulgare* L.)	*Bacillus amyloliquefaciens*	Pot	Increased plant growth and tolerate salt stress	Seed treatment	Kasim et al. (2016)
	Curtobacterium flaccumfaciens	Pot (under salinity stress)	Increased plant growth	Seed treatment (10^7–10^8 cfu/ml)	Cardinale et al. (2015)
Chickpea (*Cicer arietinum* L.)	*Rhizobium pusense*, *Paraburkholderia*, *Stenotrophomonas maltophilia*	Greenhouse	Improved nodulation, N fixation, PGP, and yields	Seed treatment (10^8 cfu/ml) and soil drench (5 ml of 10^8 cfu/ml)	Gopalakrishnan et al. (2018)
	Streptomyces	Pot	Increased growth and host-plant resistance induction against *Botrytis cinerea* (when co-inoculated with *Mesorhizobium ciceri*)	Seed treatment (10^8 cfu/ml) and soil drench (5 ml of 10^8 cfu/ml)	Vijayabharathi et al. (2018)
	Pseudomonas geniculate, *Chryseobacterium indologenes*, *Stenotrophomonas*, *Pantoea dispersa*	Field	Improved nitrogen fixation, plant growth, and yield enhancements	Seed treatment (10^8 cfu/ml) and soil drench (5 ml of 10^8 cfu/ml)	Gopalakrishnan et al. (2017)

(continued)

Table 3.2 (continued)

Crops/plants	Microbe	Growth condition	Effect on plant	Application mode	References
Phaseolus vulgaris	*Cellulosimicrobium funkei* KM032184	Greenhouse (under heavy metal toxicity (chromium VI)	Enhanced seed germination, increased shoot and root length, total biomass, chlorophyll a, b, total chlorophyll, and carotenoid content	Seed treatment (10^8 cells/ml)	Karthik et al. (2016)
Oat *(Avena sativa)*	*Klebsiella* sp.	Hydroponic condition (under salt stress)	Improved shoot and root length, dry weight (root and shoot), and relative water content	Seed treatment (10^8 cfu/ml)	Sapre et al. (2018)
Soybean	Arbuscular mycorrhizal (AM) fungi	Pot (drought condition)	Increased growth	20 g of soybean root fragments, spores, and mycelia	Salloum et al. (2017)
	Pseudomonas putida H-2-3	Greenhouse (salt and drought stress)	Enhanced shoot length and fresh weight, chlorophyll content, abscisic acid, salicylic acid	Bacterial culture suspension (5 ml of 10^8 cfu/ml)	Kang et al. (2014)
Groundnut *(Arachis hypogaea L.)*	*Pseudomonas* sp.	Field	Increased in pod yield and reduces disease	Bacterial culture suspension	Le et al. (2018)
	Bacillus licheniformis A2	Pot (under saline soil condition)	Increased fresh biomass, shoot and root length	Soil treatment (talc-based bioformulation 0.5 g/kg)	Goswami et al. (2014)
Mung bean (*Vigna radiata*)	*Pseudomonas aeruginosa* and *Bacillus subtilis*	Pot	Enhanced root and shoot length, fresh and dry weight of root and shoot, leaf area, and chlorophyll content	Seed treatment	Kumari et al. (2018)
	Pseudomonas fluorescens MC46	Pot	Plant growth and health restoration, enhanced soil enzyme activities, rhizoremediation (triclocarban)	Formulated bacterial inoculant (saw dust) (4×10^9 cfu/g)	Sipahutar et al. (2018)

	Candida sp. AVGB4	In vitro	Increased growth, shoot and root biomass, rhizoremediation	Soil mixing (10^8 cfu/ml)	Silambarasan and Vangnai (2017)
Apple	*Pseudomonas putida*, *P. fluorescens*, *P. aeruginosa*	Field	Promoted plant growth, number of nodes, branches and chlorophyll content	Bacterial culture suspension (10^9 cfu/ml)	Sharma et al. (2017a, b)
Vigna radiata and *Glycine max* (L.)	*Ochrobactrum* sp. MC22	Growth chamber	Restored the damage to plant structure and root system, rhizoremediation (triclocarban)	Bacterial culture suspension ($2.9 \pm 0.3 \times 10^8$ cfu/ml)	Sipahutar and Vangnai (2017)
Alfalfa (*Medicago sativa*)	*Bacillus megaterium*	Growth chamber	Promoted plant growth	Bacterial culture suspension	Chinnaswamy et al. (2018)
Rapeseed (*Brassica napus*)	*Bacillus*, *Serratia*, *Arthrobacter*, and *Pantoea*	Field	Increased plant growth and yield	Bacterial culture suspension (10^9 cfu/ml)	Valetti et al. (2018)
Fenugreek (*Trigonella foenum-graecum*)	*Sinorhizobium meliloti*, *Pseudomonas fluorescens*	Pot	Improved leaf area, shoot fresh and dry weight, nitrogen, phosphorus and potassium content, and water use efficacy (WUE)	Seed treatment	Bolandnazar et al. (2018)
Amaranth (*Amaranthus hypochondriacus*)	*Bacillus* sp.	Field	Enhancement of essential amino acid (methionine, lysine, and tryptophan) contents and other nutritive chemical constituents' grains	Seed treatment (talc formulation 8.0 g/kg seeds)	Pandey et al. (2018)
Lentil (*Lens culinaris* L.)	*Bacillus cereus* and *Bacillus safensis*	Pot	Increased plant growth and suppresses *Alternaria* leaf spot and blight diseases	Bacterial culture suspension (5×10^6 cfu/ml)	Roy et al. (2018)
Ginger (*Zingiber officinale* Rosc.)	*Bacillus amyloliquefaciens* and *Serratia marcescens*	Green house and field	Enhanced sprouting, suppressed soft rot incidence and increased rhizome yield	Seed treatment (10^{10} cfu/ml) and soil drench (10^8 cfu/ml)	Dinesh et al. (2015)

(continued)

Table 3.2 (continued)

Crops/plants	Microbe	Growth condition	Effect on plant	Application mode	References
Pepper (*Capsicum annuum* L.)	*Bacillus velezensis*	Greenhouse	Promoted seedling growth and induced systemic resistance (ISR) pepper gray mold disease	Bacterial culture suspension (10^8 cfu/ml)	Jiang et al. (2018)
	Serratia nematodiphila PEJ1011	Growth chamber (under cold temperature stress)	Increased endogenous ABA levels	Bacterial culture suspension (40 ml of 10^8 cfu/ml twice)	Kang et al. (2015)
	Bacillus amyloliquefaciens	Greenhouse	Increased plant growth and suppresses anthracnose disease	Seed treatment (25 ml of 1×10^8 cfu/ml)	Gowtham et al. (2018)
Tomato (*Lycopersicon esculentum*)	*Pseudomonas stutzeri* E25 and *Stenotrophomonas maltophilia*	Pot	Higher shoot and root length, chlorophyll content, and total fresh weight of tomato	Bacterial culture suspension applied every week	Rojas-Solís and Santoyo (2018)
	Bacillus amyloliquefaciens subsp. *plantarum* strain 32a	Growth chamber	Promoted plant growth and reduces crown gall symptoms	Bacterial culture suspension (10 ml of 10^8 cfu/ml)	Abdallah et al. (2018)
	Bacillus altitudinis, Bacillus velezensis	Greenhouse	Increased plant growth and biocontrol against multiple plant diseases	Seed treatment (1 ml of 10^6 cfu/ml)	Liu et al. (2018a, b)
	Pseudomonas strains	Pot	Stimulate plant growth, increase yield, enhanced antioxidant enzyme activities and proline concentrations	Seed treatment (10^7–10^8 cells/ml)	Egamberdieva et al. (2017)
	Sphingomonas sp. LK11	Pot	Increased growth attributes	Bacterial culture suspension (50 ml of 10^8 cfu/ml)	Khan et al. (2017)
	Bacillus subtilis LK14	Pot	Increased shoot and root biomass and chlorophyll (a and b) contents	Bacterial culture suspension (5 ml of 10^8 cfu/ml)	Latif Khan et al. (2016)

Radicchio (*Cichorium intybus* L.)	*Pseudomonas* sp. and *Bacillus* sp.	Pot	Increased aerial parts and root dry biomass	Bacterial culture suspension (50 ml of 10^6 cfu/ml)	Stanojković-Sebić et al. (2018)
Lettuce	*Pseudomonas fluorescens*	Greenhouse	Increased shoot dry mater (SDM) and shoot N, P, Ca, Mg, Mn, and Na uptake rates	Seed treatment (2 ml of 10^7 cfu/ml)	Khosravi et al. (2018)
	Pseudomonas nitroreducens strain IHB B 13561	Growth chamber	Improved plant growth	Bacterial culture suspension (5 ml of 10^5 cfu/ml)	Trinh et al. (2018)
	AM fungi	Pot	Improved plant growth	Seed treatment (720 propagules/g)	Konieczny and Kowalska (2016)
Alemow (*Citrus macrophylla*)	*Pseudomonas putida* KT2440 or *Novosphingobium* sp. HR1a	Pot (under salinity stress)	Decreased the production of abscisic acid and salicylic acid, better plant performance	–	Vives-Peris et al. (2018)
Tobacco (*Nicotiana tabacum*)	*Pseudomonas fluorescens SS101*	In vitro and in planta	Enhanced plant growth	Bacterial culture suspension (10^7 cfu/ml)	Park et al. (2015)
Carrot (*Daucus carota*)	*Pseudomonas fluorescens* Pf	Field	Increased yield and suppress *Meloidogyne hapla* (root-knot nematode)	Seed treatment (100 ml/kg)	Seenivasan (2018)

3.7.2 Selection of Carrier Materials

In bioformulation development, carrier comprises the major portion of the inoculant (by volume or weight). It is used to deliver the PGPM (or active ingredient) in suitable physiological condition. The carriers include the following categories (Bashan et al. 2014): *soils* (coal, clays, peat, and inorganic soil), *plant waste materials* (composts, farmyard manure [FYM], wheat bran, press mud, spent mushroom compost, plant debris, etc.), *inert carrier materials* (ground rock phosphate, talc, vermiculite, perlite, etc.), *lyophilized microbial cultures and oil-dried bacteria* (these can be used as such or can be incorporated into a solid carrier), and *liquid inoculants* (like emulsions, oils, and broth). The carrier helps in protection and stabilization of cells during storage and transportation to the target site. These can be organic, inorganic, or synthesized from specific molecules. The desirable characteristics of an ideal carrier with organism (bioformulation) include (Bashan et al. 2014; Sahu and Brahmaprakash 2016):

1. Increased shelf life and stability (5–30 °C).
2. Deliver appropriate number of viable cells.
3. Cheaply and nearly sterilized to deliver the appropriate microbe.
4. It should be chemically and physically uniform.
5. It should be suitable for numerous microbes and must have high water-holding capacity.
6. It should be eco-friendly, i.e., nonpolluting, biodegradable, and nontoxic.
7. It should not be phytotoxic to the crop plants.
8. It should be well dissolved and release active component in water.
9. It should be able to tolerate adverse environmental conditions.
10. It should be able to work in diverse field conditions and soil types.
11. It should be cost-effective and compatible with agrochemicals.
12. It should be easily manufactured, and carrier material must be cheap and easily available.
13. It should be able to improve soil properties and resist pH changes during storage.
14. Its release in entrapped formulation should not be too fast or too slow.
15. It should complete the BIS norms for biofertilizers.

3.7.3 Application/Delivery Methods

The bioformulations come in various dispersal forms such as dry products (dusts, granules, and wettable powders), liquid products (oil, water, and emulsions), and slurry and microencapsulation (in polymeric matrix). The use of different bioformulations depends on the need of the type of crop, choice of farmers, market availability, and cost (Bashan et al. 2014). They can be readily delivered through *soil*, *seed*, *rhizomes*, *setts*, and *foliage* or through the combination of these methods (Nakkeeran et al. 2005). The seed inoculation/treatment uses the cell suspensions of specific microbe or the bacteria incorporated in dry products that can grow in

Table 3.3 Some of the commercially available PGPM products used in different countries

Product	Microbe	Crop	Company/country	Function	References
BioYield	*Bacillus amyloliquefaciens* IN937a and *B. subtilis* GB03	Tomato	Gustafson Inc. USA	Management of soilborne pathogens and suppression of *Meloidogyne incognita*	Xiang et al. (2018)
Serenade®	*Bacillus amyloliquefaciens* QST713	Maize and soybean	Bayer Crop Science, Thane, India	Increase in plant growth and protection against pathogens/pests	Mendis et al. (2018)
VOTiVO®	*Bacillus firmus* I-1582	Maize and soybean	Bayer Crop Science, Thane, India	Promote plant growth and offer protection against pathogens/pests	Mendis et al. (2018)
Organo,	*Bacillus* spp., *Pseudomonas, Enterobacter* spp., *Stenotrophomonas, Rhizobium*	Wide variety of crops	Amka Products (Pty) Ltd, South Africa	Solubilize P and K, produce phytohormones and siderophore	Raimi et al. (2017)
RhizoVital42	*Bacillus velezensis* FZB42	Beet, carrot, cucumber, pepper, potato, radish, squash, tomato, and turnip	ABiTEP GmbH Inc., Berlin, Germany	Promoting plant growth	Meng et al. (2016)
Onix®	*Bacillus methylotrophicus*	Carrot	Farroupilha's group Patos de Minas, Brazil	Increased plant growth and production	Clemente et al. (2016)
Rizos®	*Bacillus subtilis*	Carrot	Farroupilha's group Patos de Minas, Brazil	Increased plant growth and production	Clemente et al. (2016)
Quartz®	*Bacillus methylotrophicus*	Carrot	Farroupilha's group Patos de Minas, Brazil	Increased plant growth and production	Clemente et al. (2016)
Mazospirflo-2	*Azospirillum brasilense* AL	Soybean and maize	Soygro Ltd, South Africa	Enhanced N uptake in maize	Laditi et al. (2012)
PHC Biopak	*Bacillus* spp. and *Paenibacillus azotofixans*	Soybean and maize	Plant Health Product (Pty) Ltd. South Africa	Enhanced nodule mass	Laditi et al. (2012)
Ecomonas	*P. fluorescens* @ 10 g/l	Rice	PJ Margo Pvt. Ltd	Increase in yield and management of sheath blight (caused by *Rhizoctonia solani*)	Kumar et al. (2009)
Florezen P	*P. fluorescens* @ 2.5 g/l	Rice	Bab India Private Limited, Hyderabad, India	Increase in yield and management of sheath blight	Kumar et al. (2009)

association with plant roots. For example, the seed treatment with *Pseudomonas fluorescens* at the rate of 100 ml/kg of carrot (*Daucus carota* subsp. *sativus*) seeds led to increase in yield and suppress root-knot nematode (Seenivasan 2018). The soil inoculation with solid or liquid bioformulations is more convenient because of the less time required for application. In this regard, direct soil delivery of PGPM will elevate the population dynamics of augmented microbes in plant rhizosphere.

3.8 Conclusion

The conventional agriculture depends on the use of agrochemicals which is mainly exploited to increase the crop yield. It has a profound negative effect on the environment leading to pollution and degradation of natural habitats. The use of PGPM is a promising approach for sustainable and eco-friendly agriculture. The field application of bioformulation to crop plants is much less effective, mainly due to the varying climatic conditions and the type of carrier material. Therefore, the bioformulation efficacy needs to be enhanced through the usage of compatible mixture of PGPM rather than using a single agent. The development of bioformulation with more than one PGPM will ensure at least one of the mechanisms to function under field conditions. The bioformulation containing multiple strains will have the enhanced efficacy, reliability, and broad spectrum of action and can operate under variable environmental conditions. They are also involved in the remediation of pollutants and heavy metals from the soil and have a great potential to improve plant and soil health (Shelake et al. 2018).

The worldwide market for bioformulation has many products that have been commercialized for use in different crops. The development of new microbial bioformulations is a complex process. It requires competence and strong collaboration of experts in various fields. The product must be produced on a large scale, preserved, and formulated to ensure the biocompatibility. The production processes are patented before commercial use of the product. However, despite a huge number of patents, there are only a few products which have been registered for agricultural application (Timmusk et al. 2017). The future challenge is to produce more economic and improved mixed bioformulations at industrial scale with longer shelf life, increased effectiveness, and higher microbial count in varying field conditions.

Acknowledgments Authors gratefully acknowledge financial support from the National Research Foundation of Korea, Republic of Korea (Grant #2017R1A4A1015515).

References

Abdallah DB, Frikha-Gargouri O, Tounsi S (2018) Rizhospheric competence, plant growth promotion and biocontrol efficacy of *Bacillus amyloliquefaciens* subsp. *plantarum* strain 32a. Biol Control 124:61–67

Ahkami AH, White RA III, Handakumbura PP (2017) Rhizosphere engineering: enhancing sustainable plant ecosystem productivity. Rhizosphere 3:233–243

Ahmad M, Pataczek L, Hilger TH (2018) Perspectives of microbial inoculation for sustainable development and environmental management. Front Microbiol 9:2992

Ansari FA, Ahmad I (2018) Plant growth promoting attributes and alleviation of salinity stress to wheat by biofilm forming *Brevibacterium* sp. FAB3 isolated from rhizospheric soil. Saudi J Biol Sci. https://doi.org/10.1016/j.sjbs.2018.08.003

Badri DV, Vivanco JM (2009) Regulation and function of root exudates. Plant Cell Environ 32 (6):666–681

Barea JM (2015) Future challenges and perspectives for applying microbial biotechnology in sustainable agriculture based on a better understanding of plant-microbiome interactions. J Soil Sci Plant Nutr 15(2):261–282

Barnawal D, Bharti N, Pandey SS et al (2017) Plant growth promoting rhizobacteria enhance wheat salt and drought stress tolerance by altering endogenous phytohormone levels and TaCTR1/ TaDREB2 expression. Physiol Plant 161(4):502–514

Bashan Y, de-Bashan LE, Prabhu SR (2014) Advances in plant growth-promoting bacterial inoculant technology: formulations and practical perspectives (1998–2013). Plant Soil 378 (1–2):1–33

Bhogal A, Nicholson F, Rollett A et al (2018) Improvements in the quality of agricultural soils following organic material additions depend on both the quantity and quality of the materials applied. Front Sustain Food Syst 2:9

Bolandnazar S, Sharghi A, Badhi HN et al (2018) The impact of *Sinorhizobium meliloti* and *Pseudomonas fluorescens* on growth, seed yield and biochemical product of fenugreek under water deficit stress. Adv Hortic Sci 32(1):19–26

Bulgarelli D, Schlaeppi K, Spaepen S et al (2013) Structure and functions of the bacterial microbiota of plants. Annu Rev Plant Biol 64:807–838

Cardinale M, Ratering S, Suarez C et al (2015) Paradox of plant growth promotion potential of rhizobacteria and their actual promotion effect on growth of barley (*Hordeum vulgare* L.) under salt stress. Microbiol Res 181:22–32

Chaparro JM, Badri DV, Bakker MG (2013) Root exudation of phytochemicals in *Arabidopsis* follows specific patterns that are developmentally programmed and correlate with soil microbial functions. PLoS One 8(2):e55731

Cheng W, Gershenson A (2007) Carbon fluxes in the rhizosphere. In: The rhizosphere. Academic, San Diego, pp 31–56

Chinnaswamy A, Coba de la Peña T, Stoll A et al (2018) A nodule endophytic *Bacillus megaterium* strain isolated from *Medicago polymorpha* enhances growth, promotes nodulation by *Ensifer medicae* and alleviates salt stress in alfalfa plants. Ann Appl Biol 172(3):295–308

Clemente JM, Cardoso CR, Vieira BSE et al (2016) Use of *Bacillus* spp. as growth promoter in carrot crop. Afr J Agric Res 11(35):3355–3359

Corral-Lugo A, Daddaoua A, Ortega A (2016) Rosmarinic acid is a homoserine lactone mimic produced by plants that activates a bacterial quorum-sensing regulator. Sci Signal 9(409):ra1-ra1

Dessaux Y, Grandclément C, Faure D (2016) Engineering the rhizosphere. Trends Plant Sci 21 (3):266–278

Dinesh R, Anandaraj M, Kumar A et al (2015) Isolation, characterization, and evaluation of multi-trait plant growth promoting rhizobacteria for their growth promoting and disease suppressing effects on ginger. Microbiol Res 173:34–43

Egamberdieva D, Davranov K, Wirth S et al (2017) Impact of soil salinity on the plant-growth–promoting and biological control abilities of root associated bacteria. Saudi J Biol Sci 24 (7):1601–1608

Elekhtyar NM (2015) Efficiency of *Pseudomonas fluorescence* as plant growth-promoting rhizobacteria (PGPR) for the enhancement of seedling vigor, nitrogen uptake. Int J Sci Res Agric Sci 2:57–67

García JE, Maroniche G, Creus C et al (2017) *In vitro* PGPR properties and osmotic tolerance of different *Azospirillum* native strains and their effects on growth of maize under drought stress. Microbiol Res 202:21–29

Glick BR (2018) Soil microbes and sustainable agriculture. Pedosphere 28(2):167–169

Gopalakrishnan S, Sathya A, Vijayabharathi R et al (2016) Formulations of plant growth-promoting microbes for field applications. In: Microbial inoculants in sustainable agricultural productivity. Springer, New Delhi, pp 239–251

Gopalakrishnan S, Srinivas V, Samineni S (2017) Nitrogen fixation, plant growth and yield enhancements by diazotrophic growth-promoting bacteria in two cultivars of chickpea (*Cicer arietinum* L.). Biocatal Agric Biotechnol 11:116–123

Gopalakrishnan S, Srinivas V, Vemula A et al (2018) Influence of diazotrophic bacteria on nodulation, nitrogen fixation, growth promotion and yield traits in five cultivars of chickpea. Biocatal Agric Biotechnol 15:35–42

Goswami D, Dhandhukia P, Patel P et al (2014) Screening of PGPR from saline desert of Kutch: growth promotion in *Arachis hypogea* by *Bacillus licheniformis* A2. Microbiol Res 169 (1):66–75

Gowtham HG, Murali M, Singh SB et al (2018) Plant growth promoting rhizobacteria *Bacillus amyloliquefaciens* improves plant growth and induces resistance in chilli against anthracnose disease. Biol Control 126:209–217

Haldar S, Sengupta S (2015) Plant-microbe cross-talk in the rhizosphere: insight and biotechnological potential. Open Microbiol J 9:1

Havlicek E, Mitchell EA (2014) Soils supporting biodiversity. In: Interactions in soil: promoting plant growth. Springer, Dordrecht, pp 27–58

Holečková Z, Kulhánek M, Hakl J et al (2018) Use of active microorganisms of the *Pseudomonas* genus during cultivation of maize in field conditions. Plant Soil Environ 64(1):26–31

Hossain MM, Sultana F, Islam S (2017) Plant growth-promoting fungi (PGPF): phytostimulation and induced systemic resistance. In: Plant-microbe interactions in agro-ecological perspectives. Springer, Singapore, pp 135–191

Jiang CH, Liao MJ, Wang HK et al (2018) *Bacillus velezensis*, a potential and efficient biocontrol agent in control of pepper gray mold caused by *Botrytis cinerea*. Biol Control 126:147–157

Kang SM, Radhakrishnan R, Khan AL et al (2014) Gibberellin secreting rhizobacterium, *Pseudomonas putida* H-2-3 modulates the hormonal and stress physiology of soybean to improve the plant growth under saline and drought conditions. Plant Physiol Biochem 84:115–124

Kang SM, Khan AL, Waqas M et al (2015) Gibberellin-producing *Serratia nematodiphila* PEJ1011 ameliorates low temperature stress in *Capsicum annuum* L. Eur J Soil Biol 68:85–93

Karthik C, Oves M, Thangabalu R et al (2016) *Cellulosimicrobium funkei*-like enhances the growth of *Phaseolus vulgaris* by modulating oxidative damage under Chromium (VI) toxicity. J Adv Res 7(6):839–850

Kasim WA, Gaafar RM, Abou-Ali RM et al (2016) Effect of biofilm forming plant growth promoting rhizobacteria on salinity tolerance in barley. Ann Agric Sci 61(2):217–227

Khan AL, Waqas M, Asaf S et al (2017) Plant growth-promoting endophyte *Sphingomonas* sp. LK11 alleviates salinity stress in *Solanum pimpinellifolium*. Environ Exp Bot 133:58–69

Khosravi A, Zarei M, Ronaghi A (2018) Effect of PGPR, Phosphate sources and vermicompost on growth and nutrients uptake by lettuce in a calcareous soil. J Plant Nutr 41(1):80–89

Konieczny A, Kowalska I (2016) The role of arbuscular mycorrhiza in zinc uptake by lettuce grown at two phosphorus levels in the substrate. Agric Food Sci 25(2):124–137

Kroll S, Agler MT, Kemen E (2017) Genomic dissection of host–microbe and microbe–microbe interactions for advanced plant breeding. Curr Opin Plant Biol 36:71–78

Kumar KVK, Raju SK, Reddy MS et al (2009) Evaluation of commercially available PGPR for control of rice sheath blight caused by *Rhizoctonia solani*. J Pure Appl Microbiol 2:485–488

Kumar A, Maurya BR, Raghuwanshi R (2014) Isolation and characterization of PGPR and their effect on growth, yield and nutrient content in wheat (*Triticum aestivum* L.). Biocatal Agric Biotechnol 3(4):121–128

Kumar P, Thakur S, Dhingra GK et al (2018) Inoculation of siderophore producing rhizobacteria and their consortium for growth enhancement of wheat plant. Biocatal Agric Biotechnol 15:264–269

Kumari P, Meena M, Gupta P et al (2018) Plant growth promoting rhizobacteria and their biopriming for growth promotion in mung bean (*Vigna radiata* (L.) R. Wilczek). Biocatal Agric Biotechnol 16:163–171

Laditi MA, Nwoke C, Jemo M et al (2012) Evaluation of microbial inoculants as biofertilizers for the improvement of growth and yield of soybean and maize crops in savanna soils. Afr J Agric Res 7(3):405–413

Latif Khan A, Ahmed Halo B, Elyassi A et al (2016) Indole acetic acid and ACC deaminase from endophytic bacteria improves the growth of *Solarium lycopersicum*. Electron J Biotechnol 19 (3):58–64

Le CN, Hoang TK, Thai TH et al (2018) Isolation, characterization and comparative analysis of plant-associated bacteria for suppression of soil-borne diseases of field-grown groundnut in Vietnam. Biol Control 121:256–262

Lebeis SL, Paredes SH, Lundberg DS (2015) Salicylic acid modulates colonization of the root microbiome by specific bacterial taxa. Science 349(6250):860–864

Levy A, Gonzalez IS, Mittelviefhaus M (2018) Genomic features of bacterial adaptation to plants. Nat Genet 50(1):138

Liu K, McInroy JA, Hu CH et al (2018a) Mixtures of plant-growth-promoting rhizobacteria enhance biological control of multiple plant diseases and plant-growth promotion in the presence of pathogens. Plant Dis 102(1):67–72

Liu X, Jiang X, Zhao W et al (2018b) Colonization of phosphate-solubilizing *Pseudomonas* sp. strain P34-L in the wheat rhizosphere and its effects on wheat growth and the expression of phosphate transporter gene TaPT4 in wheat. BioRxiv. 294736

Mendes R, Garbeva P, Raaijmakers JM (2013) The rhizosphere microbiome: significance of plant beneficial, plant pathogenic, and human pathogenic microorganisms. FEMS Microbiol Rev 37 (5):634–663

Mendis HC, Thomas VP, Schwientek P et al (2018) Strain-specific quantification of root colonization by plant growth promoting rhizobacteria *Bacillus firmus* I-1582 and *Bacillus amyloliquefaciens* QST713 in non-sterile soil and field conditions. PLoS One 13(2):e0193119

Meng Q, Jiang H, Hao JJ (2016) Effects of *Bacillus velezensis* strain BAC03 in promoting plant growth. Biol Control 98:18–26

Mishra J, Singh R, Arora NK (2017) Plant growth-promoting microbes: diverse roles in agriculture and environmental sustainability. In: Probiotics and plant health. Springer, Singapore, pp 71–111

Mukhtar S, Shahid I, Mehnaz S et al (2017) Assessment of two carrier materials for phosphate solubilizing biofertilizers and their effect on growth of wheat (*Triticum aestivum* L.). Microbiol Res 205:107–117

Nakkeeran S, Fernando WD, Siddiqui ZA (2005) Plant growth promoting rhizobacteria formulations and its scope in commercialization for the management of pests and diseases. In: PGPR: biocontrol and biofertilization. Springer, Dordrecht, pp 257–296

Pandey C, Bajpai VK, Negi YK et al (2018) Effect of plant growth promoting *Bacillus* spp. on nutritional properties of *Amaranthus hypochondriacus* grains. Saudi J Biol Sci 25 (6):1066–1071

Park YS, Dutta S, Ann M et al (2015) Promotion of plant growth by *Pseudomonas fluorescens* strain SS101 via novel volatile organic compounds. Biochem Biophys Res Commun 461 (2):361–365

Raimi A, Adeleke R, Roopnarain A (2017) Soil fertility challenges and Biofertiliser as a viable alternative for increasing smallholder farmer crop productivity in sub-Saharan Africa. Cogent Food Agric 3(1):1400933

Rais A, Shakeel M, Malik K et al (2018) Antagonistic *Bacillus* spp. reduce blast incidence on rice and increase grain yield under field conditions. Microbiol Res 208:54–62

Rodriguez A, Sanders IR (2015) The role of community and population ecology in applying mycorrhizal fungi for improved food security. ISME J 9(5):1053
Rojas-Solís D, Santoyo G (2018) Data on the effect of *Pseudomonas stutzeri* E25 and *Stenotrophomonas maltophilia* CR71 culture supernatants on the mycelial growth of *Botrytis cinerea*. Data Brief 17:234–236
Roy T, Bandopadhyay A, Sonawane PJ et al (2018) Bio-effective disease control and plant growth promotion in lentil by two pesticide degrading strains of *Bacillus* sp. Biol Control 127:55–63
Sahu PK, Brahmaprakash GP (2016) Formulations of biofertilizers–approaches and advances. In: Microbial inoculants in sustainable agricultural productivity. Springer, New Delhi, pp 179–198
Salloum MS, Menduni MF, Luna CM (2017) A differential capacity of arbuscular mycorrhizal fungal colonization under well-watered conditions and its relationship with drought stress mitigation in unimproved vs. improved soybean genotypes. Botany 96(2):135–144
Sapre S, Gontia-Mishra I, Tiwari S (2018) *Klebsiella* sp. confers enhanced tolerance to salinity and plant growth promotion in oat seedlings (*Avena sativa*). Microbiol Res 206:25–32
Sarkar A, Ghosh PK, Pramanik K et al (2018) A halotolerant *Enterobacter* sp. displaying ACC deaminase activity promotes rice seedling growth under salt stress. Res Microbiol 169(1):20–32
Seenivasan N (2018) Effect of concomitant application of *Pseudomonas fluorescens* and *Purpureocillium lilacinum* in carrot fields infested with *Meloidogyne hapla*. Arch Phytopathol Plant Protect 51(1–2):30–40
Shahzad R, Waqas M, Khan AL et al (2016) Seed-borne endophytic *Bacillus amyloliquefaciens* RWL-1 produces gibberellins and regulates endogenous phytohormones of *Oryza sativa*. Plant Physiol Biochem 106:236–243
Shahzad R, Khan AL, Bilal S et al (2017) Inoculation of abscisic acid-producing endophytic bacteria enhances salinity stress tolerance in *Oryza sativa*. Environ Exp Bot 136:68–77
Sharma IP, Chandra S, Kumar N (2017a) PGPR: heart of soil and their role in soil fertility. In: Agriculturally important microbes for sustainable agriculture. Springer, Singapore, pp 51–67
Sharma P, Verma PP, Kaur M (2017b) Comparative effect of *Pseudomonas aeruginosa*, *Pseudomonas fluorescens* and *Pseudomonas putida* on the growth of replanted apple. J Pure Appl Microbiol 11(2):1141–1148
Shelake RM, Waghunde RR, Morita EH et al (2018) Plant-microbe-metal interactions: basics, recent advances, and future trends. In: Plant microbiome: stress response. Springer, Singapore, pp 283–305
Shelake RM, Waghunde RR, Verma et al (2019) Carbon sequestration for soil fertility management: microbiological perspective. In: Soil fertility management for sustainable development. Springer, Singapore, pp 25–42
Silambarasan S, Vangnai AS (2017) Plant-growth promoting *Candida* sp. AVGB4 with capability of 4-nitroaniline biodegradation under drought stress. Ecotoxicol Environ Saf 139:472–480
Singh BK, Trivedi P (2017) Microbiome and the future for food and nutrient security. Microb Biotechnol 10(1):50–53
Sipahutar MK, Vangnai AS (2017) Role of plant growth-promoting *Ochrobactrum* sp. MC22 on triclocarban degradation and toxicity mitigation to legume plants. J Hazard Mater 329:38–48
Sipahutar MK, Piapukiew J, Vangnai AS (2018) Efficiency of the formulated plant-growth promoting *Pseudomonas fluorescens* MC46 inoculant on triclocarban treatment in soil and its effect on *Vigna radiata* growth and soil enzyme activities. J Hazard Mater 344:883–892
Stanojković-Sebić A, Pivić R, Dinić Z et al (2018) Effect of indigenous *Pseudomonas* sp. and *Bacillus* sp. strains on yield and main chemical growth parameters of Radicchio. Contemporary Agric 67(1):20–26
Timmusk S, Behers L, Muthoni J (2017) Perspectives and challenges of microbial application for crop improvement. Front Plant Sci 8:49
Trabelsi D, Mhamdi R (2013) Microbial inoculants and their impact on soil microbial communities: a review. Biomed Res Int. https://doi.org/10.1155/2013/863240

Trinh CS, Lee H, Lee WJ et al (2018) Evaluation of the plant growth-promoting activity of *Pseudomonas nitroreducens* in *Arabidopsis thaliana* and *Lactuca sativa*. Plant Cell Rep 37 (6):873–885

Trivedi P, Schenk PM, Wallenstein MD et al (2017) Tiny microbes, big yields: enhancing food crop production with biological solutions. Microb Biotechnol 10(5):999–1003

Valetti L, Iriarte L, Fabra A (2018) Growth promotion of rapeseed (*Brassica napus*) associated with the inoculation of phosphate solubilizing bacteria. Appl Soil Ecol 132:1–10

Vandenkoornhuyse P, Quaiser A, Duhamel M (2015) The importance of the microbiome of the plant holobiont. New Phytol 206(4):1196–1206

Venturi V, Keel C (2016) Signaling in the rhizosphere. Trends Plant Sci 21(3):187–198

Verbon EH, Liberman LM (2016) Beneficial microbes affect endogenous mechanisms controlling root development. Trends Plant Sci 21(3):218–229

Verma PP, Thakur S, Kaur M (2016) Antagonism of *Pseudomonas putida* against *Dematophora necatrix* a major apple plant pathogen and its potential use as a biostimulant. J Pure Appl Microbiol 10(4):2717–2727

Vijayabharathi R, Gopalakrishnan S, Sathya A et al (2018) Deciphering the tri-dimensional effect of endophytic *Streptomyces* sp. on chickpea for plant growth promotion, helper effect with *Mesorhizobium ciceri* and host-plant resistance induction against *Botrytis cinerea*. Microb Pathog 122:98–107

Vives-Peris V, Gómez-Cadenas, Pérez-Clemente RM (2018) Salt stress alleviation in citrus plants by plant growth-promoting rhizobacteria *Pseudomonas putida* and *Novosphingobium* sp. Plant Cell Rep 37(11):1557–1569

Wallenstein MD (2017) Managing and manipulating the rhizosphere microbiome for plant health: a systems approach. Rhizosphere 3:230–232

Wang B, Adachi Y, Sugiyama S (2018) Soil productivity and structure of bacterial and fungal communities in unfertilized arable soil. PLoS One 13(9):e0204085

Weidner S, Latz E, Agaras B (2017) Protozoa stimulate the plant beneficial activity of rhizospheric pseudomonads. Plant Soil 410(1–2):509–515

Xiang N, Lawrence KS, Donald PA (2018) Biological control potential of plant growth-promoting rhizobacteria suppression of *Meloidogyne incognita* on cotton and *Heterodera glycines* on soybean: a review. J Phytopathol 166:449–458

Zayed MS (2016) Advances in formulation development technologies. In: Microbial inoculants in sustainable agricultural productivity. Springer, New Delhi, pp 219–237

Zhang X, Huang Y, Harvey PR (2012) Enhancing plant disease suppression by *Burkholderia vietnamiensis* through chromosomal integration of *Bacillus subtilis* chitinase gene chi113. Biotechnol Lett 34(2):287–293

Microorganisms Improving Food Quality and Safety

4

Manpreet Kaur and Vijay Kumar

Abstract

Food quality and safety depends upon many factors, including various microbial properties. Microorganisms have been used for the production as well as for the quality and safety of the various food items. Antimicrobial and other properties of some selected microorganisms are being used to prevent food spoilage and food preservation. Some of the recent examples are use of bacteriocins and probiotics in food or by increasing the shelf life of food items by using microbial interactions or by reducing the pathogenic microorganisms by competitive microorganism. So the potential of microorganisms can be used as a tool for upgrading food safety and quality. The present work provides the summary on the use of various microbial systems, their modes of action, and application in various types of food systems.

Keywords

Bacteria · Bacteriocin · Food quality · Food safety · Probiotics

4.1 Introduction

Foodborne diseases are globally spread, and they are the main concern in today's era for creating health and economic problems. There has been much advancement in the field food industries related to processing, production, and packaging related to meet the changing requirements of changing society and our food habits. These have been the major health issue in developed as well as developing countries. This problem becomes more important due to the evolution of the microbes and their

M. Kaur · V. Kumar (✉)
Department of Microbiology, Kurukshetra University, Kurukshetra, Haryana, India
e-mail: vijaykuk@kuk.ac.in

S. G. Sharma et al. (eds.), *Microbial Diversity, Interventions and Scope*,
https://doi.org/10.1007/978-981-15-4099-8_4

microbial environments, which plays a crucial role in the foodborne diseases. Food safety is a never-ending issue.

With the revolution in area of food sanitation and hygiene, there has been increase in the use of thermally pasteurized milk products which has led to improvements in food safety. There is also increase in the sophistication of new technologies which have contributed to advances in microbiological food safety with bringing healthier quality of foods and superior nutritional value.

4.2 Microorganism in Improving Food Quality and Safety

Food quality along with food safety has become a major concern in the food business since evolving food-associated pathogens have been associated with human infections and diseases. The contamination of bacteria in ready-to-eat products is of emerging concern to human health. This gives emphasis to the importance of developing active packaging, which prevents the growth and spread of food-associated pathogenic microorganisms. There has been surge in the consumer demand for reduced food additives and, mainly, chemical preservatives. So, there is currently extreme need for carrying out research in food applications of biopreservation and the natural antimicrobials.

Among various microorganisms, lactic acid bacteria provide an edge as they are observed as GRAS (generally recognized as safe). They produce various organic acids, enzymes, hydrogen peroxide, lytic agents, and antimicrobial peptides as well as bacteriocins which inhibit the growth of various microorganisms. More than 100 species of the *Lactobacillus* genus are used commercially as probiotics.

4.3 Use of Bacteriocin as Biopreservative

Among antimicrobial agents produced by bacteria, bacteriocins have acknowledged a significant attention to be used in food preservation. Bacteriocins are antimicrobial peptides produced in ribosomes by many bacteria. There have been various studies conducted on the role of bacteriocins in food preservation synthesized by lactic acid bacteria (LAB). These antimicrobial peptides inhibit the growth of several pathogenic bacteria such as *Bacillus cereus, Listeria monocytogenes*, and *Clostridium botulinum* by challenging their need for nutrients or producing different antimicrobial substances like bacteriocins, acetic acid, and lactic acid.

The means of action of bacteriocins could be bactericidal, with or without lysis of cell, or it may be bacteriostatic. They are often confused with antibiotics, but the major difference is in the bacteriocins that restrict their activity to particular species. Some of the major differences are given in Table 4.1.

Table 4.1 Comparison between bacteriocins and antibiotics

Characteristics	Bacteriocins	Antibiotics
Area of use	Food	Clinical
Synthesis	Ribosomal	Secondary
Range of activity	Narrow	Broad
Immunity to host cell	Yes	No
Mechanism of target cell	Usually adaptation affecting cell	Usually genetically transferable
Resistance or tolerance	Membrane composition	Determinant affecting different sites depending on the mode of action
Interaction requirements	Docking molecules are required sometimes	Specific target required
Mode of action	Mainly cause formation of pores, but in a few cases possibly cell wall biosynthesis	Cell membrane or intracellular targets
Toxicity/side effects	Unknown	Yes

Adapted from Cleveland et al. (2001)

4.4 Classification of Bacteriocins

The universally accepted classification of bacteriocins is still not there (Cleveland et al. 2001; Garcia et al. 2010; Zacharof and Lovitt 2012). Heng and Tagg in 2006 proposed the classification of bacteriocins by taking the nature of colicins into account. Therefore, they have grouped bacteriocins into four main classes. Majority of class Ia, II, and IV bacteriocins obtained from lactic acid bacteria have been applied in food biopreservation.

Class I (lantibiotics): They are posttranslationally modified peptides; they have varying lengths of amino acids from 19 to more than 50, typical thioether-based intramolecular rings of lanthionine and b-methyl-lanthionine. Class I bacteriocins are divided into two more categories, namely, class Ia and class Ib.

- **Class Ia** bacteriocins: Composed of cationic and hydrophobic peptides; they are flexible in their structure contrasting to the rigid structure of class Ib and form pores in target membranes.
- **Class Ib** bacteriocins: The structure of these peptides is globular in nature, with no net charge or a net negative charge. The antimicrobial activity of these bacteriocins is characterized by preventing the synthesis of specific enzymes (Cleveland et al. 2001; Garcia et al. 2010; Zacharof and Lovitt 2012).

Class II includes small (< 10 KDa) heat-stable, membrane-active peptides and is identified as the broad group of Gram-positive bacteriocins. Generally, they are short peptides with cationic nature and high isoelectric points. They are categorized into three subgroups.

Class IIa comprises pediocin-like *Listeria* active peptides with a conserved N-terminal sequence Tyr–Gly–Asn–Gly–Val and two cysteines forming an S–S bridge in the N-terminal half of the peptide.

Class IIb is a complex of two different peptides for its activity. Both peptides have different sequences in their primary amino acid. However, both peptides are encoded by their own adjoining genes, and solely one immunity gene is needed.

Class IIc includes those bacteriocins which are secreted by bacterium in the growth medium with general secretory pathway.

Class III comprises of large (> 30 KDa) heat-labile proteins which have uncertain potential of being used as food biopreservatives. Only two Gram-negative bacteriocins, colicin V and microcins, also fall in this class.

Class IV includes complex bacteriocins which include peptides distinguished by a peptidyl bond connecting the C- and N-termini that are clustered and essential carbohydrate and lipid moieties.

4.5 Commercially Available Bacteriocins

Bacteriocins are the preservatives synthesized by bacteria and have antibiotic attributes that are not the same as therapeutic antibiotics. Nisin and pediocin PA-1 are the only bacteriocins which are produced commercially. Bacteriocins were not used in food products until 1951. The search for new types of bacteriocins has been motivated in order to maintain the FDA's zero tolerance policy toward contamination of food by *Listeria monocytogenes*, a pathogenic bacterium common in the environment (Settanni and Corsetti 2008).

4.5.1 Nisin

Nisin is a class Ia bacteriocin or lantibiotic and is produced mainly by *Lactococcus lactis* subsp. *lactis*. It was developed in the early 1960s and is one of the most characterized and researched and also the most commercially important of all the bacteriocins. It is recognized to be safe by the Joint Food and Agriculture Organization/World Health Organization (FAO/WHO) Expert Committee on Food Additives and has been licensed as a food preservative (E234). It has a molecular mass of 3510 daltons, and it is a 34-amino-acid-long, posttranslational modified polypeptide. Nisaplin™ is the most commercially available form of nisin from Danisco's for food preservative uses (Deegan et al. 2006). Eight different forms of nisin have been discovered and characterized by O'Connor et al., viz., nisins A, Z, F, and Q produced by *Lactococcus lactis* and nisins U, U2, P, and H produced by some *Streptococcus* strains. It has a wide range of activity against various lactic acid bacteria and other Gram-positive bacteria. It is also effective against heat-resistant bacterial spores of

Clostridium botulinum and against foodborne pathogens such as *L. monocytogenes, S. aureus,* or *B. cereus.* Its activity can be enhanced by using chelating agents (such as EDTA), sublethal heat, osmotic shock, and freezing, as these processes make the cell wall of Gram-negative microorganisms more permeable, which further makes them susceptible to the nisin (Galvez et al. 2007; Lucera et al. 2012).

4.5.2 Pediocin

Pediocin PA-1 is the class IIa bacteriocins and is produced by *Pediococcus acidilactici,* and it is marketed as Alta 2341™ or Microgard™. The pediocin PA-1 inhibited the growth of *L. monocytogenes* in ready-to-eat meat products (Deegan et al. 2006; Rodríguez et al. 2002). This bacteriocin has antimicrobial activity against a wide spectrum of Gram-positive bacteria such as *L. monocytogenes* and *S. aureus* and also against Gram-negative bacteria such as *Pseudomonas* and *Escherichia coli,* which are considered to be responsible for food spoilage or foodborne diseases. Pediocin is stable in a range of aqueous solutions and pH and resistant to heating and freezing. Recently, Verma et al. have produced the semi-purified pediocin which was effective in reducing the number of *S. aureus* and also increased the shelf life of buffalo milk.

4.6 Food Systems Application of Bacteriocins as Biopreservative

LAB and their bacteriocins have been taken up by humans from the past many years. The bacteriocins are proteinaceous in nature assumed to be destroyed by digestive proteases, with no adverse effect on the gut microbes, normal pH, and heat tolerance. Moreover, they are used as starter cultures for fermented foods. Three basic methods adopted for applying bacteriocins in food are:

1. Directly applied as purified and semi-purified preparation
2. The inoculation of lactic acid bacteria (LAB) that will synthesize bacteriocin in the food product
3. The use of an ingredient previously fermented with bacteriocin-producing bacteria (Garcia et al. 2010)

4.6.1 Milk and Milk Products

The spoilage of milk and its products is mainly caused by Gram-negative bacteria which are aerobic psychrotrophic in nature, heterofermentative lactobacilli, yeasts, molds, and spore-forming bacteria (Aneja et al. 2008; Ledenbach and Marshall 2009). LAB has a prolonged and harmless history to be used as preservatives in dairy fermentations where they are generally used as starter cultures, chiefly in the production of cheese. Those cultures which produce bacteriocin during cheese manufacture have been related to stop the growth of *Clostridium* which causes gas blowing in Swiss. Nisin-producing lactococci were found effective against clostridial spoilage when applied in these situations. So these primal examples explained the use of nisin-producing strains for inhibition of *Clostridium* sp., but nisin decreases the starter performance and ripening of cheese which diminished its use as biopreservative in cheese making. However, nisin-producing strains have shown slower rates of acid development and limited proteolytic activity and have the ability to ferment sucrose. Also, they have been reported as more sensitive to bacteriophage, which is an important consideration in commercial-scale cheese manufacture. The main aim of nisin is to provide protection against contamination with *L. monocytogenes* owing to create a problem during cheese manufacture and ripening. Many researchers have also carried out research on other bacteriocins as possible biopreservatives in various dairy products to control various pathogenic and spoilage microorganisms.

4.6.2 Meat and Meat Products

Fresh meat and fermented meat products make a perfect environment for the growth of pathogenic and spoilage microorganisms. The microorganisms associated with meat products generally belong to Enterobacteriaceae family, lactic acid bacteria, *Brochothrix thermosphacta*, and pseudomonads. The spoilage of refrigerated meat and meat products is caused by proliferation of *Listeria monocytogenes* during storage (Nychas et al. 2008). Nitrite is frequently used to stabilize red meat color and to prevent the food spoilage and poisoning organisms such as *C. botulinum* in curing meats. However, the adverse health effect of nitrite such as reaction with secondary amines in meats to form carcinogenic nitrosamines has encouraged the researchers to find the potential of using bacteriocins as an alternative to nitrite.

The mainly studied bacteriocins in meat and meat products are nisin, enterocin AS-48, enterocins A and B, sakacin, leucocin A, and especially pediocin PA-l/AcH. They have also been tested alone or in combination with various physicochemical treatments, modified atmosphere packaging, high hydrostatic pressure (HHP), heat, and chemical preservatives, as an additional hurdle to control the growth of *L. monocytogenes* and other pathogens (Ananou et al. 2007).

4.6.3 Fruits and Vegetables

Fruits and vegetables have short life span due to their high moisture content. After harvesting, their quality also decreases due to microbial growth, environmental factors, and maturity. They are made up of polysaccharide such as cellulose, hemicelluloses, and pectin. Spoilage microorganisms, particularly fungi, produce extracellular pectinases and hemicelluloses that can degrade these polysaccharides and cause spoilage in fruits and vegetables (Barth et al. 2009). LAB is also used as starter culture in various fermented vegetables, including table olives (*Enterococcus faecium* BFE 900, *Lactobacillus plantarum* LPC010), sauerkrauts (*Leuconostoc mesenteroides* NCK293, *L. lactis*), and pickles (Settanni and Corsetti 2008). Along with fermentation, starter cultures produce an extensive range of antimicrobial and proteinaceous substances which can inhibit the growth of flora in food products (Ross et al. 2002). The antimicrobial activity of bacteriocin has also been reported in literature against pathogenic microorganisms in fruits and vegetables. The effectiveness of nisin has been studied in combination with sodium lactate, EDTA, and potassium sorbate against pathogenic microorganisms in fresh cut fruits and found to be effective in reductions of 1 and 1.4 log cfu/g in the *Salmonella* population in fresh cut cantaloupe (Raybaudi-Massilia et al. 2009).

4.6.4 Seafood Products

The microflora on seafood constitutes microorganisms existing on the live animal and the microorganisms contaminating during processing. Only few microorganisms survive and proliferate under the product-specific conditions during storage, for example, *Vibrio* sp., *Aeromonas* sp., *Shewanella* sp., *Photobacterium phosphoreum*, and *Pseudomonas* cause fish-like smell in seafood (Gram 2009; Chahad et al. 2012). Many pathogenic microorganisms such as *Aeromonas, Clostridium botulinum, C. perfringens, Escherichia coli, L. monocytogenes, Salmonella, Staphylococcus aureus, Vibrio parahaemolyticus,* and *V. cholerae* serovar O1 and O139 have also been reported from seafood products (Ghanbari et al. 2013). Although successful studies in marine products have been carried out to prevent the growth of *Listeria* sp. by different species of bacteriocin-producing LAB, mainly from the *Carnobacterium* genus (Katla et al. 2001; Brillet et al. 2005; Vescovo et al. 2006; Pinto et al. 2009).

Antimicrobial Packaging Film The bacteriocins can be used in the active biopolymer films for applying in food packaging. Over-conventional synthetic packaging edible films provide several advantages. They exhibit biodecomposibility and are considered environmentally friendly by consumers, so they present as interesting substitute to prevent microorganism in foods.

To retard moisture, oxygen, aroma, and solute transport, the abilities of edible films or coatings have been examined. This is moreover improved by film-carrying food additives such as antioxidants, antimicrobials, flavors, and colorants. To enhance antimicrobial activity of bacteriocins, films should be used instead of being incorporated directly to the products. Many researchers have developed the antimicrobial films using other semi-purified bacteriocins. Out of numerous bacteriocins, enterocins produced by *Enterococcus faecium* have been determined to be effective for managing the growth of *L. monocytogenes* in meat products (Ananou et al. 2005). Looking into the advantage of adding bacteriocins in biofilms, various biofilms have been developed by various researchers.

4.7 Conclusion

With the growing awareness and demand of chemicals or additive free foods of consumers, there is a demand of new alternative methods of improving food quality and safety. To increase the shelf life and safety of dairy products, bacteriocins can be used as a beneficial approach. Bacteriocins are assorted groups of antimicrobial peptides/proteins which have a large spectrum of applications in foods and can be used against a wide range of bacteria with narrow specificity, although the efficiency of bacteriocins is little due to various reasons like food components, their solubility, and their degradation. If LAB-synthesized bacteriocins are used proportional to that present in cultured foods, then they can be termed as consumable. They can be combined with other technologies for food preservation. More research is needed to discover more bacteriocins which can be commercially used and have GRAS status.

References

Ananou S, Maqueda M, Martínez Bueno M, Gálvez A, Valdivia E (2005) Control of *Staphylococcus aureus* in sausages by enterocin AS48. Meat Sci 71:549–576

Ananou S, Maqueda M, Martínez-Bueno M, Valdivia E (2007) Biopreservation, an ecological approach to improve the safety and shelf-life of foods. Commun Curr Res Edu Topics Trends Appl Microbiol 2007:475–486

Aneja KR, Jain P, Aneja R (2008) A textbook of basic and applied microbiology. New Age International Publishers, New Delhi, pp 479–480

Barth M, Hankinson TR, Zhuang H, Breidt F (2009) Microbiological spoilage of fruits and vegetables. In: Sperber WH, Doyle MP (eds) Compendium of the microbiological spoilage of foods and beverages, food microbiology and food safety. Springer Science and Business Media, New York, pp 135–180

Brillet A, Marie-France P, Hervé P, Mireille C, Leroi F (2005) Effect of inoculation of *Carnobacterium divergens* V41, A biopreservative strain against *Listeria monocytogenes* risk, on the microbiological, chemical and sensory quality of cold-smoked salmon. Int J Microbiol 104:309–324

Chahad OB, El Bour M, Calo-Mata P, Abdellatif B, Barros-Velàzquez J (2012) Discovery of novel biopreservation agents with inhibitory effects on growth of food-borne pathogens and their application to seafood products. Res Microbiol 163:44–54

Cleveland J, Montville TJ, Nes IF, Chikindas ML (2001) Bacteriocins: safe, natural antimicrobials for food preservation. Int J Food Microbiol 71:1–20

Deegan LH, Cotter PD, Hill C, Ross P (2006) Bacteriocins: biological tools for bio-preservation and shelf-life extension. Int Dairy J 16:1058–1071

Galvez A, Abriouel H, López RL, Ben Omar N (2007) Bacteriocin based strategies for food biopreservation. Int J Food Microbiol 120:51–70

Garcia P, Rodríguez L, Rodríguez A, Martínez B (2010) Food biopreservation: promising strategies using bacteriocins, bacteriophages and endolysins. Trends Food Sci Technol 21:373–382

Ghanbari M, Jami M, Domig KJ, Kneifel W (2013) Seafood biopreservation by lactic acid bacteria – a review. LWT – Food Sci Technol. https://doi.org/10.1016/j.lwt.2013.05.039

Gram L (2009) Microbiological spoilage of fish and sea food products. In: Sperber WH, Doyle MP (eds) Compendium of the microbiological spoilage of foods and beverages, food microbiology and food safety. Springer Science and Business Media, New York, pp 87–120

Heng NCK, Tagg JR (2006) What is in a name? class distinction for bacteriocins. Nat Rev Microbiol 4. https://doi.org/10.1038/nrmicro1273-c1

Katla T, Moretro T, Aasen I, Holck A, Axelsson L, Naterstad K (2001) Inhibition of *Listeria monocytogenes* in cold smoked salmon by addition of sakacin P and/or live *Lactobacillus sakei* cultures. Food Microbiol 18:431–439

Ledenbach LH, Marshall RT (2009) Microbiological spoilage of dairy products. In: Sperber WH, Doyle MP (eds) Compendium of the microbiological spoilage of foods and beverages, food microbiology and food safety. Springer Science and Business Media, New York, pp 41–45

Lucera A, Costa C, Conte A, Nobile MAD (2012) Food applications of natural antimicrobial compounds. Front Microbiol, Antimicrob, Resistance Chemother 3(287):1–13

Nychas GJE, Skandamis PN, Tassou CC, Koutsoumanis KP (2008) Meat spoilage during distribution. Meat Sci 78:77–89

Pinto AL, Fernandes M, Pinto C, Albano H, Castilho F, Teixeira P, Gibbs PA (2009) Characterization of anti-Listeria bacteriocins isolated from shellfish: potential antimicrobials to control non-fermented seafood. Int J Food Microbiol 129:50–58

Raybaudi-Massilia RM, Mosqueda-Melgar J, Soliva Fortuny R, Martin-Belloso O (2009) Control of pathogenic and spoilage microorganisms in fresh cut fruits and fruit juices by traditional and alternative natural antimicrobials. Comprehensive Rev Food Sci Food Saf 8(3):157–180

Rodríguez JM, Martínez MI, Kok J (2002) Pediocin PA-1, a wide-spectrum bacteriocin from lactic acid bacteria. Crit Rev Food Sci Nutr 42(2):91–121

Ross RP, Morgan S, Hill C (2002) Preservation and fermentation: past, present and future. Int J Food Microbiol 79:3–16

Settanni L, Corsetti A (2008) Application of bacteriocins in vegetable food biopreservation. Int J Food Microbiol 121:123–138

Vescovo M, Scolari G, Zacconi C (2006) Inhibition of *Listeria innocua* growth by antimicrobial producing lactic acid cultures in vacuum-packed cold-smoked salmon. Food Microbiol 23:689–693

Zacharof MP, Lovitt RW (2012) Bacteriocins produced by lactic acid bacteria. APCBEE Proc 2:50–56

Microbes in Soil and Their Metagenomics

5

Suruchi Jindal

Abstract

Metagenomics refers to the study of microbial architecture in a community using next-generation sequencing platforms. Soil microbes play an important role in maintenance of ecosystem. The study of metagenomes of soil microbial community helped in identifying the microbes which play an important role under different environmental conditions. Various tools and techniques are now available for the study of soil metagenomes effectively.

Keywords

Next-generation sequencing · Rhizosphere metagenomics · Taxonomic units · Antarctic soil metagenomics · Saline soil metagenomics

5.1 Introduction

Microbes are the critical components of the environments they live in and provide ecosystem services (Arrigo 2005; van der Heijden et al. 2008). They reside everywhere ranging from human gut to geothermal hot springs. Microbial communities have a huge reservoir of genetic and metabolic diversity. So far, mostly efforts are made in the direction of culturing the microbes, but success rate is low that is up to 1% only (Amann et al. 1995). Handelsman et al. in 1998 made an *E. coli* genomic library using BACs (bacterial artificial chromosomes) and screened it for new enzymatic activities. With the aid of next-generation sequencing technologies available at cheaper rates, it is possible to characterize microbes without the need of

S. Jindal (✉)
School of Agricultural Biotechnology, Punjab Agricultural University, Ludhiana, India
e-mail: suruchi-coasab@pau.edu

S. G. Sharma et al. (eds.), *Microbial Diversity, Interventions and Scope*,
https://doi.org/10.1007/978-981-15-4099-8_5

culturing. An omics approach called metagenomics is being widely used to study and characterize microbes from different environments.

5.2 Definition

Metagenomics could be defined as a technique that requires common microbiology/ molecular biology methodological approaches for assessing the genetic information from environmental microbial consortia, including uncultured microorganisms representing the vast majority of the total, without the need of previous isolation techniques (Handelsman 2004).

Metagenomics refers to the use of DNA sequencing to determine the phylogenetic and functional gene complement of a sample, such as microbial community DNA in soil, human gut, geothermal hot springs, sea water, and glacial ice (Jansson 2013).

Metagenomics comprises the culture-independent and DNA-based analysis of entire microbial communities and complement cultivation-based analysis of microorganisms (Nacke and Daniel 2014).

Metagenome is the entire collection of genetic material of a microbial community Ye 2012. In metagenomics, species diversity is often estimated as operational taxonomic units (OTUs). Species diversity would be defined as the number of species in the community.

Metagenomics provides access to previously hidden genetic information in genomes from uncultured organisms, to the isolation of novel genes and proteins, and to the analysis of genomes and metabolic pathways from uncultured microorganisms, paving the way to elucidate the functions of microbial communities (Riesenfeld et al. 2004; Daniel 2004).

In soil, the amount of present microorganisms is the number of billions (10^9–10^{10}) of cells by milliliter (Daniel 2004). Soil microbial biology is often treated as "black box" (Tiedje et al. 1999) because of its complexity. There are different classes of soils depending on the texture and other geochemical characteristics (Jansson 2013). Different soils have different mineral compositions, redox reactions, and different potential electron acceptors. Metagenome data can provide the information regarding the genes for the reduction of electron acceptors that is methanogenesis, denitrification, sulfate reduction, etc. So far, it is not possible to extract the information representing all the orders of soils through metagenomics because soil habitat is complicated because of partitioning of resources into different microscopic niches. So far, various different platforms have been used for studying metagenomes (Table 5.1 illustrates various soil metagenomes decoded using shotgun sequencing approach using NGS platforms; Jansson 2013). Tringe et al. (2005) carried out the metagenomic analysis of the Wisconsin farm soil using the Sanger sequencing. Yergeau et al. in 2010 used 454 sequencing to compare an Arctic soil active and permafrost to find out the functional genes.

Actinobacteria are abundant in 16S surveys of soil samples because of their resistance to desiccation, and they can withstand long-term starvation conditions

Table 5.1 Examples of soil metagenomes obtained using a shotgun metagenome sequencing approach in the published literature

Study site	Sequencing platform and sequence data	General analyses
Permafrost and active layer samples from a single core, Canadian high Arctic soils: 2 samples total	Roche 454 GS FLX	Assembled using Phrap
	Titanium sequencing (454 Life Sciences, Branford, CT)	Software and annotation
		Using MG-RAST server
	DNA amplified by MDA	Combined assembled and unassembled reads for downstream analyses
	prior to sequencing	
Permafrost and active layer samples from 2 replicate cores, 3 time points (before and after 2 and 7 days of thaw): 12 samples total	Illumina GAII (40 Gb total)	Assembled using Velvet
	1776 million reads	9.7 Mb assembly
	DNA amplified by emPCR prior to sequencing	3700 contigs >1 Kb
		Longest contig 67 Kb
		Draft genome 1.9 Mb
Waseca farm soil: 1 sample	Sanger sequencing	Assembled using Phrap software
100 Mbp	ABI Genome Analyzer (Applied Biosystems, Life Technologies Corporation, Carlsbad, CA)	Combined assembled and unassembled reads
Rothamsted park grass	Roche 454 GS FLX Titanium (13 runs)	Newbler assembly on 454 GS de novo assembler software (Newbler v2.0.00.22)
10 samples. Different DNA extraction methods	12 million reads	Sequences annotated on MG-RAST and reads distributed into metabolic subsystems
FACE sites (biorust and creosote bush root zones – ambient and elevated CO_2), 4 metagenomes total: 1 sample per condition	180 Mb (480,000 reads)	Trimmed unassembled
		Reads analyzed on MG-RAST
Contaminated Arctic soils	450 Mb	Using MG-RAST. Hydrocarbon degradation genes identified by BLAST
4 samples: before treatment, after 1-month treatment, 1-year treatment, and uncontaminated control (approx. 450 Mbp total)	Roche 454 GS FLX	
	Titanium (approx. 1 milj reads)	

(Fierer et al. 2007, Mackelprang et al. 2011). The other most prevalent soil microbes are *Chloroflexi*, *Fibrobacter*, *Acidobacteria*, *Planctomycetes*, and *Synergistetes* (Delmont et al. 2015).

5.3 Basic Methodology of Soil Metagenomics

5.3.1 Soil Sample Collection and DNA Extraction

Soil samples from different environments depending upon the need of experiment are collected in sterile containers. Sample size of soil is crucial as it reveals the structure of different microbial communities. About 10 g of soil is found to be suitable for retrieving optimal diversity (Penton et al. 2016). Metagenomic DNA is extracted from the collected samples using modified CTAB method (Saghai-Maroof et al. 1984). High molecular weight DNA is purified using 0.7% agarose and collected from the gel for library construction (Chauhan et al. 2009). However, the major bottleneck in soil metagenomics is variable activity of microbes of soils, which is reflected during DNA isolations of several microorganisms, and subsequently analysis of metagenomes becomes difficult. In particular, the cell lysis step is very crucial in DNA extraction. Depending on the lysis step methods used, a variety of the microbial communities are either overrepresented or underrepresented. There may be two types of lysis: one is soft lysis and other one is harsh lysis. Soft lysis includes enzymatic lysis or lysis by chemical means; it generally leads to the isolation of high-quality DNA but little quantity. Cells are mechanically disrupted during harsh lysis which involves bead beating, sonication, freeze-thawing, and grinding; it leads to a large quantity of DNA isolation but with poor quality as DNA is highly sheared. Various commonly used DNA extraction protocols were compared by Delmont et al. (2011), and they found the significant representation of different members of soil community. One way to rectify this problem is through fractionation of soil, according to the physiological status of microbial community. For example, in a technique called stable isotope probing (SIP), ^{13}C-labeled substrate is added to nutrient medium, and microbes that incorporate the ^{13}C-labeled during metabolism of the substrate can be fractioned on density gradients. This approach was used by Dumont et al. in 2006 to enrich methanotrophs in a forest soil. Delmont et al. in 2011 used PAGE (polyacrylamide gel electrophoresis) to fractionate DNA representing different microbial communities on the basis of their weights. Yield of DNA is another factor that influences the metagenomics studies. Some soil environments result in low-yield DNA (nanogram quantities), for example, Antarctic soils and Arctic permafrost soils. One of the ways to get the better yields of DNA is to amplify it, which can be done using the molecular displacement amplification approach (MDA) and emulsion PCR (emPCR) approach in which DNA dilution is done to represent single DNA molecule per bead. Both the approaches have been used to amplify the DNA from permafrost layer prior to sequencing (Yergeau et al. 2010 and Mackelprang et al. 2011). Figure 5.1 illustrates the flow diagram showing different steps involved in soil metagenomic analysis.

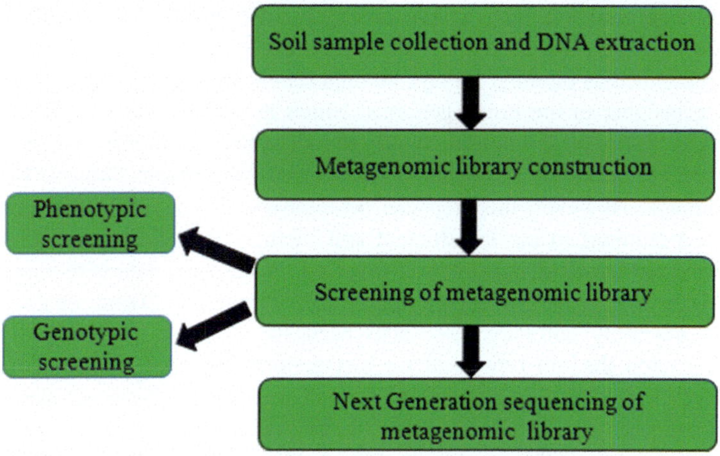

Fig. 5.1 Flow diagram showing different steps involved in soil metagenomic analysis

5.3.2 Metagenomic Library Construction

To construct the metagenomic library, the following steps are being used; high molecular weight DNA is digested partially using restriction enzymes and fractionation of partially digested fragments using gel, followed by elution of fragments (2–10 Kb length) from the gel. Then partially digested DNA fragments are cloned into vector followed by transformation in *E. coli* DH10B. Selection of recombinant transformants is done on LB agar medium supplemented with antibiotic.

5.3.3 Screening of Metagenomic Library

Once the metagenomic clones are prepared in the form of library, selection of clones is done expressing a particular phenotype. The selection of metagenomic clones can be done either through "phenotype-based" approach or functional metagenomics (Daniel 2004) or "genotypic" or "sequence-driven metagenomics." The later approach is useful when there is less opportunity of phenotypic screening; however, a little information of sequence is necessary to design primers and probes.

5.3.4 Next-Generation Sequencing of Metagenomic DNA

Sequencing of metagenomic DNA is done with the help of available NGS platforms, for example, Roche 454, Illumina HiSeq Genome analyzer, etc. The sequencing data is analyzed with the help of QIIME (Quantitative Insights into Microbial Ecology) 1.9.0 pipeline which includes assembly of sequencing data and finding of operational taxonomic units (OTUs) in sequencing reads. The OTUs are then classified taxonomically into phylum, class, order, family, genus, and species using greengenes as reference. This approach of classification based on some referential data is also called as binning. It plays an important role in metagenome analysis by providing insights to the novel genomes which are otherwise difficult to reveal, number and type of taxa; it provides a way to reduce the complexity of metagenomic analyses. The assembly step after sequencing is the major bottleneck in soil metagenomics. However, this step can be improved by adjusting the amount of sequencing required to cover sufficient soil metagenome. For example, the first shotgun metagenome sequence of soil revealed that 2–5 Gb data is required to get eightfold coverage of most of the members of soil community (Tringe et al. in 2005). Different assembly software have also been implemented in different studies depending upon the soil type used for metagenome studies, for example, Velvet was used to assemble a draft genome of novel methanogen from permafrost soils (Mackelprang et al. 2011). Annotation of the genome sequence data to find the functional units like OTUs is also challenging due to lack of databases. The most commonly used server is MG-RAST (metagenomics.anl.gov) for annotation. Genes (coding sequences) in sequencing reads are annotated to decipher their functions using various bioinformatics tools. Both homology-based searches and de novo gene predictions can be done using tools like transeq (Rice et al. 2000), USEARCH (Edgar 2010), RAPsearch (Zhao et al. 2012), MetaGene (Noguchi et al. 2006), FragGeneScan (Rho et al. 2010), and MetaGun (Liu et al. 2013). Another strategy that can be used effectively to find novel genes is to use codon usage (CU) information. Lucić et al. in 2014 analyzed 11 distinct metagenomes and showed that microbial communities belonging to same environments exhibit codon usage (CU) bias similar to single microbial species.

To find out the similarities and dissimilarities at genic level between microbial communities, metagenomic sequence data can be used to decipher information of 16S locus. Through shotgun metagenomic sequences, it is possible to retrieve 16S rRNA sequences which is both taxonomically and phylogenetically informative marker (Pace et al. 1986). However, there are various lacunas in this approach like experimental bias associated with PCR, sequencing errors, and chimeras (incorrectly assembled amplicons), which provides information of only taxonomic composition, not biological function information of taxa and overestimation of community diversity due to horizontal gene transfers of 16S locus. There are several tools that are available to analyze the sequence data like MetaPhyler and MetaPhlAn (homology based, which depends on sequence comparison between the read and taxonomic marker). AMPHORA and PhylOTU are the tools that rely on phylogenetic information and use hidden Markov models (HMM).

5.3.5 Forest Soil Metagenomics

The total microbial genome from forest soils represents forest soil metagenome (Lee 2013). Forest soils have huge microbial diversity in comparison to any other ecosystem generated through the accumulation of large amounts of organic matter.

The forest soils can be classified into four major horizons: O-horizon (organic horizon), carrying debris of plants and partially decomposed matter; A-horizon (alluvial horizon), rich in organic matter with high microbial activity; B-horizon, rich in minerals with relatively low microbial activity; and C-horizon, the mineral horizon having no microbial activity.

Amplification of 16S rRNA genes from the metagenomes of forest soils revealed a huge microbial diversity (Handelsman 2004). The most abundant bacterial phyla present in forest soils are *Acidobacteria*, *Proteobacteria*, *Actinobacteria*, and *Firmicutes* (Nacke et al. 2011). Out of these, *Acidobacteria* is the most abundant phylum in forest soils because of soil pH and holds potential degradation properties (Jones et al. 2009).

Many novel enzymes such as lipases have been obtained from forest soils using small size insert libraries (Faoro et al. 2012), and pigments like indirubin and indigo along with antifungal compounds such as polyketide synthase have been obtained using large insert size metagenomic libraries.

5.3.6 Antarctic Soil Metagenomics

The Antarctic soil area is one of the richest areas considering the bacterial diversity. Bacterial growth and bacterial metabolism have been detected at even at −20 °C. The microorganisms found in Antarctic soils are considered to be important for functioning of global ecosystems. Few adaptations are required to survive in cold environment like enzymatic fitness, presence of least amount of liquid water, high viscosity, ability to modulate membrane fluidity, production of cryoprotectants, and resistance to oxidative stress (Margesin and Miteva 2011).

Many novel cold-adapted enzymes can be found in Antarctica. Lipases and esterases are the most frequent Antarctic enzymes studied by metagenomic approaches. The MHlip enzyme is a 262-amino-acid lipase highly adapted to cold environment because of discrete structural modifications. Enzyme CHA3 belonging to esterase has been shown to be active over a wide range of temperatures. The cellulase RB cell having endocellulolytic activity is an enzyme detected by functional metagenomics (Berlemont et al. 2009). Table 5.2 illustrates the list of enzymes screened through phenotypic screening from the Antarctic soil metagenomes.

Metagenomic sequencing of Antarctic soils revealed the presence of *Proteobacteria*, *Actinobacteria*, and *Cyanobacteria* as the most abundant bacterial

Table 5.2 Enzymatic activities screened in Antarctic soil metagenomic libraries by phenotypic-based approaches

Target	Screening	Vector	Total clones
Lipase/esterase	Phenotypic	BAC	32,000
Amylase	Phenotypic	BAC	27,500
Cellulase	Phenotypic	BAC	8800
Protease	Phenotypic	BAC	16,000
MTA phosphorylase	Phenotypic	Plasmid	85,000
Alkaliphilic esterase	Phenotypic	Fosmid	10,000

classes where *Cyanobacteria* are more prevalent. Recently, *Flavobacterium* was found as the major genus isolated from various sites of Russian Antarctic zone. A considerable diversity of CRISPR spacer type II-C was observed among the flavobacterial isolates which establish the fact that CRISPR spacer content can act as a major tool to study bacterial populations (Lopatina et al. 2016).

5.3.7 Rhizosphere Metagenomics

Rhizosphere is defined as the one of the zones in soil where microbial and plant root interaction occurs. Rhizosphere is regarded to be the ecosystem which has greater impacts on the functioning of the biosphere due to its extension and to the energy consumption of living organisms. The root exudates, mucilage, and dead root cells influence the microbial activity in this zone, which further poses effect on root growth, water and nutrient uptake, respiration, and rhizodeposition. Biotic as well as abiotic stresses such as soil temperature and water content affect the microbial activities and composition. Rondon et al. in 2000 constructed metagenomic libraries using BACs and found the microbial communities actively involved in antibacterial, lipase, amylase, nuclease, and hemolytic activities. Many studies have been carried out in a variety of crops like maize (Chauhan et al. 2011), rice (Somenahally et al. 2011), and biofuel-dedicated crops like soybean, canola, and switchgrass (Jesus et al. 2010) to learn about rhizosphere microbial diversity. The most common genera were *Proteobacteria* and *Acidobacteria* in the rhizosphere of all the crops, which indicates their important role in plant growth and stress response. Rhizosphere metagenomics studies have also been carried out in forest soils, and it was revealed that the trees could select the microorganisms that proliferate under their influence (Uroz et al. 2010). Oak forest rhizosphere showed significant proportion of the *Proteobacteria* followed by *Acidobacteria* and *Actinobacteria* phyla as the most abundant. Kanokratana et al. (2011) found high proportion of aerobic microorganisms in peat swamp forest soils with high metabolic activity to degrade plant polysaccharides. Metagenomic analysis of oil-contaminated soil planted with barley and alfalfa suggested the enrichment of known oil-degrading genera, such as *Alcanivorax* and *Aequorivita* along with *Thermi* and *Gemmatimonadetes* (Kumar et al. in 2018). Romero et al. (2019) studied rhizosphere metagenomics of mine tailings colonizing plants to enhance bioremediation. They sequenced the metagenome of *Enterobacter* sp., Nacozari in Mexico and predicted coding genes for direct and indirect plant growth promotion along with adhesion and oxidative stress-related proteins.

5.3.8 Metagenomic Analysis of Earthquake-Affected Soil Microbes

Hiraoka et al. in 2016 carried out metagenomic analysis of the soil samples collected from earthquake-affected area in Eastern Japan in 2011 using general low-nutrient and seawater-based media, and it was found that genus *Arthrobacter* was in high proportion in both the conditions. Also there were loss of siderophore synthesis genes and overrepresentation of denitrification genes. It was concluded that microbial communities got adapted to drastic environments created by tsunami.

5.3.9 Saline Soil Metagenome

Ahmed et al. in 2018 studied the metagenomes of soil in saline environments to find the genes tolerant to saline environments. Saline soil microbiome metagenomics studies revealed novel genes playing role in osmoadaptation like *GSDH*, *STK_Pknb*, and *duf3445*. These genes are probably involved in the accumulation of amino acids in cytosol and decreased osmolytes, thus increasing osmotolerance. 16S locus studies gave insights about the dominance of halophilic/halotolerant phylotypes belonging to *Proteobacteria*, *Actinobacteria*, *Gemmatimonadetes*, *Bacteroidetes*, *Firmicutes*, and *Acidobacteria*. These studies can be useful for crop improvement in saline environments.

5.4 Conclusions

The advent of next-generation platforms at a cost-effective manner and their ease of availability have advanced the metagenomics research in a considerable manner. So far, many metagenomics studies are successfully accomplished to decipher the role of diverse microbial communities. Soil metagenomics has gained interest and vast area of research these days, but because of soil complexity and presence of huge diversity of microbes at various zones of soil, it has impeded a whole lot of knowledge of soil microbial community. There is constant need of improving the computing facility and developments in algorithms for assembly of bulky data and bioinformatics tools to analyze the data.

Since the sequencing of metagenomes reveals identity of microbial communities and information about function of genes, it only provides estimation of community functional potential. Soil metaphenome is one branch which is gaining importance in which metaphenome could be defined as the outcome of functions encoded in microbial genomes (metagenome) and resources available (biotic and abiotic constraints, spatial and environment; Jansson and Hofmockel 2018). The study of metaphenome along with metagenome will allow the elucidation of complex metabolic networks, and the knowledge gained will be helpful to study the effect of environment on key functions carried out by soil microbiome which will be helpful to design new approaches like carbon cycling and sustainable crop production.

References

Ahmed V, Verma MK, Gupta S, Mandhan V, Chauhan NS (2018) Metagenomic profiling of soil microbes to mine salt stress tolerance genes. Front Microbiol 9. https://doi.org/10.3389/fmicb.2018.00159

Amann RI, Ludwig W, Schleifer KH (1995) Phylogenetic identification and in situ detection of individual microbial cells without cultivation. Microbiol Rev 59:143–169

Arrigo KR (2005) Marine microorganisms and global nutrient cycles. Nature 437:349–355. https://doi.org/10.1038/nature04159

Berlemont R, Delsaute M et al (2009) Insights into bacterial cellulose biosynthesis by functional metagenomics on Antarctic soil samples. ISME J 3(9):1070–1081

Chauhan NS, Ranjan R, Purohit HJ, Kalia VC et al (2009) Identification of genes conferring arsenic resistance to Escherichia coli from an effluent treatment plant sludge metagenomic library. FEMS Microbiol Ecol 67:113–130

Chauhan P, Chaudhry V, Mishra S et al (2011) Uncultured bacterial diversity in tropical maize (Zea mays L.) rhizosphere. J Basic Microbiol 51:15–32

Daniel R (2004) The soil metagenome – a rich resource for the discovery of novel natural products. Curr Opin Microbiol 15:199–204

Delmont T, Robe P, Clark IN, Vogel TM (2011) Metagenomic comparison of direct and indirect soil DNA extraction approaches. J Microbiol Methods 86(3):397–400

Delmont TO, Eren AM, Maccario L, Prestat E, Esen ÖC, Pelletier E, Paslier DL, Simonet P, Vogel TM (2015) Reconstructing rare soil microbial genomes using in situ enrichments and metagenomics. Front Microbiol 6:358. https://doi.org/10.3389/fmicb.2015.00358

Dumont MG, Radajewski SM, Miguez CB, McDonald IR, Murrell JC (2006) Identification of a complete methane monooxygenase operon from soil by combining stable isotope probing and metagenomic analysis. Environ Microbiol 8:1240

Edgar RC (2010) Search and clustering orders of magnitude faster than BLAST. Bioinformatics 26:2460–2461. https://doi.org/10.1093/bioinformatics/btq461

Faoro H, Glogauer A, Couto GH et al (2012) Characterization of a new Acidobacteria-derived moderately thermostable lipase from a Brazilian Atlantic forest soil metagenome. FEMS Microbiol Ecol 81:386–394

Fierer N, Bradford MA, Jackson JA (2007) Toward an ecological classification of soil bacteria. Ecology 88:1354–1364

Handelsman J (2004) Metagenomics: applications of genomics to uncultured micro-organisms. Microbiol Mol Biol Rev 68:669–685

Handelsman J, Rondon MR, Brady SF, Clardy J, Goodman RM (1998) Molecular biological access to the chemistry of unknown soil microbes: a new frontier for natural products. Chem Biol 5: R245–R249

Hiraoka S, Machiyama A, Ijichi M, Inoue K, Oshima K, Hattori M, Yoshizawa S, Kogure K, Iwasaki W (2016) Genomic and metagenomic analysis of microbes in a soil environment affected by the 2011 Great East Japan Earthquake tsunami. BMC Genomics 17:53. https://doi.org/10.1186/s12864-016-2380-4

Jansson J (2013) Soil metagenomics. In: Encyclopedia of metagenomics. https://doi.org/10.1007/978-1-4614-6418-1_701-4

Jansson JK, Hofmockel KS (2018) The soil microbiome-from metagenomics to metaphenomics. Curr Opin Microbiol 43:162–168. https://doi.org/10.1016/j.mib.2018.01.013

Jesus E, Susilawati E, Smith SL et al (2010) Bacterial communities in the rhizosphere of biofuel crops grown on marginal lands as evaluated by 16S rRNA gene pyrosequences. Bio Energy Res 3:20–27

Jones RT, Robeson MS, Lauber CL et al (2009) A comprehensive survey of soil acidobacterial diversity using pyrosequencing and clone library analyses. ISME J 3:442–455

Kanokratana P, Uengwetwanit T, Rattanachomsri U et al (2011) Insights into the phylogeny and metabolic potential of a primary tropical peat swamp forest microbial community by metagenomic analysis. Microb Ecol 61:518–528
Kumar V, AlMomin S, Al-Aqeel H, Al-Salameen F, Nair S, Shajan A (2018) Metagenomic analysis of rhizosphere microflora of oil-contaminated soil planted with barley and alfalfa. PLoS One 13 (8):e0202127
Lee SW (2013) Forest soil metagenomics. In: Encyclopedia of metagenomics. https://doi.org/10.1007/978-1-4614-6418-1_622-3
Liu Y, Guo J, Hu G, Zhu H (2013) Gene prediction in metagenomic fragments based on the SVM algorithm. BMC Bioinformatics 14(Suppl.5):S12. https://doi.org/10.1186/1471-2105-14-S5-S12
Lopatina A, Medvedeva S, Shmakov S, Logacheva MD, Krylenkov V, Severinov K (2016) Metagenomic analysis of bacterial communities of Antarctic surface snow. Front Microbiol 7:398. https://doi.org/10.3389/fmicb.2016.00398
Lucić V, Roller M, Nagy I, Vlahoviček K (2014) Environmental shaping of codon usage and functional adaptation across microbial communities. In: Nelson K (ed) Encyclopedia of metagenomics. Springer, New York, NY
Mackelprang R, Waldrop MP, DeAngelis KM, David MM, Chavarria KL, Blazewicz SJ, Rubin EM, Jansson JK (2011) Metagenomic analysis of a permafrost microbial community reveals a rapid response to thaw. Nature 480(7377):368–371. https://doi.org/10.1038/nature10576
Margesin R, Miteva V (2011) Diversity and ecology of psychrophilic microorganisms. Res Microbiol 162(3):346–361
Nacke H, Daniel R (2014) Approaches in metagenome research: progress and challenges. In: Nelson K (ed) Encyclopedia of metagenomics. Springer, New York
Nacke H, Thürmer A, Wollherr A et al (2011) Pyrosequencing-based assessment of bacterial community structure along different management types in German forest and grassland soil. PLoS One 6:e17000
Noguchi H, Park J, Takagi T (2006) Meta gene: prokaryotic gene finding from environmental genome shotgun sequences. Nucleic Acids Res 34:5623–5630. https://doi.org/10.1093/nar/gkl723
Pace NR, Stahl DA, Lane DJ, Olsen GJ (1986) The analysis of natural microbial populations by ribosomal RNA sequences. Adv Microb Ecol 9:1–55
Penton CR, Gupta VVSR, Yu J, Tiedje JM (2016) Size matters: assessing optimum soil sample size for fungal and bacterial community structure analyses using high throughput sequencing of rRNA gene amplicons. Front Microbiol 7:824. https://doi.org/10.3389/fmicb.2016.00824
Rho M, Tang H, Ye Y (2010) Frag GeneScan: predicting genes in short and error-prone reads. Nucleic Acids Res 38:e191. https://doi.org/10.1093/nar/gkq747
Rice P, Longden I, Bleasby A (2000) EMBOSS: the European molecular biology open software suite. Trends Genet 16:276–277. https://doi.org/10.1016/S0168-9525(00)02024-2
Riesenfeld CS, Schloss PD, Handelsman J (2004) Metagenomics: genomic analysis of microbial communities. Annu Rev Genet 38:525–552
Romero M, Gallego D, Blaz J, Lechuga A, Martínez JF, Barajas HR, Hayano-Kanashiro C, Peimbert M, Cruz-Ortega R, Molina-Freaner FE, Alcaraz LD (2019) Rhizosphere metagenomics of mine tailings colonizing plants: assembling and selecting synthetic bacterial communities to enhance in situ bioremediation bioRxiv. https://doi.org/10.1101/664805
Rondon MR, August PR, Bettermann AD et al (2000) Cloning the soil metagenome: a strategy for accessing the genetic and functional diversity of uncultured microorganisms. Appl Environ Microbiol 66(6):2541–2547
Saghai-Maroof MA, Soliman KM, Jorgensen RA, Allard RW (1984) Ribosomal DNA space length polymorphism in barley: Mendelian inheritance chromosomal location and population dynamics. Proc Natl Acad Sci U S A 81:8014–8018

Somenahally AC, Hollister EB, Loeppert RH et al (2011) Microbial communities in rice rhizosphere altered by intermittent and continuous flooding in fields with long-term arsenic application. Soil Biol Biochem 43:1220–1228

Tiedje JM, Asumming-Brempong S, Nusslein K, Marsh TL, Flynn SJ (1999) Opening the black box of soil microbial diversity. Appl Soil Ecol 13:109–122

Tringe SG, von Mering C, Kobayashi A, Salamov AA, Chen K, Chang HW, Podar M, Short JM, Mathur EJ, Detter JC et al (2005) Comparative metagenomics of microbial communities. Science 308:554–557

Uroz S, Buée M, Murat C et al (2010) Pyrosequencing reveals a contrasted bacterial diversity between oak rhizosphere and surrounding soil. Environ Microbiol Rep 2:281–288

Van der Heijden MG, Bardgett RD, van Straalen NM (2008) The unseen majority: soil microbes as drivers of plant diversity and productivity in terrestrial ecosystems. Ecol Lett 11:296–310. https://doi.org/10.1111/j.1461-0248.2007.01139

Ye Y (2012) Metabolic and species diversity analysis for metagenomics. In: Encyclopedia of metagenomics. https://doi.org/10.1007/978-1-4614-6418-1_88-4

Yergeau E, Hogues H, Whyte LG, Greer CW (2010) The functional potential of high Arctic permafrost revealed by metagenomic sequencing, qPCR and microarray analyses. ISME J 4:1206–1214. https://doi.org/10.1038/ismej.2010.41

Zhao Y, Tang H, Ye Y (2012) RAP Search2: a fast and memory-efficient protein similarity search tool for next-generation sequencing data. Bioinformatics 28:125–126. https://doi.org/10.1093/bioinformatics/btr595

Part II

Microbes in Nanotechnology

Microbial Cell Factories in Nanotechnology 6

Deepak Sharma, Reena Sharma, and Abhishek Chaudhary

Abstract

The nanotechnology is the fast-growing field that offers a huge application in various disciplines of science and technology. The nanoscale materials can be synthesized by physical, chemical, physicochemical, or biological methods. All the synthesis processes except biological process have some environmental and operational constraints. The biological synthesis process or green synthesis of these nanomaterials is an eco-friendly and cost-effective approach which utilizes bacteria, fungi, and plant sources. Biological systems are a good producer of nanoparticles such as magnetotactic bacteria that are capable of producing magnetite (Fe_3O_4), while diatoms are capable of producing siliceous materials. Magnetotactic bacteria produce magnetosomes which are greatly used for the immobilization of enzymes, antibodies, DNA, and RNA. Metal and microbial interactions are greatly involved in the processes like biomineralization, bioremediation, bioleaching, and microbial corrosion. *Pseudomonas stutzeri* AG259 is a metal-accumulating bacterium that has the capability to produce silver nanoparticles; fungi like *Candida glabrata* and *Schizosaccharomyces pombe* have the potential to produce cadmium sulfide particles. *Schizosaccharomyces pombe* has been well studied for its potential to detoxify cadmium from the environment by active intracellular uptake of cadmium and its bioconversion to small iso-peptides. In a summarized way, we can say microbes are the living factories for the generation of advanced materials.

D. Sharma · A. Chaudhary (✉)
Department of Biotechnology and Bioinformatics, Jaypee University of Information Technology, Solan, Himachal Pradesh, India
e-mail: abhishek.chaudhary@juit.ac.in

R. Sharma
School of Basic Sciences, Indian Institute of Technology Mandi, Kamand, Himachal Pradesh, India

S. G. Sharma et al. (eds.), *Microbial Diversity, Interventions and Scope*,
https://doi.org/10.1007/978-981-15-4099-8_6

Keywords

Nanoparticle · Cell factory · Nanosynthesis · Bacteria · Fungi · Actinomycetes

6.1 Introduction

Nanotechnology is an art of fabrications and manipulations of nanoscale objects especially in the range from 1 to 100 nm. At this scale, nanoscale material or nanoparticles have unique physical, chemical, and biological properties. Due to these distinctive properties, nanoscale materials have a wide range of applications in the fields of mechanics, catalysis, optics, electronics, agriculture, and biomedical sciences. Huge applications of nanomaterials in various fields can be explained on the basis of their peculiar properties such as greater surface area-to-volume ratio, unique optical, mechanical properties, ease in uptake, high diffusion rates in the tissues, and advantages in cell signaling (Curtis et al. 2006; Jiang et al. 2008). Nanoparticles are easy to functionalize by various ligands and can be used for bio-imaging, biosensing, medicinal therapy, drug delivery, and gene therapy (Villaverde 2010). Nanoparticles can be synthesized by chemical, physical, and physicochemical methods, but most of the time these methods are not cost-effective and harmful to the surroundings or environments. The biological synthesis of nanoparticles can be used as an alternative to overcome the limitations of chemical and physical methods. Biologically synthesized nanomaterials are eco-friendly, safer, energy efficient, and cost-effective as they utilize the microbial system, plant, and animal sources to generate nanomaterials. In biological methods, microbes are the choice of system for the synthesis of nanoscale material. Microbes due to their diverse physiological activity, size, ease in genetic manipulations, culturing, and downstream processing are greatly explored for the fabrication of nanocreatures. The nanomaterials produced by microbes like polymers, magnetosomes, and engineered systems like for proteins, peptides, and customized metallic nanoparticles are greatly recognized in the field of nanotechnology (Sarikaya et al. 2003).

This chapter gives a detailed outline of the biological synthesis of nanoparticles, especially microbial synthesis of nanoparticles, mechanism, and applications in various fields such as nanomedicine, sensing, and agriculture that are also discussed.

6.1.1 Microbes in Nanotechnology

The group of microorganisms engaged in the manufacturing process of nanoparticles includes bacteria, yeast, actinomycetes, fungi, and algae. Microbial synthesis of metal nanoparticles like silver, gold, platinum, and iron and the oxides of metals like titanium, zinc, copper, etc., have been well documented. Various microbes have been used to synthesize the different sizes and shapes of nanoparticles depending on the applications (Fig. 6.1).

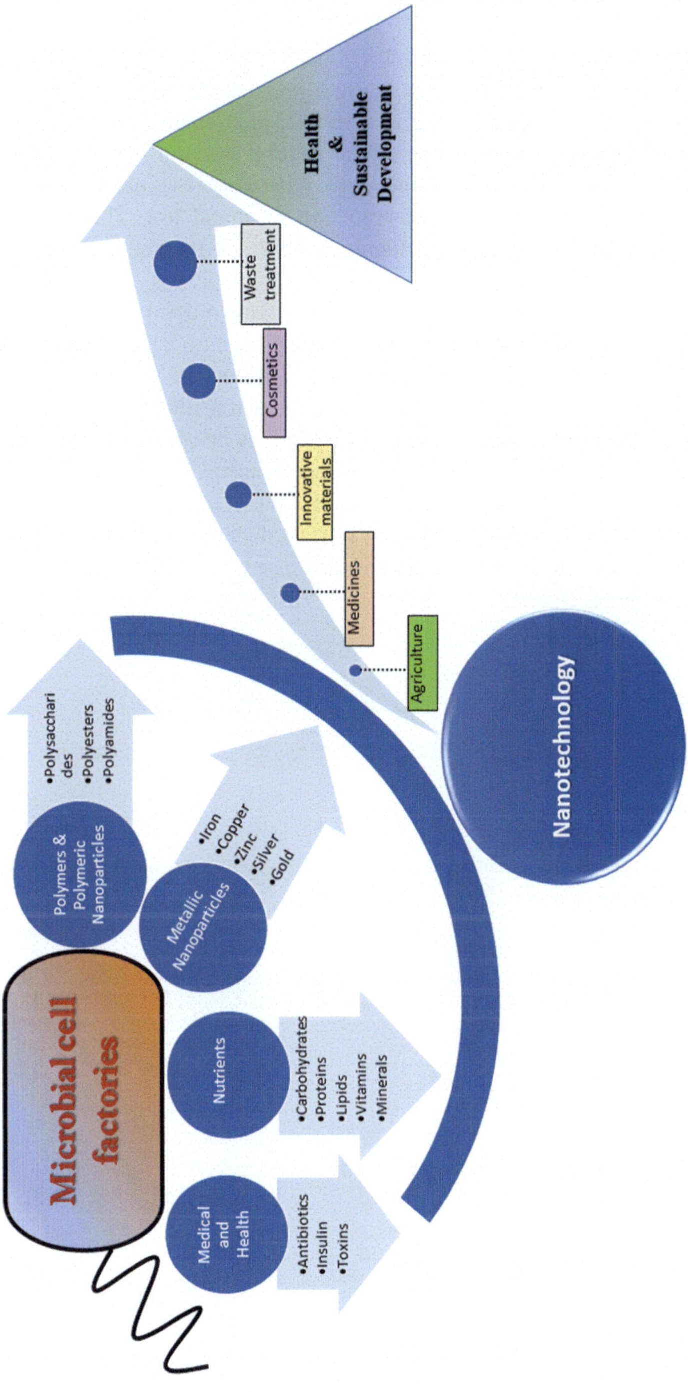

Fig. 6.1 Microbial cell factories and its applications

6.2 Mechanism of Synthesis

The exact mechanism of nanomaterial synthesis through microbial agencies is not yet clear. This is due to different microorganisms that react differently with the substrate material leading to the synthesis of nanomaterials. Microbial synthesis of nanomaterials can be intracellular or extracellular. Intracellular mode of synthesis involves electrostatic interaction between the negatively charged cell wall and positively charged substrate ion or metal ion. The proteins and enzymes of the cell wall act as reducing agents for the metal ion and form nanoparticles.

Positively charged metal ion is reduced by an enzyme present in the cell wall, and nanoparticles come out through the cell wall; after synthesis, the nanoparticles are thrown out from the cell wall by diffusion process (Mukherjee et al. 2001a; Nair and Pradeep 2002; Ahmad et al. 2003a). Whereas in the extracellular mode of synthesis, extracellular enzyme or protein secreted by the cells is involved in the reduction of substrate ion or metal ion, e.g., synthesis of silver nanoparticles using an extracellular nitrate-reductase enzyme secreted by fungi (Kumar et al. 2007a, b; He et al. 2007; Ottoni et al. 2017), similarly NADH- and NADH-dependent enzymes secreted by *Rhodopseudomonas capsulata* bacterium are used in the synthesis of gold nanoparticles (Fig. 6.2).

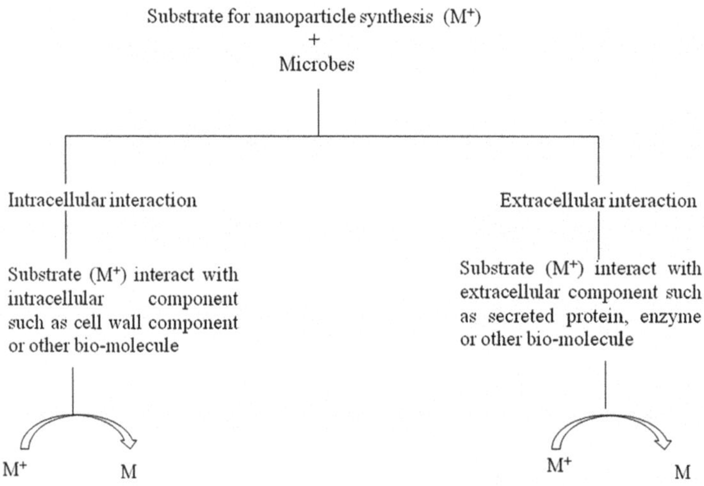

Fig. 6.2 Microbes as a factory for the synthesis of nanoparticles

6.3 Important Microbes Used for Nanoparticle Synthesis

6.3.1 Bacteria

Bacteria are one of the most promising candidates to form nanoparticles. Some bacterial species have the ability to convert metal ion to metal nanoparticles at extreme conditions such as high salt or high temperature, e.g., *Pseudomonas stutzeri and aeruginosa* species have the ability to survive at high salt concentration. Sometimes, the bacterial system utilizes metal ion or another substrate as an energy source and converts them to nanoparticles (Iravani 2014). The rate of nanoparticle synthesis depends on the rate of oxidation and reduction process, and variation in nanoparticle size depends on the microorganism used and the conditions provided during the synthesis process.

Pseudomonas stutzeri AG259 is a Gram-negative metal-accumulating bacterium that has the capability to produce silver nanoparticles of 200 nm in size. This capability of *Pseudomonas stutzeri* helps in the eradication of toxic silver from the environment along with the production of silver nanoparticles in the periplasmic space of bacteria (Klaus et al. 1999). Similarly, *Corynebacterium* sp. has the potential to form silver nanoparticles of about 10–15 nm (Fu et al. 2006). The bacterium *Rhodopseudomonas capsulata* has the potential to form gold nanoparticles in the range of 10–20 nm at neutral pH. The size and morphology of microbial nanoparticles can be regulated by some physiological parameters such as pH and temperature of the system. Deplanche et al. have reported that acidic pH is mainly responsible for the synthesis of spherical gold nanoparticles which are less than 10 nm in size and alkaline pH is responsible for the production of different shape nanoparticles such as hexagons, triangles, and rods with 50 nm in size (He et al. 2007). Similarly, lactate has been reported to reduce $PtCl_6^{2-}$ to Pt^0 in marine bacterium *Shewanella* at neutral pH (Konishi et al. 2007). This bacterium also has the potential to form gold nanoparticles (about 10–20 nm) at neutral pH conditions (Konishi et al. 2006). Similarly, *Acinetobacter schindleri* has been used for the production of zinc oxide nanoparticles, and *Pseudomonas stutzeri* has been used for the production of copper nanoparticles. The mechanism involved in the synthesis process may be the variation in solubility of compound and the extent of toxicity through redox process, absence of transport system for the metals, biological absorption, extracellular complexities, precipitation of metals, accumulation, and the efflux system, greatly depending on the microorganism and condition at the time of nanomaterial synthesis (Husseiny et al. 2007). Magnetotactic bacteria produce magnetosomes which are greatly used for the immobilization of enzymes, antibodies, DNA, and RNA. The microbial systems are frequently exposed to extreme conditions like high metal content, high salt content, etc. Thus, these systems adopt the capability to grow in these conditions and become resistant. Metal and microbial interactions are greatly involved in the processes like biomineralization, bioremediation, bioleaching, and microbial corrosion.

6.4 Actinomycetes as a Microbial Cell Factory

Actinomycetes, another important class of microorganisms, are involved in the production of nanoparticles. Actinomycetes use both the intracellular and extracellular modes of synthesis, and cell wall membrane proteins play a crucial role in nanoparticle synthesis. Several reports are available, indicating the use of actinomycetes for the production of nanoparticles. Kannibaran et al. have reported the synthesis of nanoparticles using actinomycetes (Hulkoti and Taranath 2014; Ahmad et al. 2003a). Similarly, *Thermomonospora* sp. is an alkalothermophilic actinomycete, which synthesizes gold nanoparticles of about 8 nm size by biological reduction (Ahmad et al. 2003a). Alkalo-tolerant actinomycetes *Rhodococcus* sp. can synthesize the gold nanoparticles of 5–15 nm range (Ahmad et al. 2003b). These findings highlighted the potential of actinomycetes in nanoparticle synthesis.

6.5 Fungi as Microbial Cell Factories

Fungi are an excellent source to synthesize metal and metal-sulfide nanoparticles. Diversity of enzymes in the cells of fungi acts as biological reducing agents to generate nanoparticles of various metals. Other advantages of using fungi as microbial cell factory are its easy and fast growth at the laboratory as well as industrial scale. The fungi are most efficient biological agents for nanoparticle synthesis. High intracellular metal uptake capacities, presence of reductase enzyme, and easy culturing on a large scale by fermentation (solid state) make them an efficient candidate for nanoparticle synthesis (Mohanpuria et al. 2008; Volesky and Holan 1995). A variety of fungal species have been used for the synthesis of nanoparticles, e.g., *Verticillium* sp. are a fungi that have the potential to synthesize nano-size particles of precious metals (Ag and Au) intracellularly. Cell wall embedded enzymes of fungi help in the reduction of silver ions into silver nanoparticles. Some of the silver ions may penetrate the fungal cell and are reduced by cytoplasmic proteins to form silver nanoparticles (Mukherjee et al. 2001b). *Fusarium oxysporum* is another fungus having the potential of forming highly stable Au–Ag alloy of size between 8 and 14 nm. Mechanistically, cofactor NADH and enzyme reductase secreted by fungi cells help in the production of Au–Ag alloy (Senapati et al. 2005). Similarly, *Fusarium oxysporum* has been used for the synthesis of semiconductor CdS nanoparticles with the help of reductase enzymes released by fungus (Ahmad et al. 2002). Similarly, luminescent CdSe nanoparticles have been synthesized extracellularly at room temperature conditions (Kumar et al. 2007b). In the same pipeline, *Aspergillus fumigatus* has been documented to form silver nanoparticles in the range of 5–25 nm in just 10 min extracellularly (Bhainsa and D'Souza 2006). *Volvariella volvacea*, an edible mushroom, has the unique capabilities to synthesize of silver, gold, and silver-gold alloy nanoparticles (Vigneshwaran et al. 2007). *Candida glabrata* and *Schizosaccharomyces pombe* have potential to produce cadmium sulfide particles (Krumov et al. 2009).

6.6 Other Microorganisms as Microbial Cell Factories

The algae and the yeasts are also equally important candidates for the synthesis of nanoparticles for various applications. Both of them are well documented as cell factories to synthesize Ag and Au nanoparticles. *Chlorella vulgaris* can be used for the synthesis of gold nanoparticles in the range of 9–20 nm by accumulating within themselves (Hosea et al. 1986). Similarly, *Sargassum wightii Greville*, a marine alga, has the potential to generate gold nanoparticles of 8–12 nm size in the extracellular medium (Singaravelu et al. 2007). The literature review suggested that yeast has the ability to produce semiconductor cadmium sulfide (CdS) crystallite (Dameron et al. 1989). Similarly, gold nanoparticles have been synthesized by *Pichia jadinii* (Gericke and Pinches 2006). *Schizosaccharomyces pombe* has been well studied for its potential to detoxify cadmium from the environment by active intracellular uptake of cadmium and its bioconversion to small iso-peptides. These iso-peptides form cadmium sulfide microcrystallites having application as quantum semiconductor (Bhainsa and D'Souza 2006).

They have a great contribution in the fields of industrial biotechnology and pharmaceutical industry. In the forthcoming section, we will discuss traditional microbial cell factories and their significance.

6.6.1 *Escherichia coli*

Escherichia coli are a favorite microorganism in the field of biotechnology and microbiology because of its fast-growing capabilities, ease in fermentation, and comparatively low cost. Secondly, the *E. coli* also provides excellent genetic tools for genetic manipulations. In the past few decades, lots of manipulations and modification have been done to utilize *E. coli* for various purposes. Easy genetic modification of *Escherichia coli* makes them a potential candidate for nanomaterial synthesis, e.g., *Choi* et al. have used recombinant *E. coli* as a potential candidate for the production of 60 different nanomaterials. They have explored the synthesis of nanomaterial by in vivo methods, where they have used live microbial cells and in vitro methods in which they used microbial cell extract (Choi et al. 2018). Similarly, Vu et al. have used *E. coli* for the synthesis of Ag nanoparticles and explored its antimicrobial activity (Vu et al. 2018). Due to well-known genetics and metabolism, *E. coli* was used as a traditional cell factory for the production of proteins and enzymes (Rueda et al. 2016). Recent applications of *E. coli* in nanomaterial synthesis make them a potential candidate for microbial cell factories for nanomaterial.

6.6.2 *Saccharomyces cerevisiae*

Saccharomyces cerevisiae is the most intensively studied unicellular eukaryotic microorganism and is an important microorganism to produce various products.

Traditionally, it was used in the processes like alcoholic fermentations, bakery product, and ethanol production and for the production of pharmaceutically important proteins (Adrio and Deman 2006). In addition to the traditional application, Jha et al. have proved the role of *S. cerevisiae* in the biosynthesis of Sb_2O_3 nanoparticles; similarly, Niknejad et al. have used *S. cerevisiae* for the production of Ag nanoparticles (Jha et al. 2009; Niknejad et al. 2015).

6.6.3 *Bacillus subtilis*

Bacillus subtilis is a Gram-positive bacterium and stands in the category of the well-characterized microorganisms. The genetic amenability and good fermentation characteristics are key properties of this bacterium making it industrially valuable (Van Dijl and Hecker 2013). Kirthi et al. have reported the role of *B. subtilis* in the synthesis of titanium nanoparticles (Kirthi et al. 2015). Priyadarshani et al. have reported the role of bacillus in the production of anisotropic nanoparticles (Priyadarshinia et al. 2013). Thus, in a summarized way, we can say that the widespread use of microbial sources allows them to pronounce as "the living factories" for the generation of advanced materials. Synthesized microbial nanoparticles have shown great potential in the fields of biomedical, agriculture, and environmental sciences.

6.7 Conclusion

The applications of microbes for the production of nanomaterials seem a promising alternative to physical and chemical approaches in the current time. The strong combination of biotechnology and nanotechnology in the twenty-first century is providing a technological advantage in the various fields of science. The use of microbes for the production of various types of nanoparticles with diversification in shapes and sizes is very much promising. The genetic engineering of microbes can help to increase the production of nanomaterials along with desired properties. Thus, conclusively, we can say that microbes can be utilized as manufacturing factories of nanomaterials for human health and sustainable development.

References

Adrio JL, Deman AL (2006) Genetic improvement of processes yielding microbial products. FEMS Microbiol Rev 30:187

Ahmad A, Mukherjee P, Mandal D, Senapati S, Khan MI, Kumar R, Sastry M (2002) Enzyme mediated extracellular synthesis of CdS nanoparticles by the fungus *Fusarium oxysporum*. J Am Chem Soc 124:12108

Ahmad A, Senapati S, Khan MI, Kumar S, Sastry M (2003a) Extracellular biosynthesis of monodisperse gold nanoparticles by a novel extremophilic Actinomycete, *Thermomonospora* sp. Langmuir 19:3550

Ahmad A, Senapati S, Khan MI, Kumar R, Ramani R, Shrinivas V, Sastry M (2003b) Intracellular synthesis of gold nanoparticles by a novel alkalotolerant Actinomycete, *Rhodococcus* species. Nanotechnology 14:824

Bhainsa KC, D'Souza SF (2006) Extracellular biosynthesis of silver nanoparticles using the fungus *Aspergillus fumigatus*. Colloids Surf B: Biointerfaces 47:160

Choi Y, Parke TJ, Leed DC, Leea SY (2018) Recombinant Escherichia coli as a biofactory for various single- and multi-element nanomaterials. Proc Natl Acad Sci 115:5944

Curtis AS, Dalby M, Gadegaard N (2006) Cell signaling arising from nanotopography: implications for nanomedical devices. Nanomedicine 1:67

Dameron CT, Reeser RN, Mehra RK, Kortan AR, Carroll PJ, Steigerwaldm ML, Brus LE, Winge DR (1989) Biosynthesis of cadmium sulphide quantum semiconductor crystallites. Nature 338:596

Fu M, Li Q, Sun D, Lu Y, He N, Deng X, Wang H, Huang J (2006) Rapid preparation process of silver nanoparticles by bioreduction and their characterizations. Chin J Chem Eng 14:114

Gericke M, Pinches A (2006) Microbial production of gold nanoparticles. Gold Bull 39:22

He S, Guo Z, Zhang Y, Zhang S, Wang J, Gu N (2007) Biosynthesis of gold nanoparticles using the bacteria *Rhodopseudomonas capsulata*. Mater Lett 61:3984

Hosea M, Greene B, McPherson R, Henzl M, Alexander MD, Darnall DW (1986) Accumulation of elemental gold on the alga *Chlorella vulgaris*. Inorg Chim Acta 123:161

Hulkoti NI, Taranath TC (2014) Biosynthesis of nanoparticles using microbes—a review. Colloids Surf B: Biointerfaces 121:474

Husseiny MI, Ei-Aziz MA, Bad RY, Mahmoud MA (2007) Biosynthesis of gold nanoparticles using *Pseudomonas aeruginosa*. Spectrochim Acta A 67:1003

Iravani S (2014) Bacteria in nanoparticle synthesis: current status and future prospects. Int Sch Res Not 1:18

Jha AK, Prasad K, Prasad K (2009) A green low-cost biosynthesis of Sb2O3nanoparticles. Biochem Eng J 43:303

Jiang W, Kim BY, Rutka JT, Chan WC (2008) Nanoparticle-mediated cellular response is size-dependent. Nat Nanotechnol 3:145

Kirthi AV, Rahuman AA, Rajakumar G, Marimuthu S, Santhoshkumar T, Jayaseelan C, Elango G, Zahir GA, Kamaraj C, Bagavan A (2015) Biosynthesis of titanium dioxide nanoparticles using bacterium Bacillus subtilis. Mater Lett 65:2745

Klaus T, Joerger R, Olsson E, Granqvist C (1999) Silver-based crystalline nanoparticles, microbially fabricated. Proc Natl Acad Sci 96:13611

Konishi Y, Tsukiyama T, Ohno K, Saitoh N, Nomura T, Nagamine S (2006) Intracellular recovery of gold by microbial reduction of $AuCl_{4}$ _ ions using the anaerobic bacterium *Shewanella* algae. Hydrometallurgy 81:24

Konishi Y, Ohno K, Saitoh N, Nomura T, Nagamine S, Hishida H, Takahashi Y, Uruga T (2007) Bioreductive deposition of platinum nanoparticles on the bacterium *Shewanella* algae. J Biotechnol 128:648

Krumov N, Perner-Nochta I, Oder S, Gotcheva V, Angelov A, Posten C (2009) Production of inorganic nanoparticles by microorganisms. Chem Eng Technol 32:1026

Kumar SA, Abyaneh MK, Gosavi SW, Kulkarni SK, Pasricha R, Ahmad A, Khan MI (2007a) Nitrate reductase-mediated synthesis of silver nanoparticles from $AgNO_3$. Biotechnol Lett 29:439

Kumar SA, Ayoobul AA, Absar A, Khan MI (2007b) Extracellular biosynthesis of CdSe quantum dots by the fungus, *Fusarium oxysporum*. J Biomed Nanotechnol 3:190

Mohanpuria P, Rana NK, Yadav SK (2008) Biosynthesis of nanoparticles: technological concepts and future applications. J Nanopart Res 10:507

Mukherjee P, Ahmad A, Mandal D, Senapati S, Sainkar SR, Khan MI, Ramani RR, Parischa R, Ajayakumar PV, Alam M, Sastry M, Kumar R (2001a) Bioreduction of $AuCl_4^-$ ions by the fungus, *Verticillium* sp. and surface trapping of the gold nanoparticles formed. Angew Chem Int Ed 40:3585

Mukherjee P, Ahmad A, Mandal D, Senapati S, Sainkar SR, Khan MI, Parishcha R, Ajay PV, Alam M, Kumar R, Sastry M (2001b) Fungus-mediated synthesis of silver nanoparticles and their immobilization in the mycelial matrix: a novel biological approach to nanoparticle synthesis. Nano Lett 1:515

Nair B, Pradeep T (2002) Coalescence of nano-clusters and formation of submicron crystallites assisted by *Lactobacillus* strains. Cryst Growth Des 2:293

Niknejad F, Nabili M, Ghazvini D, Moazeni M (2015) Green synthesis of silver nanoparticles: advantages of the yeast *Saccharomyces cerevisiae* model. Curr Med Mycol 1:17

Ottoni CA, Simões MF, Fernandes S, Santos JG, Silva ES, De Souza RFB, Maiorano AE (2017) Screening of filamentous fungi for antimicrobial silver nanoparticles synthesis. AMB Express 7:31

Priyadarshinia S, Gopinatha V, Priyadharsshinia NM, MubarakAli D, Velusamy P (2013) Synthesis of anisotropic silver nanoparticles using novel strain, *Bacillus flexus*, and its biomedical application. Colloids Surf B: Biointerfaces 102:232

Rueda F, Céspedes MV, Chardi AS, Franzoso JS, Pesarrodona M, Miralles NF, Vázquez E, Rinas U, Unzueta U, Mamat U, Mangues R, Fruitos EG, Villaverde A (2016) Structural and functional features of self-assembling protein nanoparticles produced in endotoxin-free Escherichia coli. Microb Cell Factories 15:59

Sarikaya M, Tamerler C, Jen AK, Schulten K, Baneyx F (2003) Molecular biomimetics: nanotechnology through biology. Nat Mater 2:577

Senapati S, Ahmad A, Khan MI, Sastry M, Kumar R (2005) Extracellular biosynthesis of bimetallic Au–Ag alloy nanoparticles. Small 1:517

Singaravelu G, Arockiamary JS, Kumar VG, Govindaraju K (2007) A novel extracellular synthesis of monodisperse gold nanoparticles using marine alga, *Sargassum wightii* Greville. Colloids Surf B: Biointerface s57:97

Van Dijl JM, Hecker M (2013) *Bacillus subtilis*: from soil bacterium to super-secreting cell factory. Microb Cell Factories 3:1

Vigneshwaran N, Ashtaputre NM, Varadarajan PV, Nachane RP, Paralikar KM, Balasubramanya RH (2007) Biological synthesis of silver nanoparticles using the fungus *Aspergillus flavus*. Mater Lett 61:1413

Villaverde A (2010) Nanotechnology, bionanotechnology, and microbial cell factories. Microb Cell Factories 9:53

Volesky B, Holan ZR (1995) Biosorption of heavy metals. Biotechnol Prog 11:235

Vu XH, Tra Duong TT, Pham TTH, Trinh DK, Nguyen XH, Dang VS (2018) Synthesis and study of silver nanoparticles for antibacterial activity against Escherichia coli and Staphylococcus aureus. Adv Nat Sci Nanosci Nanotechnol 09:025019

Nanotechnology and Food Microbiology 7

Gagandeep Kaur and Shivani Sharma

Abstract

Nanotechnology has emerged as a safer and more effective alternative to conventional food microbiology techniques used to assess pathogens or enhance shelf life of foods. Nanoparticles of silver, zinc oxide, and titanium are already being used as effective antimicrobial agents and are being incorporated in various food-related equipment and packaging materials. Nanoparticles can also be used in combination with polymers leading to formation of nanocomposites. These nanocomposites have enhanced antimicrobial properties and, when embedded in packaging material, provide better shelf life. Nanoemulsions formed from essential oils offer controlled release of these oils. This leads to prolonged antimicrobial activity from the oils without the risk of overpowering aroma or thermal destruction. Nanosensors offer a faster way to detect pathogens with minimum amount of sample. However, despite their manifold advantages, nanostructures are yet to be fully understood in terms of their biological safety. Many food safety agencies across the world are still in process in developing protocols to ensure better and safer use of nanotechnology-based products.

Keywords

Nanotechnology · Nanosensors · Nanoparticles · Nanocomposites · Nanoemulsions · Food safety

G. Kaur (✉) · S. Sharma
Punjab Agricultural University, Ludhiana, Punjab, India

S. G. Sharma et al. (eds.), *Microbial Diversity, Interventions and Scope*,
https://doi.org/10.1007/978-981-15-4099-8_7

7.1 An Overview of Nanotechnology

With his classic talk "There Is Plenty of Room at the Bottom," Richard Feynman envisioned the development of nanostructures with *atom-by-atom* design (Feynman 1960). The transformation of nanotechnology into a separate branch of science has been a rapid one and has led to its applications in all the major disciplines of science. According to the National Nanotechnology Initiative (Arlington, VA, USA), nanotechnology is defined as "the understanding and control of matter at dimensions of roughly 1-100 nm, where unique phenomena enable novel applications." In the nanodimensions (10^{-9} m), matter acquires some unique properties compared to the bulk counterparts due to increased surface area-to-volume ratio. The physical, optical, magnetic, thermodynamic properties of nano-materials show significant variations compared to bulk materials (Rai et al. 2009). Manufacturing of nanomaterials is accomplished by two approaches: "the top-down" approach and "the bottom-up" approach. The top-down approach involves breaking down of bulk material into nanodimensions through techniques like nano-lithography, milling, and precision engineering. The bottom-up approach deals with synthesis of nanomaterials from atoms or molecules, accumulated using crystallization, self-assembly, or microbial synthesis (Iqbal et al. 2012). Nanostructures can vary in their composition (monometallic, bimetallic, composite, hybrid, magnetic, metal oxides, semiconductors, etc.) and shapes and sizes (nanoparticles, nano-rods, nano-wires, nanofibers, nanotubes, nanofluids, nanocapsules, quantum dots, nanosheets, nanoribbons, etc.) (Nasrollahzadeh et al. 2019). These nanostructures have found remarkable applications in food industry, medicine, defense, electronics, textiles, agriculture, and cosmetics (Ozimek et al. 2010).

7.2 Role of Nanotechnology in Food Microbiology

As consumers are becoming more and more aware of food safety, focus of research and development in food sector has shifted toward advanced techniques for food preservation and quality enhancement.

The aim is to have safe, pathogen-free, and high-quality food products with zero defects. Traditional microbiological approaches are time and labor intensive. Therefore, many large food companies like Nestlé, Unilever, and Kraft have already started experimenting with the use of nanotechnology to prepare foods with enhanced quality and safety levels (Ozimek et al. 2010).

7.3 Use of Nanoparticles as Antimicrobial Agents

Nanoparticles can act as potent antibacterial agents and reduce the spoilage of food products. Silver nanoparticles are most commonly used in food products as antimicrobials. One of the major issues encountered by food microbiologists is development of biofilms. Microbial biofilms are impenetrable to most antimicrobial agents. Nano-silver can easily penetrate biofilms and is readily ionized into chemically active form (Zarei et al. 2014). Silver is a stable metal, and within FDA-recommended limit, it can be safely incorporated into packaging materials without posing any health risks. Silver nanoparticles have higher bactericidal effects toward Gram-negative bacteria than Gram-positive ones due to the thinner cell wall of the former. The mechanism of action of silver involves the disruption of ribosomal activity, thereby inhibiting the production of many important microbial proteins and enzymes. Other than silver, titanium dioxide, zinc dioxide, and chitosan nanoparticles can be used as antimicrobial agents in foods. Zinc oxide nanoparticles can be incorporated in polymeric matrices for preparation of antimicrobial packaging materials (Xie et al. 2011). ZnO nanoparticles have been identified as GRAS (generally recognized as safe) material by FDA. Titanium dioxide nanoparticles show photocatalytic activity and are only bactericidal under UV irradiation (Weir et al. 2012). Another way of using nanoparticles as antimicrobials is by loading them with antimicrobial agents. For example, pectin nanoparticles loaded with natural broad-spectrum antimicrobial agent, nisin, can be used as food additives to enhance microbiological safety of the foods (Krivorotova et al. 2016). Cadmium, telluride, selenium, copper, and copper oxide nanoparticles and carbon nanotubes have also shown antimicrobial activity.

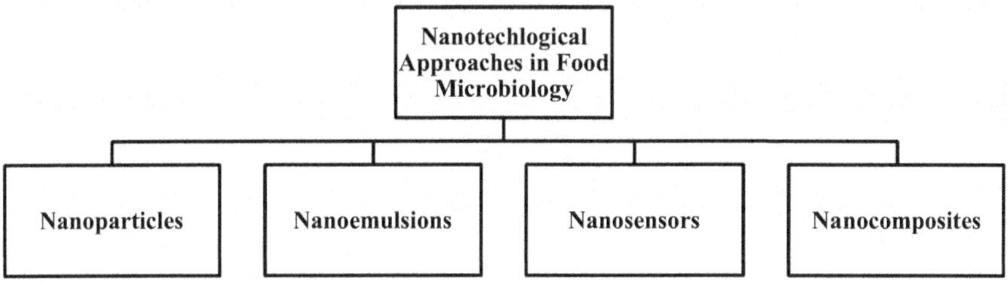

7.4 Polymer-Based Nanocomposites for Barrier Applications

Nanocomposites are formed by a combination of polymers and nanoparticles. Nanocomposite formation enhances the activity of the polymer used. Nanocomposites are used in packaging materials (Llorens et al. 2012). They prevent the microbial infestation of food and do so in a controlled rate over a prolonged duration of time (De Azeredo 2009). Nano-laminates made up of polymers and nano-metals are used to coat meats, fruits, and vegetables and prevent microbial infestations. Nanoforms of zinc oxide, magnesium oxide, and silver have been successfully used in antimicrobial coatings made up of nanocomposites (Garcia et al. 2018). Bio-based nanocomposites offer a biodegradable and environmentally efficient way of food packaging. These are commonly made up of starch and cellulose derivatives (Rhim et al. 2013).

7.5 Nanoemulsions

Emulsions are created by dispersing two immiscible phases. Based on their physical and structural properties, emulsions can be microemulsions or nanoemulsions. While microemulsions are thermodynamically stable, nanoemulsions are only kinetically stable. Microemulsions are transparent, while nanoemulsions are opaque. Functional efficacy of nanoemulsions is increased owing to their decreased particle size. Due to their higher stability, hydrophobic active agents can be dispersed reliably through them (Salvia-Trujillo et al. 2014).

Nanoemulsions can be made through two kinds of techniques: high-energy and low-energy techniques. High-energy techniques are used when energy acts as a disruptive agent to create smaller particle size. These involve techniques such as high-pressure homogenization, microfluidization, and sonication. Low-energy approaches involve techniques such as spontaneous emulsification, emulsion inversion point, phase inversion composition, and phase inversion temperature (PIT). Nanoemulsions made up of essential oils have been used as antimicrobials to control foodborne pathogens like *Listeria monocytogenes*, *Salmonella typhi*, and *Escherichia coli* O157:H7 (Amaral and Bhargava 2015). A soybean oil-based nanoemulsion has recently been of interest due to its antagonistic properties against Gram-positive pathogens and enveloped viruses. It has also been found to be fungistatic.

7.6 Nanosensors

Nanosensors are nanoscale-based sensors that characterize chemical, mechanical, and optical properties of nanoparticles. These nanosensors are being developed with numerous food-based applications. In agri-food sector, it is used as gas sensors for detecting variation in temperature, pressure, and other processing parameters in food samples (Joyner and Kumar 2015). It can also be used for detecting foodborne pathogens, dangerous microorganisms, toxins, contaminants, and chemicals present in foods. These nanosensors have potential advantages over conventional sensors in terms of enhanced sensitivity, specificity, increased speed analysis, low cost, increased number of sample analysis, and less complexity in sample assay. These sensors require a very small amount of analyte in order to get a response that is largely due to small sensing surface area (Driskell and Tripp 2009). The small sensing surface areas lead to fabricate higher-density arrays. Therefore, it analyzes a maximum number of analytes (toxins, contaminants) which also reduces the complexity and cost by reducing the sample processing steps. Nanosensors are classified on the basis of external and internal food conditions. For external conditions of the food products, these sensors are able to detect atmospheric effect, while in internal conditions, these are able to detect certain foodborne pathogens and chemicals (Ramachandraiah et al. 2015). The crucial application of nanosensors is to reduce the pathogen detection time from days to minutes (Bhattacharya et al. 2007). These sensors would also act as "electronic tongue" or "noses" in the packaging material by detecting chemicals during food spoilage (García et al. 2006). On the basis of microfluidic devices, these are having potential applications in detecting food pathogens in less time with high sensitivity (Baeumner 2004). The nanosensor research has led to the development of scientific advancements in the field of nanotechnology that forms new generation nanomachines.

7.7 Risks Related to Use of Nano-products

Nanostructures can cross the cellular barriers and lead to complications at the molecular level. Some nanoparticles can attach to cellular receptors in cells of immune system and lead to their disruptions, cause degradation of cellular proteins, or increase oxidative stress in cells. They can also cause genotoxic and cytotoxic effects by disrupting the DNA of the cells. Consumption of nanoemulsions at a large scale can cause health-related problems as large amounts of surfactants are used to stabilize them. Nanoparticles can also adversely affect the environment. They can get washed away into the ocean where they can enter the plankton, thereby entering the entire aquatic food web.

Internationally, several agencies are in place to regulate the use of nanotechnology-based products in food sector. Some of these include the European Food and Safety Authority (EFSA), Environmental Protection Agency (EPA), Food and Drug Administration (FDA), National Institute for Occupational Safety and Health (NIOSH), Occupational Safety and Health Administration (OSHA), US Department of Agriculture (USDA), Consumer Product Safety Commission (CPSC), and US Patent and Trademark Office (USPTO).

7.8 Conclusion

An increased demand for biologically safe and eco-friendly methods of food safety is causing a surge in production of nanotechnology-based products in the food industry. Some of these products and technologies have successfully been able to control the foodborne pathogens and enhance the shelf life and safety of the food products. However, most government policies related to nanotechnology are still in their nascent phase as the field has grown at a rate much faster than anticipated. Lack of regulations can put consumers at risk and defeat the purpose in the process.

References

Amaral DMF, Bhargava K (2015) Essential oil nanoemulsions and food applications. Adv Food Technol Nutr Sci Open J 1(4):84–87

Baeumner A (2004) Nanosensors identify pathogens in food. Food Technol (Chicago) 58(8):51–55

Bhattacharya S, Jang J, Yang L, Akin D, Bashir R (2007) BioMEMS and nanotechnology-based approaches for rapid detection of biological entities. J Rapid Methods Autom Microbiol 15 (1):1–32

De Azeredo HM (2009) Nanocomposites for food packaging applications. Food Res Int 42 (9):1240–1253
Driskell JD, Tripp RA (2009) Emerging technologies in nanotechnology-based pathogen detection. Clin Microbiol Newsl 31(18):137–144
Feynman RP (1960) There's plenty of room at the bottom. California Institute of Technology, Engineering and Science Magazine
Garcia CV, Shin GH, Kim JT (2018) Metal oxide-based nanocomposites in food packaging: applications, migration, and regulations. Trends Food Sci Technol 82:21–31
García M, Aleixandre M, Gutiérrez J, Horrillo MC (2006) Electronic nose for wine discrimination. Sensors Actuators B Chem 113(2):911–916
Iqbal P, Preece JA, Mendes PM (2012) Nanotechnology: the "top-down" and "bottom-up" approaches. In: Supramolecular chemistry: from molecules to nanomaterials. Wiley, Chichester
Joyner JJ, Kumar DV (2015) Nanosensors and their applications in food analysis: a review. Int J Sci Technol 3(4):80
Krivorotova T, Cirkovas A, Maciulyte S, Staneviciene R, Budriene S, Serviene E, Sereikaite J (2016) Nisin-loaded pectin nanoparticles for food preservation. Food Hydrocoll 54:49–56
Llorens A, Lloret E, Picouet PA, Trbojevich R, Fernandez A (2012) Metallic-based micro and nanocomposites in food contact materials and active food packaging. Trends Food Sci Technol 24(1):19–29
Nasrollahzadeh M, Issaabadi Z, Sajjadi M, Sajadi SM, Atarod M (2019) Types of nanostructures. In: Interface science and technology, vol 28. Elsevier, Amsterdam, pp 29–80
Ozimek L, Pospiech E, Narine S (2010) Nanotechnologies in food and meat processing. Acta Sci Pol Technol Aliment 9(4):401–412
Rai M, Yadav A, Gade A (2009) Silver nanoparticles as a new generation of antimicrobials. Biotechnol Adv 27(1):76–83
Ramachandraiah K, Han SG, Chin KB (2015) Nanotechnology in meat processing and packaging: potential applications—a review. Asian Australas J Anim Sci 28(2):290
Rhim JW, Park HM, Ha CS (2013) Bio-nanocomposites for food packaging applications. Prog Polym Sci 38(10–11):1629–1652
Salvia-Trujillo L, Rojas-Graü MA, Soliva-Fortuny R, Martín-Belloso O (2014) Impact of microfluidization or ultrasound processing on the antimicrobial activity against Escherichia coli of lemongrass oil-loaded nanoemulsions. Food Control 37:292–297
Weir A, Westerhoff P, Fabricius L, Hristovski K, Von Goetz N (2012) Titanium dioxide nanoparticles in food and personal care products. Environ Sci Technol 46(4):2242–2250
Xie Y, He Y, Irwin PL, Jin T, Shi X (2011) Antibacterial activity and mechanism of action of zinc oxide nanoparticles against campylobacter jejuni. Appl Environ Microbiol 77(7):2325–2331
Zarei M, Jamnejad A, Khajehali E (2014) Antibacterial effect of silver nanoparticles against four foodborne pathogens. Jundishapur J Microbiol 7(1):e8720

Part III

Microbes in Genetic Engineering

Bioengineered Microbes in Disease Therapy

8

Rahul Mehta

Abstract

The developments in biotechnology, genomics, proteomics, immunology, and metabolic engineering have provided with unprecedented applications of microbes to improve human health. Disease therapies are experimenting with new regimes to alleviate the problems of diseases. Bioengineered microbes present with ample opportunities to uniquely address the problem of ailments by increasing the length and breadth of the available options to tackle diseases. Microbial immunomodulation of the host, targeted killing of the tumor cells using microbes, and specific delivery of drugs into the host tissues are some of the examples of application of bioengineered microbes. Although history of the concept of possibility of using the microbes for tweaking the disease progression and therefore promoting human health is an old one, with recent research studies, the field has received a much needed impetus. In the future, the need is to establish more in vivo evidences for wide acceptance of therapies using bioengineered microbes as well as to demonstrate the extent of the possible horizontal transfer of such bioengineered genes in order to undertake its practical applications.

Keywords

Bioengineered microbes · Therapeutic · Immunomodulation · Tumor cells

R. Mehta (✉)
Department of Microbiology, DAV University, Jalandhar, India

S. G. Sharma et al. (eds.), *Microbial Diversity, Interventions and Scope*,
https://doi.org/10.1007/978-981-15-4099-8_8

8.1 Introduction

The relationship between humans and microorganisms has been well established and studied for over a century with microbes used in various applications for the benefit of human race (Paton and Paton 2012). The major areas in which microbes have been a key player include but are not limited to model organisms in laboratory (for development of molecular biotechnology and studying gene functions), industrial microbiology (development and production of newer products), and the pathogenic microbes (studying infections and potential way of controlling them). However, the discovery of advanced techniques in molecular biology studies, genetic tweaking, and immunology has brought unforeseen avenues to develop newer and better bioengineered microbes with multidimensional applications in disease therapy (Das et al. 2010; Chikazu D et al. 2017). Bioengineered microbes can be helpful in disease management and therapies (Yu H et al. 2011) in one of the following possible ways:

- Immunomodulation by probiotics (Culligan EP et al. 2009)
- Targeted delivery of drugs and gene therapy vectors with a higher site specificity in comparison with traditionally employed methods (Paton et al. 2012)
- Use of anaerobic bacteria to specifically target the killing of tumor cells (Roberts et al. 2014)
- Use of bioengineered microbes like *Salmonella* sp. to target specific viral mRNAs with RNase P (Al-Ramadi B.K. et al. 2009)

Thus, it becomes quite evident that bioengineered microbes have a variety of applications (Table 8.1) and they can significantly reduce the global disease burden by helping to control morbidity and mortality brought about by various diseases.

Table 8.1 Examples of bioengineered microbes, their disease targets, and the reported mechanism of action

S. no.	Bacterial species	Type of cancer	Mechanism of action
1.	*Mycobacterium bovis* BCG	Bladder cancer	Immune stimulation, enhances the pool of pro-inflammatory cytokines (Shintani Y et al. 2007)
2.	*Streptococcus pyogenes* OK-432 (Chikazu et al. 2017)	Bone sarcoma (Mccarthy EF 2006), lymphangiomas, intraoral ranula	Immune sensitization, increase in neutrophils, macrophages, and lymphocytes (Miwa S et al. 2019)
3.	*Clostridium novyi*	Leiomyoma	Mechanism unknown (Roberts et al. 2014; Barbé et al. 2006)
4.	*Salmonella typhimurium*NVP2009	Pancreatic cancers	Blocks angiogenesis as it has an expression plasmid encoding VEGFR2 (Al-Ramadi B.K. et al. 2009; Gajda et al. 2007)

8.2 History of Bioengineered Microbes

The therapeutic application of microbes predates even the experimental establishment of Koch's "germ theory of disease" and is as old as two centuries. It was a peculiar and fine observational prowess of Vautier in 1813 that captured the fact that those persons who had a progressing tumor and developed gas gangrene experienced tumor regression. It was later established that gas gangrene results from *Clostridium perfringens.* This ultimately indicated that some bacteria may naturally promote tumor regression or growth of some bacteria in tumor cells may result in such cells getting killed. In the late nineteenth century, it was also observed that *Streptococcus pyogenes* natural infection also accelerates the killing of tumor cells. This led to the use of Coley's toxins (killed extracts of *S. pyogenes & Serratia* sp.) for inoperable sarcomas showing potential curative effects. The use of BCG strain of *M. bovis* was also employed to cure bladder cancer, and this practice still finds its use in modern times. The success of BCG strain to cure cancer has been attributed to the fact that this strain in particular is able to induce the production antitumor cytokines (IL-2), TNF-α, and interferon-Υ. By extension, it may be possible to predict that along with bacterial cytotoxicity and immunomodulation, these microbes may be affecting the survival of cancer cells.

Probiotics, especially the natural, native microorganisms, had been used for promoting disease therapies in many cases since a century ago or so. A well-suited example would be the use of *Lactobacillus* sp., *Bifidobacterium* sp., and *Leuconostoc* sp. for the treatment of CDAD (*Clostridium difficile*-associated diarrhea). It is an encouraging observation that standard treatment for inflammatory bowel disease (IBD) and Crohn's disease is probiotics. These naturally available probiotic organisms can be better engineered so as to enhance their efficacy by putting in resistance mechanisms to host's tissue environment such as stomach acidity, bile, and increased osmolarity in intestinal tissue.

8.3 Areas Where Bioengineered Microbes Are Currently Used for Cancer Therapy

The currently used bacterial cultures for cancer treatment can be viewed as a supplementary treatment to the standard treatment, enabling complete recovery of the patient (Sun et al. 2008).

The major advantage in using the group of anaerobic organisms in cancer therapy is the fact that unlike chemotherapeutic agents that can spread all over the body tissues, these anaerobic microbes can only grow in tumor microenvironment due to the presence of hypoxic conditions in the latter (Forbes NS 2010).

8.4 Immunomodulatory Recombinant Probiotics

While a lot of natural probiotic organisms are used for therapeutic interventions in many diseases, using the advancement of genetic manipulations, molecular biology can increase the efficiency of such probiotics by making them more suitable for disease therapies (Culligan EP et al. 2009). An example could be the use of such recombinant probiotics to deliver immunomodulatory proteins at sites which are otherwise difficult to access because of physiological constraints such as colon tissue, e.g., excessive inflammatory responses to normal intestinal bacteria are a characteristic symptom in Crohn's disease; IL-10 is a cytokine that can regulate mucosal immune system and thus prevent excessive inflammatory response in such patients. However, systemic IL-10 treatment is not very effective and shows significant side effects. To overcome this, a genetically engineered *Lactococcus lactis* (expressing a mature human IL-10 & made thymine dependent to prevent its growth outside body) was made and showed significant improvement in disease reduction, thereby offering a greater hope with extended therapeutic potential against Crohn's disease.

8.5 Advantages and Limitations

Advantages of such therapeutic interventions include the specificity and highly selective effects of such therapy interventions that make using bioengineered microbes in disease therapy less burdensome and convenient to use (Table 8.1). Other advantages include the capacity to use such bioengineered microbes as in situ production strategies for targeted antitumor and recombinant proteins.

Like every technology, this approach also has some limitations that can be addressed in subsequent times to make it a more robust and gradually increase the acceptance of such therapeutic interventions; few of the limitations for consideration are as follows: most of the studies involving such bioengineered microbes are largely based on animal models of cancer, infection, etc. In vivo stability of the constructs as well as the prevention of horizontal transmission of such vectors from bioengineered microbes needs to be ascertained before being used in actual therapy interventions. (Łukasiewicz and Fol 2018)

8.6 Future Directions

Cancer, infectious diseases, and inflammatory disorders are a rising cause for morbidity and mortality due to such pathologies, especially in the light of increasing antimicrobial resistance; it is imperative to look out and establish the reliability along with validity of alternative strategies to combat such disease modalities. Microbial-based therapies can provide an answer to such problems as demonstrated by ongoing research in microbial bioengineering for disease therapy such as immunomodulatory probiotics, targeted drug and vaccine delivery, and targeted killing of tumor cells.

The future course for this field will depend on the successful transition of such therapies from laboratories to the clinic by accumulating extensive data for safety and efficacy of such bioengineered microorganisms.

8.7 Conclusion

Infectious diseases, cancers, and chronic inflammatory conditions are leading causes of morbidity and mortality all over the world and need immediate attention for newer and more reliable and robust therapeutics especially in the era of the emergence of multidrug resistance. The idea of bioengineered microbes for therapeutic intervention is not a new one, and with recent advancement in many areas of biological sciences, such therapies can offer promising results. However, extensive clinical trials need to be validated before such interventions reach the patient. The major issues of biological containment, preventing horizontal transmission to other species, and greater safety for human use need urgent attention to translate such therapies from laboratory into clinics. The attitude of the regulatory authorities also needs to be changed to be more acceptable toward using genetically modified organisms. The general public should also be made more aware about the potential benefits of such bioengineered microbes even if in the capacity of a complimentary treatment options along with standard chemotherapeutics and radiation therapies.

References

Al-Ramadi BK et al (2009) Potent anti-tumor activity of systemically-administered IL2-expressing *Salmonella* correlates with decreased angiogenesis and enhanced tumor apoptosis. Clin Immunol 130:89–97

Barbé S et al (2006) The use of clostridial spores for cancer treatment. J Appl Microbiol 101:571–578

Chikazu D et al (2017) Evaluation of OK-432 injection therapy as possible primary treatment of intraoral ranula. J Oral Maxillofac Surg 75:336–342

Culligan EP, Hill C, Sleator RD (2009) Probiotics and gastrointestinal disease: Successes problems and future prospects,Gut Pathog 1(1):19. https://doi.org/10.1186/1757-4749-1-19. Published 23 Nov 2009

Das BK et al (2010) Bacteria in cancer therapy: a novel experimental strategy. J Biomed Sci 17:21

Forbes NS (2010) Engineering the perfect (bacterial) cancer therapy. Nat Rev Cancer 10:785–794

Gajda M et al (2007) Improving tumor targeting and therapeutic potential of Salmonella VNP20009 by displaying cell surface CEA-specific antibodies. Vaccine 25:4183–4192

Łukasiewicz K, Fol M (2018) Microorganisms in the treatment of cancer: advantages and limitations. J Immunol Res 2018:1–8

Macarthy EF (2006) The toxins of William B. Coley and the treatment of bone and soft-tissue sarcomas. Iowa Orthop J 26:154–158

Miwa S et al (2019) Current and emerging targets in immunotherapy for osteosarcoma. J Oncol 2019(11):1–8

Paton JC et al (2012) Bioengineered microbes in disease therapy. Trends Mol Med 18:417–425

Roberts NJ, Zhang L, Janku F et al (2014) Intratumoral injection of clostridium novyi-NT spores induces antitumor responses. Sci Transl Med 6(249):249ra111. https://doi.org/10.1126/scitranslmed.3008982

Shintani Y et al (2007) Intravesical instillation therapy with Bacillus Calmette–Guérin for superficial bladder cancer: study of the mechanism of *Bacillus Calmette*–Guérin immunotherapy. Int J Urol 14:140–146

Sun J-B, Duan J-H, Dai S-L, Ren J, Lin G, Jiang W, Li Y (2008) Preparation and anti-tumor efficiency evaluation of doxorubicinloaded bacterial magnetosomes: Magnetic nanoparticles as drug carriers isolated from Magnetospirillum gryphiswaldense. Biotechnol Bioeng. https://doi.org/10.1002/bit.22011

Yu H et al (2011) Bacteria-mediated disease therapy. Appl Microbiol Biotechnol 92:1107–1113

Benefits and Biohazards of Microbial Recombinants

9

Alla Singh, Mehak Sethi, Diksha Verma, Ritu Naliath, and D. P. Chaudhary

Abstract

Recombinant DNA technology has brought paradigm changes in the field of biological interventions in modern society. Be it pharmaceuticals, nutraceuticals, or other products of commercial use, recombinant proteins have brought new opportunities. Way back in 1974, a paper in the journal *Science*, commented on the acquisition of new capabilities in recombinant DNA (rDNA) technology and the immense opportunities in biological sciences created thereby. At the same time, potential biohazards of the technology were also pointed out, which had earlier been discussed by a group of scientists in the 1973 Gordon Research Conference. The scientists called for a voluntary deference of experimentation using rDNA technology. Today, strict guidelines exist regarding the use of rDNA technology for basic and applied sciences. This chapter discusses the various benefits as well as biohazards of microbial recombinants, i.e., rDNA technology using microbial strains and viral vectors. The known and upcoming recombinant products are documented, along with a discussion on the biohazard issues.

Keywords

rDNA technology · Microbial recombinants · Genetic engineering · Protein purification · Biohazard

A. Singh (✉) · D. Verma · R. Naliath · D. P. Chaudhary
ICAR – Indian Institute of Maize Research, Ludhiana, Punjab, India
e-mail: alla.singh@icar.gov.in

M. Sethi
Punjab Agricultural University, Ludhiana, Punjab, India

S. G. Sharma et al. (eds.), *Microbial Diversity, Interventions and Scope*,
https://doi.org/10.1007/978-981-15-4099-8_9

9.1 Benefits of Recombination: The First Recombinant

Utilization of the basic process of recombination for expression of desired proteins in simple organisms like *Escherichia coli* opened new vistas in bio-based industries. The first sector to use this technology was the pharmaceutical industry, which designed engineered insulin protein(s) for management of diabetes (Quianzon and Cheikh 2012). rDNA technology helped design new pharmaceuticals with desired biological activities. But before the advent of rDNA technology, insulin derived from animals was the only product available for treating diabetes. For the discovery of insulin and establishment of its blood glucose reduction effects, Frederick Banting, Charles Best, and John MacLeod got the Nobel Prize. The story of pharmaceutical insulin aptly explains the enhancement of research repertoire, achieved by rDNA technology (Quianzon and Cheikh 2012). The company Eli Lilly started insulin production from animal pancreas, but challenges of various dimensions were encountered soon. Due to the use of animals, meeting the demand was an issue. Potency of the product varied to levels as high as 25%. Animal insulin also suffered from the drawback of allergic reactions and toxicities in humans. Later years saw improvements in methodology, incorporating protamine and zinc to improve peak activities and further refinement in purification. It was in 1978 that Goeddel et al. expressed the two chains A and B of human insulin in *E. coli* to produce rDNA insulin. Genentech and Eli Lilly joined hands to commercially produce rDNA insulin, named Humulin. Two versions R (rapid) and N (intermediate-acting) were produced and marketed.

Cloning of the insulin gene and its successful production opened newer ways for modification to further upgrade the product. Site-directed mutagenesis of key amino acids in the insulin protein was used to achieve desired pharmacokinetics, absorption level, peak, and duration of action. This led to more products in the market, namely, lispro, aspart, glulisine, glargine, detemir, etc. rDNA insulin is a wonderful example that demonstrates addressing of challenges using nonconventional tools. Talking to The New York Times at the announcement of approval of Humulin on 30th October 1982, Dr. Henry Miller, the medical officer in charge of Humulin at the FDA, commented that the development of Humulin was a major step forward in the scientific and commercial viability of recombinant DNA techniques and that "We have now come of age," referring to the dawn of new tools in the hands of medical research community to offer required products and solve existing challenges (http://members.tripod.com/diabetics_world/humulin_history.html). Thus, cloning and successful expression of a human protein in a simple bacterium not only upscaled production but also offered avenues of further improvements in the basic product and its quality control.

9.2 Overview of the Technology and Advantage with Microbial Hosts

The immense potential inherent in rDNA technology is mainly because of the ability to join/delink/modify DNA. The specific advantage with microbial recombinants is the flexibility offered due to simplicity of prokaryotic organisms. *E. coli* has been successfully used as an experimental organism (Cronan 2014). *E. coli*, also known as the workhorse of molecular biology, has simple nutrient requirements and a rapid generation time of 18 min, which brings both cost- and time-effectiveness during expression of a desired protein. Protein expression in any host utilizes the basic fundamental of molecular biology, also referred to as central dogma, which outlines the flow of information in cellular systems (Fig. 9.1a). DNA, being the defining molecule, is used for directing the microorganism to produce a particular protein

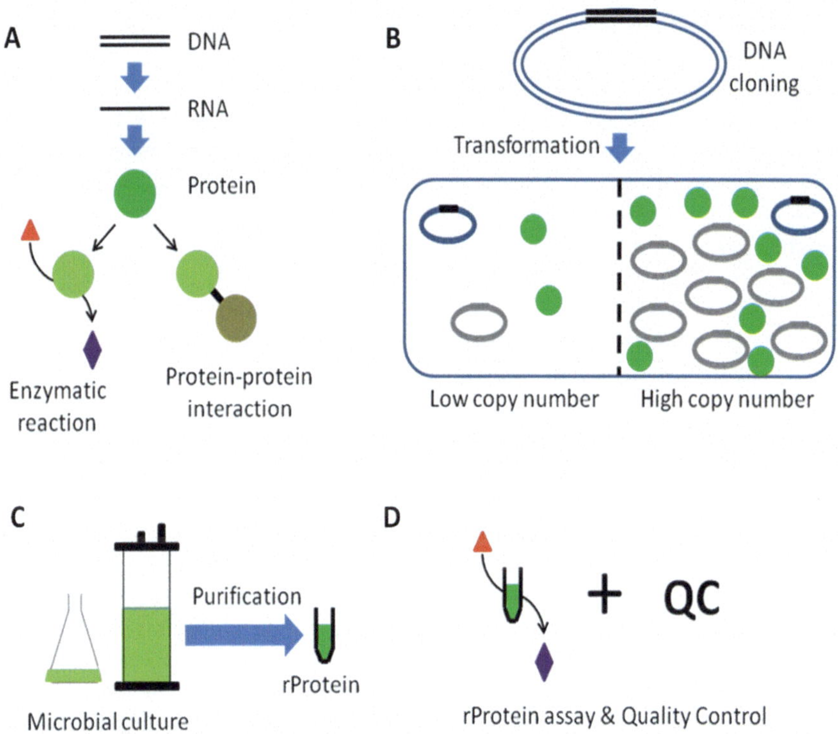

Fig. 9.1 Overview of rDNA technology in microbial hosts. (**a**) Central dogma of molecular biology. (**b**) Cloning of the DNA encoding a desired protein in plasmid and transformation of the plasmid in a compatible microbial host. Depending on the stringency/regulations of application of a particular protein, a choice is made to use either a low- or high-copy-number plasmid, among other factors. (**c**) Protein expression in culture flasks or fermentor based on scale of production and its downstream processing to yield recombinant protein purified to homogeneity. (**d**) Determination of activity of the recombinant product and its quality control

(Fig. 9.1b). Many factors like codon usage, choice of promoter to drive gene expression, and type of plasmid in terms of copy number to be used for protein expression need to be optimized for successful protein expression (Rosano and Ceccarelli 2014). For example, while high-copy-number plasmids warrant more protein expression, the same may also result in mutations. This situation can be avoided by using low-copy-number plasmids. In practical scenario, the choice would be determined in part by the type and usage of protein being expressed and the effect of mutations on it. In addition to the many tools for manipulating DNA, a number of expression systems have been characterized and standardized in model microorganisms like *E. coli, Saccharomyces cerevisiae, Pichia pastoris*, etc. Inducible expression systems have been designed for cloning of proteins that are toxic to the host. Advances in fermentation technology and downstream processing have resulted in optimized protein purification protocols capable of purifying desired protein to homogeneity (Fig. 9.1c). The purified protein is finally tested for its activity and other quality control parameters (Fig. 9.1d).

Many factors influence the choice of microbial host system, which include the necessity of posttranslational modifications like glycosylation, secretion of the overexpressed protein in media for easier purification, etc. Although many advances have been made in insect cell and mammalian cell cultures and they have proved their potential as upcoming expression platforms, microbial recombinants still top in terms of number of products made in them. *E. coli* continues to enjoy its position as the topmost protein expression platform in terms of number of pharmaceuticals (Sanchez-Garcia et al. 2016).

9.3 Applications of Recombinant Proteins in Medical and Nonmedical Industry

While most of the recombinant pharmaceuticals have been designed for treating metabolic disorders, products for treating hematological disorders and cancers also abound (Sanchez-Garcia et al. 2016). Prokaryotic microbial host *E. coli* and eukaryotic microbial host *S. cerevisiae* together express around three-fourths of the total recombinant proteins. Many products like Protropin, Roferon A, Intron A, Recombivax, Humatrope, filgrastim, pegfilgrastim, and many antitumor drugs have been expressed in micro-organisms. This indicates the utilization of microbial recombinants in medical sector.

Similar to changes in the pharmaceutical paradigm brought, new markets have resulted due to increased opportunities in rDNA technology. An example of creation of a new market is the fully vegetarian cheese and dairy products. Many populations of the world stringently follow vegetarianism. In dairy industry, rennin was widely used for cheese making. Rennin was isolated from young calves. Nowadays, microbial strains are used to express recombinant coagulants. Many companies have advertised this strategy in order to accomplish better sales from the customer base concerned for vegetarianism (https://economictimes.indiatimes.com/amuls-worlds-biggest-vegetarian-cheese-brand/articleshow/1288927.cms). Other important products include those needed in cosmetics (Sunar et al. 2016), detergent industry (Wang et al. 2018; Fariha et al. 2010), animal feed and additives industry (Claudia et al. 2013), food industry (Claudia et al. 2013), and vitamin industry (Jose-Luis and Arnold 2010). The portfolio of industrial proteins is wide and includes phytase, laccase, proteases, antibiotics, lipases, etc., tailored to various applications. Cellulase and β-glucosidase have been manufactured for biofuel industry (Puetz and Wurm 2019). For nonmedical applications, protein function is of utmost function, while for those intended for medical use, both protein function and its reception by human/animal immune system assume significance.

9.4 Viral Vectors and Applications

Viruses have the natural ability of introducing DNA into specific host cells. They have been modified for use in recombinant DNA technology. Viral vectors have gained much importance as they are widely used as tool for delivery of genetic material in molecular biology experiments, both in vivo and in vitro. The process of gene transfer by viral vector is known as transduction, and this process was harnessed in molecular biology experiments in the late 1970s when modified form of SV40 viruses was used to infect in vitro cultures of monkey kidney cell lines (Vanncci et al. 2013).

The major aim behind development of viral vector was to increase the probability of DNA uptake by host cells without affecting viability and viral vector proved to be good alternative for gene delivery as compared to traditional methods of DNA uptake by chemical treatment. The added benefit of using viral vector is that certain viral vectors integrate the delivered DNA into host cell genome which ensures stable expression of transferred gene. Advantages associated with viral vectors have increased their applicability in basic molecular biology experiments as well as clinical studies, which gives avenues for construction of novel viruses used in gene therapy and for vaccine development (Perricaudet and Stratford-Perricaudet 2012).

There are two broad categories of viral vectors, depending on whether their genome integrates into host chromatin or remains in the nucleus majorly in the form of extrachromosomal episomes. Lentiviruses and oncoretroviruses are the two subgroups which integrate their genome into host DNA, whereas adenoviruses, adeno-associated viruses, and herpes viruses come under the category of vectors that exist as independent entity in the host nucleus. The aforementioned classification is important to ascribe the applicability of viral vector for a specific purpose. For example, integrating viral vectors can be used for the dividing host cell, whereas non-integrating vectors are of major choice when dealing with non-proliferating stable cell lines. As there is no single multipurpose viral vector available, there is need to develop specific viral vectors as per demand of a particular target (Bouard et al. 2009). The choice of a particular viral vector is established by several parameters which include stability constraints, production processes, demand of their transient or long-term expression, and the need of regulating the transgene expression.

9.5 Biohazards of Microbial Recombinants

One of the initial concerns about the nature of microbial recombinants was expressed by the Committee on Recombinant DNA molecules (Berg et al. 1974). While the technology has come a long way since then and many products have been commercialized, still combating the biological obstacles posed by cell factories and standardization of methodology is a huge challenge for rDNA pharmaceuticals in the development of protein-based molecular medicine. The major reason for some of apprehensions regarding the recombinant enzymes and pharmaceuticals from the market is low product quality, which is a common problem associated with biological synthesis of target proteins in host systems (Ferrer-Miralles et al. 2009). Poor protein quality of expressed recombinant protein in microbial cell factories is mainly due to altered folding, changes in codon preference, formation of inclusion bodies, and triggering of the consequent cell responses in host cell (Gasser et al. 2008). A reason for the potential hazardous nature of recombinant proteins is distortion in the protein structure, also known as misfolding. At the atomic level, a protein is composed of secondary structure elements α-helices and β-sheets. The spatial arrangement of secondary elements imparts a unique three-dimensional conformation to the protein, which is important for its biological activity. In the case of protein enzymes, active site (where the enzymatic reaction takes place) and allosteric sites (which are influenced by cofactors) are important. In case of protein-protein interactions, hotspot regions whereby two or more proteins recognize each other to manifest biological functions are essential. Any changes at the important sites that disrupt the normal activity will render the protein ineffective. In case of pharmaceutical proteins, misfolding of protein structure has the additional drawback of possibility of immunological reactions, in case the misfolded site adopts a local conformation recognized as "foreign" by the body's immune system. The recombinant proteins used presently for the treatment of several human disorders, including malignancies, endocrine, infectious, inflammatory diseases, etc., are similar to naturally occurring human proteins. In human body, these recombinant proteins prove to be immunogenic, which is a major setback for their frequent use in medicine development, as autoimmune response towards these self-antigens is observed by the administration of these recombinant proteins. The antibody production represents a system response against an exogenously administrated recombinant drug which leads to unwanted immune response causing loss of efficacy, which distorts the ability to distinguish self from non-self-antigens and serious side effects. The major obstacle for developing stable recombinant drug is to analyze, characterize, and finally quantify the amount of anti-recombinant drug antibodies induced by it in patients, and further major challenge is to overcome these side effects. Figure 9.2 provides a schematic representation of the protein structure along with secondary structure elements.

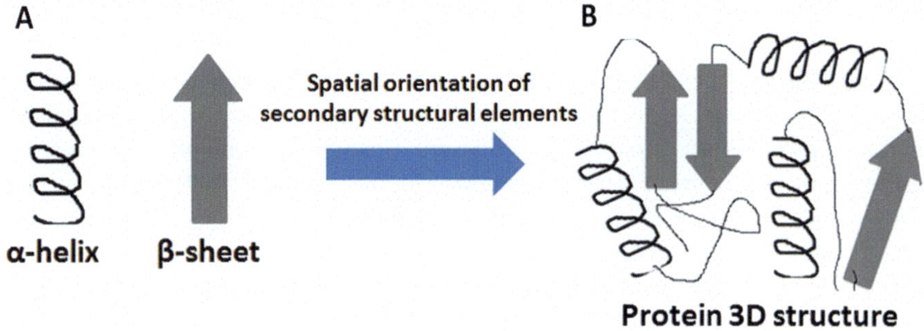

Fig. 9.2 Overview of the protein structure. (**a**) Secondary structure elements. (**b**) Spatial arrangement of secondary structural elements with loops that result in distinct three-dimensional conformation of a particular protein, important for its biological function

Hence, due to issues described above, recombinant proteins may suffer from certain post-application hazardous effects, some of which are described below:

1. **Hypersensitive Reactions Against Recombinant Protein in Host System**

With consistent use of recombinant pharmaceuticals, there are more evidences for induction of immune response in patients which seems to be a major issue not only during development of recombinant drugs but also during marketing these drugs after several clinical trials.

(a) *Insulin*

Hypersensitive reactions have been observed since early times due to the administration of insulin produced in beef or pork for treatment of type I diabetes, mainly due to antibody production. Anti-insulin antibody, particularly IgE, is ascribed to induce skin and anaphylactic reactions which necessitated increased insulin doses over time to gain its primary benefits, i.e., treatment of diabetes. rDNA technology came with vision of developing human insulin by gene technologies that was assumed to eliminate the immunogenic reactions, but such benefits were not fully attained as reports regarding insulin allergies to human insulin exist (Durand-Gonzalez et al. 2003). It was observed that amount of antibodies with cross-reactivity was greater when treatment was dependent on application of subparts of insulin protein compared to natural form of insulin, but clinically the information about these antibodies is obscure. The antibodies against insulin may potentially lead to autoimmune type I diabetes. The need of comparing the levels of antibodies against insulin before and after administrating insulin and correlating it with diabetes induction has been stressed.

(b) *Interferon*

Naturally occurring interferons have many defense-related benefits. For example, alpha-interferons are known to inhibit viral replication. Beta-interferon has been used to treat multiple sclerosis (MS) (Limmroth et al. 2011). Immunogenic reactions against these recombinant interferons have been observed due to the production of neutralizing antibodies (NAbs). Fibroblast interferon is the only beta-interferon present naturally in humans. The relapse rate of multiple sclerosis was observed to be reduced by the administration of recombinant beta-interferon named interferon-beta-ser (Betaferon®). Later on, it was observed that neutralizing antibody production was more in multiple sclerosis-affected patients administrated with (Betaferon®) essentially due to variation in the molecular domains of artificially produced interferon-beta-ser and its natural counterparts, i.e., beta-interferon (Limmroth et al. 2011). Earlier, it was reported that one-third of MS patients treated with recombinant interferon developed antibodies within the first year of therapy. In some cases, antibodies even appear to stimulate the function of interferon. Hence, it was recommended that treatment decisions should be made on case-to-case basis.

(c) *Erythropoietin*

Erythropoietin (Epo) is an activator protein for erythroid progenitor cells through receptors known as Epo-receptors. Epo mainly induces the erythroid progenitor cell for production of red blood cells. In normal person, the kidney is the target organ for production of erythropoietin that is produced in sufficient quantities by the kidneys under normal conditions, but due to renal failure erythropoietin content decreases drastically which subsequently affects RBC production through erythroid progenitor cells which generates anemic condition in affected patient (Fisher 2003). The treatment of erythropoietin deficiency anemia was treated using recombinant form of human erythropoietin known as r-HuEpo since the late 1980s, and it was found that r-HuEpo do not induce immunogenic reactions in in vivo systems as the recombinant erythropoietin had nearly identical protein structure as that of its naturally existing counterparts with minute difference in pattern of sugar moieties associated with protein part of erythropoietin (Skibeli et al. 2001). Recombinant erythropoietin was found to induce a disease condition known as pure red cell aplasia (PRCA) in rare cases at initial period of its use, and just three cases were reported until 1998 by Eckardt and Casadevall (2003). However, with time, the recombinant erythropoietin-mediated PRCA was increased in receptive patients of recombinant erythropoietin drugs due to generation of neutralizing antibodies (Casadevall et al. 2002). Thereafter, the number of patients diagnosed with PRCA has increased substantially (Verhelst et al. 2004). This data reveals regular need of monitoring the immunogenicity of recombinant proteins even after its approval for clinical use.

2. **Autoimmune Response and Pharmacokinetics**

Autoimmune antibodies mainly affect the immune-complex formation, hypersensitivity reactions, and pharmacokinetics. The reason behind production of autoimmune antibody is elevated subjection of unnatural epitopes to the immune system, newness of recombinant drugs to the host immune systems, the presence of adjuvants, varied carbohydrate patterns in glycosylated protein products, drug aggregate formation, the predisposition of genetic system to autoantibody formation, and the previous exposure of identical compounds in an individual immune system. Established examples of antibodies produced against recombinant drugs were antibodies against factors VIII and IX. The clinical management of hemophilia A and B is mainly managed by acting as inhibitors (Oldenburg et al. 2004). There should be prevention of antidrug antibodies which reduces benefit of a given drug and increases the risk of autoimmune reactions. Recombinant drug development comes with additional step of detecting and monitoring antibodies before, during, and after drug application through special clinical trials. There is need to develop highly specific assays and technologies to completely understand the immunogenicity of a particular recombinant protein in a particular person. The autoimmune status of a particular patient has to be included into drug development program for overall assessment of antibodies produced due to autoimmune response. The antidrug antibody production can be reduced by advancement in recombinant technologies, by establishing the underlined mechanism of autoimmunity, along with the role of human genetics and environmental factors that affects immunogenicity.

3. **Biohazards of Virus-Derived Vectors**

Viruses have a natural tendency of caring genetic material into the host cell and introducing there genetic material into host genome; this natural process has gained much importance in rDNA technologies by exploiting virus-based systems as a vector for transferring required genetic information in host cell. Virus vectors are formed by genetically modifying the natural viral genome by removing virulence sequences while keeping sequence important for gene delivery, integration, and functional effectiveness. Virus vectors differ on the basis of insert size, duration of gene expression, targeted host, and the wild-type virus from which they are derived. Clinical trials are conducted prior to approval of genetically modified organism as pharmaceutical products, so there is need to contemplate that a precise boundary exists between deliberate use and contained use of recombinant proteins that contained GMO, as patients who accrued recombinant drugs are in long-term exposure to natural environment which increases the risk of spreading these GMO in the environment.

Genetically modified viral vector used for medical benefits may add adverse effect to the environment due to alteration in the properties compared to the parental virus, such as changes in aspects of the viral life cycle, the viral vector interactions with host, change in the gene products encoded by the insert sequence within host system, increased virulence in comparison to the parental virus in the human nontarget population, increased pathogenic effects including toxic and allergenic effects in animals, effects on the population dynamics in the natural environment, e.g., effects on the dynamics of populations of species in the receiving environment, and the genetic diversity of each of these populations (Baldo et al. 2013). The gene product expressed in virus vector may have intrinsic hazardous properties by acting toxic or allergenic in host system, and combating these hazardous effects in clinical trials solely depends on genetic and physiological context of parental virus, its interaction with host system, and condition of use.

Some of the biohazards associated with virus vectors are as follows:

(a) *Altered virulence*: Recombinant viral vectors can undergo inadequate virulence changes which increase the risk to both environment and human. Major reason behind virulent virus vector formation is recombination of the viral vector with wild-type viruses or with related viruses having similar sequences.
(b) *Host range and tissue tropism-associated alterations*:

    The viral vector should ideally transfer the genetic material to the target cells with showing any effect on the adjacent cells of healthy tissues (Gogev et al. 2003). Several efforts have been made in direction of developing viral vectors with limited tropism towards the targeted host cell population. As a result, pseudotyped virus vectors are developed, where replacement of viral envelope proteins is done by either proteins encoding envelope from other viruses or by using chimeric proteins. These chimeric vectors will have necessary information for incorporation into virion along with sequence required for interacting specific host cell proteins, for example, the vesicular stomatitis virus glycoprotein (VSV-G) was coated on HIV virus, which is the host range of virus as compared to its parental counterparts which have binding tendency towards CD4 receptors of T helper cells. Reconstruction of viral surface ligands may have several benefits with binding specificity for the target cells, but it comes with the huge concern of mediated side effects, so while developing such pseudotyped viral vectors, there is need of environmental risk assessment (ERA) to ensure that all modifications in the pseudotyped viruses are safe for long-term clinical use and environmental release because these modifications can act in opposite direction by increasing the tropism or the host range pseudotyped vector, which can lead to more severe and novel disease symptoms in humans or animals.
(c) *Changes in susceptibility towards immune system*: A competition has been observed between the host immune system and viruses where host immune responses work in the removal of virus-borne infections from the body by affecting long-term expression and clinical efficacy of viral vectors, whereas viruses evolve mechanisms for invasion of host immune system. This host immune system and virus interaction defines the need of either deleting immune evasion determinants from virus or by encoding immune modulatory function in viral vectors. Attenuation of viral vectors leads to loss immune evasion property, such as E3 deletion from adenovirus or the interleukin (Il)-18 binding protein from poxviruses, which leads to more efficient clearing of the viral vector during an infection and also leads to inflammation as a result of acute immune response

(CHMP 2007). Pathogenic responses may be observed due to addition of genes encoding immune attuning functions which is rare property of wild-type virus. For example, more pathogenic form of poxviruses was formed due to expression of Il-4 which led to inhibition of immune response required for effective clearance of viral infection (Jackson et al. 2001). Enhanced proliferation is one of the possible effects of a viral vector with impaired immune evasion systems in immunosuppressed patients. This aspect should be critically considered during treatment.

9.6 Conclusion

There is no doubt the recombinant biologicals have brought revolution in medical and nonmedical field. They have brought many diseases and other tasks under management by offering new solutions. However, particularly in the field of pharmaceuticals, the complexity of the human body continues to pose new challenges based on the immunological profile of patients. At the core of this are the changes in protein structure that result when they are produced through recombinant DNA technology. While for nonmedical applications, protein's function is the most important parameter, in medical applications, the ability to remain "natural" and evade immune responses is a challenge that persists. Nonetheless, new improvements in protein purification and characterization along with framing of guidelines for quality control of recombinants and the lessons learned during the past years are being put to good use. This will further expand the repertoire of the areas covered by microbial recombinants and offer advantages of new functionalities with cost-effective deployment.

References

Baldo A, Akker EV, Bergman HE et al (2013) General consideration on biosafety of virus-derived vector used in gene therapy and vaccination. Cur Gen Ther 13:385–394

Berg P, Baltimore D, Boyer HW et al (1974) Letter: Potential biohazards of recombinant DNA molecules. Science 185(4148):303

Bouard D, Alazard-Dany N, Cosset F-L et al (2009) Viral vectors: from virology to transgene expression. Brit J Pharm 157:153–165

Casadevall N, Nataf J, Viron B et al (2002) Pure red cell aplasia and antierythropoietin antibodies in patients treated with recombinant erythropoietin. N Engl J Med 346:469–475

Claudia R, Daniel W, Adalberto P Jr, Jorge GF (2013) Recombinant protein for Industrial uses: utilization of *Pichia pastoris* expression system. Braz J Microbiol 44(2):351–356

Committee for the Medicinal Product for Human Use (CHMP) (2007) Guideline on scientific requirements for the environmental risk assessment of gene therapy medicinal products. EMA. http://www.ema.europa.eu/docs/en_GB/document_library/Scientific_guideline/2009/09/WC500003

Cronan JE (2014) Escherichia coli as an experimental organism. In: eLS. Wiley, Chichester. https://doi.org/10.1002/9780470015902.a0002026.pub2

Durand-Gonzalez KN, Guillausseau N, Anciaux ML et al (2003) Allergy to insulin in a woman with gestational diabetes mellitus: transient efficiency of continuous subcutaneous insulin lispro infusion. Diab Metab 29:432–434

Eckardt KU, Casadevall N (2003) Pure red-cell aplasia due to anti-erythropoietin antibodies. Nephrol Dial Trans 18:865–869

Fariha H, Aamer Ali S, Sundus J, Abdul H (2010) Enzymes used in detergent: lipases. Afr J Biotechnol 9(3):4836–4844

Ferrer-Miralles N, Domingo-Espin J, Corchero JL et al (2009) Microbial factories for recombinant pharmaceuticals. Microb Cell Factories 8:17–19

Fisher JW (2003) Erythropoietin: physiology and pharmacology update. Exp Biol Med 228:1–14

Gasser B, Saloheimo M, Rinas U et al (2008) Protein folding and conformational stress in microbial cells producing recombinant proteins: a host comparative overview. Microb Cell Factories 7:11–18

Gogev S, Schynts F, Meurens F et al (2003) Biosafety of Herpesvirus vectors. Curr Gene Ther 6:597–611

http://members.tripod.com/diabetics_world/humulin_history.htm

https://economictimes.indiatimes.com/amuls-worlds-biggest-vegetarian-cheese-brand/articleshow/1288927.cms

Jackson RJ, Ramsay AJ, Christensen CD et al (2001) Expression of mouse Interleukin-4 by a recombinant ectromelia virus suppresses cytolytic lymphocyte responses and overcomes genetic resistance to mousepox. J Virol 75:1205–1210

Jose-Luis A, Arnold LD (2010) Recombinant organisms for production of industrial products. Bioeng Bugs 1(2):116–131

Limmroth V, Putzki N, Kachuck NJ (2011) The interferon beta therapies for treatment of relapsing–remitting multiple sclerosis: are they equally efficacious? A comparative review of open-label studies evaluating the efficacy, safety, or dosing of different interferon beta formulations alone or in combination. Therapeutic Adv in Neurolog Dis 4(5):281–296

Oldenburg J, Schröder J, Brackmann HH et al (2004) Environmental and genetic factors influencing inhibitor development. Semin Hematol 41:82–88

Perricaudet M, Stratford-Perricaudet LD (2012) Adenovirus-mediated *In Vivo* gene therapy. J Vir Hum Gen Ther 5:1–32

Puetz J, Wurm FM (2019) Recombinant proteins for industrial versus pharmaceutical purposes: a review of process and pricing. Processes 7:476

Quianzon CC, Cheikh I (2012) History of insulin. J Commun Hosp Intern Med Perspect 2(2). https://doi.org/10.3402/jchimp.v2i2.18701

Rosano GL, Ceccarelli EA (2014) Recombinant protein expression in *Escherichia coli*: advances and challenges. Front Microbiol 5:172

Sanchez-Garcia L, Martín L, Mangues R et al (2016) Recombinant pharmaceuticals from microbial cells: a 2015 update. Microb Cell Fact 15:33

Skibeli V, Nissen-Lie G, Torjesen P (2001) Sugar profiling proves that human erythropoietin differs from recombinant human erythropoietin. Blood 98:3626–3634

Sunar K, Kumar U, Deshmukh S (2016) Recent application of enzymes in personal care products. https://doi.org/10.1016/B978-0-12-802392-1.00012-5

Vanncci L, Lia M, Chiuppesi F et al (2013) Viral vectors: a look back and ahead on gene transfer technology. New Microbiol 36:1–22

Verhelst D, Rossert J, Casadevall N et al (2004) Treatment of erythropoietin-induced pure red cell aplasia: a retrospective study. Lancet 363:1768–1771

Wang X, Kan G, Ren X et al (2018) Molecular cloning and characterization of a novel α-amylase from Antarctic Sea ice bacterium *Pseudoalteromonas* sp. M175 and its primary application in detergent. https://doi.org/10.1155/2018/3258383

Part IV

Microbial Enzymes & Applications

Extremophile Microorganisms and Their Industrial Applications

10

Kamla Malik, Nisha Kumari, Sushil Ahlawat, Upendra Kumar, and Meena Sindhu

Abstract

Microorganisms are ubiquitous in nature and are the backbones of all ecosystems. The halophilic microbes are extremely diverse that survive in high-saline environments. Most halophile (salt-loving) microbes belong to Bacteria, Archaea, and Eukarya, which are found in ponds, lake, deep-sea brines, etc. These microorganisms have adapted their surface and membrane structures to their highly ionic environments. They use two fundamentally different strategies in medium for balancing of their cytoplasm osmotically. The first strategy includes the accumulation of molar concentrations of KCl that requires adaptation mechanism of the intracellular enzymatic activity in the presence of salt to maintain proper conformation and activity of the proteins. The second strategy includes where salt is to be excluded from their cytoplasm and synthesized as well as accumulated organic solutes that could not be interfered with enzyme machinery. In recent years, halophiles have been attractive among researchers due to their adaptation mechanism in a wide range of salinities with potentially promising applications and a source of various bioactive compounds. Therefore, the saline and hypersaline environments could be more explored in the field of ecological

K. Malik (✉) · M. Sindhu
Department of Microbiology, Chaudhary Charan Singh Haryana Agricultural University, Hisar, India

N. Kumari
Department of Biochemistry, Chaudhary Charan Singh Haryana Agricultural University, Hisar, India

S. Ahlawat
Department of Chemistry, Chaudhary Charan Singh Haryana Agricultural University, Hisar, India

U. Kumar
Department of Molecular Biology, Biotechnology and Bioinformatics, Chaudhary Charan Singh Haryana Agricultural University, Hisar, India

S. G. Sharma et al. (eds.), *Microbial Diversity, Interventions and Scope*,
https://doi.org/10.1007/978-981-15-4099-8_10

niches which have been used for the discovery of bioactive compounds/ metabolites. However, these halophiles are potential sources for a wide range of new therapeutic compounds. The aim of this introductory chapter is to provide brief perspectives on microbial ecology and highlight the applications in various fields and also understand the function and ecological roles of microbial community in saline environment. It also concludes putting the current status and future prospective with potential applications using advanced approaches derived from genomics and proteomics.

Keywords

Microbial ecology · Halophiles · Bioactive metabolites · Bacteria · Archaea · Fungi · Enzymes

10.1 Introduction

On Earth, the most diverse and abundant organisms are microorganisms. They are ubiquitous in nature and prevalent in air, water, soil, etc., as well. The study of interactions of microorganisms and their relationship with one another and with their environment, plant, and animal species is called microbial ecology which also includes the study of symbiosis, interaction of the microorganisms with anthropogenic effects like pollution and climate changes, and microbes also involved in biochemical cycles. Presently, major types of microorganisms remain unknown; indeed, only 1% of the microbial species is known. Without microbes in the environment, life on Earth may have never come into existence and certainly not as we know it today. Microbes are responsible for cycling nutrients through the environment, creating important symbiotic relationships, providing energy in the absence of sunlight, and digesting the food we eat. According to antropocentric view, these microorganisms are required extreme optimum conditions for growth and metabolism (Vauclare et al. 2014). Approximately 70% of the surface of planet Earth has been covered by seawater that contains 35 g/l of total dissolved salts and 78% of sodium chloride. The marine environment could not be considered extreme habitat; however, many microbes are unable to cope up or survive at seawater salinity, and the seas also contained a tremendous diversity of micro- and macroorganisms. However, these environments contain much higher salt concentration than those found in the sea. As the concentration of soluble salts increased, the biological diversity decreased. On Earth, extreme ecosystems are one of the most dynamic environments. Such ecosystems are characterized on the basis of physico-chemical conditions which make them suitable for maximum survival of life. Some microorganisms could survive at different extreme environmental conditions that varied from pH, temperature, salinity, pressure, and radiation; therefore, they are known as extremophiles (Rampelotto 2013; Selvarajan et al. 2017).

The saline and hypersaline enviroments are well known for their unique properties such as microbial population, biogeochemical characteristics, and aesthetic appeal. In these environments, microbial activities and diversity are distinct properties with many novel species that have been reported and identified. Different characteristics such as adaptation, evolutionary studies, morphology, and environmental and various applications of these organisms have been studied (Paul and Mormile 2017). However, these unique environments are critical to conserve and limit the damage due to anthropogenic influences. These conditions have been increased salinization due to water diversion, undersirable freshening, sewage effluents, high extraction of mineral, pollution, and global climate changes are the important factors that adversely affect their surrounding environments such as hypersaline ponds, lakes, rivers, etc. If these negative effects continue to proceed at the faster rates, irreparable consequences for the particular environments would occur that results in losing the possible opportunities to gain about the new concepts and knowledge about the biogeochemistry and role of beneficial microbial population in saline and hypersaline environments.

Saline environments are one of the most important niches for different microorganisms with significant industrial applications. The landmass contains about one-quarter of salt deposits in saline ponds, lakes, seas, and oceans that represented a huge part of the Earth's surface (Jackson et al. 2001). Hypersaline and saline habitats are a widely distributed environment, such as a variety of terrestrial lakes, ponds, and basins of deep sea with high salt concentrations that exceeded about three times greater than seawater saturation and salt flats. The environmental factors, viz., higher salt concentration, alkalinity, and low oxygen concentration, may also limit their biodiversity (Ventosa 2006). Moreover, high saline concentration represents an extreme environment in which only few organisms have been able to survive. On the basis of salt (sodium chloride) requirement, these organisms are categorized for survival and growth, which are basically Archaea, Bacteria, and Eukarya domains of life possessing a great diversity of microorganisms, for example, halophilic aerobic bacteria (moderately), anaerobic bacteria, cyanobacteria (BGA), sulfur-oxidizing bacteria, heterotrophic bacteria, archaea, protozoa, algae, fungi, and other multicellular eukaryotes. These microorganisms are widely distributed in hypersaline environments in various geographical ecosystems on Earth, such as saline ponds, lakes, salt pans or salt marshes, hypersaline brines in deep sea, arid and artificial saltern, and coastal locations. Halophilic microorganisms grow vigorously in highly saturated brines and are tolerant to several stresses such as temperature, pH, toxic metals, desiccation, radiation, etc. (Baliga et al. 2004; Kottemann et al. 2005; Mc Cready et al. 2005; Whitehead et al. 2006), and space conditions (Moissel et al. 2008). These extreme halophiles are robust organisms that are confined in fluid inclusions within salt crystals. Therefore, these organisms have been represented by manageable models to know about how much they are fast and to what extent these organisms had adapted to different conditions according to environmental changes. In high salinity, these halophiles employ differing strategies such as prevention to desiccation through osmotic movement of water to come out of their cytoplasm. Strategies involved the increase in the internal osmolarity of cell. The halophilic microorganisms which survive well at high salt concentrations are bacteria, archaea, yeasts, algae, fungi, and viruses. These microorganisms were isolated and screened from different saline environments, viz., natural brines, artificial solar salterns, hypersaline lakes, ponds, submarine pools and ponds, and deep salt mines (DasSarma and DasSarma 2015; Waditee Sirisatha et al. 2016). This chapter provides brief perspectives on classification, mechanisms, and diversity of halophilic microorganisms and its industrial and environmental applications using recent molecular approaches.

10.2 Classification of Halophilic Microorganisms

In hypersaline ecosystems, the salinity is greater than seawater. Microorganisms that survive in these ecosystems are called "halophiles." The physiological basis of their responses to salty environment in various groups could be distinguished. Based on optimized growth conditions and media, different categories of microorganisms were studied (Kushner and Kamekura 1988). Halophiles are classified based on their optimum salt concentration for growth as slight, moderate, and extreme halophiles that have undertaken various physiological adaption mechanisms for survival in higher salt concentration (De Lourdes et al. 2013), and different products are produced that are suitable for various applications in different industries. Slight halophiles prefer 0.3–0.8 M (1.7–4.8%), moderate halophiles survive at 0.8–3.4 M (4.7–20%), and extreme halophiles require 3.4–5.1 M (20–30%) salt concentration (Ventosa 2006; De La Haba et al. 2011; Das Sarma and Das Sarma 2012) (Fig. 10.1).

10.3 Ecology and Characteristics of Saline and Hypersaline Microorganisms

The hypersaline environment in which salinity is greater than 1.5 M or $>$ 10% concentration mainly includes two predominate groups of microorganisms, i.e., extreme and moderate halophilic bacteria and archaea. Among normal inhabitants with non-extreme environments, archaea have constituted a large proportion of hypersaline environments. In these environments, the most common inhabitant is Haloarchaea, and only a few species of methanogens have also been found. Haloarchaea are found in saturated ponds of salterns and salt lakes. Most microbial inhabitants of hypersaline environment are Halobacteriales (Archaea). Species of the genera *Haloarcula*, *Haloferax*, *Halobacterium*, *Natronobacterium*, and *Natronococcus* and bacterial species *Salinibacter ruber* were commonly isolated from solar saltern and alkaline lakes. Proteobacteria (*Halomonas* sp. and *Salinivibrio* sp.), Firmicutes (*Bacillus* and *Flavobacterium*), and Cyanobacteria, Gram-negative and Gram-positive bacteria, and aerobic and anaerobic microorganisms including fermentative, phototrophic, sulfate-reducing, homo-acetogenic, and methanogenic species were also observed (Ventosa 2006) (Table 10.1).

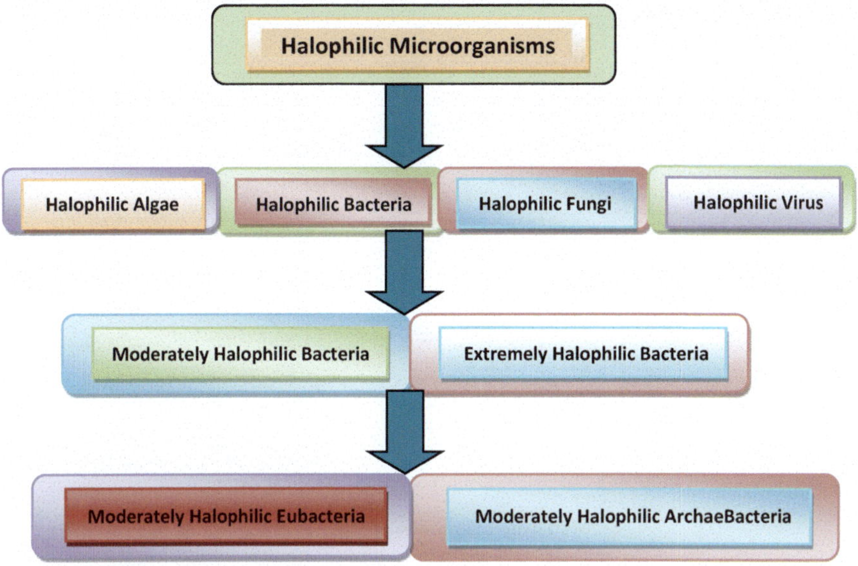

Fig. 10.1 Various types of halophilic microorganisms

Different microorganisms, e.g., archaea, bacteria, single-celled algae *Dunaliella salina*, and halophytes (salt-loving plants), are capable of surviving under high-saline environment. Similarly, brine flies and brine shrimps *Artemia salina* could survive under extreme NaCl concentrations along with the broad range of salinity tolerance between 6.0 and 19 M or 35 and 110% NaCl (Vanhaecke et al. 1984). Hypersaline environments with their unique characteristics, metabolic activity or adaptations, diversity, and potential economic and scientific benefits have been studied. In these environments, the limited biodiversity is found due to higher salt concentration, along with other parameters like availability of oxygen, low nutrients, higher UV radiation, heavy metals, toxic compounds, and alkalinity (Rodriguez-Valera 1988; Ventosa 2006). The haloalkaline *Halomonas campisalis* bacteria that were isolated from Soap Lake, Washington survived at a wide range of pH (6–12) and salinity (1.0–26.5% or 0.2–4.5 M) level (Mormile et al. 2003). In these environments, organisms that exhibited in various conditions with complex adaptation approaches have been studied by researchers (Madern et al. 2000; Roberts 2005). In some studies, a number of polymorphic black yeasts (*Hortaea, Phaeotheca, Aureobasidium, Cladosporium, and Trimmatostroma*) have been retrieved from saltern hypersaline water (3–30 NaCl) and play an important ecological role in hypersaline ecosystems. Alkaline pH of some haloalkaliphiles was required for optimum growths in soda lakes. Ecological studies demonstrated that they might be reached at high cell densities ($>10^7$ cells/ml). On the basis of conventional studies, the cultivation of viable cells of halophiles was predominant genera belonging to the *Haloferax, Halobacterium, Halorubrum*, and *Haloarcula* found in most neutrophilic hypersaline environments (Rodriguez-Valera et al. 1985;

Table 10.1 Classification of halophilic organisms

Types of halophiles	Species
Slight halophiles	*Chromatium buderi*
	Chloroherpeton thalassium
	Ectothiorhodospira mobilis
	Rhodobacter sulfidophilus
	Pelodictyon phaeum
	Rhodopseudomonas marina
	Ectothiorhodospira vacuolata
	Prosthecochloris phaeoasteroidea
	Thiorhodovibrio winogradskyi
	Chlorobium chlorovibrioides
	Chromatium purpuratum
	Rhodobacter adriaticus
	Prosthecochloris aestuarii
	Chromatium vinosum HPC
	Lamprobacter modestohalophilus
Moderate halophiles	*Rhodospirillum mediosalinum*
	Rhodospirillum salexigens
	Ectothiorhodospira marismortui
	Thiocapsa halophila
	Chromatium salexigens
	Ectothiorhodospira abdelmaiekii
	Rhodospirillum salinarum
Extreme halophiles	*Ectothiorhodospira halophila*
	Ectothiorhodospira halochloris

Rodriguez-Valera 1988; Benlloch et al. 2002). Therefore, ecological and advanced molecular studies have been based on cultivation; characterizations using individual methods indicated that the members of the particular genera that constituted a small proportion of the microbial community have also been observed in these environments. In haloarchaeal group, many environmental clones were observed which did not show close relationship tophylogenetically with *Halobacterium* species. *Halorubrum*, *Haloarcula*, *Natronobacterium*, and *Natronomonas* clones were also obtained (Benlloch et al. 2002; Burns et al. 2004).

The literature suggested that the bacteria community (haloarchaeal) was dominated in the crystallizer habitat by the two groups of *Halorubrum* and related to environmental phylotypes. Additionally, to the environmental phylotypes (two groups), members of haloarchaeal genera (four) were also observed (Pasic et al. 2005). This predominant member belongs to the *Halobacterium* and *Haloarcula* which were not closely related to Haloarchaea. Other extremophile Archaea required specific cultural conditions, e.g., Haloarchaea grown better in complex media but in some cases grown in aerobic conditions in minimal growth media with standard procedures for growing other prokaryotes. Microorganisms were genetically manipulated, and their genetically exchangeable mechanisms were also well known. At laboratory scale, the methods and process for genetic manipulation were standardized (Robb et al. 1995). The complete genome sequencing of some Haloarchaea has been investigated, e.g., *Haloarcula marismortui* ATCC 43049 T (4274 kb); *Halobacterium salinarum* NRC-1 (2571 kb) (Ng et al. 2000); *Halobacterium* species NRC-1 (Baliga et al. 2004); *Haloarcula marismortui*, a halophilic archaea isolated from the Dead Sea (Baliga et al. 2004); and *Natronomonas pharaonis* DSM 2160 T (2749 kb) (Falb et al. 2005). Genome sequences of other halophiles, e.g., *Halobacterium salinarum* and *Haloferax volcanii*, *Halobiforma lacisalsi*, *Halobaculum gomorrense*, *Halorubrum lacusprofundi*, and *Natrialba asiatica* strains, have been studied (Bolhuis et al. 2006; Bakke et al. 2009; Hartman et al. 2010). The genomic information and diversity were found within the Haloarchaea. Different species of *halophytica* were isolated from the Dead Sea, Great Salt Lake, solar lake, and artificial solar ponds (Baronio et al. 2010). *Dactylococcopsis salina* (planktonic cyanobacterium) was isolated from the Great Salt Lake. Different genera of cyanobacteria (filamentous), belonging to the order Oscillatoriales, *Oscillatoria neglecta, Oscillatoria limnetica, Oscillatoria salina*, and *Phormidium ambiguum* were identified. These microbes developed second layer of green mats in hypersaline lakes. Mostly, these were moderate halophiles that grew at 1.0–2.5 M NaCl concentration, formed heterocyst, and fixed nitrogen. In hypersaline environments, the cyanobacterial diversity has not yet been reported (Caumette et al. 1994).

Phototrophic hypersaline anaerobic bacteria produced microbial mats beneath the cyanobacterial layers in light zones, and some heterotrophs were grown aerobically as well. They also used reduced sulfur (elemental sulfur and H_2S), organic compounds, and hydrogen ions as electron donors. On the basis of bacteriochlorophyll pigments, these microorganisms were characterized as non-sulfur bacteria and purple and green sulfur bacteria. The bacteria (green sulfur) *Chlorobium limicola* and *C. phaeobacteriales* were slightly moderate halophiles that deposit elemental sulfur granules outside their cells which have the capability to fix nitrogen. The filamentous halophile green non-sulfur bacteria (*Chloroflexus aurantiacus*) were slight thermophiles, whereas *Chromatiaceae* were moderate halophilic (purple sulfur bacteria) and deposit sulfur granules inside the cells. Photoorganotrophic bacteria *Chromatium glycolicum* were grown well using glycolate. Similarly, *Chromatium violescens* and *Chromatium salexigens* also used glycerol for their growth. The sulfate-reducing bacteria (SRB) such as *Desulfovibrio salexigens*, *Desulfovibrio desulfuricans*, *Desulfomicrobium baculatus*, *Desulfococcus multivorans*, *Desulfotomaculum nigrificans*, and *Desulfomonile* sp. were isolated from the salt pans of Ribandar, Goa, and required higher salinity level >4.0 M for growth (Kerkar 2004). The osmoregulations of sulfate-reducing bacteria have not been studied, but initial studies indicated that these were not synthesized compatible solutes but internally accumulated salts. In hypersaline environments, generally methanogens were strict anaerobes and used methylotrophic substrates rather than CO_2, acetate, and hydrogen. The *Methanohalophilus (*halophilic methanogens) were isolated from microbial mat, *Methanohalophilus mahii* from the Great Salt Lake, and *Methanohalophilus portucalensis* from a saltern. *Methanosalsus zhilinaeae* was alkaline thermophilic and slight halophile. *Methanohalobium evestigatum* was extremely thermophilic methanogen halophile that required 4.5 M NaCl and 50 °C temperature for its optimum growth. Methanogens which grew at moderately high salinity were isolated from brine pools of Deep Sea in the Gulf of Mexico. The intracellular salt concentration in the bacteria was higher than those required about 0.6 M KCl and lower than the halophilic Archaea (Das Sarma and Arora 2001).

Gram-positive aerobic and facultative anaerobic bacteria were moderate halophiles. Other halophilic genera were *Salinivibrio, Alcaligenes, Alteromonas, Flavobacterium, Dichotomicrobium, Acinetobacter, Pseudomonas, Spirochaeta*, and *Arhodomonas* (Ventosa et al. 1998b). *Arhodomonas aquaeolei* was isolated from an oil field of subterranean brine, which exhibited nitrate reduction capability (MacDonald et al. 1990). Similarly, *Chromohalobacter marismortui* was isolated from the Dead Sea and has the ability for nitrate reduction; strain of *Pseudomonas beijerinckii* was isolated from salted beans and preserved in brine solution. However, *Pseudomonas halophila* was isolated from the Great Salt Lake and *Salinivibrio costicola* from Australian bacon. Species of *Halomonas (H. elongate)* were isolated from a solar saltern and have also shown nitrate reduction capability. Some aromatic compounds were degraded by *H. halodurans*. *H. panteleriense* was isolated from alkaline saline soil, *H. halophila* and *H. salina* were isolated from saline soil and grew at pH 9.0, and *Halomonas subglaciescola* was isolated from beneath the ice layer of organic lake in Antarctica. Primarily, organisms used compatible solutes glycine, betaine, and ectoine for growth (Das Sarma and Arora 2001). Psychrotolerant halophile flavobacteria, *Flavobacterium gondwanese* and *F. salegens*, were isolated from Antarctic lakes. Some species of the genera *Bacillus, Halobacillus, Marinococcus, Salinococcus, Nesterenkonia*, and *Tetragenococcus* were Gram-positive moderate halophilic bacteria. *Salinococcus* sp. was isolated from salterns. *Bacillus diposauri* was isolated from a dessert iguana (nasal cavity), *Bacillus haloalkaliphilus* from Wadi Natrun, and *Bacillus denitrificans* from a solar saltern located in Southern France. Two species of *H. trueperi* and *Halobacillus litoralis* were observed in the Great Salt Lake. From saline soils, actinomycete *Actinopolyspora halophila* was isolated and grown at moderate NaCl concentrations, and few heterotrophic bacteria such as *Norcardopsis halophila* could synthesize compatible organic solutes such as glycine betaine and use a beta-glutamate and hydroxyl group derivative of ectoine as compatible solutes (Ventosa et al. 1998). The halotolerant or halophilic fungi and yeasts have been also observed. The *Hortaea werneckii* was black yeasts grown up to 5 M NaCl concentration. A truly halophilic yeast, *Wallemia ichthyophaga*, required low concentration of 1.5 M NaCl and grew up to saturation level, and *Aureobasidium pullulans* required 3.0 M NaCl for its growth, and these were isolated in polar ice, hypersaline lakes, domestic dishwashers, and spider webs in desert caves. These fungi used glycerol, mannitol, erythritol, and arabitol polyols as osmotic solutes, and their cytoplasm retained lower salt concentrations (Gunde et al. 2009). However, molecular studies on osmotic adaptation of *Hortaea werneckii* and *Wallemia ichthyophaga* were reported by Vaupotic et al. (2007). *Siphoviridae* and halophilic virus BJ2 of the *Myoviridae* that infect *Halorubrum kocurii* bacterium were observed from a salt lake (Pietila et al. 2010). On the basis of dinucleotide frequency and G+C content with bioinformatics analysis, it was observed that 24% of viral sequences were retrieved which belongs to the *Salinibacter* phages. The phages infecting *Salinibacter* were more prominent in hypersaline environment as compared to *Haloquadratum* infected by phages (Santos et al. 2010).

10.4 Recent Advances in Microbial Diversity of Saline and Hypersaline Environments

Saline and hypersaline microorganisms are categorized on the basis of their requirements and tolerances to salt as belonging to the Eukarya, Bacteria, and Archaea. The microbial diversity has been studied in hypersaline environments, and DNA extracted from various samples isolated from different environment were analyzed by 16S rRNA genes and amplified by PCR (polymerase chain reaction). Phylogenetic and physiological diversity of hypersaline Archaea has been studied by using molecular phylogenetic techniques. The halophilic Archaea that belongs to the Halobacteriaceae family were aerobic and produced red colorations in hypersaline lakes and saltern crystallizer ponds. Archaea belongs to the family Halobacteriacea were produced large amount of pigments C-50 carotenoid in the membrane. Eukaryotic microorganism halotolerant green alga *Dunaliella* was present in high salt environments. The different branches of the Proteobacteria, Cyanobacteria, Firmicutes, Actinobacteria, and actinomycetes were halophilic organisms which are closely relative to non-halophilic organisms. However, there are a few types of Archaea homophile that belongs to the family Halobacteriaceae according to salt requirements and tolerance capabilities (Oren 2007).

On the basis of small subunit rRNA gene sequencing, the phylogenetic tree of life of halophilic microorganisms is shown in Fig. 10.2.

The halophile belongs to Bacteria, Archaea, and Eukarya domains. The phylogenetic tree indicated that halophilic and non-halophilic organisms and its relatives were found jointly with genera, families, and orders and have represented with diverse salt requirement and tolerance. The salt-requiring Archaea belongs to the family Halobacteriaceae (*Halobacterium*) and required 100–150 gl^{-1} salt for their growth conditions and structural stability. Other halophiles or highly halotolerant Archaea belongs to the class Methanothermea (Methanococci), order Methanosarcinales, genera *Methanohalophilus* and *Methanohalobium*, and phylum Euryarchaeota. To date, halophilic microorganisms have not been identified within the Crenarchaeota. A halophilic bacterium mainly belongs to the different phyla Cyanobacteria, Proteobacteria, Actinobacteria, Firmicutes, *Bacteroidetes*, and Spirochaetes (Allers and Mevarech 2005).

Saline and hypersaline environments offer a broad range of microbial diversity and have potential to produce various natural products. The cultural and molecular techniques have been employed for characterization of halophilic bacteria isolated from saltpan water samples and their applications in biotechnological field. The water samples were analyzed for physicochemical parameters, and it was observed that they grew at alkaline pH (8.8) with 12.8% salinity. A 16S ribosomal RNA gene-

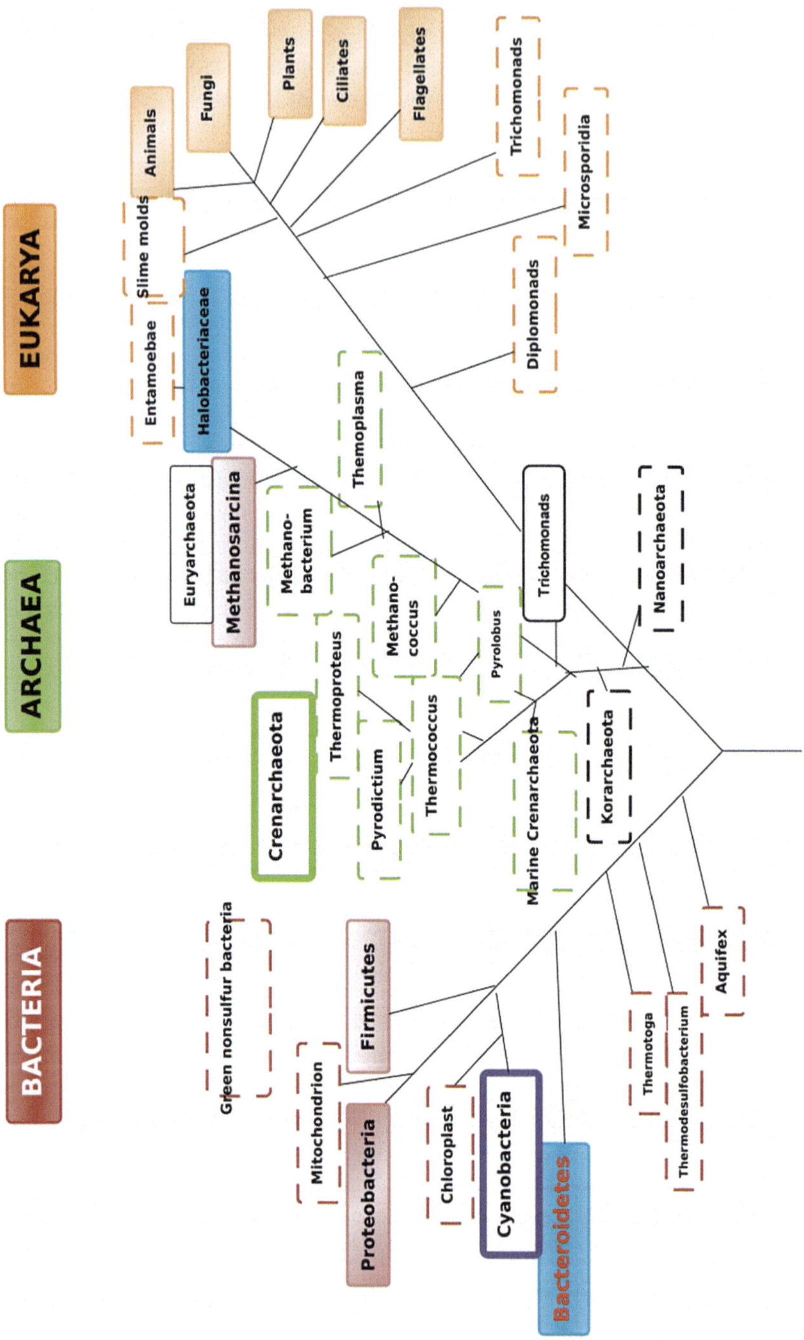

Fig. 10.2 Phylogenetic tree of life of halophilic microorganisms

targeted amplicon analysis has been produced by ten bacterial phyla that constitute of 30.57% Bacteroidetes, 15.27% Proteobacteria, 9.05% Actinobacteria, 5.52% Planctomycetes, and 3.18% Cyanobacteria. On the basis of gene sequencing analysis of the culturable bacterial strains, only 18 strains were identified. Only two strains, namely, SP7 and SP9, were positive for cellulase production, while SP4, SP8, and SP22 strains were positive for lipase production. Quantitatively enzyme assays showed that the extracellular cellulase and lipase activities were 1.95 U/mL and 3.71 U/mL by the isolate SP4 and SP9, respectively. Moreover, the SP9 isolate, i.e., *Salinivibrio* has been potentially used in the bioremediation of hydrocarbon pollution due to high activity against benzanthracene (70% 2, 6-dicholophenolindophenol reduction). These isolates also produced secondary metabolites such as 2, 3-butanediol, hexahydro-3(2-methyl propyl), pyrrole [1,2a]pyrazine-1,4-dione, aziridine, dimethylamine, and ethyl acetate detected by gas chromatography-mass spectrometry (GC-MS) and oxypurinol and 5-hydroxydecanoic acid by LC-MS (liquid chromatography mass spectrometry). These secondary metabolites have possessed various industrial and pharmaceutical applications (Selvarajan et al. 2017). Recently, two strains of methyl-reducing halophilic methanogenic species were isolated from hypersaline soda lake sediments and belong to a new euryarchaeal class within SA1 called as *Methanonatronarchaeia,* which have deep hypersaline anoxic brines (DHABs). These SA1 strains were dissimilar up to 10% at the 16S ribosomal RNA gene sequencing level but associated with one of the two discrete phylogenetic clusters which have been isolated from the Shaban deep brine pool in the Red Sea and completed single-cell genome that has been sequenced (SCG-AAA382-B04) (Sorokin et al. 2016; Ngugi and Stingl 2018; Shibl et al. 2018).

On the basis of 16S ribosomal RNA sequencing, the microbial community structures of halophile microorganisms are not sufficient to explain about the habitat-specific functional pathways. However, meta-genomics is a recent technique which helps to mark functional profiling at the microbial community level and utilizes taxonomical and functional binning to cluster with similar nucleotide patterns together and also allocate taxonomic and metabolic functions in the sample population (Droge and McHardy 2012). Various techniques have been employed such as GC content, tetramer frequency, short oligomer (kmer) frequency, and also read coverage to assign sequence fragments of the metagenome into microbial community (Alneberg et al. 2014). Molecular techniques and metagenomics analysis have provided the detailed information of the microbial diversity and metabolic activities/function of microorganisms in hypersaline and saline environments. The recent model of community structure was identified as square Archaea *Haloquadratum walsbyi*, the bacteroidetes *Salinibacter ruber*, and Nanohaloarchaea. They were dominant members at higher salt concentrations, while more diverse Archaea and bacterial taxonomy were also observed in hypersaline habitats with intermediate salt concentrations. However, metagenomics studies characterized gamma proteobacterium and identified *Spiribacter salinus* found in hypersaline environment (Ventosa et al. 2015). Therefore, advanced or recent technology has facilitated a significant knowledge of saline and hypersaline microorganisms at the molecular level. This is an emerging field, and a lot of important issues should be considered when analyzing these ecosystems. Generally, it is an ideal technique that utilizes multiple approaches and facilitates a more elaborated view of complex microbial population or diversity; therefore, we are expanding our knowledge about the interactions, mechanisms, and adaptations in hypersaline ecosystems. Emerging or combined approaches such as niche differentiation (Vigneron et al. 2014), single-cell genomics (SCG) (Brown et al. 2015), multiple "omics," and metatranscriptomics (Chen et al. 2015) have been used for microbial diversity in hypersaline environment. Considering these tremendous microbial diversity of unculturable microorganisms, various omics approaches have potentially identified a large numbers of microorganisms, niche-specific gene, and protein profile among localized microbial communities in saline and hypersaline environment. It can also help the relationship between hypersaline microbial communities and their growth environments and different strategies for adaptation and metabolic activity that might be employed. Database-dependent analyses (bioinformatics tools) would be used to their inherent biases in hypersaline ecosystem.

The unique characteristics of hypersaline environment are revealed not only in extreme environment with fluctuating growth conditions but also in the microbial diversity and biogeochemical metabolic activities. There is no doubt that we need a continous effort towards identification of indigenous microorganisms which have unknown characteristics. However, saline and hypersaline environments have provided a number of beneficial effects to human beings (Table 10.2).

Keeping this in view, this chapter elaborated on distinct characteristics of microbial community in saline and hypersaline environments that protect and strengthen these ecosystems and enlisted the significant factors or conditions that have adverse effects on these ecosystems. Changes in various factors such as the salinity of these unique networks and the loss of microbial communities are still not described. Thus, research efforts should be made for protection of hypersaline environments worldwide so that such interaction could be further studied for their diverse microbial communities with numerous scientific and commercial values (Moissl-Eichinger 2011).

Table 10.2 Numbers of benefits and impacts are provided by the hypersaline environments

Benefits provided by saline environments	Negative impacts on saline environments
Mineral extraction	Climate change
Brine shrimp production	Increased salinization
Beta-carotene	Water diversions
Hydrolytic enzymes	Extensive mineral extraction
Attraction for shore birds	Sewage and industrial effluents
Biotechnological advances	Agricultural runoff
Clues to ancient life on Earth	Urbanization
Astrobiological implications	Mining impacts
Potential bioremediation solutions	
Considered natural spas	

10.5 Mechanisms of Halophilic and Hypersaline Microorganisms

The cell structures and metabolic activity of halophilic and halotolerant microorganisms that survived in hypersaline environments have been well developed. All halophiles and halotolerant microorganisms have been used strategy to balance their cytoplasm osmotically (salt concentration) with their medium. Biological membranes of these microbes have permeability to water and active energy-dependent processes that carry the transport of water and compensates the loss of water due to osmotic processes, which are not feasible energetically. However, cells maintained a turgor for balancing their osmotic pressure intracellularly which is greater than that of their surrounding environment (Vreeland et al. 2002). Microorganisms have the ability to adapt a broad range of sodium chloride concentrations by three mechanisms. The first mechanism would be a passive in which the cytoplasmic ions are always equal to the medium. In the second mechanism, many organisms produced compatible solute that creates an osmotic balance between the cytoplasm and external environment. In third mechanisms icludes when changes in cell physiology of halophiles, the control movement of water is under control and exist the ions or diluted in the cytoplasm. Loss of water due to high osmolarity and accumulation of potassium ions within their cells was prevented by strict halophilic microorganisms (Madern et al. 2000). By these strategies, microbes produced various enzymes and proteins that are more stable under extreme saline and hypersaline conditions. Some alternative strategies for accumulation of compatible organic solutes, osmolytes, sugars, amino acids, and polyols have been produced by the halophilic and halotolerant microbes. Additionally, these also served as osmotic regulators and osmolytes that could also help in the maintenance of protein stability of halophiles (Roberts 2005). However, due to evaporation and rainfall could be lead to change in salt concentration that would be adverse effect on the survival of halophilic bacteria (Oren 2010b).

Basically, halophilic and halotolerant microorganisms used two fundamentally different strategies for balancing of their cytoplasm osmotically in the medium. These organisms accumulate the molar concentration of K^+ and Cl^- ions and require a wide adaptation/survival mechanism for their intracellular machinery (enzymatic activity) in the availability of salt. Their proteins have proper conformation which maintained their activity at nearby saturated salt concentrations. Halophilic organism's proteome is extremely acidic, and mostly proteins are denaturing in lower salt suspensions. Generally, "high-salt-in strategy" of microorganisms could not survive in low salt concentration medium. Halotolerant and halophilic microorganisms are survived at high salt concentration in outside the medium have been maintained in their cytoplasm isotonic osmotically. The halophilic Archaea belongs to the family Halobacteriaceae. *Salinibacter* was aerobic and Halanaerobiales was anaerobic. They used the "high-salt-in" strategies which maintain higher intracellular ionic concentrations; instead of sodium ions, the potassium ions are the dominant cation and adapted the entirely intracellular machinery in the high salt concentration (Dennis and Schimmin 1997). The second strategy includes the biosynthesis and accumulation of organic osmotic solutes that do not interfere in normal enzyme machinery. But some time the organisms require salt and proteins in the membrane to survive in the saline medium. Much possibility in this strategy is that the cells used excluded salt from their cytoplasm and organic "compatible" solutes are in high concentrations which do not interfere with normal enzymatic machinery. Therefore, adaptation mechanisms of the cell proteome are required by microorganisms. Such microorganisms could be adapted to a wide range of salt concentration. Halotolerant and halophilic Eukarya, Bacteria, and Archaea methanogenic species were involved in the maintenance of their cytoplasm during lowering salt concentration and accumulated osmotic "compatible" solutes that play important role to achieve the osmotic equilibrium without inhibitory or side effect to enzymatic machinery (Gallinski 1995; Ventosa et al. 1998b). These compatible osmotic solute concentrations were regulated accordingly to the variable salt concentration; the cells survived and adjusted according to the requirement whenever salinity was changed outside. Compatible organic solutes were synthesized de novo and accumulated in the medium with constant growth. It may be possible when intracellular Na + concentrations are lower than outside concentrations that directed sodium pumps in the cytoplasmic membranes; in both cases, they have maintained proper intracellular ionic concentration environment as well as pH during regulation.

Growth and adaptability of halophilic bacteria depends on the synthesis of regulation of organic osmolytes such as ectoine (1, 4, 5, 6-tetrahydro-2-methyl-4-pyrimidine carboxylic acid), glycine betaine, glucosylglycerol, etc. (Galinski and Louis 1999). In addition, halophilic methanogens (*Methanohalophilus* sp.) containing glycine, betaine, β-amino acids, and derivatives and other groups such as β-glutamate, β-glutamine, and N-acetyl-β-lysine have also been reported (Oren 2010b). Sulfotrehalose has been observed only in a few alkaliphilic and served as the osmotic solute. A bacterium that belongs to the family Halobacteriaceae was accumulated in 1.0 M concentrations in addition to potassium chloride when sulfotrehalose was used as osmotic solute. The osmotic adaptation mechanisms have been studied in model organisms *Halobacterium salinarum* (Archaea) and *Salinibacter ruber* (Bacteria) for KCl accumulation and *Halomonas elongata* that synthesized organic osmotic solutes (Edbeib et al. 2016). Unicellular eukaryotic *Dunaliella salina* (green alga), *Hortaea werneckii* (black yeasts), and the *Wallemia ichthyophaga* (basidiomycetes) use glycerol and other compatible organic solutes for growth (Cimerman and Oren 2018).

10.6 Applications of Halophiles

Hypersaline and saline environments are habitat in everywhere, and halophiles could be adopted or survive in the environments which have limited the growth conditions. Earlier studies indicated that halophilic microorganisms (bacteria) spoil salted foods, i.e., salted fish and salted hides. Moreover, halophilic bacteria have possessed many beneficial properties and produced stable, unique biomolecules that have practical utility. Pigment producing archae and microalgae have absorbed light energy from saltern ponds when increase in temperature and water evoration rate and hasten the salt deposition process. Moreover, halophilic microorganisms have produced various hydrolytic enzymes, viz., amylases, proteases, gelatinases, DNAses, and lipases, that have capabilities for functioning under optimized conditions and also lead to denaturing and precipitation of proteins. Production of halophilic protein is effective with hydration of salts that provide resistance to lower water activity due to the availability of organic solvents. *Dunaliella* strains have rich in carotenoid and produced β-carotene which is used as food supplement and food additive (Ben-Amotz et al. 1989). Ectoine extracted from moderately halophilic bacteria were also used as moisturizer in the cosmetic industry and enzyme protectant (Galinski and Louis 1999). Many novel biomolecules produced by halophiles have also been used for different applications, for example, bacteriorhodopsin used for biocomputing, gas vesicles used for bioengineering floating particles, and food coloring pigments and compatible solutes used as stress protectants. Halophilic microbes possess a commercial potential due to its unique and novel characteristics which denoted the biotechnological applications in different fields (Rodriguez-Valera 1992), for example, *Bacillus* sp. produced proteases enzymes which were used in hydrolysis of starch and food, brewing, textile, and distilling industries and *Micrococcus* strains were produced DNase, which further used in pulp and paper-making industry and increase loaf volume in baking industry. These microbes were isolated and produced many halophilic enzymes and organic osmotic solutes, and they have adapted and are more stable in saline conditions (Coker 2016). These microorganisms have a broad range of applications and are used as nutrient supplements, cosmetics, antioxidant activity, food preservation, stress protectants, immune system boosters, coloring agents, etc. (Oren 2010a). Commercially available products were listed such as β-carotenoid produced from halophilic green algae, ectoine produced by *Dunaliella* which is an organic solute that can be used as enzyme protectant from denaturation of protein and produced in extreme conditions, and another nuclease H, ribonucleic acid-degrading enzyme that was isolated from *Micrococcus varians* (*halophilus* subsp.) (Kamekura 2002; Kamekura et al. 1982). The halophiles also produced various hydrolytic novel enzymes such as proteases, lipases, amylases, etc. (de Lourdes Moreno et al. 2013). These enzymes have been used for bioenergy and biofuel production (Begemann et al. 2011). Halophiles have also been potentially used as alternative biofuels, such as hydrogen production (Begemann et al. 2011, 2012), and also used as microbial fuel cells for electricity generation (Paul et al. 2014). The retinal pigments from halorhodopsin and bacteriorhodopsin were extracted from *Halobacterium salinarum*, extremely halophilic Archaea, and have potential bioelectric response when exposed to light. The uniqueness and natural phenomenon of halophiles have been studied for holography, artificial neural network systems (computing), and retinas (Margesin and Schinner 2001).

Worldwide, halophilic microbes had gained more importance for environmental applications. Several industries are producing a large amount of wastewater that contains high amounts of heavy metals, salt, and toxic compounds. Microorganisms were used for wastewater treatment by conventional methods and could not function properly due to the cell membrane disruption, enzymes, protein denaturation, and osmotic stress. In this direction, halophilic microorganisms have been considered as suitable and potential alternatives due to their unique characteristics, i.e., resistance and potential to grow in harsh conditions. Halophiles metabolized various organic substrates that possess some essential characteristics and produced extracellular polymeric substances that have been used as electron acceptors required for wastewater treatments (Peyton et al. 2004). The wastewater with salinity is released from various industries such as food; biodiesel production and gas and oil recovery have been treated by halophilic bacteria (Kargi 2002; Ulrich et al. 2009). A few members of halophilic bacteria and archaea have degraded hydrocarbons analyzed from oil fields. Wang et al. (2010) isolated a moderate halophile *Amycolicicoccus subflavus* from Daqing Oilfield, China and had the ability to degrade crude oil with 0.2–2.0 M NaCl concentration. Mailem et al. (2010) investigated halophilic bacteria and their potential use as bioremediation agents in hypersaline environments (Al-).

PHA (poly-β-hydroxyalkanoate) is a polyester and produced in nature by many microorganisms (Bacteria and Archaea). These were used for the production of biodegradable plastics (biological polyesters), and their properties resemble with the polypropylene (thermoplastic polymer). *Haloferax mediterranei* (Archaea) grew in simple media containing starch as cheap carbon sources and accumulated these compounds (38% DW) (Fathepure 2014; Koller 2017).

10.7 Conclusions

Uniqueness of hypersaline and saline ecosystems is widely known for their biogeochemical characteristics, microbial populations/community, and aesthetic appeal. Microbial metabolism and diversification in saline and hypersaline ecosystems are distinct with various novel genus and species that have been isolated and identified. Many distinguishing characteristics of halophilic organisms like morphology, adaptation, evolutionary, environmental, and industrial applications in biotechnological fields have been investigated. Halophiles are extremophile organisms that have survived to adverse environment. In these environments, the microbial diversity would be the area of interest among researchers. In recent years, halophiles have gained attraction among the researchers/scientists due to their adaptability or survival to a broad range of salt concentration as well as their potential applications as bioactive compounds, enzymes, polymers, β-carotene, etc. Thus, the saline and hypersaline ecosystems could be largely unexplored ecological ecosystem/niches for the discovery of various bioactive compounds, and these halophiles are potentially important sources for a wide range of newly identified therapeutic important compounds. Recent techniques such as transcriptomics, genetic engineering, proteomics, and metabolomics and bioinformatics analyses should be used for the investigation of morphological and physiological mechanisms involved in the adaptation of halophilic microbes. The multidisciplinary functions and opportunities would be explored for understanding the potential uses of saline and hypersaline microorganisms that hold promising potential solutions for the present and future prospective at world issues.

References

Allers T, Mevarech M (2005) Archaeal genetics-the third way. Nat Rev Genet 6:58–66

Al-Mailem DM, Sorkhoh NA, Al-Awadhi H (2010) Biodegradation of crude oil and pure hydrocarbons by extreme halophilic archaea from hypersaline coasts of the Arabian Gulf. Extremoph Godalming 14:321–328

Alneberg J, Bjarnason BS, de Bruijn I, Schirmer M, Loman NJ, Quince C (2014) Binning metagenomics contigs by coverage and composition. Nat Methods 11:1144–1150

Bakke P, Carney N, Deloache W, Gearing M, Campbell AM (2009) Evaluation of three automated genome annotations for *Halorhabdus utahensis*. PLoS One 4:e6291

Baliga NS, Bonneau R, Facciotti MT (2004) Genome sequence of *Haloarcula marismortui*: a halophilic archaeon from the Dead Sea. Genome Res 14:2221–2234

Baronio M, Lattanzio VMT, Vaisman N, Oren A, Corcelli A (2010) The acylhalocapnines of halophilic bacteria: structural details of unusual sulfonate sphingoids. J. Lipid Res 51:1878–1885

Begemann MB, Mormile MR, Paul VG (2011) Potential enhancement of biofuel production through enzymatic biomass degradation activity and biodiesel production by halophilic microorganisms. In: Ventosa A, Oren A, Ma Y (eds) Halophiles and hypersaline environments. Springer, Berlin/Heidelberg, pp 341–357

Begemann MB, Mormile MR, Sitton OC (2012) A streamlined strategy for biohydrogen production with Halanaerobium hydrogeniformans, an alkaliphilic bacterium. Front Microbiol 3:93–100

Ben-Amotz A, Shaish A, Avron M (1989) Mode of action of the massively accumulated β-carotene of *D. bardawil* in protecting the alga against damage by excess irradiation. Plant Physiol 91:1040–1043

Benlloch S, Lopez-Lopez A, Casamayor EO (2002) Prokaryotic genetic diversity throughout the salinity gradient of a coastal solar saltern. Environ Microbiol 4:349–360

Bolhuis H, Palm P, Wende A, Farb M, Rodriguez-Valera F, Oesterhelt D (2006) The genome of the square archaeon "*Haloquadratum walsbyi*": life at the limits of water activity. BMC Genomics 7:169

Brown CT, Hug LA, Thomas BC, Sharon I, Castelle CJ, Singh A, Banfield JF (2015) Unusual biology across a group comprising more than 15% of domain Bacteria. Nature 523:208–211

Burns DG, Camakaris HM, Janssen PH, Dyall-Smith ML (2004) Combined use of cultivation-dependent and cultivation-independent methods indicates that members of most haloarchaeal groups in an Australian crystallizer pond are cultivable. Appl Environ Microbiol 70:5258–5265

Caumette P, Matheron R, Raymond N, Relexans JC (1994) Microbial mats in the hypersaline ponds of Mediterranean salterns (Salins-de-Giraud, France). FEMS Microbiol Ecol 13:273–286

Chen LX, Hu M, Huang LN, Hua ZS, Kuang JL, Li SJ, Shu WS (2015) Comparative metagenomics and metatranscriptomic analyses of microbial communities in acid mine drainage. ISME J 9:1579–1592

Cimerman NG, Oren A (2018) Strategies of adaptation of microorganisms of the three domains of life to high salt concentrations. FEMS Microbiol Rev 42:353–375

Coker JA (2016) Extremophiles and biotechnology: current uses and prospects. Life 2:3–10

DasSarma S, Arora P (2001) A general review on Halophiles. In: Encyclopedia of life sciences. Nature Publishing Group. www.els.netf

DasSarma S, DasSarma P (2012) Halophiles. https://doi.org/10.1002/9780470015902.a0000394.pub3

DasSarma S, DasSarma P (2015) Halophiles and their enzymes: negativity put to good use. Curr Opin Microbiol 25:120–126

De la Haba R, Sanchez-Porro C, Marquez MC (2011) Taxonomy of halophiles. In: Horikoshi K, Antranikian G, Bull A (eds) Extremophiles handbook. Springer, Berlin/Heidelberg, pp 344–361
De Lourdes MM, Perez D, Garcia MT, Mellado E (2013) Halophilic bacteria as a source of novel hydrolytic enzymes. Life 3:38–51
Dennis PP, Shimmin LC (1997) Evolutionary divergence and salinity-mediated selection in halophilic archaea. Microbiol Mol Biol Rev 61(1):90–104
Droge J, McHardy AC (2012) Taxonomic binning of metagenome samples generated by next-generation sequencing technologies. Brief Bioinform 13:646–655
Edbeib MF, Wahab RA, Huyop F (2016) Halophiles: biology, adaptation, and their role in decontamination of hypersaline environments. World J Microbiol Biotechnol 32:135
Falb M, Pfeiffer F, Palm P, Rodewald K, Hickmann V, Tittor J, Oesterhelt D (2005) Living with two extremes: conclusions from the genome sequence of *Natronomonas pharaonis*. Genome Res 15:1336–1343
Fathepure BZ (2014) Recent studies in microbial degradation of petroleum hydrocarbons in hypersaline environments. Front Microbiol 5:173. https://doi.org/10.3389/fmicb.2014.00173
Galinski EA (1995) Osmoadaptation in bacteria. Adv Microb Physiol 37:273–328
Galinski EA, Louis P (1999) Compatible solutes: ectoine production and gene expression. In: Oren A (ed) Microbiology and biogeochemistry of hypersaline environments. CRC Press, Inc. Boca Raton, pp 187–202
Gunde-Cimerman N, Ramos J, Plemenitas A (2009) Halotolerant and halophilic fungi. Mycol Res 113:1231–1241
Hartman AL, Norais C, Badger JH, Allers T, Eisen JA (2010) The complete genome sequence of *Haloferax volcanii* DS2, a model archaeon. PLoS One 5:e9605
Jackson RB, Carpenter SR, Dahm CN, McKnight DM, Naiman RJ, Postel SL, Running SW (2001) Water in a changing world. Ecol Appl 11:1027–1045
Kamekura M (2002) Halosimplex carlsbadense gen. nov., sp. nov., a unique halophilic archaeon, with three 16S rRNA genes, that grows only in defined medium with glycerol and acetate or pyruvate. Extremophiles 6:445–452
Kamekura M, Hamakawa T, Onishi H (1982) Application of halophilic nuclease H of Micrococcus varians subsp. halophilus to commercial production of flavoring agent 5'-GMP. Appl Environ Microbiol 44:994–995
Kargi F (2002) Enhanced biological treatment of saline wastewater by using halophilic bacteria. Biotechnol Lett 24:1569–1572
Kerkar S (2004) Ecology of hypersaline microorganisms. In: Ramaihan (eds) Marine microbiological facts and opportunities. NIO, Goa, pp 53–67
Koller M (2017) Production of poly hydroxyalkanoate (PHA) biopolyesters by extremophiles. Pol Sci 1(2):69–85
Kottemann M, Kish A, Iloanusi C, Bjork S, Diruggiero J (2005) Extremophiles 9:219–227
Kushner DJ, Kamekura M (1988) Physiology of halophilic eubacteria. In: Rodriguez-Valera F (ed) Halophilic bacteria. CRC Press, Boca Raton, pp 109–138
Mac Donald IR, Reiley JF II, Guinasso NL (1990) Chemosynthetic mussels at a brine-filled pockmark in the northern Gulf of Mexico. Science 248:1096–1099
Madern D, Ebel C, Zaccai G (2000) Halophilic adaptation of enzymes. Extremophiles 4:91–98
Margesin R, Schinner F (2001) Potential of halotolerant and halophilic microorganisms for biotechnology. Extremophiles 5:73–83
Mc Cready S, Muller JA, Boubriak I, Berquist BR, Ng WL, DasSarma S (2005) Saline systems. BioMed Central 1:3
Moissl C, Bruckner JC, Venkateswaran K (2008) Archaeal diversity analysis of spacecraft assembly clean rooms. ISME J 2:115–119
Moissl-Eichinger C (2011) Archaea in artificial environments: their presence in global spacecraft clean rooms and impact on planetary protection. ISME J 5:209–219
Mormile MR, Biesen MA, Gutierrez MC (2003) Isolation of *Halobacterium salinarum* retrieved directly from halite brine inclusions. Environ Microbiol 5:1094–1102

Ng WV, Kenned SP, Mahairas GG (2000) Genome sequence of *Halobacterium* species NRC-1. Proc Natl Acad Sci U S A 97:12176–12181

Ngugi David K, Stingl U (2018) High-quality draft single-cell genome sequence belonging to the archaeal candidate division SA1, isolated from Nereus Deep in the Red Sea. Genome Announc 6:e00383-18

Oren A (2007) Biodiversity in highly saline environments. In: Gerday C, Glansdorff N (ed) Physiology and biochemistry of extremophiles. ASM Press, Washington, DC, pp 223–231. https://doi.org/10.1128/9781555815813.ch17

Oren A (2010a) Industrial and environmental applications of halophilic microorganisms. Environ Technol 31:825–834

Oren A (2010b) Microbial life at high salt concentrations: phylogenetic and metabolic diversity. Saline Syst 4:2–8

Pasic L, Galan Bartual S, Poklar Ulrih N, Grabnar M, Herzog-Velikonja B (2005) Diversity of halophilic Archaea in the crystallizers of an Adriatic solar saltern. FEMS Microbiol Ecol 54:491–498

Paul VG, Minteer SD, Treu BL (2014) Ability of a haloalkaliphilic bacterium isolated from Soap Lake, Washington to generate electricity at pH 11.0 and 7% salinity. Environ Technol 35:1003–1011

Paul VG, Mormile MR (2017) A case for the protection of saline and hypersaline environments: a microbiological perspective. FEMS Microbiol Ecol 98:8–16

Peyton BM, Mormile MR, Alva V (2004) Biotransformation of toxic organic and inorganic contaminants by halophilic bacteria. In: Ventosa A (ed) Halophilic microorganism. Springer, Berlin/Heidelberg, pp 315–331

Pietila MK, Laurinavicius S, Sund J, Roine E, Bamford DH (2010) The single-stranded DNA genome of novel archaeal virus *Halorubrum* pleomorphic virus 1 is enclosed in the envelope decorated with glycoprotein spikes. J Virol 84:788–798

Rampelotto PH (2013) Extremophiles and extreme environments. Life 3:482–485

Robb FT, Place AR, Sowers KR, Schreier HJ, Dasarma S, Fleischmann EM (1995) *Archaea.* A laboratory manual, Cold Spring Harbor (eds). Cold Spring Harbor Laboratory, New York

Roberts MF (2005) Organic compatible solutes of halotolerant and halophilic microorganisms. Saline Syst 4:1–5

Rodriguez-Valera F (1988) Characteristics and microbial ecology of hypersaline environments. In: Rodriguez-Valera F (ed) Halophilic bacteria. CRC Press, Boca Raton, pp 3–30

Rodriguez-Valera F (1992) Biotechnological potential of halobacteria. In: Danson MJ, Hough DW, Lunt GG (eds) The archaebacteria: biochemistry and biotechnology. Portland Press, London, pp 135–147

Rodriguez-Valera F, Ventosa A, Juez G, Imhoff JF (1985) Variation of environmental features and microbial populations with salt concentrations in a multi-pond saltern. Microb Ecol 11:107–115

Santos F, Yarza P, Parro V, Briones C, Anton J (2010) The metavirome of a hypersaline environment. Environ Microbiol 10:1–10. https://doi.org/10.1111/j.1462-2920.2010.02273x

Selvarajan R, Sibanda T, Tekere M, Nyoni H, Taylor SM (2017) Diversity analysis and bioresource characterization of halophilic bacteria isolated from a south african saltpan. Molecules 22:657. https://doi.org/10.3390/molecules22040657

Shibl AS, Ngugi DK, Talarmin A, Thompson LR, Blom J, Stringl U (2018) The genome of a novel isolate of *Prochlorococcus* from the Red Sea contains transcribed genes for compatible solute biosynthesis. FEMS Micrb Ecol 94(12). https://doi.org/10.1093/femsec/fiy182

Sorokin DY, Kublanov IV, Yakimov MM (2016) *Halanaeroarchaeumsulfurireducens* gen. nov., sp. nov., a first obligately anaerobic sulfur-respiring haloarchaeon from hypersaline lakes. Int J Syst Evol Microbiol 66:2377–2381

Ulrich AC, Guigard SE, Foght JM, Semple KM, Pooley K, Armstrong JE, Biggar KW (2009) Effect of salt on aerobic biodegradation of petroleum hydrocarbons in contaminated groundwater. Biodegradation 20:27–38

Vanhaecke P, Siddall SE, Sorgeloos P (1984) International study on *Artemia* XXXII. Combined effects of temperature and salinity on the survival of *Artemia* of various geographical origin. J Exp Mar Biol Ecol 80:259–275

Vauclare P, Madern D, Girard E, Gabel F, Zaccai G, Franzetti B (2014) New insights into microbial adaptation to extreme saline environments. BIO Web of Conferences published by EDP Sciences, https://doi.org/10.1051/bioconf/20140202001

Vaupotic T, Gunde-Cimerman N, Plemenitas A (2007) Novel 3-phosphoadenosine-5-phosphatases from extremely halotolerant *Hortaeawerneckii*reveal insight into molecular determinants of salt tolerance of black yeasts. Fungal Genet Biol 44:1109–1122

Ventosa A (2006) Unusual microorganisms from unusual habitats: hypersaline environments in prokaryotic diversity, mechanisms and significance. In: Logan NA, Lappin-Scott HM, Oyston PCF (eds) Society for general microbiology symposium 66. Cambridge University Press, Cambridge, pp 223–253

Ventosa A, Nieto J, Oren A (1998) Biology of moderately halophilic aerobic bacteria. Microbiol Mol Biol Rev 62:504–544

Ventosa A, Nieto JJ, Oren A (1998b) Biology of moderately halophilic aerobic bacteria. Microbial Mol Biol Rev 62:504–544

Ventosa A, Rafael RH, Porro CS, Papke RT (2015) Microbial diversity of hypersaline environments: a metagenomic approach. Curr Opin Microbiol 25:80–87

Vigneron A, Cruaud P, Roussel EG, Pignet P, Caprais JC, Callac N (2014) Phylogenetic and functional diversity of microbial communities associated with subsurface sediments of the Sonora Margin, Guaymas Basin. PLoS One 9:e104427

Vreeland RH, Rosenzweig WD, Powers DW (2002) Isolation of a 250 million-year-old halotolerant bacterium from a primary salt crystal. Nature 407:897–900

Waditee Sirisattha R, Kageyama H, Takabe T (2016) Halophilic microorganism resources and their applications in industrial and environmental biotechnology. AIMS Microbiol 2:42–54

Wang YN, Chi CQ, Cai M (2010) *Amycolicicoccussubflavus* gen. nov., sp. nov., an actinomycete isolated from a saline soil contaminated by crude oil. Int J Syst Evol Microbiol 60:638–643

Whitehead K, Kish A, Pan M, Kaur A, Reiss DJ, King N, Hohmann L, DiRuggiero J, Baliga NS (2006) An integrated systems approach for understanding cellular responses to gamma radiation. Mol Syst Biol 2:47–55

Innovative Techniques for Improving Microbial Enzyme Production

11

Abhishek Thakur, Chayanika Putatunda, Rashmi Sharma, Rahul Mehta, Preeti Solanki, and Kavita Bhatia

Abstract

Enzymes are biocatalysts which have a central role in the biochemical, physiological, and metabolic functioning of all the organisms from micro to macro level. These catalytic proteins or metalloproteins have wonderful applications in a number of processes for conversion of substrates to useful products and are used in pharmaceutical, nutraceutical, cosmetic, food and beverage, and other industries. Different sources of these enzymes have been explored for commercial production. The reservoir of microbial world has not been investigated to a greater extent. The microorganisms have proved as an effective source of these industrially important enzymes. The reason for exploring enzymes from microorganisms is their stability at different physiological conditions like high and low temperature, pH, salinity, and others like high catalytic activity and ease of standardization and production. In recent times, the enhancement of production of these microbial enzymes can be achieved using different genetic modulatory methodologies, physiological parameter redesigning, and protein bioengineering. These techniques are the focus of researchers for industrial-friendly hyperproduction of microbial enzymes for generating various formulations.

Keywords

Purification · Solid-state fermentation · Submerged fermentation · Hyperproduction

A. Thakur (✉) · C. Putatunda · R. Sharma · R. Mehta
Department of Microbiology, DAV University, Jalandhar, Punjab, India
e-mail: abhishek10379@davuniversity.org

P. Solanki
Department of Biotechnology, DAV University, Jalandhar, Punjab, India

K. Bhatia
Faculty of Applied Sciences and Biotechnology, Shoolini University, Solan, India

S. G. Sharma et al. (eds.), *Microbial Diversity, Interventions and Scope*,
https://doi.org/10.1007/978-981-15-4099-8_11

11.1 Introduction to Microbial Enzymes

Enzymes are macromolecules of biological origin which are generated by all living organisms. These act as catalysts to enhance the rate of mostly all the biochemical reactions. These proteinaceous macromolecules speed up the rate of metabolic reactions. These are commonly termed as "biocatalysts." Enzymes found in nature have been used widely since old times in the manufacture of different products like leather, linen, and indigo. Production of these products involves either enzymes produced by microbes or enzymes present in added preparations such as papaya fruit or rumen of cattles. The development and refinement of fermentation industry is aimed at the production of microbial enzymes using specifically picked hyperproducer strains, which has led to the recovery of enzymes in purified forms at bulk scale along with full molecular characterization. More than 500 industrial products are being produced using different enzymes (Kumar and Singh 2013; Johannes and Zhao 2006). The demand for industrial enzymes has increased continuously in order to meet the growing need for sustainable solutions. Microorganisms have served as one of the biggest and important sources of many enzymes (Demain and Adrio 2008). Many industrial processes, like chemical synthesis of pharmaceuticals and chemicals, have many drawbacks: less catalytic efficiency, absence of enantiomeric specificity for chiral synthesis, requirement of high temperature, high pressure, and low pH. Also, using organic solvents for chemical synthesis leads to the formation of organic waste and pollutants. Enzymes are suitable aptly for these processes and applications as these require gentle and mild reaction parameters like temperature, pH, and atmospheric pressure and do not even require protection of side chains and functional groups of substrate molecules, show a longer half-life, and work on unnatural substrate molecules (Johnson 2013). Furthermore, enzymes can be chemically modified to elevate their key properties like specific activity, stability, and substrate specificity. The recombinant DNA technology has further upgraded production processes and aided in producing enzymes at commercial level that could not be produced earlier. The evolution of applied sciences like biotechnology in the areas like genetic and protein engineering have further revolutionized the commercialization and wide penetration of these industrial important microbial enzymes. Most of the currently used industrial enzymes are hydrolytic in action, as they are used for the breakdown of different natural substances. Proteases are the most prevalent enzyme type, because of their ample use in the dairy as well as detergent industries. Other different enzymes like amylases, lipases, and cellulases, used starch, textile, detergent, brewing, and baking industries represent the second biggest group of microbial enzymes (Underkofler et al. 1957). Thus we can say that enzymes have a wide range of applications in various industries like food, beverage, textile, pharmaceutical, dairy, and antiaging formulations (Kumar et al. 2018; Bhatia et al. 2018). With the advancement and refinement of modern biotechnology in the field of protein engineering, genetic engineering, and metabolic engineering, we can design innovative strategies to produce these novel enzymes from microorganisms.

11.2 Conventional Methods of Enzyme Production and Purification

Industrial enzymes may be derived from a variety of sources, viz., plants, animals, and microbial sources; however, majority of the production is based and derived from microorganisms for industrial enzymes. The enzymes thus produced majorly can be divided into two categories of extracellular and intracellular enzymes. Majority of the industrial enzymes are extracellular like proteases and carbohydrate-degrading enzymes. Intracellular enzymes, like glucose oxidase, remain associated with the cell and need to be released from the cell by disruption techniques unless the microbe itself is used as a catalyst. Thus, it becomes quite evident that the production and purification techniques to be employed for industrial enzymes will depend upon the source of the enzyme as well as the location of the enzyme, i.e., site of enzyme production. The specific procedure followed for a particular enzyme may vary from manufacturer to manufacturer, but the production methods can principally be divided into two main categories:

- Solid-state fermentation
- Submerged fermentation

Majority of the industrial enzymes, wherever possible, are preferred to be produced using aerobic submerged fermentation process that allows for a greater process control and a better monitoring of growth regime of producer strain of the enzyme in consideration. The production process is invariably followed by recovery and purification process which again depends upon the production method used. Some of the recovery and purification techniques (commonly referred to as "downstream processing") that are applicable to almost all fermentation processes are cell disruption, precipitation, and solid-liquid separation including ultracentrifugation. These techniques are not used in isolation; rather a well-designed combination of these techniques can be used to produce a range of products with varied specifications.

11.2.1 Solid-State Fermentation (SSF)

SSF essentially refers to a process of growth of microorganism growing on a solid surface with moisture content present in the adsorbed form only in the solid matrix and without any presence of a liquid phase. There are a majority three ways in which SSF is used, namely, oriental preparations, mold ripening of cheese, and waste composting. The major organism that can be used to carry out SSF is usually the fungi as opposed to bacterial species because in comparison fungi exhibit a larger capacity to tolerate low water availability. This method of fermentation has the following advantages:

1. Owing to air spaces present in the media, aeration can be achieved easily.
2. Agitation needs to be done intermittently rather than continuously.
3. Low moisture content and absence of a liquid phase too have some additional benefits like the space needed to set up SSF is less in comparison to submerged fermentation, inhibits growth of most bacterial species, and requires less or no effluent treatment processes.

A number of enzymes are produced at industrial scale using SSF, viz., alpha-amylase (*Bacillus* sp.), proteases (*Lactococcus lactis*), lactase (*Aspergillus oryzae*), cellulase (*Trichoderma viride*), and pectinase (*Aspergillus niger*).

11.2.2 Submerged Fermentation (SMF)

SMF is the most familiar method of fermentation for the production of majority of microbial products. Stainless steel vessels of different grades and volumes are designed and operated for growth of microbial producer organisms depending upon the scale of operation, which is usually one of the following three kinds:

- Laboratory scale
- Pilot scale
- Industrial scale

The two major running conditions of SMF may be batch mode or continuous mode of submerged fermentation. The chemical media can be very much expensive for operation at such large scale, and thus SMF generally uses crude media which can easily be procured as waste from other industries like corn-steep liquor, whey, etc. The range of organisms that can be used in SMF included both molds and varied bacterial species. The SMF offers a tight and well-monitored control over various process parameters and thereby results in a number of industrial enzymes being produced using submerged fermentation (Meers and Lambert 1983). After downstreaming, purification of these enzymes is done using different chromatographic techniques. These techniques involve column chromatography (gel exclusion, ion exchange, affinity), HPLC, and UFPLC followed by gel electrophoresis for analysis of purity and homogeneity of enzymes.

11.3 Enhancement of Microbial Enzyme Production by Exploring Natural Hyperproducers

Nature provides an ample supply of biomolecules in the form of enzymes from microbial sources. The capability to explore such huge reservoir relies on the various tools that are accessible to us. We can enhance our search for new enzymes by the following different approaches.

(i) Metagenomic screening (Rondon et al. 1999; Uchiyama and Miyazaki 2009)
(ii) Genome mining in various microbes whose genomes have been sequenced (Ahmed 2009; Kaul and Asano 2012)
(iii) Exploring diverse extremophiles (Schiraldini and De Rosa 2002; Kumar et al. 2011)

11.3.1 Metagenomic Screening

(i) Screening based on metagenomes is mostly dependent on two approaches either function based or sequence based (Rondon et al. 1999; Uchiyama and Miyazaki 2009). Function-based screening is a direct method to isolate genes that exhibit the desired function, and this can be achieved by heterologous complementation along with direct detection on the basis of phenotypes and induced gene expression (Li et al. 2012). The sequence-based screening is done using polymerase chain reaction (PCR) or hybridization methods. Normally, the popular approach includes use of a set of degenerate primers that have the designing based on consensus sequences of amino acid. Different habitats, like volcanic vents, rumen of cow (Hess et al. 2011), and oceans (Kennedy et al. 2008), have provided microbial enzymes that have great potential for industrial applications such as superoxide dismutase (Kumar et al. 2018; Bhatia et al. 2018), lipase (Jeon et al. 2009), amylase (Rondon et al. 2000), nitrilase (Bayer et al. 2011), and decarboxylase (Jiang et al. 2009).

11.3.2 Genome Mining

In the case of genome mining, two methodologies have being practiced to discover novel enzymes (Luo et al. 2012). The first approach involves genome searching which is based on finding open reading frames in the genome of a specific microorganism. Sequences that are determined as presumed enzymes are exposed to further cloning strategies, activity screening, and overexpression. Other approach known as data mining is based on homology alignment of all sequences that are found in databases. Different bioinformatics tools like BLAST help in searching conserved regions between sequences which yield homologous protein sequences that are recognized as feasible candidates for further studies and characterization.

11.3.3 Extremophiles

Extremophiles have inherent capacity to survive under extreme environments, which includes factors like temperature (−2 to 12 °C, 60–110 °C), high pressure, pH (less than 2 and more than 9), and high salinity, and these are a great source of enzymes with extreme stability. Thermophilic superoxide dismutases, proteases, amylases, as well as cellulases are being used in different industries for various applications

(Kumar et al. 2011; Atomi et al. 2011). Another example of extremophiles is psychrophiles (microorganisms that can survive and proliferate at low temperatures), and these are potential sources of enzymes that are stable at very low temperatures. These cold-active enzymes have found application in various fields like paper industry (xylanases), detergent and reagent industry (lipases), and pharmaceutical industry (SODs). Halophilic enzymes tolerate very high salt concentrations. These proteins have gotten the adaptation to this system by attaining a high number of negatively charged amino acid residues on their surface to prevent denaturation by precipitation. Xylanases, amylases, proteases, and lipases from *Halobacterium*, *Halobacillus*, and *Halothermothrix* have been isolated, and their applications have been explored thoroughly (Gomes et al. 2003; Van den Burg 2003).

11.4 Strategies for Enhancing Microbial Enzyme Production Using Genomic Alterations

The significance of enzymes in the metabolic activities as well as their innumerable applications in almost all spheres of life cannot be overemphasized. The microorganisms are known to provide many advantages over the plant and animal sources especially in terms of rapid production rates, diversity, and ease of cultivation (Rao et al. 1998). Another major advantage associated with the utilization of microbes is their amenability to genetic improvement (Adrio and Demain 2014). There are ample examples of application of conventional genetic improvement techniques as well as novel genetic engineering techniques for enhancement of enzyme production by various microorganisms (Baweja et al. 2016). Nowadays, even metabolic engineering is being utilized for hyperproduction of microbial enzymes.

11.4.1 Conventional Genetic Improvement Approaches

Initially the improvement strategies focused on the optimization of cultural conditions for enhanced enzyme yields by the microbes. However, it was realized that even at the most suitable conditions, the maximum possible output is going to depend on the genetic potential of the producing microbial strain (Stanbury et al. 2016). So, in order to overcome this obstacle, several approaches have been used. Some of these strategies are being discussed below.

11.4.1.1 Mutagenesis

Mutations are known to provide raw material for evolution by inducing genetic variation. These may appear spontaneously or may be induced with the help of mutagenic agents. For commercial enzyme purposes, mainly the induced mutants obtained by the treatment of physical or chemical mutagens have been used. However, in the literature, one can encounter some cases of spontaneous hyperproducing mutants also. A spontaneous mutant *Beauveria bassiana* P2 has been isolated by Borgi and Gargouri (2014), which was reported to show a ninefold higher protease production with a wider substrate specificity. The conventional chemical mutagens which have been used by many workers include base analogues (like bromouracil), direct DNA reacting chemicals (like ethyl methanesulfonate or EMS), and intercalating agents (like acridine orange) (Errol et al. 2006). The physical mutagens include the ionizing radiations and non-ionizing radiations (Maloy et al. 1994). While the former includes X-rays and gamma-rays, the classical example of the latter is the ultraviolet (UV) radiations. These mutagens vary in their modus operandi of the DNA damage, e.g., UV radiation leads to pyrimidine dimerization and EMS leads to DNA alkylation (Auerbach 1976; Rowlands 1984a). The ultimate outcome of mutagenesis is dependent not only on the type and extent of DNA damage caused by the mutagenic agent but also on the microbial DNA repair mechanisms. For example, photoreactivation repair simply breaks the thymine-thymine dimers, thus offsetting the damage resulting from UV radiations and reducing the chances of mutations, while the SOS repair system is known to be an error-prone repair system and often leads to multiple mutations (Kenyon 1983). Another important observation associated with the mutagens was the fact that mutagens usually affected particular regions of the genome with relatively fewer effects on other parts (Auerbach 1976). Also, while planning for mutagenic experiments, it is imperative to standardize the required dosage of the mutagen, as lower dosage may result into no or very less number of mutants and very high dosage may lead to killing of the microbe and production of unwanted secondary mutations (Rowlands 1984a). The classical method of random mutagenesis involves exposing the appropriately grown microbial culture to the appropriate dosage of mutagen, followed by screening of the survivors and assessment of the enzyme activity. The hyperproducers thus obtained may further be subjected to subsequent round(s) of mutagenesis, till the desired level of production is obtained. The screening process may also be random, or some kind of selective pressure may be applied to isolate the desired mutant (Rowlands 1984b; Kostyleva et al. 2017).

There are ample examples where mutagenesis has led to the development of microbial strains with enhanced enzyme activity and/or better enzymatic characteristics. Ali and Mahmood (2019) carried out mutagenesis of *Alkalibacillus flavidus* with EMS and were able to obtain a mutant EMS-cys2 with a 6.8-fold higher production of acetyl xylan esterase. Ameri et al. (2019) applied UV radiations for developing a mutant strain of *Bacillus atrophaeus* capable of overproducing thermoalkalophilic lipase. Kamalambigeswari et al. (2018) obtained several mutants with higher pectinase production capability by treating the *Aspergillus niger* with UV radiations, ethyl methanesulfonate, etc. They also reported that combined treatment with UV radiations and ethidium bromide yielded a mutant strain of *A. niger* with 1.69 times higher production of polygalacturonase. Similarly, Aleem et al. (2018) obtained a mutant strain of *A. oryzae* M100(6) with the capability of producing higher levels of alpha-amylase. The enzyme was also observed to display greater thermal stability and had immense potential for commercial application in koji fermentation. The aforementioned mutant was obtained by gamma radiation-mediated random mutagenesis followed by a five-step screening protocol. Khedr et al. (2017) applied UV radiations for the development of several α-amylase hyperproducers of *Bacillus licheniformis* MK90. On comparing the amylase gene (*amy* E) in the parental strain and one of the mutants UV-5-M-12, it was observed that the mutant gene had five nucleotide substitutions. Suribabu et al. (2014) were able to obtain ten mutants of *Brevibacillus borstelensis* R1 which showed enhanced production of α-amylase. These mutants were obtained by using different mutagens like UV rays, nitrous acid, EMS, ethidium bromide, acrylamide, and 5′-fluorouracil. Haq et al. (2010) tried UV radiations for improving α-amylase yield of *B. amyloliquefaciens*, however were not able to get any hyperproducing mutant. However, with the EMS treatment, the workers were successful in obtaining a mutant with 1.4 times higher production than the original bacterial strain. Azin and Noroozi (2001) tried various mutagens like *N*-methyl-*N*′-nitro-*N*-nitrosoguanidine, UV radiations, and nitrous acid and 2-deoxy-D-glucose as anti-metabolite for screening α-amylase-overproducing mutants of *Aspergillus oryzae* and reported the maximum saccharogenic and dextrinizing activity in a mutant obtained by nitrous acid treatment.

Another very significant category of microbial enzymes includes the proteases. These enzymes have been employed in a very diverse range of products and processes and hence form significant part of many industries including detergent, food processing, pharmaceutical, etc. (Adrio and Demain 2014). Hence a lot of effort has been directed for the improvement of protease production (Contesini et al. 2018). de Paiva et al. (2019) obtained several mutants of *B. subtilis* LFB-FIOCRUZ 1266 with the help of ethyl methanesulfonate, which displayed 1.4–2.4 times higher keratinolytic activity. Mehtani et al. (2017) applied acridine orange, ethidium bromide, and UV rays sequentially and developed a mutant strain of *Streptomyces* with 2.2 times higher protease production. Wang et al. (2016) have reported about the mutant of *B. subtilis* S1-4 designated as UMU4 which produced a 2.5-fold enhanced proteolytic activity as well as better feather-degrading activity. They were able to obtain this mutant by combined treatment of UV radiations and N-methyl-N-nitro-N-nitrosoguanidine (NTG). Karn and Karn (2014) were able to isolate a mutant designated as *Bacillus* sp. RS1, capable of showing higher protease activity, with the help of UV ray mutagenesis. Mukhtar and Haq (2013) reported 11.28 ± 0.45 U/mL of alkaline protease production from an EMS-treated mutant *Bacillus subtilis* IH72EMS8 on soybean meal containing medium as compared to 5.74 ± 0.26 U/mL protease production by the wild-type strain. Nadeem et al. (2010) tried UV rays and the chemical mutagens NTG and methyl methanesulfonate (MMS) on *Bacillus licheniformis* N-2 and finally screened out a mutant UV-9 with 1.4 times increase in alkaline protease production. Dutta and Banerjee (2006) subjected an extracellular protease-producing *Pseudomonas* sp. RAJR 044 to UV mutagenesis and obtained a hyperproducing mutant JNGR 242. Sunitha et al. (1999) tried random mutagenesis with the combined application of UV rays and NTG on alkaline protease-producing *Thermoactinomyces* sp. E79 and obtained a mutant which not only was capable of producing double the amount of enzyme but also was reported to be resistant to catabolite repression by glucose.

Microbial especially fungal cellulases are another set of very important enzymes. Apart from the conventional roles in paper and pulp industries, food industries, textile industries, etc., these are emerging as a major player in the production of biofuels from lignocellulosic biomass (Shin et al. 2000; Bischof et al. 2016). However, a main obstacle in their widespread application is their cost. So, in order to bring down the cost, the scientists are trying for the development of hyperproducing strains. Sadhu et al. (2014) were able to develop a mutant strain of *Bacillus* with higher cellulase production by treating the parental bacterial culture with NTG. Agrawal et al. (2013) targeted the β-glucosidase enzyme for improving the overall cellulolytic activity of *B. subtilis* and reported the isolation of a mutant, designated as PS-CM5-UM3, obtained by combined treatment of UV radiations and EMS. Dillon et al. (2006) reported higher cellulase production from mutant obtained by treating *Penicillium echinulatum* spores with hydrogen peroxide. Chand et al. (2005) have suggested a new method of mutagenesis by growing fungal isolates on selective media incorporated with sub-lethal concentrations of chemical mutagens like 1-methyl-3-nitro-1-nitrosoguanidine and EMS for obtaining higher yields of cellulase. Bakare et al. (2005) were able to obtain two catabolite repression resistant hyper-cellulase producing mutants by EMS treatment of *Pseudomonas fluorescens*.

There are a few reports where the mutagenesis resulted in hyperproduction of more than one enzyme. For example, Yoneda and Maruo (1975) isolated several mutants of a *B. subtilis* strain, using NTG, which were reported to produce 2–3 times higher protease and 5–16 times higher amylase as compared to the parent strain. Kostyleva et al. (2018) achieved eightfold and fivefold enhancement in the production of xylanase and endoglucanase, respectively, by treating *Trichoderma reesei* with gamma radiations.

Apart from the conventional physical and chemical mutagens which are very commonly employed, another novel type of mutagen, known as atmospheric and room temperature plasma (ARTP) technology, is gaining popularity nowadays. The alterations arising in the microbial cell and biomolecules including the genetic material, proteins, cell wall, etc. due to ARTP treatment have been attributed to the combined effects of electromagnetic field, charged particles, neutral reactive particles, as well as heat and UV radiations (Laroussi and Leipold 2004; Zhang et al. 2014). Zou et al. (2018) reported higher cellulase production by a mutant *T. reesei* obtained by ARTP mutagenesis. Ma et al. (2016) have applied the ARTP mutagenesis for enhancing alkaline amylase production from a recombinant *B. subtilis*. Xu et al. (2011) tried a combined approach using ARTP, UV rays, and genome shuffling for obtaining higher yield of cellulase from a mutant *T. viride*.

Another approach of mutagenesis makes use of a so-called biological mutagen, i.e., transposons. The transposons or transposable elements are those DNA segments which can change their location and move from one place to another within the genome of a cell. Their insertion may lead to overexpression, insertional inactivation, or alteration in the gene expression (Muñoz-López and García-Pérez 2010). These have played a very significant role in identifying the functions of various genes (Hu and Coates 2005; Ebert et al. 2013), but there are some cases where transposon-based mutagenesis has resulted into higher production of microbial enzymes. *Stenotrophomonas maltophilia*, subjected to a transposon-mediated mutagenesis, showed overexpression of extracellular serine protease and better control of phytopathogen *Pythium ultimum* due to this (Dunne et al. 2000). Htway et al. (2018) also reported higher expression of cellulolytic activity in nitrogen-fixing bacteria with the help of transposon-mediated mutagenesis.

Apart from a few cases of spontaneous mutations and a plethora of cases of random induced mutagenesis, another category of mutagenesis referred to as site-directed mutagenesis has also been used by several workers for obtaining higher production of enzymes from microbial sources. The greatest advantage of this technique is that one can alter the exact desired sequences without any possible undesirable effects arising in the case of random mutagenesis (Chung et al. 2017). However, in the case of site-directed mutagenesis, it is imperative to possess the knowledge about the particular gene(s) to be targeted (Goraya et al. 2017). One of the conventional approaches of site-directed mutagenesis involved targeting the actively transcribing genes during continuous cultures. This approach relied on the fact that the genes undergoing transcription tend to be more prone to mutagenesis, so if mutagenesis is carried out during the period when the microbes are carrying out the biosynthesis of the desired product, the chances of obtaining the desired mutants will be drastically enhanced (Rowlands 1984a, b). Another approach relied on the fact that NTG specifically mutated the replication fork. So, if one is in possession of a proper time-dependent mutational map, the desired gene(s) can be mutated (Rowlands 1984a, b). Apart from that, the recombinant DNA technology and PCR-mediated techniques are also used very commonly for site-directed mutagenesis (Shortle et al. 1981; Goraya et al. 2017). The site-directed mutagenesis or rational mutagenesis can be used to construct hyperproducing microbial strains by enhancing the expression of key enzyme genes or alleviating the inhibition and/or repression of particular key biosynthetic enzyme or preventing the synthesis of unwanted by-products (Xu and Zhang 2016). The site-directed mutagenesis (SDM) can either target a single site (known as single SDM) or multiple sites (known as multi-SDM). Although several techniques of single SDM are available, single SDM most commonly employs amplification of plasmid using complementary primer pairs, which contain the desired mutation.

There are several examples where the higher production of microbial enzymes and/or improvement in their desirable characteristics has been achieved with the help of SDM, a few of which are discussed below. Sriprang et al. (2006) were able to enhance the thermostability of xylanase from *A. niger* by replacement of serine and threonine residues with arginine using SDM. Similarly, Korman et al. (2013) reported the improvement in methanol tolerance of lipase obtained from *Proteus mirabilis*. Zhang et al. (2015) reported 87% improvement in the β-endoglucanase activity and better thermostability with the help of SDM based on homology modeling and rational design in *Thermotoga maritima*. Weng et al. (2015) carried out single site SDM and double site SDM of nattokinase (serine protease produced by *B. subtilis* var. *natto*) and observed better fibrinolytic properties. Gai et al. (2018) reported the improvement in properties like thermostability, acid tolerance, as well as production of α-amylase in *B. stearothermophilus* by carrying out SDM resulting in the deletion of arginine (179) and glycine (180) amino acid residues. Fang et al. (2019) were able to enhance the extracellular keratinolytic protease activity of *E. coli* with the help of site-directed mutagenesis. Apart from this, another type of mutagenesis, known as site saturation mutagenesis, is gaining importance nowadays. Liu et al. (2014) were able to improve the activity of the nitrilase enzyme of *Alcaligenes faecalis* with the help of gene site saturation method, based on homology modeling. Zheng et al. (2016) made use of this technique for improving the activity of lipase being synthesized by *Thermomyces lanuginosus*.

11.4.1.2 Genetic Recombination

Recombination allows the bridging of two or more desirable traits in a single organism. The conventional recombination was possible only by crossing two sexually compatible organisms. However, there are many fungi which lack sexual reproduction, or the sexual reproduction has not yet been observed. Also, the bacteria do not show the sexual reproduction in the conventional terms. So, in order to deal with such situations, two main techniques have been reported traditionally, viz., recombination via parasexual cycle and protoplast fusion. While the former is possible within the same or closely related organisms, the latter does permit greater levels of phylogenetic distance among the organisms used in protoplast fusion. Both these processes have been tried in case of fungi and involve development of heterokaryon. In a heterokaryotic cell, more than one type of genetically different nuclei co-exist (Weichert and Fleibner 2015).

A conventional parasexual cycle in fungi involves the formation of a heterokaryotic stage by fusion of hyphal cells of compatible mycelia, followed by karyogamy in some cases (Pontecorvo 1956; Read et al. 2010). Then such cell may undergo mitotic crossing overs leading to a diploid recombinant, or it may undergo chromosomal losses resulting into formation of aneuploid or haploid cells (Raper 1966; Roper et al. 2013). In the case of protoplast fusion, the cell walls are broken (most commonly) by enzymatic treatments leading to the formation of protoplasts, which are allowed to fuse with each other with the help of chemical or physical agents leading to heterokaryons. Subsequently, like a heterokaryon obtained by parasexual cycle, one may end up with a recombinant diploid, haploid, or aneuploid (Shoji et al. 2015; Strom and Bushley 2016). These types of recombination experiments have gone a long way in deciphering the intricacies of the fungal genetics including the concept of sexual incompatibility, the concept of dominance, the mapping of fungal genes, etc. (Marek et al. 2003; Ma et al. 2010). It has also been suggested that parasexual cycles have played a very significant role in the evolution and adaptations of fungi especially the ones where an apparent sexual stage is absent or unknown (James et al. 2008).

Apart from this, such recombination experiments have also shown the possibility of improving the production of several metabolites as well as enzymes from various fungi. Ushijima and Nakadai (1987) were able to develop heterozygous diploids by protoplast fusion of two parental strains of *Aspergillus sojae*, one of which was a good protease producer and the other was a good glutaminase producer. Many of the heterozygous diploids obtained from them produced higher levels of both the enzymes. Khattab and Bazaraa (2005) carried out protoplast fusion of various mutant *A. niger* strains and reported that 19 fusants had enhanced glucose oxidase activity as compared to the mutant parents. Prabavathy et al. (2006) undertook self-fusion of *Trichoderma harzianum* and observed that the self-fusant SFTh8 showed a twofold higher chitinase activity as compared to the parent. Varavallo et al. (2007) have reported that a recombinant haploid generated by protoplast fusion of two different species of *Penicillium*, viz., *P. expansum* and *P. griseoroseum*, showed higher levels of pectinase production as compared to the parental cultures. Solís et al. (2009) observed better hydrolysis of orange peel by the activity of pectin lyase-overproducing hybrid which was obtained by protoplast fusion between a mutant strain of *Aspergillus flavipes* and *A. niveus* CH-Y-1043. Hassan (2014) reported that some of the fusants obtained from fusion of *T. harzianum* and *T. viride* showed a threefold higher chitinase activity and a twofold enhanced β-glucanase activity. Adeleye et al. (2019) reported that overall the enzyme production by fusants, obtained by fusing the protoplasts of two amylase-producing *Aspergillus* sp., was found to be more favorable than the parental strains.

The utilization of the aforementioned techniques is not solely restricted to fungi, as there are cases where bacterial cultures, especially the filamentous bacteria, have also been subjected to protoplast fusions (Okanishi et al. 1974; Illing et al. 1989). Sivakumar et al. (2004) reported that a fusant, designated as F_4, had higher laccase activity as compared to the parents *Streptomyces* sp. RK1 and *S. lividans*. Anitha and Rebeeth (2009) have reported higher chitinase activity as well as biocontrol activity against *Fusarium oxysporium* in self-fusants of *Streptomyces griseus*. Li et al. (2018) have reported about an intergeneric protoplast fusion between *Streptomyces albulus* and *Bacillus subtilis*, with the aim of transferring the proteolytic activity of *B. subtilis* into *S. albulus* so that the hybrid can produce ε-poly-L-lysine and utilize the proteins in corn starch residues (CSR). They have concluded that the fusant LS-84 produced significantly higher product formation rate on CSR due to proteolytic activity from the parental strain.

11.4.2 Using Genetic Engineering and rDNA Technology

The present applications of microbial enzymes in the development of various bioprocesses are focused on the industries including leather, pharmaceuticals, pulp and paper, biofuels, chemical agricultural, cosmetics, and food and beverage, among others. With the increased market demand of products and their versatility, there is also a growing need to develop novel and improved enzymes to sustain the competitiveness by the industries. Present-day approaches to develop microbial biocatalysts not only include identification of the enzymes from the available microorganisms already present in the environment but also utilize new age biotechnology and genetic engineering approaches like metagenomic screening for potent proteins and enzymes present in microorganisms and other species that may help in improving the industrial products, identifying and isolating their genes, and developing vector and microbial host expression systems for these proteins/enzymes (Adrio and Demain 2014).

11.4.2.1 Computer-Aided Molecular Designing (CAMD) and Molecular Docking

The availability of online resources, including chemical and biomolecule data banks, has significantly decreased the time-consuming processes required in the discovery and development of enzymes used in industries. These techniques have the leverage of chemical and biological information about enzymes and new enzyme-target molecular docking to identify and design/optimize ideal protein candidates with improved biocatalytic activity (Frushicheva et al. 2014; Mulholland 2008; Warshel 1991). The approach toward molecular designing and docking may be structure based or ligand based. If the molecular structure and chemical properties of the target molecules for the enzymes are known, then the CAMD and molecular docking techniques will provide the structure of the potential enzyme candidate, which can now be screened from the available protein data banks or created through protein engineering. Early approaches in the computational analysis of enzyme catalysis were based on mutation studies (Hwang and Warshel 1987; Rao et al. 1987). Computer-simulated enzyme structure-catalysis models have substantially reduced the exhaustive work of isolating and testing potential microbes and/or their enzymes. Successful "hits" generated enlist the ideal enzyme candidates which may or may not be from microbial source or may not even be a naturally occurring protein. There are however limitations to the computer-simulated models due to insufficient proficiency in the knowledge of physical chemistry, and thus only consistent results of CAMD or specifically computer-aided enzyme design (CAED) must be relied upon (Frushicheva et al. 2014). A general example of reference for CAED may be the designing of Kemp eliminases (Frushicheva et al. 2010; Khersonsky et al. 2012; Rothlisberger et al. 2008) which demonstrated small but important improvements in biocatalysis primarily based on modulation of electromotive forces via computational directed evolution models. Combinatorial approaches having both computational and mathematical modelling simulations have further enhanced the knowledge in selecting the enzymes for the required catalysis (Alexandrova et al. 2008; Kiss et al. 2010). Hence, we may say that the approaches utilizing the predictive computational and mathematical tools must be must be optimized on the basis of core concepts of physical chemistry for developing enzyme candidates with reproducible biocatalysis effects and well-defined active sites for the targeted molecule. The future of CAED holds great promise ahead with more reliable structure-catalysis modelling and production of designer enzymes.

11.4.2.2 Protein Engineering

When a native microbial enzyme is needed for large-scale applications in the industry, it generally may not meet the requirements for the needed biocatalytic process, and hence its properties must be modified and optimized. The main advantage of using protein engineering in the field of industrial microbiology is to overcome the any limitations of native microbial enzyme systems and create process-specific microbial enzymes. The approaches toward designing new proteins or enzymes may be based on a either a rational design/structure or directed evolution (Verma et al. 2012). A rational design approach is based on the knowledge of protein/enzyme structure and employs a more precise technique of saturation mutagenesis or site-directed mutagenesis for creating changes on the protein folding and scaffolds and hence the active sites of the biocatalyst. This technique is primarily based upon our knowledge of the protein structure, folding, and molecular dynamics (Gerlt and Babbitt 2009). Directed evolution, on the other hand, is based on repeated rounds of random mutagenesis and reshuffling of genes (Tracewell and Arnold 2009). While the former technique requires in silico prescreening, the latter is followed by screening of the mutant library. Although directed evolution is a powerful approach in redesigning a protein structure, it however suffers the limitation of the generated diversity in sequence, hence reducing the efficiency in sampling of the sequence space having functional importance (Tracewell and Arnold 2009; Wong et al. 2007). Recent developments in computer-aided protein-directed evolution (CAPDE) that involve a combinatorial approach involving computational designing and methods of directed evolution have further strengthened the techniques employed in protein engineering (Verma et al. 2012).

Natural selection and evolution of proteins has taken place by minor changes in the structure of active sites, and based on these facts, the goals of protein engineering focus on both designing binding sites to fit the targeted substrate and also constructing new catalytic polypeptide residues leading to modifications in the functions and binding mechanisms of the enzymes (Cedrone et al. 2000). The rational design approach leads to the production of a small number of variants when compared to the directed evolution approach, but improvements in the availability of detailed three-dimensional structure of proteins through protein data banks and CAED technology has paved the way to significant success (Porter et al. 2016).

11.4.2.3 Genetic Recombination

The possible mechanisms for recombination in microorganisms considering primarily parasexual events are conjugation, transduction, and transformation. Besides, there may also occur genetic rearrangements internally by the translocation of DNA segments, i.e., transposons or insertion sequences. Transfer of DNA takes place during conjugation by cell-to-cell contact, transduction is mediated by bacteriophage, and transformation involves the uptake of DNA molecules by the bacteria from its environment which may be mediated by physical or chemical triggers naturally as well as in vitro (Adrio and Demain 2010). The interest in application of recombinant DNA technology in improving the microbial products of industrial value has risen since the 1980s, especially after the knowledge of protoplast fusion (Ryu et al. 1983). Programs have been initiated to improve microbial strains using protoplast fusion between different mutant lines. Yoneda (1980) demonstrated that developing a strain by recombining five individual mutations, each of which led to an increase in the production of α-amylase in *Bacillus subtilis* by two- to sevenfold, was able to produce 250 times more of α-amylase than the native strain. Similarly, genes encoding for various microbial enzymes have been cloned and used to increase the expressed enzyme levels hundreds of times more than the expression in naturally occurring strains, and presently more than 60% enzymes used in industries like food processing, agriculture, pharmaceuticals, etc. are leading recombinant proteins (Cowan 1996).

The focuses of recombinant DNA technology in the industries employing the use of enzymes are primarily to (a) create genetically modified microorganisms that can produce industrially beneficial enzymes isolated from other microbes that are difficult to grow and maintain or genetically modified for industrial purposes, (b) design efficient open reading frames for use in expression of multiple gene copies to increase the enzyme production, (c) create safe host-vector system for expression of enzymes produced naturally in pathogenic or toxin-producing organisms, and (d) increase environmental stability, catalytic efficiency, stereospecificity, half-life, and reaction kinetics (Adrio and Demain 2010).

The genes for industrially beneficial enzymes are identified and isolated from naturally occurring organisms and expressed in microorganisms employed for industrial production like *Trichoderma* spp., *Aspergillus* spp., *Yarrowia lipolytica*, *S. cerevisiae*, *K. lactis*, and *Bacillus licheniformis*. Recombinant mold and filamentous fungi are able to express recombinant proteins at levels increased to 4.6 g/L. Genetic engineering techniques that are employed in the development of these recombinant strains and proteins are enlisted in Table 11.1.

11.4.2.4 Applications of Genetic Engineering and Recombinant DNA Technology in Production of Microbial Enzymes

Saturation mutagenesis and screening have been successfully employed in producing enantioselective hydantoinase with fivefold more productivity for the production of l-Met (l-amino acids) in *Arthrobacter* sp. (May et al. 2000). Dumon et al. (2008) used gene site-saturation mutagenesis to improve Tm of xylanase by 25 °C employed in degradation of hemicellulose. A 79.4-fold increment in activity and a 6.3–79-fold enhanced in thermostability of lipases used for transesterification reactions and produced by *Geobacillus* sp. NTU 03 were shown by Shih and Pan (2011). Endoglucanase CelA used in bioconversion of cellulosic biomass was produced in *Clostridium thermocellum* with a tenfold increase in half-life of inactivation at 86 °C (Yi et al. 2011).

Site-directed mutagenesis has been used in the modulation of cyclizing activity and thermostability of cyclodextrin glucanotransferase and amylase produced by *Bacillus stearothermophilus* ET and *Bacillus* sp. US149 in the bread industry (Ben Mabrouk et al. 2011; Lee et al. 2002); increasing specific activity 1.5-fold higher of protease BYA in *Bacillus* sp. Y for use in detergent industry (Tobe et al. 2006); increased alkali stability of endo-1,4-β-xylanase II produced by *Trichoderma reesei* for use in sulfate pulp bleaching (Fenel et al. 2006); increasing thermostability and 4-fold increase in k_{cat} of lipase *Bacillus pumilus* for use in chemical, food, leather, and detergent industries (Bustos-Jaimes et al. 2010); improving catalytic activity and thermostability of β-agarase AgaA in *Zobellia galactanivorans* for the production of functional neo-agarooligosaccharides (Lee et al. 2011) and cholesterol oxidase in *Brevibacterium* sp. for the detection and conversion of cholesterol (Sun et al. 2011);

Table 11.1 Common genetic technologies successfully employed for increasing industrial productivity by microorganisms

Genome-based strain reconstruction	Creating a superior strain which only has the mutations critical for increased productivity and no other unknown mutations assimilated through brute-force mutagenesis followed by screening.	Ohnishi et al. (2002)
Metabolic engineering and reverse (inverse) metabolic engineering	Improvement in the formation of product or the properties of the cells by modifying the existing cellular metabolic reactions or creating new ones which are more desirable using recombinant DNA technology.	Nielsen (2001), Stephanopoulos (1999)
	Reverse or inverse metabolic engineering is carried out by first selecting a strain with desirable cellular phenotype, followed by evaluating and characterizing environmental and/or genetic factors that produce that phenotype, and finally transferring the phenotype to another strain through modifications in the identified environmental and/or genetic factors.	Bailey et al. (1996), Santos and Stephanopoulos (2008)
Molecular breeding and DNA shuffling	This technique mimics the strategies of natural recombination possible in microorganisms via in vitro recombination of homologous genes. It creates not only recombinant genes but also point mutation at controlled low rates. It holds the advantage of multi-species crossing over by DNA shuffling and creating recombinant pools of entire genomes.	Hou (2009), Ness et al. (2000), Zhang et al. (2002), Zhao and Arnold (1997)
Protoplast fusion	Two genetically different protoplasts are experimentally fused to produce parasexual hybrid protoplasts containing heteroplasmic cytoplasm and fused parent nuclei. This technique induces genetic recombination in several of prokaryotic and eukaryotic species. It produces inter-specific and even inter-generic hybrids and is hence an important tool of genetic manipulation as it breaks down barriers of genetic exchange imposed by natural mechanisms.	Power and Davey (1990), Ryu et al. (1983)

improving oxidative stability of alkaline amylase produced in *Alkalimonas amylolytica* for use in detergent and textile industries (Yang et al. 2012); 4-fold increase in k_{cat} and 2.5-fold improvement in hydrolytic activity on cellulosic substrates of endoglucanase produced in *Thermoascus aurantiacus* for bioethanol production (Srikrishnan et al. 2012); increasing substrate specificity of D-glucose 1-dehydrogenase isozymes expressed in *Bacillus megaterium* for measurements of blood glucose level (Nishioka et al. 2012); making superoxide dismutase that scavenges of O_2^- thermostable and expressed in *Potentilla atrosanguinea* (Kumar et al. 2012); and enhanced k_{cat}/K_m and k_{cat} values by 5.3- and 6.9-fold of β-glucosidase expressed in *Trichoderma reesei* for hydrolysis of cellobiose and cellodextrins (Lee et al. 2012).

Random mutagenesis, gene shuffling, directed evolution, and screening have been successfully employed in *Thermotoga neapolitana* (high activity on glucose at low temperature and low pH for xylose isomerase and 2.3-fold increases in catalytic efficiency used in preparation of high fructose syrup, Sriprapundh et al. 2003), *Neisseria polysaccharea* (fivefold increased activity of amylosucrase for synthesis or the modification of polysaccharides, Van der Veen et al. 2004), *P. aeruginosa* (twofold increase in amidase activity of lipase in understanding lipase inability to hydrolyze amides, Fujii et al. 2005), *Pyrococcus horikoshii* (improving thermostability of prolidase for detoxification of organophosphorus nerve agents, Theriot et al. 2011), and *Symbiobacterium toebii* (improved thermal stability and activity of tyrosine phenol-lyase in industrial production of l-tyrosine and its derivatives, Rha et al. 2009).

Directed evolution and site-directed mutagenesis have been used to improve thermostability and catalytic efficiency of fructosyl peptide oxidase (*Coniochaeta* sp.) in clinical diagnosis, subtilase (*Bacillus* sp.) in detergent additives and food processing, and xylanase XT6 (*Geobacillus stearothermophilus*) in degradation of hemicellulose (Hirokawa et al. 2008; Zhang et al. 2010; Zhong et al. 2009).

11.4.3 Metabolic Pathway Engineering for Enhanced Enzyme Production

Metabolic pathway engineering is a combination of two words: metabolism and engineering. It may be referred to the specific alteration of a metabolic pathway for better understanding of cellular network or for enhanced production of cellular metabolites (Lessard 1996). The pathways are modified either by deliberative change in specific biochemical reactions or by addition of new genes by genetic engineering (Stephanopoulos 1999).

Strain improvement is a fundamental process in the development of various industrial metabolites. Traditional methods employed for strain improvement include induced and site-directed mutagenesis, improvement of cultural conditions, and genetic recombination. The other less exploited but an equally efficient process is metabolic engineering where product yield is enhanced by recombinant DNA technology (Yang et al. 1998). Metabolic pathway engineering is a multidisciplinary science that involves use of various biological fields such as biochemistry, chemical engineering, computational biology, bioinformatics, and molecular biology with the ultimate goal of increasing the production of a particular biological metabolite. Various microbes act as miniature cell factories for industrial production of these metabolites, viz., enzymes, biosurfactants, antibiotics, and other therapeutics. These microbial cell factories can be modified by pathway engineering for overproduction of industrial metabolites (Farmer and Liao 2000; Gupta 2007; Yazdani and Gonzalez 2008).

In natural conditions, microbes produce metabolites as per cellular need, and the yield is significantly less to meet industrial requirement. In such conditions, the metabolite yield can be enhanced by detailed biochemical pathway analysis to identify the steps or genes where the alteration can increase the metabolite yield (Fig. 11.1). The technology involved is robust and uses various methods of pathway modification.

The microbial enzymes have widespread applications in various fields such as food, pharmaceutical, leather, detergent, paper-pulp, chemical protectant, and other manufacturing industries. Their production processes are being modified since long using conventional as well as advanced methods.

If the genes involved in enzyme production pathway are unknown, the conventional methods are used such as mutagenesis or modification of cultural conditions (Adrio and Demain 2006), whereas, if the genes are known, the enzymes can be overproduced by gene upregulation or downregulation using rDNA technology (Stephanopoulos et al. 1998; Adrio and Demain 2010). If producer microorganism cannot be efficiently modified, the gene of interest can be overexpressed in other microbes in which genetic alteration can be easily achieved (Stephanopoulos et al. 1998; Keasling 2012).

Metabolic pathway engineering for enhanced enzyme production involves the identification of coding genes. For gene identification, genomic data of the producer

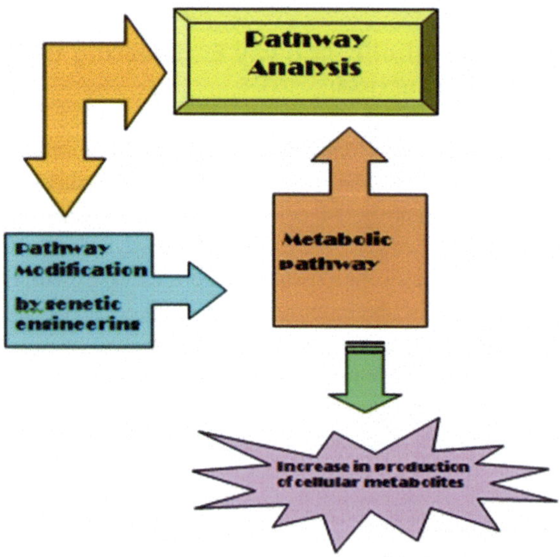

Fig. 11.1 Overview of metabolic pathway modification

microorganism should be known in addition to the molecular weight and N-terminal amino acid sequence of enzyme. Genomic data is mined from databases and molecular weight; amino acid sequence of the purified protein is worked. After gene identification, the gene is overexpressed in either producer microorganism or another host microbe (Demain and Vaishnav 2009).

Metabolic pathway engineering involves four strategies for overproduction of enzymes and other biomolecules:

(i) The first and most common strategy is to increase expression of genes involved in metabolite production. However, the results are always not satisfactory; this may be due to the involvement of multiple genes coding for multiple enzymes in a pathway. One enzyme with minimum reaction rate regulates whole of the pathway, thus acting as bottleneck step. If a gene is upregulated that does not code for enzyme at bottleneck step, the overall effect will be negligible. Therefore, it is required to increase expression of each gene involved in the biosynthesis of enzyme. However, this is difficult to achieve because it is time-consuming and labor-intensive.

(ii) The second strategy involves either knocking down or downregulating a gene coding for enzymes that degrades or transforms the required enzyme into another form. In *E. coli*, large amount of fatty acid production was achieved by using the first strategy of overexpression of metabolic pathway genes and also by knockout of genes involved in degradation (Steen et al. 2010).

(iii) The third strategy involves increased production of various coenzymes such as NADH, NADPH, FAD, and Co-A which are actively involved in the biosynthesis of metabolites. If these coenzymes are not produced in sufficient concentration, the final end product will not be produced in significant concentration. Therefore, coenzyme production is increased by change in genetic makeup of producer microorganism. An increase in the final yield of fatty acids was observed by increasing the intracellular NADPH molecules used in fatty acid biosynthesis. This was achieved by overexpression of the malic enzyme (ME) gene in *Mucor circinelloides* (Zhang et al. 2007).

(iv) The fourth strategy involves release of bio-metabolites extracellularly. Increased intracellular accumulation results in stressing of producer strain; moreover, it can pose growth-inhibiting effect on the host microbe. Therefore, if microbe is made to discharge bio-metabolite extracellularly by change in genetic makeup or by improved cultural conditions, the target biomolecule will continue to be produced, because the cell is free from stress posed by higher intracellular accumulation. Various enzyme-producing microbes have been modified by pathway engineering and overexpression of genes. Overexpression is achieved in same most microbes, or if it is difficult to achieve in similar host, heterologous overexpression is used. Bacteria such as *Bacillus subtilis* and *Lactobacillus lactis*; yeasts such as *Saccharomyces cerevisiae*, *Pichia pastoris*, *Hansenula polymorpha*, and *Yarrowia lipolytica*; and fungi such as *Aspergillus* and *Trichoderma* are commonly used for heterologous overexpression for increased metabolite production (Liu et al. 2013).

11.5 Conclusion and Future Perspectives

It has now been vividly established that microorganisms contain a vast repertoire of enzymes that have a variety of applications in different industries like household products, food, animal feed, leather processing industry, technical industries, fine chemicals, healthcare, and pharmaceuticals. The novel properties of enzymes like their very high specificity toward their substrate, high turnover number, high catalytic efficiency, and biodegradability and reusability authorize enzyme-aided processes in these industries to operate under mild reaction parameters, with refined yields and reduction in toxic by-products or waste generation.

The naturally occurring enzymes in their native forms are not appropriate for these biocatalytic reactions without further modification or restructuring of the enzymes so as to refine their specific activity or other key enzymatic attributes.

Recent advancements in genomics, metagenomics, proteomics, efficient expression systems, and rDNA technologies along with metabolic pathway engineering have stimulated the mining of new microbial enzymes from nature or by designing enzymes with refined catalytic features.

The current progress in enzyme technology will definitely pave a way for success in the arena of industrial biocatalysis and biotransformation. The coming years will show a lot of exciting and interesting developments in the area of biotransformations. The future investigations will use combinations of both engineered and de novo designed enzymes associated with chemistry to develop more substrates and materials from cheaper as well as renewable resources that will definitely contribute to establish a bio-based economy and to achieve sustainable development for the betterment of mankind and other lives on this planet.

References

Adeleye T, Kareem S, Dairo O, Atanda O (2019) Studies on amylase from protoplast fusants of *Aspergillus* species using response surface methodology. Access Microbiol 1(1A)

Adrio JL, Demain AL (2006) Genetic improvement of process yielding microbial products. FEMS Microbiol 30(2):187–214

Adrio JL, Demain AL (2010) Recombinant organisms for production of industrial products. Bioeng Bug 1(2):116–131

Adrio JL, Demain AL (2014) Microbial enzymes: tools for biotechnological processes. Biomol Ther 4(1):117–139

Agrawal R, Satlewal A, Verma AK (2013) Development of a β-glucosidase hyperproducing mutant by combined chemical and UV mutagenesis. 3 Biotech 3:381

Ahmed N (2009) A flood of microbial genomes—do we need more? PLoS One 4:1–5

Aleem B, Muhammad HR, Zeb N, Saqib A, Ihsan A, Iqbal M, Ali H (2018) Random mutagenesis of super Koji (*Aspergillus oryzae*): improvement in production and thermal stability of α-amylases for maltose syrup production. J BMC Microbiol 18:1471–2180

Alexandrova AN, Rothlisberger D, Baker D, Jorgensen WL (2008) Catalytic mechanism and performance of computationally designed enzymes for Kemp elimination. J Am Chem Soc 130:15907–15915

Ali S, Mahmood S (2019) Mutagenesis of a thermophilic *Alkalibacillus flavidus* for enhanced production of an extracellular acetyl Xylan esterase in semi-solid culture of linseed meal. Waste Biomass Valoriz. https://doi.org/10.1007/s12649-019-00665-2

Ameri A, Shakibaie M, Soleimani-Kermani M, Faramarzi MA, Doostmohammadi M, Forootanfar H (2019) Overproduction of thermoalkalophilic lipase secreted by *Bacillus atrophaeus* FSHM2 using UV-induced mutagenesis and statistical optimization of medium components. Prep Biochem Biotechnol 49:184–191

Anitha A, Rebeeth M (2009) Self-fusion of *Streptomyces griseus* enhances chitinase production and biocontrol activity against *Fusarium oxysporum* F. Sp. *lycopersici*. Biosci Biotechnol Res Asia 6:175–180

Atomi H, Sato T, Kanai T (2011) Application of hyperthermophiles and their enzymes. Curr Opin Biotechnol 22:618–626

Auerbach C (1976) Mutation research: problems, results and perspectives. Chapman and Hall, London

Azin M, Noroozi E (2001) Random mutagenesis and use of 2-deoxy-D-glucose as an antimetabolite for selection of α-amylase-overproducing mutants of *Aspergillus oryzae*. World J Microbiol Biotechnol 17:747

Bailey JE, Sburlati A, Hatzimanikatis V, Lee K, Renner WA, Tsai PS (1996) Inverse metabolic engineering: a strategy for directed genetic engineering of useful phenotypes. Biotechnol Bioeng 52(1):109–121

Bakare MK, Adewale IO, Ajayi A, Shonukan OO (2005) Purification and characterization of cellulase from the wild-type and two improved mutants of *Pseudomonas fluorescens*. Afr J Biotechnol 4:898–904

Baweja M, Nain L, Kawarabayasi Y, Shukla P (2016) Current technological improvements in enzymes toward their biotechnological applications. Front Microbiol 7:965

Bayer S, Birkemeyer C, Ballschmiter M (2011) A nitrilase from ametagenomic library acts regioselectively on aliphatic dinitriles. Appl Microbiol Biotechnol 89:91–98

Ben Mabrouk S, Aghajari N, Ben Ali M, Ben Messaoud E, Juy M, Haser R, Bejar S (2011) Enhancement of the thermostability of the maltogenic amylase MAUS149 by Gly312Ala and Lys436Arg substitutions. Bioresour Technol 102:1740–1746

Bhatia K, Mal G, Bhar R, Jyoti AC, Seth A (2018) Purification and characterization of thermostable superoxide dismutase from *Anoxybacillus gonensis* KA 55 MTCC 12684. Int J Biol Macromol 117:1133–1139

Bischof RH, Ramoni J, Seiboth B (2016) Cellulases and beyond: the first 70 years of the enzyme producer *Trichoderma reesei*. Microb Cell Factories 15:106

Borgi I, Gargouri A (2014) Investigations on a hyper-proteolytic mutant of *Beauveria bassiana*: broad substrate specificity and high biotechnological potential of a serine protease. FEMS Microbiol Lett 351:23–31

Bustos-Jaimes I, Mora-Lugo R, Calcagno ML, Farres A (2010) Kinetic studies of Gly28:Ser mutant form of *Bacillus pumilus* lipase: changes in kcat and thermal dependence. BiochimBiophysActa 1804:2222–2227

Cedrone F, Ménez A, Quéméneur E (2000) Tailoring new enzyme functions by rational redesign. Curr Opin Struct Biol 10(4):405–410

Chand P, Aruna A, Maqsood AM, Rao LV (2005) Novel mutation method for increased cellulase production. J App Microbiol 98:318–323

Chung D, Potter SC, Tanomrat AC, Ravikumar KM, Toney MD (2017) Site-directed mutant libraries for isolating minimal mutations yielding functional changes. Protein Eng Des Sel 30:347–357

Contesini FJ, de Melo RR, Sato HH (2018) An overview of *Bacillus* proteases: from production to application. Crit Rev Biotechnol 38:321–334

Cowan D (1996) Industrial enzyme technology. Trends Biotechnol 14:177–178

de Paiva DP, de Oliveira SS, Mazotto AM et al (2019) Keratinolytic activity of *Bacillus subtilis* LFB-FIOCRUZ 1266 enhanced by whole-cell mutagenesis. 3 Biotech 9:2

Demain AL, Adrio JL (2008) Contributions of microorganisms to industrial biology. Mol Biotechnol 38:41–45

Demain AL, Vaishnav P (2009) Production of recombinant proteins by microbes and higher organisms. Biotechnol Adv 27:297–306

Dillon AJP, Zorgi C, Camassola M, Henriques JA (2006) Use of 2-deoxyglucose in liquid media for the selection of mutant strains of *Penicillium echinulatum* production increased cellulase increased cellulase and β-glucosidase activities. Appl Microbiol Biotechnol 70:740–746

Dumon C, Varvak A, Wall MA, Flint JE, Lewis RJ, Lakey JH, Morland C, Luginbuhl P, Healey S, Todaro T et al (2008) Engineering hyperthermostability into a GH11 xylanase is mediated by subtle changes to protein structure. J Biol Chem 283:22557–22564

Dunne C, Moënne-Loccoz Y, de Bruijn F, O'Gara F (2000) Overproduction of an inducible extracellular serine protease improves biological control of *Pythium ultimum* by *Stenotrophomonas maltophilia* strain W81. Microbiology 146:2069–2078

Dutta JR, Banerjee R (2006) Isolation and characterization of a newly isolated *Pseudomonas* mutant for protease production. Braz Arch Biol Technol 49:37–47

Ebert M, Laaß S, Burghartz M, Petersen J, Koßmehl S, Wöhlbrand L, Rabus R, Wittmann C, Tielen P, Jahn D (2013) Transposon mutagenesis identified chromosomal and plasmid genes essential for adaptation of the marine bacterium *Dinoroseobacter shibae to* anaerobic conditions. J Bacteriol 195:4769–4777

Errol CF, Graham CW, Wolfram S, Richard DW, Roger AS, Tom E (2006) DNA repair and mutagenesis, 2nd edn. ASM Press, Sterling

Fang Z, Sha C, Peng Z, Zhang J, Du G (2019) Protein engineering to enhance keratinolytic protease activity and excretion in Escherichia coli and its scale-up fermentation for high extracellular yield. Enzym Microb Technol 121:37–44

Farmer WR, Liao JC (2000) Improving lycopene production in Escherichia coli by engineering metabolic control. Nat Biotechnol 18:533–537

Fenel F, Zitting AJ, Kantelinen A (2006) Increased alkali stability in *Trichoderma reesei* endo-1,4-β-xylanase II by site directed mutagenesis. J Biotechnol 121:102–107

Frushicheva MP, Cao J, Chu ZT, Warshel A (2010) Exploring challenges in rational enzyme design by simulating the catalysis in artificial kempeliminase. Proc Natl Acad Sci U S A 107:16869–16874

Frushicheva MP, Mills MJ, Schopf P, Singh MK, Prasad RB, Warshel A (2014) Computer aided enzyme design and catalytic concepts. Curr Opin Chem Biol 21:56–62

Fujii R, Nakagawa Y, Hiratake J, Sogabe A, Sakata K (2005) Directed evolution of *Pseudomonas aeruginosa* lipase for improved amide-hydrolyzing activity. Protein Eng Des Sel 18:93–101

Gai Y, Chen J, Zhang S, Zhu B, Zhang D (2018) Property improvement of α-amylase from *Bacillus stearothermophilus* by deletion of amino acid residues arginine 179 and glycine 180. Food Technol Biotechnol 56:58–64

Gerlt JA, Babbitt PC (2009) Enzyme (re)design: lessons from natural evolution and computation. Curr Opin Chem Biol 13(1):10–18

Gomes I, Gomes J, Steiner W (2003) Highly thermostable amylase and pullulanase of the extreme thermophilic eubacterium *Rhodothermus marinus*: production and partial characterization. Bioresour Technol 90:207–214

Goraya A, Asghar F, Javaid I, Ali S (2017) Microbial strain improvement for overproduction of industrial products. J Biol Sci 3:10–29

Gupta PK (2007) Metabolic engineering for over production of metabolites. Elem Biotechnol:458–470

Haq I, Ali S, Javed MM, Hameed U, Saleem A, Adnan F, Qadeer MA (2010) Production of alpha amylase from a randomly induced mutant strain of *Bacillus amyloliquefaciens* and its application as a desizer in textile industry. Pak J Bot 42:473–484

Hassan MM (2014) Influence of protoplast fusion between two Trichoderma spp. on extracellular enzymes production and antagonistic activity. Biotechnol Biotechnol Equip 28(6):1014–1023

Hess M, Sczyrba A, Egan R, Kim TW, Chokhawala H, Schroth G, Luo S, Clark DS, Chen F, Zhang T (2011) Metagenomic discovery of biomass-degrading genes and genomes from cow rumen. Science 331:463–467
Hirokawa K, Ichiyanagi A, Kajiyama N (2008) Enhancement of thermostability of fungal deglycating enzymes by directed evolution. Appl Microbiol Biotechnol 78:775–781
Hou L (2009) Novel methods of genome shuffling in *Saccharomyces cerevisiae*. Biotechnol Lett 31 (5):671–677
Htway HTP, Yu SS, Latt ZK, Yi KP (2018) Improvement of cellulolytic activity in cellulolytic nitrogen-fixing bacteria by transposon mutagenesis. J Bacteriol Mycol 6:147–153
Hu Y, Coates ARM (2005) Transposon mutagenesis identifies genes which control antimicrobial drug tolerance in stationary-phase *Escherichia coli*. FEMS Microbiol Lett 243:117–124
Hwang JK, Warshel A (1987) Semiquantitative calculations of catalytic free-energies in genetically modified enzymes. Biochemistry 26:2669–2673
Illing GT, Normansell ID, Peberdy JF (1989) Protoplast isolation and regeneration in *Streptomyces clavuli-gerus*. J Gen Microbiol 135:2289–2297
James TY, Stenlid J, Olson Å, Johannesson H (2008) Evolutionary significance of imbalanced nuclear ratios within heterokaryons of the basidiomycete fungus *Heterobasidion parviporum*. Evolution 62:2279–2296
Jeon JH, Kim JT, Kim YJ, Kim HK, Lee HS, Kang SG, Kim SJ, Lee JH (2009) Cloning and characterization of a new-cold active lipase from a deep-se sediment metagenome. Appl Microbiol Biotechnol 81:865–874
Jiang C, Shen P, Yan B, Wua B (2009) Biochemical characterization of a metagenome-derived decarboxylase. Enzym Microb Technol 45:58–63
Johannes TW, Zhao H (2006) Directed evolution of enzymes and biosynthetic pathways. Curr Opin Microbiol 9:261–267
Johnson EA (2013) Biotechnology of non-*Saccharomyces* yeasts- the ascomycetes. Appl Microbiol Biotechnol 97:503–517
Kamalambigeswari R, Alagar S, Sivvaswamy N (2018) Strain improvement through mutation to enhance pectinase yield from *Aspergillus niger* and molecular characterization of *Polygalacturonase* gene. J Pharm Sci Res 10(2018):989–994
Karn N, Karn SK (2014) Evaluation and characterization of protease production by *Bacillus* sp. induced by UV mutagenesis. Enzym Eng 3:119
Kaul P, Asano Y (2012) Strategies for discovery and improvement of enzyme function: state of the art and opportunities. Microbiol Biotechnol 5:18–33
Keasling JD (2012) Synthetic biology and the development of tools for metabolic engineering. Metab Eng 14:189–195
Kennedy J, Marchesi JR, Dobson AD (2008) Marine metagenomics: strategies for the discovery of novel enzymes with biotechnological applications from marine environments. Microb Cell Factories 7:27–37
Kenyon CJ (1983) The bacterial response to DNA damage. Trend Biochem Sci 2:84–87
Khattab AA, Bazaraa WA (2005) Screening, mutagenesis and protoplast fusion of *Aspergillus niger* for the enhancement of extracellular glucose oxidase production. J Ind Microbiol Biotechnol 32:289
Khedr MA, Ewais EA, Khalil KMA (2017) Improvement of thermophilic α-amylase productivity through UV mutagenesis and AmyE gene amplification and sequencing. J Innov Pharm Biol Sci 4:58–67
Khersonsky O, Kiss G, Rothlisberger D, Dym O, Albeck S, Houk KN, Baker D, Tawfik DS (2012) Bridging the gaps in design methodologies by evolutionary optimization of the stability and proficiency of designed Kemp eliminase KE59. Proc Natl Acad Sci U S A 109:10358–10363
Kiss G, Rothlisberger D, Baker D, Houk KN (2010) Evaluation and ranking of enzyme designs. Protein Sci 19:1760–1773

Korman TP, Sahachartsiri B, David MC, Grace HL, Beauregard M, Bowie JU (2013) Dieselzymes: development of a stable and methanol tolerant lipase for biodiesel production by directed evolution. J Biotechnol Biofuel 6:1754–6834

Kostyleva EV, Sereda AS, Velikoretskaya IA (2017) Development of schemes of induced mutagenesis for improving the productivity of *Aspergillus* strains producing amylolytic enzymes. Microbiology 86:493

Kostyleva EV, Tsurikova NV, Sereda AS, Velikoretskaya IA, Veselkina TN, Lobanov NS, Shashkov IA, Sinitsyn AP (2018) Enhancement of activity of carbohydrases with endo-depolymerase action in *Trichoderma reesei* using mutagenesis. Microbiology 87:652–661

Kumar A, Singh S (2013) Directed evolution: tailoring biocatalysis for industrial application. Crit Rev Biotechnol 33:365–378

Kumar L, Awasthi G, Singh B (2011) Extremophiles: a novel source of industrially important enzymes. Biotechnology 10:1–15

Kumar A, Dutt S, Bagler G, Ahuja PS, Kumar S (2012) Engineering a thermo-stable superoxide dismutase functional at sub-zero to >50 °C, which also tolerates autoclaving. Sci Rep 2:387:1–387:8

Laroussi M, Leipold F (2004) Evaluation of the roles of reactive species, heat, and UV radiation in the inactivation of bacterial cells by air plasmas at atmospheric pressure. Int J Mass Spectrom 233:81–86

Lee SH, Kim YW, Lee S, Auh JH, Yoo SS, Kim TJ, Kim JW, Kim ST, Rho HJ, Choi JH (2002) Modulation of cyclizing activity and thermostability of cyclodextrin glucanotransferase and its application as an antistaling enzyme. J Agric Food Chem 50:1411–1415

Lee S, Lee DG, Jang MK, Jeon MJ, Jang HJ, Lee SH (2011) Improvement in the catalytic activity of β-agarase AgaA from *Zobellia galactanivorans* by site-directed mutagenesis. J Microbiol Biotechnol 21:1116–1122

Lee HL, Chang CK, Jeng WY, Wang AH, Liang PH (2012) Mutations in the substrate entrance region of β-glucosidase from *Trichoderma reesei* improve enzyme activity and thermostability. Protein Eng Des Sel 25:733–740

Lessard P (1996) Metabolic engineering, the concept coalesces. Nat Biotechnol 14:1654–1655

Li S, Yang X, Yang S, Zhu M, Wang X (2012) Technology prospecting on enzymes: application, marketing and engineering. Comput Struct Biotechnol J 2:1–11

Li S, Wang N, Du ZJ, Chen GJ (2018) Intergeneric hybridization between *Streptomyces albulus* and *Bacillus subtilis* facilitates production of ε-Poly-L-lysine from corn starch residues. Biotechnol Bioprocess Eng 23:580–587

Liu L, Yang H, Shin HD, Chen RR, Li J, Du G (2013) How to achieve high-level expression of microbial enzymes: strategies and perspectives. Bioengineered 4:212–223

Liu Z-Q, Zhang X-H, Xue Y-P, Xu M, Zheng Y-G (2014) Improvement of *Alcaligenes faecalis* Nitrilase by gene site saturation mutagenesis and its application in stereospecific biosynthesis of (R)-(−)-mandelic acid. J Agric Food Chem 62:4685–4694

Luo XJ, Yu HL, Xu JH (2012) Genomic data mining: an efficient way to find new and better enzymes. Enzym Eng 1:104–108

Ma L-J, van der Does HC, Borkovich KA, Coleman JJ, Daboussi M-J, Di Pietro A et al (2010) Comparative genomics reveals mobile pathogenicity chromosomes in *Fusarium*. Nature 464:367–373

Ma Y, Shen W, Chen X (2016) Significantly enhancing recombinant alkaline amylase production in *Bacillus subtilis* by integration of a novel mutagenesis-screening strategy with systems-level fermentation optimization. J Biol Eng 10:13

Maloy SR, Cronan JE, Freifelder D (1994) Microbial genetics, 2nd edn. Jones and Bartlett Publishers, Burlington

Marek SM, Wu J, Glass NL, Gilchrist DG, Bostock RM (2003) Nuclear DNA degradation during heterokaryon incompatibility in *Neurospora crassa*. Fungal Genet Biol 40:126–137

May O, Nguyen PT, Arnold FH (2000) Inverting enantioselectivity by directed evolution of hydantoinase for improved production of l-methionine. Nat Biotechnol 18:317–320

Meers JL, Lambert PW (1983) The production of industrial enzymes. Philos Trans R Soc Lond:263–282
Mehtani P, Sharma C, Bhatnagar P (2017) Strain improvement of halotolerant Actinomycete for protease production by sequential mutagenesis. Int J Chem Sci 15:109
Mukhtar H, Haq I (2013) Comparative evaluation of Agroindustrial byproducts for the production of alkaline protease by wild and mutant strains of *Bacillus subtilis* in submerged and solid state fermentation. Sci World J, Article ID 538067:1–6
Mulholland AJ (2008) Computational enzymology: modelling the mechanisms of biological catalysts. Biochem Soc Trans 36:22–26
Muñoz-López M, García-Pérez JL (2010) DNA transposons: nature and applications in genomics. Curr Genomics 11(2):115–128
Nadeem M, Qazi JI, Baig S (2010) Enhanced production of alkaline protease by a mutant of *Bacillus licheniformis* N-2 for dehairing. Braz Arch Bio Technol 53:1015–1025
Ness JE, Del Cardayré SB, Minshull J, Stemmer WP (2000) Molecular breeding: the natural approach to protein design. Adv Protein Chem 55:261–292
Nielsen J (2001) Metabolic engineering. Appl Microbiol Biotechnol 55(3):263–283
Nishioka T, Yasutake Y, Nishiya Y, Tamura T (2012) Structure-guided mutagenesis for the improvement of substrate specificity of *Bacillus megaterium* glucose 1-dehydrogenase IV. FEBS J 279:3264–3275
Ohnishi J, Mitsuhashi S, Hayashi M, Ando S, Yokoi H, Ochiai K, Ikeda M (2002) A novel methodology employing *Corynebacterium glutamicum* genome information to generate a new L-lysine-producing mutant. Appl Microbiol Biotechnol 58(2):217–223
Okanishi M, Suzuki K, Umezawa H (1974) Formation and reversion of streptomycetes protoplasts: culture? Conditions and morphological study. J Gen Microbiol 80:389–400
Pontecorvo G (1956) The parasexual cycle in fungi. Annu Rev Microbiol 10:393–400
Porter JL, Rusli RA, Ollis DL (2016) Directed evolution of enzymes for industrial biocatalysis. Chembiochem 17(3):197–203
Power JB, Davey MR (1990) Protoplasts of higher and lower plants: isolation, culture, and fusion. Methods Mol Biol 6:237–259
Prabavathy VR, Mathivanan N, Sagadevan E, Murugesan K, Lalithakumari D (2006) Self-fusion of protoplasts enhances chitinase production and biocontrol activity in *Trichoderma harzianum*. Bioresour Technol 97:2330–2334
Rao SN, Singh UC, Bash PA, Kollman PA (1987) Free-energy perturbation calculations on binding and catalysis after mutating Asn-155 in subtilisin. Nature 328:551–554
Rao MB, Tanksale AM, Ghatge MS, Deshpande VV (1998) Molecular and biotechnological aspects of microbial proteases. Microbiol Mol Biol Rev 62:597–635
Raper JR (1966) Genetics of sexuality in higher fungi. Ronald Press Company, New York
Read ND, Fleibner A, Roca MG, Glass NL (2010) Hyphal fusion. In: Borkovich KA, Ebbole DJ (eds) Cellular and molecular biology of filamentous fungi. ASM Press, Washington, DC, pp 260–273
Rha E, Kim S, Choi SL, Hong SP, Sung MH, Song JJ, Lee SG (2009) Simultaneous improvement of catalytic activity and thermal stability of tyrosine phenol-lyase by directed evolution. FEBS J 276:6187–6194
Rondon MR, Goodman RM, Handelsman J (1999) The Earth's bounty: assessing and accessing soil microbial diversity. Trends Biotechnol 17:403–409
Rondon MR, August PR, Betterman AD, Brady SF, Grossman TH, Liles MR (2000) Cloning the soil metagenome: a strategy for accessing the genetic and functional diversity of uncultured microorganisms. Appl Environ Microbiol 66:2541–2547
Roper M, Simonin A, Hickey PC, Leeder A, Glass NL (2013) Nuclear dynamics in a fungal chimera. Proc Natl Acad Sci U S A 110:12875–12880
Rothlisberger D, Khersonsky O, Wollacott AM, Jiang L, DeChancie J, Betker J, Gallaher JL, Althoff EA, Zanghellini A, Dym O (2008) Kemp elimination catalysts by computational enzyme design. Nature 453:190–U194

Rowlands RT (1984a) Industrial strain improvement: mutagenesis and random screening procedures. Enzym Microb Technol 6:3–10
Rowlands RT (1984b) Industrial strain improvement: rational screens and genetic recombination techniques. Enzym Microb Technol 6:290–300
Ryu DD, Kim KS, Cho NY, Pai HS (1983) Genetic recombination in *Micromonospora rosaria* by protoplast fusion. Appl Environ Microbiol 45(6):1854–1858
Sadhu S, Ghosh PK, Aditya G, Maiti TK (2014) Optimization and strain improvement by mutation for enhanced cellulase production by *Bacillus* sp. (MTCC10046) isolated from cow dung. J King Saud Univ Sci 26:323–332
Santos CSS, Stephanopoulos G (2008) Combinatorial engineering of microbes for optimizing cellular phenotype. Curr Opin Chem Biol 12:168–176
Schiraldini C, De Rosa M (2002) The production of biocatalysts and biomolecules from extremophiles. Trends Biotechnol 20:515–521
Shih TW, Pan TM (2011) Substitution of Asp189 residue alters the activity and thermostability of *Geobacillus* sp. NTU 03 lipase. Biotechnol Lett 33:1841–1846
Shin CS, Lee JP, Lee JS, Park SC (2000) Enzyme production of *Trichoderma reesei* Rut C-30 on various lignocellulosic substrates. Appl Biochem Biotechnol 84–86:237–245
Shoji JY, Charlton ND, Yi M, Young CA, Craven KD (2015) Vegetative hyphal fusion and subsequent nuclear behavior in *Epichloë* grass endophytes. PLoS One 10:1–21
Shortle D, DiMaio D, Nathans D (1981) Directed mutagenesis. Annu Rev Genet 15:265–294
Sivakumar U, Kalaichelvan G, Ramasamy K (2004) Protoplast fusion in *Streptomyces* sp. for increased production of laccase and associated ligninolytic enzymes. World J Microbiol Biotechnol 20:563–568
Solís S, Loeza J, Segura G, Tello J, Reyes N, Lappe P et al (2009) Hydrolysis of orange peel by a pectin lyase-overproducing hybrid obtained by protoplast fusion between mutant pectinolytic *Aspergillus flavipes* and *Aspergillus niveus* CH-Y-1043. Enzym Microb Technol 44:123–128
Srikrishnan S, Randall A, Baldi P, Da Silva NA (2012) Rationally selected single-site mutants of the Thermoascus aurantiacus endoglucanase increase hydrolytic activity on cellulosic substrates. Biotechnol Bioeng 109:1595–1599
Sriprang R, Asano K, Gobsuk J, Tanapongpipat S, Champreda V, Eurwilaichitr L (2006) Improvement of thermostability of fungal xylanase by using site-directed mutagenesis. J Biotechnol 126:454–462
Sriprapundh D, Vieille C, Zeikus JG (2003) Directed evolution of *Thermotoganea politana* xylose isomerase: high activity on glucose at low temperature and low pH. Protein Eng 16:683–690
Stanbury PF, Whittaker A, Hall SJ (2016) Principles of fermentation technology, 3rd edn. Butterworth-Heinemann, Oxford
Steen EJ, Kang Y, Bokinsky G, Hu Z, Schirmer A, McClure A (2010) Microbial production of fatty-acid-derived fuels and chemicals from plant biomass. Nature 463:559–562
Stephanopoulos G (1999) Metabolic fluxes and metabolic engineering. Metab Eng 1:1–11
Stephanopoulos GN, Aristidou AA, Nielsen J (1998) Metabolic engineering: principles and methodologies. Academic, San Diego
Strom NB, Bushley KE (2016) Two genomes are better than one: history, genetics, and biotechnological applications of fungal heterokaryons. Fungal Biol Biotechnol 3:4
Sun Y, Yang H, Wang W (2011) Improvement of the thermostability and enzymatic activity of cholesterol oxidase by site-directed mutagenesis. Biotechnol Lett 33:2049–2055
Sunitha K, Park YS, Oh TK, Lee JK (1999) Synthesis of alkaline protease by catabolite repression-resistant *Thermoactinomyces* sp. E79 mutant. Biotechnol Lett 21:155–158
Suribabu K, Lalitha Govardhan T, Hemalatha KPJ (2014) Strain improvement of *Brevibacillus borstelensis* R1 for optimization of α-amylase production by mutagens. J Microb Biochem Technol 6:123–127
Thakur A, Kumar P, Lata J, Devi N, Chand D (2018) Thermostable Fe/Mn superoxide dismutase from Bacillus licheniformis SPB-13 from thermal springs of Himalayan region: purification, characterization and antioxidative potential. Int J Biol Macromol 115:1026–1032

Theriot CM, Semcer RL, Shah SS, Grunden AM (2011) Improving the catalytic activity of hyperthermophilic Pyrococcus horikoshii prolidase for detoxification of organophosphorus nerve agents over a broad range of temperatures. Archaea 2011:565127

Tobe S, Shimogaki H, Ohdera M, Asai Y, Oba K, Iwama M, Irie M (2006) Expression of Bacillus protease (Protease BYA) from *Bacillus* sp. Y in *Bacillus subtilis* and enhancement of its specific activity by site-directed mutagenesis-improvement in productivity of detergent enzyme. Biol Pharm Bull 29:26–33

Tracewell CA, Arnold FH (2009) Directed enzyme evolution: climbing fitness peaks one amino acid at a time. Curr Opin Chem Biol 13(1):3–9

Uchiyama T, Miyazaki K (2009) Functional metagenomics for enzyme discovery: challenges to efficient screening. Curr Opin Biotechnol 20:616–622

Underkofler LA, Barton RR, Rennert SS (1957) Production of microbial enzymes and their applications. Appl Microbiol 6(3):212–221

Ushijima S, Nakadai T (1987) Breeding by protoplast fusion of koji mold, *Aspergillus sojae*. Agric Biol Chem 51:1051–1057

Van den Burg B (2003) Extremophiles as a source for novel enzymes. Curr Opin Microbiol 6:213–218

Van der Veen BA, Potocki-Veronese G, Albenne C, Joucla G, Monsan P, Remaud-Simeon M (2004) Combinatorial engineering to enhance amylosucrase performance: construction, selection, and screening of variant libraries for increased activity. FEBS Lett 560:91–97

Varavallo MA, de Queiroz MV, Lana TG, de Brito ATR, Gonçalves DB, de Araújo EF (2007) Isolation of recombinant strains with enhanced pectinase production by protoplast fusion between *Penicillium expansum* and *Penicillium griseoroseum*. Braz J Microbiol 38:52–57

Verma R, Schwaneberg U, Roccatanoa D (2012) Computer-aided protein directed evolution: a review of web servers, databases and other computational tools for protein engineering. Comput Struct Biotechnol J 2:e201209008

Wang XC, Zhao HY, Liu G, Cheng XJ, Feng H (2016) Improving production of extracellular proteases by random mutagenesis and biochemical characterization of a serine protease in *Bacillus subtilis* S1-4. Genet Mol Res 15

Warshel A (1991) Computer modeling of chemical reactions in enzymes and solutions. Wiley, New York

Weichert M, Fleibner A (2015) Anastomosis and heterokaryon formation. In: van den Berg MA, Maruthachalam K (eds) Genetic transformation systems in fungi. Springer, Cham, pp 3–21

Weng M, Deng X, Bao W, Zhu L, Wu J, Cai Y, Jia Y, Zheng Z, Zou G (2015) Improving the activity of the subtilisin nattokinase by site-directed mutagenesis and molecular dynamics simulation. Biochem Biophys Res Commun 465:580–586

Wong TS, Roccatano D, Schwaneberg U (2007) Steering directed protein evolution: strategies to manage combinatorial complexity of mutant libraries. Environ Microbiol 9(11):2645–2659

Xu JZ, Zhang WG (2016) Strategies used for genetically modifying bacterial genome: site-directed mutagenesis, gene inactivation, and gene over-expression. J Zhejiang Univ Sci B 17:83–99

Xu F, Wang J, Chen S, Qin W, Yu Z, Zhao H, Xing X, Li H (2011) Strain improvement for enhanced production of cellulase in *Trichoderma viride*. Appl Biochem Microbiol 47:53–58

Yang YT, Bennet GN, San KY (1998) Genetic and metabolic engineering. Electron J Biotechnol 1 (3):134–141

Yang H, Liu L, Wang M, Li J, Wang NS, Du G, Chen J (2012) Structure-based engineering of methionine residues in the catalytic cores of alkaline amylase from *Alkalimonas amylolytica* for improved oxidative stability. Appl Environ Microbiol 78:7519–7526

Yazdani SS, Gonzalez R (2008) Engineering *Escherichia Coli* for the efficient conversion of glycerol to ethanol and coproducts. Metab Eng 10:340–351

Yi ZL, Pei XQ, Wu ZL (2011) Introduction of glycine and proline residues onto protein surface increases the thermostability of endoglucanase CelA from *Clostridium thermocellum*. Bioresour Technol 102:3636–3638

Yoneda Y (1980) Increased production of extracellular enzymes by the synergistic effect of genes introduced into *Bacillus subtilis* by stepwise transformation. Appl Environ Microbiol 39 (1):274–276

Yoneda Y, Maruo B (1975) Mutation of *Bacillus subtilis* causing hyper-production of α-amylase and protease, and its synergistic effect. J Bac 124:48–54

Zhang YX, Perry K, Vinci VA, Powell K, Stemmer WP, del Cardayré SB (2002) Genome shuffling leads to rapid phenotypic improvement in bacteria. Nature 415(6872):644–646

Zhang Y, Adams IP, Ratledge C (2007) Malic enzyme: the controlling activity for lipid production? Overexpression of malic enzyme in *Mucor circinelloides* leads to a 2.5-fold increase in lipid accumulation. Microbiology 153:2013–2025

Zhang ZG, Yi ZL, Pei XQ, Wu ZL (2010) Improving the thermostability of *Geobacillus stearothermophilus* xylanase XT6 by directed evolution and site-directed mutagenesis. Bioresour Technol 101:9272–9278

Zhang X, Zhang X-F, Li H-P, Wang L-Y, Zhang C, Xing X-H, Bao C-Y (2014) Atmospheric and room temperature plasma (ARTP) as a new powerful mutagenesis tool. Appl Microbiol Biotechnol 98:5387–5396

Zhang J, Shi H, Xu L, Zhu X, Li X (2015) Site-directed mutagenesis of a hyperthermophilic endoglucanase Cel12B from *Thermotoga maritima* based on rational design. PLoS One 10(7): e0133824

Zhao H, Arnold FH (1997) Optimization of DNA shuffling for high fidelity recombination. Nucleic Acids Res 25(6):1307–1308

Zheng R-C, Ruan L-T, Ma H-Y, Tang X-L, Zheng Y-G (2016) Enhanced activity of *Thermomyces lanuginosus* lipase by site-saturation mutagenesis for efficient biosynthesis of chiral intermediate of pregabalin. Biochem Eng J 113:12–18

Zhong CQ, Song S, Fang N, Liang X, Zhu H, Tang XF, Tang B (2009) Improvement of low-temperature caseinolytic activity of a thermophilic subtilase by directed evolution and site-directed mutagenesis. Biotechnol Bioeng 104:862–870

Zou Z, Zhao Y, Zhang T, Xu J, He A, Deng Y (2018) Efficient isolation and characterization of a cellulase hyperproducing mutant strain of *Trichoderma reesei*. J Microbiol Biotechnol 28:1473–1481

Microbial Enzymes from In Vitro to Market 12

Ashish Vyas and Abdulhadi Yakubu

Abstract

A number of chemical transformations sustaining life are mediated by enzymes. All metabolic processes in the system are accelerated by these macromolecules. Microorganisms producing complex enzyme systems are pivotally found in all environments, where they exist in consortia with other microorganisms. In laboratory, optimum synergism amongst enzymatic systems and condition needs to be explored for complete harnessing of its application. Last four to five decades have shown exploration of research with promising insight into the working mechanism of microbial enzymes. The requirement for particular enzymatic formulations is growing more rapidly than ever before, and this demand has become the driving force for research on enzymatic systems. This chapter focuses on the applications, commercialization and market outreach of microbial enzymes from in vitro to market.

Keywords

Industrial utilization · Microbial enzymes · Sustainable environment

A. Vyas (✉)
Department of Microbiology and Biochemistry, School of Bioengineering and Biosciences, Lovely Professional University, Phagwara, Punjab, India
e-mail: ashish.vyas@lpu.co.in

A. Yakubu
Department of Microbiology and Biochemistry, School of Bioengineering and Biosciences, Lovely Professional University, Phagwara, Punjab, India

Department of Science Laboratory Technology, College of Science and Technology, Jigawa State Polytechnic, Dutse, Jigawa State, Nigeria

S. G. Sharma et al. (eds.), *Microbial Diversity, Interventions and Scope*,
https://doi.org/10.1007/978-981-15-4099-8_12

12.1 Introduction

Microbial enzymes are used in various eco-friendly purposes. These microbial enzymes can expedite and fasten up reactions by forming transitional complexes states and reduce the activation energy of the reactions. Microbial enzymes are widely used in large number of fields like medicinal products, cosmetics, manufacturing of food and feedstuffs. These microbial catalysts are applied in cleansers in detergents, for pulp and paper applications, in textile applications, for fuel generation and for the creation of pharmaceuticals and chiral substances. Microbial feed enzymes are used in degrading specific components of feed which otherwise would have been harmful or of no value to mulching animals. Microbial enzymes is also helpful in medical field by preventing the spreading of septicaemia such as in healing of wound, lysis of vein thrombosis and acute therapy for myocardial infarction, in diagnosis and curing of different types of leukaemia. This chapter is centred on three pivotal points. Primarily, the focus is given to the latest scenario about the applications of enzyme in the field of food and feed industry, cosmetics and food processing followed by characteristics of enzymes, and sources are discussed. Lastly, enzyme companies involved in large-scale production with highlights, features and commercial brand names have been discussed. In spite of the fact that enzyme formulations have been utilized by humankind over a long history, leaps forward are expected to expand their uses in more extensive applications with progressive unrivalled execution. Moreover, endeavours are made to draw a clear situation about the industrial structure of worldwide enzyme market regarding India.

12.2 History of Microbial Enzymes

Dr. Christian Hansen, a Danish chemist, in 1874 extracted the enzyme rennet from saline solution from dried calf stomach. Since then a lot of exploration and elucidation related to microbial enzymes have occurred. French scientist Louis Pasteur in the eighteenth century reported the fermentative activity of microorganisms. The main utilization of cell-free proteins was the utilization of rennin aspartic protease characterized from calf or sheep stomach in cheddar making. The principal commercial enzyme (trypsin) was set up by Rohn in Germany in 1914, isolated and separated from animals and utilized in detergents to degrade proteins. This enzyme trypsin has shown extraordinary results that original small packet size made the German housewives suspicious, so the product had to be reformulated and sold in larger packages. The innovation of microbial proteases has revolutionized the production of detergent industries. Novozymes is a global biotechnology located in Denmark, and its application to major detergents started around 1965. The company has shown remarkable performance of laundry detergents with improved stain removal, garment care, wash performance and replacing ingredients like surfactants and builders by more environment-friendly product. A 2.9 billion people's clothes were estimated to be washed per week with a detergent containing enzyme in 2015.

As world current population increased to about 7.4, the need for laundry detergent is now around 40%.

Alpha-amylases and glucoamylases have completely replaced the traditional acid hydrolysis, converting 95% of starch to yield simple monosaccharide sugar. Apart from detergent companies, starch processing industries are the second highest enzyme users. Many different extracellular enzymes such as pectinases, esterase, cellulases and many others were reported to replace conventional (chemical) method of ink removal from waste paper. These enzymatic proteins are accounted for eco-friendly to be earth well-disposed when contrasted with regular strategy (Table 12.1).

12.3 Enzyme Application Based on Fields

Microbial enzymes can be the effective mitigator for solution against set goal no 6, 11 and 12 of UN Sustainable Development Goals (SDGs) and the 2030 Agenda for sustainable development. Many industrial products and manufacturing processes were improved by microbiological enzymatic solutions, thereby saving energy, water and raw materials as well as decreasing waste and emissions. The United Nation goal no. 6 has been planned to ensure the presence and availability along with sustainable utilization and proper management of water and hygienic sanitation for all. Target 6.3 further explains that by 2030, there will be improved water quality by decreasing contamination, taking out dumping and limiting arrival of perilous synthetic concoctions and materials, dividing the extent of untreated waste water and significantly expanding reusing and safe reuse all around. A large number of Novozymes enzymatic arrangements help clients and shoppers spare water during application contrasted with regular strategies. For instance, compounds can be utilized in the material business to consolidate procedures and spare critical measures of water. Different Novozymes arrangements help clients in the mash and paper industry to address lignin lethality in effluents created during the generation procedure. It additionally offers answers for waste water treatment and slop decrease for city and mechanical applications. In China, Novozymes is attempting to understand water difficulties in the south-eastern industrialized region just as in the less grown north-west. Their microbial waste water treatment arrangements have been applied in processing plants in Ningxia, Shanxi, Xinjiang and Inner Mongolia, guaranteeing the consistent release of waste water and improved water accessibility. As a feature of the objectives in China's Water 10 Plan, the organization is as of now cooperating with accomplices to extend utilizations of bio-arrangements into recuperation and protection from dark smelly water bodies. Table 12.2 summarizes the aspects of this hydrolytic inducible enzyme and its usage in industry.

Table 12.1 Biochemical characterization of cellulases enzymes with sources and location

Sample	Location	Organism	Enzyme	pH	Temp (°C)	References
Soil	Pulp and paper industries, India	*Bacillus subtilis*	Cellulase	4.0	60	Pala et al. (2004)
Soil	Macuya rainforest, Pucallpa, Peru	*Aspergillus sp*. LM-HP32, *Penicillium* sp. LM-HP33 and 37	Cellulase	4.8–9.4	28	Vega et al. (2012)
Soil	Iguazu rainfalls, Argentina	*Penicillium sp*. CR-313 and *Penicillium sp*. CR-316	Cellulase	4.5	65	Picart et al. (2007)
Waste paper	USM campus, Penang, Malaysia	*Aspergillus niger*	Cellulase, hemicellulose	6.0	50	Lee et al. (2013)
Agricultural waste	Cairo, Egypt	*Bacillus thuringiensis* MAM-29, MAM-38	Cellulase, xylanase	3–7.6	60–80	Abo-State et al. (2013)
Waste photocopy paper	Medellin, Colombia	NA	Cellulase, amylase	7.0	40	Gil et al. (2013)
Wild herbivore, rain deer	Wayanad, Kerala, India	*Escherichia coli* SD5	Cellulase, xylanase	NA	37–39	Vinod Kumar et al. (2018)
Soil, compost, animal waste slurry	Jeju Island, South Korea	*Bacillus subtilis* C5–16 and S52–2	CMCase, avicelase, xylanase	5.0	50	Kim et al. (2012)
Waste paper	NA	NA	Cutinase, amylase	9–11	50	Wang et al. (2018)
Water	Lonar Lake, Buldhana, Maharashtra, India	Many haloalkaliphilic bacteria	Lipase, amylase, caseinase, cellulose	10.5	23	Kanekar et al. (2008)
Soil	Vellore, Tamil Nadu, India	*Streptomyces sp.*	Xylanase	7.5	37	Kalpana and Rajeswari (2015)

Old newsprint, magazine, inkjet, xerox	Chandigarh, Punjab, India	*Bacillus halodurans* FNP135	Xylanase	8–9.5	65	Virk et al. (2013)
Soil	Ambala Cantt., Haryana, India	*Bacillus pumilus*	Xylanase	6–11	60	Nagar et al. (2012)
Soil	Tianshan Xinjiang, China	*Streptomyces rameus* L2001	Xylanase	5–8	70	Li et al. (2010)
Industrial effluents	Shreyan paper industry, Ahmedgarh, Punjab India	*Aspergillus nidulans* KK-99	Xylanase	8–8.5	55	Taneja et al. (2002)
Compost pit	BREC Sadra, Gujarat, India	*Bacillus altitudinis* DHN8	Xylanase	8.0	45–55	Adhyaru et al. (2017)
Waste paper	Chandigarh, Punjab, India	*Bacillus halodurans*	Xylanase and Laccase	8–9.5	65	Virk et al. (2013)
Soil	Effluents of paper industries, India	*Bacillus pumilus* AJK10414	Xylanase, Pectinase	8.5	55	Singh et al. (2012)

Table 12.2 Industrial use of enzymes

Application fields	Enzyme	Technical benefits
Pulp and paper industry	Amylases	Cleaving starch molecules to reduce the viscosity for surface sizing in coating but not used for dry strength agent additive
	Lipases	Deinking to control pitch in pulping processes
	Cellulases	Improving softness by hydrolysing cellulose fibres, breaking weak points in fibres, increasing the flexibility of fibres
	Mannanases	Degrading the residual glucomannan to increase brightness
	Laccases	Bleaching to increase brightness
	Beta-xylanases	Pulp bleaching process efficiency
Laundry detergents	Proteases	Hydrolysis of protein-based stains in fabrics into soluble amino acids
	Lipases	Collar and cuff cleaning by removing the thick spots of oils
	Cellulases	Separation of the microfibres and fuzz to give glossier appearance
	Amylases	Removing resistant starch residues
Cosmetics industry	Oxidases, peroxidases, polyphenyl oxidases	Hair dyeing
	Protein disulphide isomerases, glutathione sulfhydryl oxidases, transglutaminases	Hair waving
	Papain, bromelain, subtilisin	Gentle peeling effect in skin care
	Amyloglucosidase, glucose oxidases	Toothpastes and mouthwashes
Dairy industry	Chymosin, lipases, lysozymes	Cheese manufacturing
	Beta-galactosidases, lactases	Conversion of lactose to glucose and galactose for lactose intolerance
Juice industry	Amylases and glucoamylases	Breaking down starch into glucose
	Pectinases	Acting on soluble pectin hydrolysis, increase in overall juice production
	Cellulases and hemicellulases	Acting on soluble pectin hydrolysis and on cell wall components of pectinases
	Laccases	Increase the susceptibility of browning during storage
	Naringinases and limoninase	Hydrolysis of narginin, resulting in decreases in bitterness of compounds
Starch processing	Beta-amylases	Cleaving alpha 1, 4 glyosidic bond from non-reducing ends amylose, amylopectin and glycogen

(continued)

Table 12.2 (continued)

Application fields	Enzyme	Technical benefits
	Pullulanases	Attacking alpha 1,6 linkages, liberating straight chain residues of oligosaccharides of glucose residues linked by alpha 1,4 bonds
	Neopullulanases and amylopullulanases	Attacking on both alpha 1,6 and 1,4 linkages
	Beta-amylases	Cleaving alpha 1, 4 glyosidic bond from non-reducing ends amylose, amylopectin and glycogen

12.4 Enzyme-Producing Microbes

12.4.1 Cellulase and Its Industrial Applications

Cellulases of microbial origin are found to be engaged as the significant enzymes based on its multi-isoenzymatic as well as broad applications. Bacteria and fungi mainly reported for this enzyme are *Trichoderma, Thermomonospora, Fusarium* and *Aspergillus*. Research on cellulase enzymes complex and related hydrolases is mainly confined to animal feed, food application, brewery and wine industry. The last 50 years of research witnessed progress in isolation, purifying, characterizing and elucidating the mechanism of action of cellulases from microbes. Recent use of these enzymes in textile, laundry, pulp and paper industries has given a new spurt for screening of new microbes having extremophilic, alkaliphilic and thermotolerent properties for novel industrial applications from lab to market utilization.

12.4.1.1 Industrial Application of Xylanase

This enzyme is a hydrolytic compound which breaks the β-1, 4 spine of the multiplex plant cell wall. Apart from cellulose, xylan is the second largest polysaccharide. Among different clusters of microorganisms that are engaged with breaking down of xylanase include fungi, bacteria, yeast and actinomycetes. In this process, bark removal, grinding and screening of wood usually take place. At that point a chip experiences boiling procedure with the goal that the amount of microbes will reduce drastically. This will be followed by cooling the chips and then inoculated with biopulping organism. The biopulping procedure is savvy and mechanically possible. The fundamental bit of leeway is the abatement in the utilization of vitality just as the expansion in factory utilization. These procedures additionally allow an upgrade of paper and decreased ecological problems. Based on previous research, it was assumed that pre-dyeing strategy of this enzyme is less expensive and eco-accommodating than chemicals. It additionally diminishes the critical measure of synthetic compounds that enjoyed request to get splendour in synthetically blanching procedure. In a customary strategy for paper-making process, the makers utilize unsafe synthetic compounds which confer negative effect to the earth.

A large quantity of xylanase was reported from *Bacillus pumilus* SV-205 under an optimized fermentation conditions. The bacterium secretes maximum amount of cellulase-free xylanase in combination with yeast and peptone which also enhanced highest xylanase production that differ from other combinations. The enzyme maintained a thermal stability of 65% activity after incubation at 60 °C for 2 h (Nagar et al. 2012). Biobleaching capability was also reported from xylanase produced by *Streptomyces* L2001 at a maximum temperature of 70 °C and 5.3 pH (Li et al. 2010) as well as *Bacillus altitudinis* DHN8 (Adhyaru et al. 2017). Using response surface technology, enzyme yield was improved by optimizing submerged fermentation conditions which include incubation time, temperature, agitation speed, sorghum straw, inoculum size and gelatin. Improvement of enzyme production was expressed using response surface methodology (RSM) which gives twofold increase in activity compared to conventional method used for biodeinking and biobleaching.

12.4.1.2 Industrial Application of Laccase

In 1883, Yoshida discovered laccase from Japanese lacquer tree called *Rhus vernicifera* and reported as one of the oldest and most widely reported industrial enzymes. A high quantity of this enzyme was present in fungi and plants including basidiomycetes, white rot and ascomycetes. Among these, white rot and basidiomycetes are found to be involved in the breaking of lignin using enzymes like manganese-dependent peroxidases, peroxidases and laccase. In different ways, this laccase usually participate in cellular process such as plant pathogenesis and sporulation. Increase in pulp strength can be achieved by woody chips pretreated with lignolytic fungi.

Some of the industrial uses of this enzyme include kraft pulp biobleaching. It was reported that 25% decrease of chloride application and 1.8 unit increase in brightness of kraft pulp were observed when SL4 lignocellulotic fungi was used (Kaur and Nigam 2014). Decrease in lignin content in the wood of eucalyptus with its application in biobleaching was found in *C. albidus*. Another vital application of this enzyme is the removal of toxic waste from pulp industries that contain high amount of phenolic compounds. Virk et al. (2013) reported an increase in ink removal of this enzyme in combination with xylanase when some physico-chemical and nutritional parameters were optimized using response surface methodology which is the first time when mediator supplements were not added in this enzyme treatment. A synthetic dye was reported to be removed by alkaline laccase produced from *Myrothecium verrucaria* 24G-4 (Sulistyaningdyah et al. 2004).

12.4.1.3 Industrial Application of Amylase

This enzyme is capable of breaking down starch into different types of products such as dextrin and glucose units. It is found in plant, animals as well as microorganisms.

Moreover, microbial amylases are now used instead of conventional chemical methods in pulp and paper industries because of its low cost. Microbial amylase is the first commercially available enzyme of fungal origin produced in 1894 with therapeutic application in digestive disorder. By the application of ethoxylated fatty acid as surfactant, amylase and cellulase were used for the deinking of waste paper. In this process, temperature of 40 C was applied with floatation consistency of 0.8% in 6 min. The result indicated an increase in brightness as well as reduction of residual ink as compared to control which contained a denatured enzyme (Gil et al. 2013).

The enzyme is basically divided in to endo- and exoamylase. The former hydrolyses oligosaccharides different lengths in a random way, while the latter hydrolyses in a non-reducing end by forming short end products. Some fungi responsible for the production of these industrial enzymes include *Aspergillus* species of *niger, flavus* and *oryzae.* These species have the ability of generating high amounts of this enzyme for commercial purposes.

12.4.1.4 Industrial Application of Lipase

In 1834, J. Eberle was the first person to discover the presence of lipase enzyme in the pancreas, which was also isolated in 1856 by C.I. Bernard. This enzyme has the ability to yield glycerol and fatty acids when working with carboxyl ester found in triacylglycerol under aqueous conditions (Gupta 2004). In the area of biotechnology, microbial lipase plays a vital role because of its versatility, excellent production of large quantity and highly use in different industries. Microorganisms responsible for the production of lipase include fungi such as *Acinotobacter radioresistens, Aeromonas hydrophila* as well as *Aspergillus oryzae* (Andualema and Gessesse 2012). Pitch described hydrophobic contents of wood which happen to be one of the cheap sources of paper. This usually brings more problems in paper processing industries. Moreover, this enzyme can remove pitch from the pulpy deposits at the point of processing paper. Almost 90% of triglycerides found in the pitch can be converted into monoglycerides, glycerol and fatty acids by this enzyme that has low stickiness and high hydrophilic activity (Jaeger and Reetz 1998). This was confirmed by Nippon, a paper industry in Japan, where 90% of woody triglycerides were hydrolysed by this enzyme produced from *Candida rugosa.* Some of the major applications of lipase are the ability to increase the rate of pulping, improve brightness and decrease the level of pollutant from waste water, time and cost of composite. Lipases isolated from *Pseudomonas* sp. (KWI-56) have been found to increase pulp brightness and reduce residual ink concentration, while a thermoalkalophilic *Bacillus coagulans* BTS-3 found to produce lipases was used to remove ink from waste paper when grown at an optimized culture conditions.

12.5 International Market Scenario

In recent year, global industrial enzyme markets are expected to create new opportunities to market players. The market has an estimate of $4.2 billion in 2014, and it was expected to reach $6.2 billion between 2015 and 2020 with almost 7% increase. The major factor for driving growth of global industrial market is obtained from beverage, paper and personal care industry. Approximately, 4000 different enzymes are known currently, and at least 200 of these enzymes are of microbial origin. Moreover, the large-scale industrial production of these microbial enzymes is just 20. However, based on the current investigations of enzyme production biochemistry, recovery methods and process fermentation, a large number of industrial enzymes can be achieved. Novozymes from Denmark, DuPont from the United States and Roche from Switzerland are three major global producers of enzymes with about 75% of the total enzyme production. In addition to these three major producers, a total of 12 major and 400 small industries supplied global enzyme needs presently. Some of the problems encountered by these companies are the market competitiveness, less profit gap as well as technology intensive (Table 12.3).

12.6 Indian Market Scenario

Some of the leading players in the India Enzymes Market are BASF SE, Associated British Foods PLC, Novozymes A/S, Dyadic International Inc., Advance Enzyme Technology Ltd., etc. As per latest survey, India Enzymes Market is anticipated to post robust growth by 2023, owing to the growing application of enzymes for fermentation of milk to produce dairy products like curd, yogurt, cheese, etc. The flourishing dairy industry in India is anticipated to have a positive impact on the demand for industrial enzyme in the coming years. Moreover, the increasing popularity of recombinant enzymes across various industries like detergents, pharmaceuticals, etc. as a result of constant innovations in R&D to increase the yield and improve enzyme specificity as well as stability will propel the growth of the market further. Additionally, the rising use of proteases in detergents coupled with the growing use in animal feed is projected to drive the growth of the market during the forecast period. Based on the type, the protease segment is expected to grow at the highest CAGR by 2023, due to widespread use in food, beverage, detergents and pharmaceuticals. Based on the application, the pharmaceutical segment is expected to lead the market during the forecast period, owing to the rising R&D in recombinant techniques that have enabled in improving the yields of enzymes through fermentation, increased stability and altered specificity. Biotech industry in India is just 2% of the global market, but as the investment opportunity is expanding, it turns more into global visibility. The Indian modern chemical showcase is taking advantage of the interest for simpler plans, helping expanded utilitarian advantages and multi-application profiles. The influx of international companies coupled with expansion of research and production in Indian companies has led to customer's choice with wider inventive products to choose.

Table 12.3 Microbial enzymes – in vitro to market outreach

Enzymes	Industries involved in enzyme production	Commercial brand product	Features	Highlights	Application
Mixed enzymes	Rossari Biotech, Mumbai	Rexsize LHT New Liquid	Mixing of hydrolytic enzymes at high temperature desizing of fabric	A unique mix of several enzymes with efficient fabric desizing	The main type of basic reactions involved in enzymatic desizing: absorption of water, pH buffering, anti-catalytic action – fibre sizes are destabilized by swelling, entry and breaking.
				Biodegrades the starch-based sizes without redeposition	
				Highly concentrated product	
				Size removal is fast	
				Operates over wide temperature range (70–110 °C) and at pH range of 6.5–8.0	Gelatinization: enzyme acts like the scissors working at molecular level – it washes off and disperses the degraded products. Rexsize LHT New Liquid degrades starch into smaller soluble fragments called disaccharides (maltose) by hydrolysis and hence can be easily removed from the fabric.
				Operational by exhaust as well as semi-continuous process	
				Economical in use	
				Eco-friendly operation and non-corrosive to the equipment	
Amylases	Maps Enzymes Limited (Formerly Maps (India) Limited)	Palkozyme	Alpha-amylase for low-medium temperature conventional desizing	–	For fabrics produced using cotton or mixes, the twist strings are covered with a sticky substance known as 'size'; to forestall the strings breaking during weaving. Albeit a wide range of mixes have been utilized to estimate textures; starch and its

(continued)

Table 12.3 (continued)

Enzymes	Industries involved in enzyme production	Commercial brand product	Features	Highlights	Application
					subsidiaries have been the most well-known measuring operator. In the wake of weaving, the size must be expelled again so as to set up the texture for colouring and finishing.
					This procedure (desizing) must be done by treating the texture with synthetic substances, for example, acids, soluble base or oxidizing operators. Anyway starch-breaking compounds (amylases) are favoured for desizing because of their high effectiveness and explicit activity. Amylases achieve total evacuation of the size with no destructive impacts on the texture. Another advantage of enzymes contrasted with solid synthetic concoctions referenced above is that chemicals are conditioned neighbourly.

Cellulases	Maps Enzymes Limited (Formerly Maps (India) Limited)	Palkofeel Palkosoft	Biological polishing with the help of cellulases with mixing of fabric and garment		Cotton and other normal filaments dependent on cellulose can be improved by an enzymatic treatment known as biopolishing. This treatment gives the texture a smoother and glossier appearance. The treatment is utilized to expel 'fluff' – the small strands of fibre that distend from the outside of yarn. A chunk of fluff is known as a 'pill' in the material exchange. After biopolishing, the fluff and pilling are decreased. Different advantages moving fluff are a milder and smoother handle, and prevalent shading brilliance.
Cellulases	Maps Enzymes Limited (Formerly Maps (India) Limited)	Palkowash	In processing of garments with process of biological stonewashing with cellulase	–	Denim finishing Numerous garments of clothing are exposed to a wash treatment to give them a somewhat worn look; model is the stonewashing of denim pants. In the conventional stonewashing process, the blue denim was blurred by the grating activity of pumice stones on the article of clothing surface. These days, denim finishers are utilizing an exceptional cellulase. Cellulase

(continued)

Table 12.3 (continued)

Enzymes	Industries involved in enzyme production	Commercial brand product	Features	Highlights	Application
					works by relaxing the indigo colour on the denim in a procedure known as 'bio-stonewashing'. A little portion of chemical can supplant a few kilograms of pumice stones. The utilization of less pumice stones brings about less harm to article of clothing, machine and less pumice dust in the clothing condition.
					Bio-stonewashing has opened up new potential outcomes in wrapping up denim by expanding the assortment of completions accessible. For instance, it is presently conceivable to blur denim to a more noteworthy degree without risking harming the article of clothing. Efficiency can likewise be expanded in light of the fact that clothing machines contain less stones or no stones and more garments. Maps scope of cellulases offers a for denim finishing, each

					with its own extraordinary properties. These can be utilized either alone or in mix with pumice stones so as to acquire a particular look.
Catalase	Maps Enzymes Limited (Formerly Maps (India) Limited)	Palkoperox	Enzymatically removing the residual hydrogen peroxide after bleaching of cotton with the help of catalase		Bleach clean-up regular textures, for example, cotton, are typically blanched with hydrogen peroxide before colouring. Fades are exceptionally receptive synthetics, and any peroxide left on the texture can meddle with the colouring procedure. A careful 'fade clean-up' is fundamental. The conventional technique is to kill the sanitizer with a lessening specialist, yet the portion must be controlled absolutely. Catalysts present an increasingly advantageous option since they are simpler and speedier to utilize. A little portion of catalase is equipped for separating hydrogen peroxide into water and oxygen. Contrasted and the customary clean-up techniques, the enzymatic procedure brings about cleaner squander water or diminished water utilization.

(continued)

Table 12.3 (continued)

Enzymes	Industries involved in enzyme production	Commercial brand product	Features	Highlights	Application
					Maps offers catalase for expelling remaining hydrogen peroxide after the fading of cotton. It diminishes the flushing important to expel dye, or it very well may be utilized to supplant synthetic medicines.
Multicomponent enzyme	Maps Enzymes Limited (Formerly Maps (India) Limited)	Palkoscour	In the case of native cellulose, bio-scouring needs to be done by partial and complete removal of non-cellulosic fraction with the help of isoenzymic form of cellulases		Bio-scouring.
					Cotton yarn or texture, preceding colouring or printing, experiences various procedures in a material handling unit. A significant procedure is scouring. Right now, cellulosic segments from local cotton are totally or incompletely removed. Scouring gives a texture with a high and even wet capacity, so it very well may be blanched and coloured effectively. Today, profoundly basic synthetic compounds scathing soft drink are utilized for scouring. These synthetic concoctions not just expel the non-cellulosic polluting

					influences from the cotton but in addition assault the cellulose prompting overwhelming quality misfortune and weight reduction in the texture. Besides, utilizing these dangerous synthetic compounds brings about high COD (substance oxygen request), BOD (organic oxygen request) and TDS, in the waste water. Recently another enzymatic scouring process known as 'bio-scouring' is utilized in material wet-preparing in which all non-cellulosic segments from local cotton are totally or halfway evacuated. After this bio-scouring process, the cotton has an unblemished cellulose structure, with lower weight reduction and quality misfortune. The texture gives better wetting and infiltration properties, making consequent blanch process simple and resultantly giving much better colour take-up.

(continued)

Table 12.3 (continued)

Enzymes	Industries involved in enzyme production	Commercial brand product	Features	Highlights	Application
Amylases	Rossari Biotech, Mumbai	Rexsize MHT-A Liquid	Exceptionally viable mix of amylase-based compound for cotton and its mixes with peak activity at 60–65 °C	Good compatibility with chemicals in desizing bath	Starch comprises of two unique polysaccharides (amylose and amylopectin) and is insoluble is water; henceforth, it must be decayed into pieces which break up more effectively. The conditions of enzymatic desizing is as per the following: wetting, pH buffering, hostile to synergist activity; swelling, infiltration, breaking and de-adjustment of size layers; gelatinization; enzyme attack – the compound plays the job of atomic scissors; wash off, scattering of the debasement items. Rexsize MHT-A Liquid corrupts starch into more little solvent pieces called disaccharides (maltose) by hydrolysis and thus can be effectively expelled from the texture.
				Applicable by exhaust as well as pad batch method	
				Highly effective for removal of size	
				Apply at 600C with pH 6–7.5	
				Biodegradable and non-corrosive in nature	
				Economical operation	
Amylases	Rossari Biotech, Mumbai	Rexsize LHT Conc Liquid	Highly effective blend of amylase-based enzyme for cotton and its blend. It has peak activity at 85C	A unique blend of several enzymes for efficient fabric desizing	
				Biodegrades the starch-based sizes without redeposition	
				Highly concentrated product	
				Operates at an optimum temperature of 60–70 ° C and at pH range of 6.5–8.0	
				Operational by exhaust as well as semi-continuous process	
				Imparts soft hand to the fabric	
				Very economical in use	
				Eco-friendly operation	

	Rossari Biotech, Mumbai	BioD –15 Plus Powder	Highly effective blend of amylase-based enzyme for cotton and its blends. It has peak activity at 98 °C	Good compatibility with chemicals in desizing bath	
				Applicable by exhaust as well as pad batch method	
				Highly effective for removal of size with partial scouring property	
				Apply at 60–98 °C with pH 6.5–7.5	
				Biodegradable and non-corrosive in nature	
				Economical operation	
Cellulases	Rossari Biotech, Mumbai	Biofast 50 Liquid	Acid cellulase enzyme for biopolishing with minimum colour loss	Anamorphic stage of *Trichoderma* cellulase having high FPAse activity	Due to the ring dyeing of warp yarn of denim, a faded fashionable look is obtained. These types of washdown effects are created due to the removal of dyes by abrasion on the garments. Washdown effects are achieved due to bio-abrasion where the enzymes act on cellulose, and hydrolysing it gives the desired look. Biofast 50 Liquid attacks the shapeless areas, creating better cutting, great puckering impacts yet with higher back recolouring look at impartial or designed cellulases.
				Hydrolytically cracks the cellulose on the surface of the garment	
				Acts only on crystalline regions ensuring minimum colour loss	
				Imparts soft feel to the fabric	
				Gives clean surface to the fabric	
				Improves drapability of the garments	
				Operates at 55 °C temperature and pH range of 4.5–5.0	
				Eco-friendly operation	

(continued)

Table 12.3 (continued)

Enzymes	Industries involved in enzyme production	Commercial brand product	Features	Highlights	Application
Cellulases	Rossari Biotech, Mumbai	Biofast FNB Conc Liquid	A unique easy-to-use cellulase that allows biopolishing to occur under neutral conditions	Hydrolyses the cellulose on the surface of the fabric	Biofast FNB Conc. Fluid is utilized for bio-cleaning or depilling of cellulosics which improve the texture quality, frequently done after overwhelming handling where pills are raised. Cellulase catalysts debilitate the strands distending from the surface by debasement, ideally of the formless structure of the fibre. The enzyme-weakens the filaments which are touchy to shear powers endless supply of adequate shear the fibre will separate from the surface. This results in improved pilling opposition, more brilliant hues, cleaner surface, improved drapability and expanded delicate quality, decrease in the measure of dead and youthful cotton.
				Imparts soft feel to the garment	
				Its high crystalline type activity cuts good amount of cotton fuzz, giving good lustre and clean surface to the fabric	
				Improves drapability of the garments	
				Operates at 55–60 C temperature and pH 6.5–7.0	
				Eco-friendly operation	

Cellulases	Rossari Biotech, Mumbai	G-Zyme BCS Conc Liquid	A cellulase enzyme preparation to produce a very fast biopolishing effect	Highly concentrated product	Cellulase chemicals are utilized for bio-cleaning or depilling of cellulosics which improve the texture quality, frequently done after substantial preparing where pills are raised. Cellulase proteins debilitate the strands distending from the surface by debasement, ideally of the formless structure of the fibre. The enzymatic action improves pilling opposition, more splendid hues, cleaner surface, improved drapability and expanded non-abrasiveness, decrease in the measure of dead and juvenile cotton. G-Zyme BCS Conc Liquid is biopolishing cellulase yet in addition reasonable for blurring where high aggression is required.
				Fast biopolishing effect	
				Prevents redeposition of fuzz by keeping it in suspension due to additional dispersing property	
				Softens the fabric, with excellent fuzz cutting	
				Imparts clean and fresh look with brightening effect	
				Gives natural and permanent finish on garments	
				Operates at a temperature of 55 °C and pH of 4.5–5.0	
				Eco-friendly operation	

(continued)

Table 12.3 (continued)

Enzymes	Industries involved in enzyme production	Commercial brand product	Features	Highlights	Application
Cellulases	Rossari Biotech, Mumbai	G-Zyme BSL Liquid	New generation, fully formulated engineered enzyme for effective biopolishing as well as fading	High aggression	Due to the ring dyeing of warp yarn of denim, a faded fashionable look is obtained. These types of washdown effects are created due to the removal of dyes by abrasion on the garments. Washdown effects are achieved due to bio-abrasion where the enzymes act on cellulose and hydrolysing it gives the desired look. G-Zyme BSL Liquid attacks the amorphous regions producing better cutting, good puckering effects and as it engineered with its CBD'S knocked off produces far less back staining compared with acid cellulases. Can call this product as high-speed enzyme with neutral looks.
				Very low back staining	
				Good pH and temperature tolerance	
				High economy as fully formulated	
				Good indigo retention with grey cast	
				Excellent fuzz cutting	
				Operates over wider range of temperature 55 to 60 °C and pH 4.5–5.5. Highest activity at 600 C	
				Eco-friendly operation and non-corrosive to the equipment	

Cellulases	Rossari Biotech, Mumbai	Genecel HEBPL CONC Liquid	High activity acid cellulase for biopolishing	Aggressive acid cellulase-based enzymes for biopolishing Hydrolyses the cellulose on the surface of the fabric High cutting giving clean look to the fabric Excellent inner softness and smooth handle Exhibits good drapability to the fabric Specially recommended for biopolishing of hosiery and defibrillation of Tencel Not recommended for use after dyeing as it gives colour loss Operates at a temperature of 55 °C and pH range of 4.5–5.0	Genecel HEBPL CONC Liquid is used for biopolishing or depilling of cellulosics which improve the fabric quality, often done after heavy processing where pills are raised. Cellulase enzymes weaken the fibres protruding from the surface by degradation, preferably, of the amorphous structure of the fibre. The enzyme-weakened fibres are sensitive to shear forces, and upon application of sufficient shear, the fibre will break from the surface. This results in improved pilling resistance, brighter colours, cleaner surface, improved drapability and increased softness, reduction in the amount of dead and immature cotton.
Cellulases	Rossari Biotech, Mumbai	Genecel JNI Liquid	An acid cellulase enzyme for washdown effect on denims	Aggressive acid cellulase-based enzymes for biopolishing Hydrolyses the cellulose on the surface of the fabric High cutting giving clean look to the fabric Excellent inner softness and smooth handle	Genecel JNI Liquid is utilized for bio-cleaning or depilling of cellulosics which improve the texture quality, regularly done after overwhelming handling where pills are raised. Cellulase chemicals debilitate the strands jutting from the surface by debasement, ideally of the nebulous structure of the fibre.

(continued)

Table 12.3 (continued)

Enzymes	Industries involved in enzyme production	Commercial brand product	Features	Highlights	Application
				Exhibits good drapability to the fabric	This enzymatic action results in improved pilling obstruction, more splendid hues, cleaner surface, improved drapability and expanded delicate quality, decrease in the measure of dead and youthful cotton.
				Specially recommended for biopolishing of hosiery	
				Operates at a temperature of 55 C and pH range of 4.5–5.0	
Cellulases	Rossari Biotech, Mumbai	Koldenz CONC Liquid	An acid cellulase for biopolishing at room temperature	Acid cellulase-based enzymes for biopolishing	Koldenz CONC Liquid is used for biopolishing or depilling of cellulosics which improve the texture quality, regularly done after overwhelming handling where pills are raised. Cellulase compounds debilitate the strands projecting from the surface by corruption, ideally of the indistinct structure of the fibre. This enzymatic action results in improved pilling obstruction, more brilliant hues, cleaner surface, improved drapability and expanded delicate quality, decrease in the measure of dead and juvenile cotton.
				Hydrolyses the cellulose on the surface of the fabric	
				High cutting giving clean look to the fabric	
				Excellent inner softness and smooth handle	
				Exhibits good drapability to the fabric	
				Specially recommended for biopolishing of hosiery and defibrillation of Tencel	
				Operates at room temperature and pH range of 4.5–5.0	

Cellulases	Rossari Biotech, Mumbai	Neutrox BDN 100 New Powde	True neutral cellulase enzyme for washdown effect on garments	Neutral cellulase enzyme for washdown effect on garments Minimum back staining Large grain size Excellent colour contrast Applicable at low MLR Operates at 50–60 °C temperature and pH 6.5–7.5	Due to the ring dyeing of warp yarn of denim, a faded fashionable look is obtained. These types of washdown effects are created due to the removal of dyes by abrasion on the garments. Washdown effects are achieved due to bio-abrasion where the enzymes act on cellulose, and hydrolysing it gives the desired look. Bio-washing improves the texture quality, regularly done after overwhelming handling where pills are raised. Cellulase compounds debilitate the strands projecting from the surface by corruption, ideally of the indistinct structure of the fibre. This enzymatic action results in improved pilling obstruction, more brilliant hues, cleaner surface, improved drapability and expanded delicate quality, decrease in the measure of dead and juvenile cotton.
Cellulases	Rossari Biotech, Mumbai	Neutrox Cool Powder	Special cold enzyme of true neutral cellulytic type for fading and biopolishing	True neutral cellulase enzyme designed for washdown effect on garments Neutral biopolish so shade change and colour loss will be less Brighter shades due to combination with surface	Due to the ring dyeing of warp yarn of denim, a faded fashionable look is obtained. These types of washdown effects are created due to the removal of dyes by abrasion on the garments. Washdown effects are achieved due to

(continued)

Table 12.3 (continued)

Enzymes	Industries involved in enzyme production	Commercial brand product	Features	Highlights	Application
				active agents for better dispersion and cleaning of fabric surface	bio-abrasion where the enzymes act on cellulose, and hydrolysing it gives the desired look. Neutrox Cool Powder attacks the amorphous regions producing better cutting, good puckering effects, and being true neutral engineered enzyme, it works at low temperature and gives less back staining compared to acid cellulases and conserves energy. Neutrox Cool Powder is used for biopolishing or depilling of cellulosics which improve the texture quality, frequently done after overwhelming handling where pills are raised. Cellulase catalysts debilitate the strands jutting from the surface by debasement, ideally of the formless structure of the fibre. This enzymatic action results in improved pilling opposition, more splendid hues, cleaner surface, improved drapability and expanded delicate quality, decrease in the measure of dead and youthful cotton.
				For the first time a biopolishing in powder form to be used at room temperature	
				Operates at 30–20 ° C temperature and pH range of 6.0–8.0, peak activity at 30 °C and 6.5pH	
				Easy to handle	
				Saves energy	
				Lower dusting	
				The enzyme bath water is harmless to operators and non-corrosive to equipment	
				Recommended mainly for higher-end garment finish	
				Eco-friendly operation	

Cellulases	Rossari Biotech, Mumbai	Neutrox MKL Liquid	Neutral cellulase enzyme for biopolishing with minimum colour loss	Hydrolyses the cellulose on the surface of the fabric	Neutrox MKL Liquid is used for biopolishing or depilling of cellulosics which improve the texture quality, regularly done after overwhelming preparing where pills are raised. Cellulase catalysts debilitate the filaments projecting from the surface by debasement, ideally of the shapeless structure of the fibre. The compound-debilitated filaments are delicate to shear powers endless supply of adequate shear the fibre will part from the surface. This results in improved pilling opposition, more brilliant hues, cleaner surface, improved drapability and expanded non-abrasiveness, decrease in the measure of dead and juvenile cotton.
				Imparts soft feel to the garment	
				Its high crystalline type activity cuts good amount of cotton fuzz giving good lustre and clean surface to the fabric	
				Improves drapability of the garments	
				Operates at 55–60+ C temperature and pH 6.5–7.0	
				Eco-friendly operation	
Cellulases	Rossari Biotech, Mumbai	Neutrox MKL Super Liquid	Neutral cellulase enzyme for biopolishing with minimum colour loss	First time telescoping of four processes in one: dyeing, scourboosting, peroxide neutralization and biopolishing	Biofast SB-NP Super Liquid has unique feature of boosting scouring efficiency, thereby enhancing penetration of dyestuff and other auxiliaries, resulting into even dyeing in a single step. It also carries out residual hydrogen peroxide removal very effectively along with excellent
				Saving in processing time, energy, water	
				Improved absorbency	
				Uniformity in dyeing with increased dye uptake	

(continued)

Table 12.3 (continued)

Enzymes	Industries involved in enzyme production	Commercial brand product	Features	Highlights	Application
				Effectively carries out residual peroxide killing and biopolishing	biopolishing or depilling of cellulosics which improve the fabric quality, often done after heavy processing where pills are raised.
				Hydrolyses the cellulose on the surface of the fabric	
				Softens the fabric, with excellent fuzz cutting	
				Gives clean surface to the fabric	
				Improves drapability	
				Operates at a temperature of 40–60 °C and pH of 4.5–7.0, peak activity at temperature 55 °C and pH 5.5	
				Eco-friendly operation	
Cellulases	Rossari Biotech, Mumbai	Rexsize Cool Liquid	A liquid desizer for efficient removal of size from fabric	Biodegrades the starch sizes very effectively	Starch consists of two different polysaccharides (amylose and amylopectin) and is insoluble in water; hence, it must be decomposed into fragments which dissolve more easily. The mechanism of enzymatic desizing is as follows: wetting, pH buffering, anti-catalytic action; swelling, penetration, cracking and destabilization of size layers; gelatinization; enzyme attack – the enzyme
				Operates in the temperature range of 60–110 ° C/steaming and at neutral pH	
				Broad temperature-sensitive amylase blend, maintains the desizing efficiency even at lower temperatures	
				Operates by exhaust, semi-continuous and steaming method	

				Eco-friendly operation and non-corrosive to the equipment	takes the role of molecular scissors; wash off, dispersion of the degradation products. Rexsize ECE Liquid degrades starch into smaller soluble fragments called disaccharides (maltose) by hydrolysis and hence can be easily removed from the fabric.
Cellulases	Rossari Biotech, Mumbai	Rexsize LHT 100 Liquid	Blend of enzymes for desizing having peak activity at 80–85 °C	A unique blend of several enzymes for efficient fabric desizing Biodegrades the starch-based sizes without redeposition Highly concentrated product Operates at an optimum temperature of 70–85 C and at pH range of 6.5–8.0 Operational by exhaust as well as semi-continuous process Very economical in use Eco-friendly operation	Starch consists of two different polysaccharides (amylose and amylopectin) and is insoluble in water; hence, it must be decomposed into fragments which dissolve more easily. The mechanism of enzymatic desizing is as follows: wetting, pH buffering, anti-catalytic action; swelling, penetration, cracking and destabilization of size layers; gelatinization; enzyme attack – the enzyme takes the role of molecular scissors; wash off, dispersion of the degradation products. Rexsize LHT 100 Liquid degrades starch into smaller soluble fragments called disaccharides (maltose) by hydrolysis and hence can be easily removed from the fabric.

12.7 Conclusion and Future Prospects

Microbial enzyme field needs more research and development expenditures. Research should also focus in and around isolating novel organisms from unexplored virgin sites across the globe. Establishment of data bank having information on enzyme production will also provide impetus to microbial enzymologist. Discovery of more enzymes from microbial technology will lead enhanced and improved enzyme products with different physiological conditions. Industrial enzymes are pivotal in current commercial status of biotechnology and may lead to more new discoveries of industrial applications in India.

Acknowledgements I thank my research colleagues for providing me valuable guidance and research papers regarding microbial enzymes. My sincere thanks to my departmental colleagues for cordial and jovial research environment.

References

Abo-State" MAM, Ghaly MF, Abdellah EM, Fadel M (2013) Studying the stability of cellulases and xylanase produced by Thermophilic and alkaliphilic bacterial strains isolated from agricultural wastes. Am-Eurasian J Agric Environ Sci 13(11):1568–1575. https://doi.org/10.5829/idosi.aejaes.2013.13.11.11262

Adhyaru DN, Bhatt NS, Modi HA, Divecha J (2017) Cellulase-free-thermo-alkali-solvent-stable xylanase from Bacillus altitudinis DHN8: over-production through statistical approach, purification and bio-deinking/bio-bleaching potential. Biocatal Agric Biotechnol 12:220–227. https://doi.org/10.1016/j.bcab.2017.10.010

Andualema B, Gessesse A (2012) Microbial lipases and their industrial application: review. Biotechnology 11(3):100–118. https://doi.org/10.3923/biotech.2012.100.118

Gil HHA, Dovale SAM, VirneyHadely CL, Munoz OA, Casas BAE, Quintana MGC, Velasquez JJA (2013) Study of the enzymatic/neutral deinking process of waste photocopy paper. O Papel 74(8):61–65

Gupta MR (2004) Bacterial lipases: an overview of production, purification and biochemical properties. Appl Microbiol Biotechnol 64:763–781. https://doi.org/10.1007/s00253-004-1568-8

Jaeger KE, Reetz MT (1998) Microbial lipases form versatile tools for biotechnology. Trends Biotechnol 16(9):396–403. https://doi.org/10.1016/S0167-7799(98)01195-0

Kalpana VN, Rajeswari VD (2015) Production of xylanase from various lignocellulosic waste materials by Streptomyces sp. and its potential role in deinking of newsprint. Asian J Biochem:1–8. https://doi.org/10.3923/ajb.2015

Kanekar PP, Joshi AA, Kelkar AS, Borgave SB, Sarnaik SS (2008) Alkaline Lonar lake, India – a treasure of alkaliphilic and halophilic bacteria. In: Sengupta M, Dalwani R (eds) Proceedings of Taal 2007: the 12th World Lake Conference, pp 1765–1774

Kaur S, Nigam V (2014) Production and application of laccase enzyme in pulp and paper Industry. Int J Res Appl Nat Soc Sci 2(4):153–158

Kim Y-K, Lee S-C, Cho Y-Y, Oh H-J, Ko YH (2012) Isolation of cellulolytic *Bacillus subtilis* strains from agricultural environments. ISRN Microbiol 2012:1–9. https://doi.org/10.5402/2012/650563

Lee CK, Ibrahim D, Che Omar I (2013) Enzymatic deinking of various types of waste paper: efficiency and characteristics. Process Biochem 48(2):299–305. https://doi.org/10.1016/j.procbio.2012.12.015

Li X, She Y, Sun B, Song H, Zhu Y, Lv Y, Song H (2010) Purification and characterization of a cellulase-free, thermostable xylanase from Streptomyces rameus L2001 and its biobleaching effect on wheat straw pulp. Biochem Eng J 52(1):71–78. https://doi.org/10.1016/j.bej.2010.07.006

Nagar S, Mittal A, Kumar D, Gupta VK (2012) Production of alkali tolerant cellulase free xylanase in high levels by Bacillus pumilus SV-205. Int J Biol Macromol 50(2):414–420. https://doi.org/10.1016/j.ijbiomac.2011.12.026

Pala H, Mota M, Gama FM (2004) Enzymatic versus chemical deinking of non-impact ink printed paper. J Biotechnol 108(1):79–89. https://doi.org/10.1016/j.jbiotec.2003.10.016

Picart P, Diaz P, Pastor FIJ (2007) Cellulases from two Penicillium sp. strains isolated from subtropical forest soil: production and characterization. Lett Appl Microbiol 45(1):108–113. https://doi.org/10.1111/j.1472-765X.2007.02148.x

Singh A, Yadav RD, Kaur A, Mahajan R (2012) An ecofriendly cost effective enzymatic methodology for deinking of school waste paper. Bioresour Technol 120:322–327. https://doi.org/10.1016/j.biortech.2012.06.050

Sulistyaningdyah WT, Ogawa J, Tanaka H, Maeda C (2004) Characterization of alkaliphilic laccase activity in the culture supernatant of Myrothecium verrucaria 24G-4 in comparison with bilirubin oxidase. FEMS Microbiol Lett 230:0–5. https://doi.org/10.1016/S0378-1097(03)00892-9

Taneja K, Gupta S, Chander Kuhad R (2002) Properties and application of a partially purified alkaline xylanase from an alkalophilic fungus Aspergillus nidulans KK-99. Bioresour Technol 85(1):39–42. https://doi.org/10.1016/S0960-8524(02)00064-0

Vega K, Villena GK, Sarmiento VH, Ludeña Y, Vera N, Gutiérrez-Correa M (2012) Production of alkaline cellulase by fungi isolated from an undisturbed rain forest of Peru. Biotechnol Res Int 2012:1–7. https://doi.org/10.1155/2012/934325

Vinod Kumar N, Rani ME, Gunaseeli R, Kannan ND (2018) Paper pulp modification and deinking efficiency of cellulase-xylanase complex from Escherichia coli SD5. Int J Biol Macromol 111:289–295. https://doi.org/10.1016/j.ijbiomac.2017.12.126

Virk AP, Puri M, Gupta V, Capalash N, Sharma P (2013) Combined enzymatic and physical deinking methodology for efficient eco-friendly recycling of old newsprint. PLoS One 8(8):1–8. https://doi.org/10.1371/journal.pone.0072346

Wang F, Zhang X, Zhang G, Chen J, Sang M, Long Z, Wang B (2018) Studies on the environmentally friendly deinking process employing biological enzymes and composite surfactant. Cellulose:1–11. https://doi.org/10.1007/s10570-018-1778-3

Part V

Microbes in Bioremediation

Microbial Clean Up Strategy for Polluted Water

13

Amaninder Kaur Riat

Abstract

The clean-up of contamination by biological agents is known as bioremediation. There are various microbes which have an ability to clear up the mess created by humans. Due to industrialization, toxic substances are added to the water inappropriately making it unfit for any other consumption. Many genetically modified organisms are used by many scientists against mercury, oil spills, radioactive wastes, etc. The microbes includes *Rhodotorula rubra*, *Deinococcus-Thermus*, *Cunninghamella elegans*, *Cyathus bulleri* and *Actinobacteria*, *Cyanobacteria*, *Flavobacteria*, etc. It is a better approach to clean the polluted water than other conventional methods as it is much cheaper, easy to handle and feasible for large water bodies. The process is very simple as the microbes will be introduced to the polluted water, and they feed on organic matter, oil spills, etc. to convert them into carbon dioxide and water. The multiplication also occurs in the same polluted water without any external source or efforts. The system does not require any construction or diversion of drainage flow; it does not require any skilled manpower. Microbes also help to restore the quality as well as self-cleaning capacity of water body; some are already existing in the same water as they are indigenous, whereas others can be introduced to the targeted sites. In a country like India, where the sewage system is not well developed, this technique will be highly beneficial as it will be carried out at a place where actually the problem is without the usage of any harmful chemicals.

Keywords

Microbial cleaners · Clean water · Bioremediation · Self-cleaning · Biological agents

A. K. Riat (✉)
Department of Zoology, School of Bioengineering and Biosciences, Lovely Professional University, Phagwara, Punjab, India

S. G. Sharma et al. (eds.), *Microbial Diversity, Interventions and Scope*,
https://doi.org/10.1007/978-981-15-4099-8_13

13.1 Introduction

There are various metals and metalloids that degrade the quality of water, such as arsenic, chromium, lead, uranium, selenium, zinc, mercury, cadmium, silver, gold and nickel. The toxicity of these metals is very high, which leads to hazardous effects on human health; apart from this, they damage the floral and faunal population because of their non-biodegradable nature. Thus, there is an alarming need to introduce the new approaches in current developing treatments to eliminate or reduce the metals present in the water and environment.

Various processes like physicochemical and biological are commonly used to eliminate the heavy metals from the wastewater discharged from industries in the ecosystem (Fomina and Gadd 2014). Methods like ion exchange, evaporation, sorption, precipitation and electrochemical treatment are very expensive, so they cannot be used at every place, and some are harmful for the environment as well (Mulligan et al. 2001; Kadirvelu et al. 2002). When such methods are not working efficiently, bioremediation can sort the need for the removal of metals, even if their concentration is in low amounts. It is an environment-friendly as well as economically feasible option. In this strategy, it is basically working with the binding of the biological agents with the metals, which will remove the heavy metals from the contaminated sites effectively. Furthermore, the microorganisms, we can say, act as a biological tool for the removal of various metals as they clean up and regenerate the degraded water quality of the contaminated aquatic system (Riggle and Kumamoto 2000).

13.1.1 Types of Organisms Used in Bioremediation

Basically, the process of bioremediation is done with the metabolic activity of one organism or a microorganism's consortium (Nicolaou et al. 2010). The food consumed as a toxic heavy metal by the organism, bioremediation works in a concept of non-beneficial biotransformation as the consumed contaminants will be of no use and biosorption of metal ions occurs (Kumar et al. 2011; Wasilkowski et al. 2012; Singh et al. 2014). Many examples of micro-organisms which are used for the process of bioremediation (Table 13.1).

The process and effectiveness of bioremediation depends upon the types of cell structures of microorganisms used for the process as there is a presence of different enzymes and molecular structures. They are mainly of two types, i.e. prokaryotic (i.e. bacteria and archaea) and eukaryotic organisms, which include fungi, protists and animals. In the case of eukaryotic cellular structure, there are various organelles present, i.e. nucleus and others. However, ribosomes in eukaryotes are larger with 80S as compared to 70S of prokaryotes (Killham and Prosser 2007). This is one of the reasons that eukaryotes are more sensitive to toxicity of heavy metals as compared to prokaryotes (Perpetuo et al. 2011). So the interaction with heavy metals depends variably with the type of the organism. It has to convert the metal into a harmless product by digesting with the enzymatic action (Sharma 2012).

Table 13.1 Different types of microorganisms used in bioremediation

Type of microorganism	Species	References
Bacteria	*Arthrobacter* spp., *Pseudomonas veronii*, *Burkholderia* spp., *Kocuria flava*, *Bacillus cereus Sporosarcina ginsengisoli* and *Pseudomonas veronii*	Roane et al. (2001), Vullo et al. (2008), Jiang et al. (2008), Achal et al. (2011) and Kanmani et al. (2012)
Algae	*Spirogyra* spp. and *Cladophora* spp., *Cladophora fascicularis* and *Spirogyra* spp. and *Spirulina* spp.	Lee and Chang (2011), Deng et al. (2007) and Mane and Bhosle (2012)
Fungi	*Penicillium canescens*, *Aspergillus versicolor* and *Aspergillus fumigatus*	Say et al. (2003), Tastan et al. (2010) and Ramasamy et al. (2011)
Yeast	*Saccharomyces cerevisiae* and *Candida utilis*	Machado et al. (2010) and Kujan et al. (2006)

13.2 Mode of Action

The mode of action of interaction has a three-way process initially:

1. Active extrusion of metal.
2. Metal binding peptides will chelate intracellularly (only in eukaryotes).
3. Conversion into other forms/products with reduced toxicity.

The higher organism and bacteria are resistant to many toxic metals and render them innocuous by developing various mechanisms in their body (Mejáre and Bülow 2001). Furthermore, aerobic and anaerobic fungi and other microbes are efficient for the enzymatic degradation process.

Aerobic conditions are more favourable for the bioremediation processes, but it can be possible for anaerobic organisms as well to degrade toxic metals with the help of recalcitrant molecules (Sharma 2012). As in a water body, the contamination can be due to many reasons like industrial wastes, human household disposable and environmental factors; so to clear and convert the harmful toxic metals into harmless products, there is a need of combination of various types of microorganisms because one type of organism can digest particular type of metal. So it depends completely on which type of microorganism is introduced in the contaminated aquatic system for bioremediation. Some microorganisms can survive in a limited range of contaminants.

The effectiveness of the Bio-degradation is the potential depends upon the type of microorganism as it will convert the the oxygen molecules into hydrocarbon compounds which will be easily degraded. Due to the chemotactic response the microorganism senses the chemical constituents of the metals and attracted towards it to bind so as to degrade it into a simpler compounds. The chemicals of heavy metals help the micro-organims to find them (Thapa et al. 2012).

Few microbial species are also helpful for the removal of trace minerals as well such as selenium, zinc and even arsenic from the consumable water used for drinking purposes. There are bacteria which are reported to bind the sulphate in water and clean-up which is available naturally named as sulphate-reducing bacteria (SRB). Sulphate-reducing bacteria (SRB) are able to binds to the sulphate group of heavy metals then degrade the heavy metals into simpler compounds. The mineral deposits formed from the bacteria surrounded by them, tiny spheres are formed as a binding substance. During the process of bioremediation the mineral deposits formation has occured due to the bacteria, which well act as binding substances. The concept of sphere formation works so efficiently that the bacteria which will bind the zinc and sulphate become million times much more concentrated in the water present in the surrounding. At the end, the spheres consisted entirely of zinc sulphide.

13.3 Genetically Engineered Microorganisms (GEMs)

The genetic material of the microorganisms which was present naturally is changed by the techniques used in genetic engineering so as to increase the performance in the field they are going to be used. The microorganisms can be modified or altered according to the requirement. This whole technique is known as recombinant DNA technology. Many reports are available, such as the work done by Jain et al. (2011) under laboratory conditions, where these microorganisms have been proved to improve the elimination of harmful waste with the help of genetic engineering technology. Such recombinant organisms are created by the exchange of genetic material between the naturally occurring organisms and the already genetically modified. These organisms are currently able to insert the correct gene to produce the appropriate enzyme which is able to degrade many pollutants and heavy metals (Jain et al. 2010). Genetically modified microorganisms (GEMs) have a good potential for the bioremediation of groundwater, soil, wastewater and activated sludge environments as they can degrade the pollutants in a wide range. The improvements have been made in the genetically engineered organisms as there are many strategies which were employed to enhance the performance of the microorganisms as their metabolism is completely modified or it is genetically manipulated as such microorganisms will degrade the pollutants better than the other native or old bacterial strains. In this case the genetically modified strains are used for the bioremediation which performs better than the original ones.

These steps need to be followed for genetically modified organisms:

1. Manipulation of enzyme specificity and affinity
2. Pathway construction and its regulation
3. Bioprocess development, monitoring and control
4. Affinity of bioreporter for sensing the chemicals, toxicity reduction and analysis of end point

Advantage of GEM in Bioremediation

1. It helps to speed up the recovery of wastewater or polluted water sites and make it suitable for the utilization of the industry.
2. It is also helpful for the degradation of substrate and emits a high catalytic capacity with very minute cell mass.
3. It created a safe and clean environmental condition by elimination or neutralization of the harmful substances.

Disadvantage of GEMs in Bioremediation

1. The major drawback is that in some of the cases, death of the cells happened, and it leads to the unsuccessful attempt to release microorganism into such areas.
2. There is another challenge to release the microorganisms in the surrounding or target area as sometimes the growth is delayed and degradation of the substrate occurs.
3. The variation in season and other abiotic factor fluctuations have also a direct and indirect impact on the activity of the microorganism.
4. Also, the introduction of foreign modified strain to the system leads to unreacted as well as unmeasurable adverse effect on the already existing natural structural and functional community of microorganism and its composition.

Successful examples of GEMs having promising approach for the bioremediation of heavy metals like *Deinococcus geothermalis* strain have been developed for the bioremediation of water and environment having mixed radioactive pollutant at high temperatures. However, the recombinant strain of *Acinetobacter baumannii* was recorded to enhance degradation rates at sites contaminated with crude oil (Paul et al. 2005). Due to the presence of metals, various higher organisms produce cysteine-rich peptides; one of them is metallothioneins (MTs), which has the ability to bind and sequester metal ions into biologically inactive forms. The enhancement of the metal accumulation occurred with the overexpression of MTs in recombinant bacterial cells, which works well for the formation of microbial-based biosorbents (Perpetuo et al. 2011).

13.4 Microorganisms Target Different Types of Pollutants

13.4.1 Oil Spills

Many microbes that use oil as their energy source have been reported for hundred millions of years. In natural places like the Gulf of Mexico, the microbial communities collectively feed on all the compounds contained in the oil which is very well established and diversified; even if the level of oil is low, a little quantity of microbes still capable of degrading oil seems to be available there. The ability of microbes varies from one another by displaying the metabolic capacity to degrade the type of oil components (Brzeszcz and Kaszycki 2018). Mostly, certain microbes prefer oil hydrocarbons over any other energy sources, and their number is rapidly increasing in response to an oil spill. However, other types of bacteria are capable of degrading and getting energy from the other type of constituent.

Environmental factors play a very important role in the case of microbes to degrade the toxic pollutants as they require an ideal condition for the process and their metabolic activity (Das and Chandran 2011). Like in the case of crops where if they get the right amount of fertilizers, water and light, they grow faster, the degradation process of oil by microbes is also enhanced during optimal environmental conditions. But various factors also participate in this case, such as the following.

13.4.1.1 Physical Nature of Oil

The nature of the oil is very much correlated with the degradation process of the oil. In a simple manner, the microbes should interact with the oil appropriately as they must coat the oil properly. If the single large slick of oil is present, less surface area is available for the microbes to gain the access of the oil; in this case the process of degradation is slow (Doshi et al. 2018). Furthermore, if heavy and viscous oil is present in the water body, the biodegradable components must first diffuse through the thick matrix to the oil-water interface so that the microbes can access them. The heavier the oil, the slower the biodegradable components to diffuse.

13.4.1.2 Chemical Nature of Oil

The rate of biodegradation is different for a hydrocarbon accordingly that makes up the oil spilled. The composition of oil includes thousands of different compounds as some have various food sources, and they will be consumed so quickly, whereas others will take time to degrade. In natural places where the synthetic components are not present, the hydrocarbons are easily degraded within days or weeks because of the presence of unbranched carbons. Furthermore, if the hydrocarbons are arranged in branched chains, it takes much longer time period for the degradation process and persists much longer in the environment.

13.4.2 Water Temperature

The degradation of heavy metals and growth of microorganisms are fast in warmer waters compared to the cold temperature water bodies. Although microbes can survive and adapt well even in the colder temperature, the only advantage is that in warmer water the metabolic rates are much faster which is the same reason that milk stays longer when kept in refrigerator than at the room temperature (Das and Chandran 2011). Genetically modified organisms adapted well to those places which remain cold throughout the year, and these microbes can degrade the oil as quickly as the normal microorganism can in the normal habitat. But the lower temperature slows down the evaporation process, so the oil is left in the environment a little longer and available for the microbes for degradation process. The viscosity of oil is more which spreads the oil and access to the microbes is less.

13.4.3 Availability of Oxygen in Polluted Water

The presence of more oxygen in the water body speeds up the enzymatic process of breaking down the oil. The availability of nutrients and oil also favours the growth of the microbial population, but due to the increase in population, the level of oxygen is exhausted from the water in the vicinity of an oil spill (Akar and Tunali 2006). Although the level of oxygen is not the limiting factor for the degradation of the heavy metals, the process rate could be much slower on those places where the rate of oxygen is slow.

13.4.4 Availability of Nutrients

Microorganisms require phosphate, nitrogen and many other nutrients as these living organisms need such nutrients to degrade the hydrocarbons present in the polluted water it depends upon the availability of nutrients. If the level of nutrients like phosphate and nitrogen is extremely low, the process of degradation of oil is also very slow. Such substances are present in nature in a very limited quantity.

13.4.5 Pressure

The process of degradation is also depending upon the pressure and temperature; if the temperature is low and pressure is high, the rate is much slower (Ahalya et al. 2003). However, many microorganisms adapted well in such environmental conditions, and the oil-degrading microbial species can survive in these extreme conditions.

13.4.6 pH and Salinity

Mostly in oceans, the variation in pH and salinity does not incorporate much difference, but in smaller water bodies, the rate of salinity is high, and also the fluctuations in salinity, oxygen and pH slow down the process of degradation by microbes in polluted water bodies.

13.4.7 Other Microbes

In water bodies, there are many microorganisms which are already existing, and they compete and cooperate well with each other. Sometimes, adding the genetically modified organisms or artificially introducing them into such environments leads to unsuccessful attempt and is quite challenging as they have a hard time breaking into the existing community structure and competing with the local species which are living well with each other in an environment as the natural microbial community is very diverse.

13.5 Mechanism

The mechanism of bioremediation is divided into two major categories for the clean-up of heavy metals:

1. Biosorption is a physicochemical interaction process between heavy metal ion and the functional group on the cell surface; metals are retained for ex. Ion exchange, precipitation, adsorption, crystallization (Gadd and White 1993; Gadd 2009; Volesky 2004; Ahalya et al. 2003; Fosso and Mulaba 2014). This process is a fast as well as reversible passive adsorption mechanism (Gadd and White 1993; Ahalya et al. 2003). Many factors affect the process of biosorption, like pH, ionic strength, temperature, biomass concentration, particle size and presence of any other ions in the solution (Volesky 2004).
2. Bioaccumulation process has a very limited part of passive uptake because mainly intracellular and extracellular processes are involved in this case.

Microorganisms contain different macromolecules on their cell wall which include proteins, polysaccharides and large amount of charged functional groups, like carboxyl, sulfhydryl, amide, carbonyl, ester sulphate, phenol, amino, thioether and hydroxyl groups (Zabochnicka and Krzywonos 2014; Bayramoglu et al. 2006; Akar and Tunali 2006). The heavy metal ions are attracted towards these groups of the cell wall microorganisms which leads to the adsorption process. The culturing of microorganism plays a very important role in this process as the more the number of functional groups, the more will be the adsorption capacity (Gadd and White 1993). The functional groups like carboxyl, aldehydes and ketones present in the cell wall of bacteria will remove the heavy metals effectively from the wastewater, and lesser amount of sludge is formed (Qu et al. 2014). The bacteria like gram-positive and gram-negative are used for the uptake of metals, whereas brown, red and green algae are also used as a biosorbant. Bacteria have some functional groups present which are capable of exchange of ions like galactans, alginic acid and xylans which are advantageous to use as compared to microorganisms as they don't produce toxic compounds unlike microorganisms (Das et al. 2008).

Whereas the other concept, bioaccumulation works in a manner as many heavy metals do not biodegrade and usually accumulate inside the microorganism (Huang et al. 2014). There are many factors which influence accumulation of metal ions (Varma et al. 2011). It is a very complex process and varies according to the metabolism pathway by the metal concentration (Fukunaga and Anderson 2011). It mainly depends upon the binding of metal ions to the surface of the cell.

13.6 Conclusion

Microorganisms play a very important role in clearing the polluted water over any other methods. It is advantageous as:

1. The ability of degradation is highly efficient.
2. The energy consumption rate is very low.
3. There is no problem of secondary pollution as it is completely a degradable process.
4. There is no need of labour, and very simple technical operations are involved.
5. It has a long-term viability, and there is no need of any construction work.

Scientists attempted the field trials and suggested that if two kinds of microorganisms are added to the polluted water which improves the water quality of heavily polluted tidal river is successfully used at many countries for various purposes to clearing the heavy metals from the environment. It is succesfully used by many countries to clear the heavy metals. This biodegradable nature and eco-friendly approach is best to break the pollutants and heavy metals which should be implemented against water pollutants.

References

Achal V, Pan X, Zhang D (2011) Remediation of copper-contaminated soil by *Kocuria flava* CR1, based on microbially induced calcite precipitation. Ecol Eng 37(10):1601–1605

Ahalya N, Ramachandra TV, Kanamadi RD (2003) Biosorption of heavy metals. Res J Chem Environ 7(4):4544–4552

Akar T, Tunali S (2006) Biosorption characteristics of *Aspergillus flavus* biomass for removal of Pb (II) and Cu (II) ions from an aqueous solution. Bioresour Technol 97:1780–1787

Bayramoglu G, Tuzun I, Celik G, Yilmaz M, Arica MY (2006) Biosorption of mercury (II), cadmium (II) and lead (II) ions from aqueous system by microalgae *Chlamydomonas reinhardtii* immobilized in alginate beads. Int J Miner Process 81(1):35–43

Brzeszcz J, Kaszycki P (2018) Aerobic bacteria degrading both n-alkanes and aromatic hydrocarbons: an undervalued strategy for metabolic diversity and flexibility. Biodegradation 29:359–407

Das N, Chandran P (2011) Microbial degradation of petroleum hydrocarbon contaminants: an overview. Biotechnol Res Int. https://doi.org/10.4061/2011/941810

Das N, Vimala R, Karthika P (2008) Biosorption of heavy metals – an overview. Indian J Biotechnol 7:159–169

Deng L, Su Y, Su H, Wang X, Zhu X (2007) Sorption and desorption of lead (II) from wastewater by green algae *Cladophora fascicularis*. J Hazard Mater 143(1–2):220–225

Doshi B, Sillanpaa M, Kalliola S (2018) A review of bio-based materials for oil spill treatment. Water Res 135:262–277

Fomina M, Gadd GM (2014) Biosorption: current perspectives on concept, definition and application. Bioresour Technol 160:3–14

Fosso-Kankeu E, Mulaba-Bafubiandi AF (2014) Implication of plants and microbial metalloproteins in the bioremediation of polluted waters: a review. Phys Chem Earth 67:242–252

Fukunaga A, Anderson MJ (2011) Bioaccumulation of copper, lead and zinc by the bivalves *Macomona liliana* and *Austrovenus stutchburyi*. J Exp Mar Biol Ecol *396*(2):244–252
Gadd GM (2009) Biosorption: critical review of scientific rationale, environmental importance and significance for pollution treatment. J Chem Technol Biotechnol 84(1):13–28
Gadd GM, White C (1993) Microbial treatment of metal pollution – a working biotechnology? Trends Biotechnol 11:353–359
Huang F, Guo CL, Lu GN, Yi XY, Zhu LD, Dang Z (2014) Bioaccumulation characterization of cadmium by growing *Bacillus cereus* RC-1 and its mechanism. Chemosphere 109:134–142
Jain PK, Gupta VK, Gaur RK et al (2010) Fungal enzymes: potential tools of environmental processes. In: Gupta VK, Tuohy M, Gaur RK (eds) Fungal biochemistry and biotechnology. LAP Lambert Academic Publishing AG and Co. KG, Germany, pp 44–56
Jain PK, Gupta VK, Bajpai V et al (2011) GMO's: perspective of bioremediation. In: Jain PK, Gupta VK, Bajpai V (eds) Recent advances in environmental biotechnology. LAP Lambert Academic Publishing AG and Co. KG, Germany, pp 6–23
Jiang CY, Sheng XF, Qian M, Wang QY (2008) Isolation and characterization of heavy metal resistant *Burkholderia* species from heavy metal contaminated paddy field soil and its potential in promoting plant growth and heavy metal accumulation in metal polluted soil. Chemosphere *72*:157–164
Kadirvelu K, Senthilkumar P, Thamaraiselvi K, Subburam V (2002) Activated carbon prepared from biomass as adsorbent: elimination of Ni (II) from aqueous solution. Bioresour Technol 81:87–90
Kanmani P, Aravind J, Preston D (2012) Remediation of chromium contaminants using bacteria. Int J Environ Sci Technol 9:183–193
Killham K, Prosser JI (2007) The prokaryotes. In: Paul EA (ed) Soil microbiology, ecology, and biochemistry. Elsevier, Oxford, pp 119–144
Kujan P, Prell A, Safár H, Sobotka M, Rezanka T, Holler P (2006) Use of the industrial yeast *Candida utilis* for cadmium sorption. Folia Microbiol 51(4):257–260
Kumar A, Bisht BS, Joshi VD, Dhewa T (2011) Review on bioremediation of polluted environment: a management tool. Int J Environ Sci 1(6):1079–1093
Lee YC, Chang SP (2011) The biosorption of heavy metals from aqueous solution by *Spirogyra* and *Cladophora* filamentous macroalgae. Bioresour Technol 102(9):5297–5304
Machado MD, Soares EV, Soares HM (2010) Removal of heavy metals using a brewer's yeast strain of *Saccharomyces cerevisiae*: chemical speciation as a tool in the prediction and improving of treatment efficiency of real electroplating effluents. J Hazard Mater 180(1–3):347–353
Mane PC, Bhosle AB (2012) Bioremoval of some metals by living Algae *spirogyra* sp. and *Spirulina* sp. from aqueous solution. Int J Environ Res 6(2):571–576
Mejáre M, Bülow L (2001) Metal-binding proteins and peptides in bioremediation and phytoremediation of heavy metals. Trends Biotechnol 19(2):67–73
Mulligan CN, Yong R, Gibbs BF (2001) Remediation technologies for metal contaminated soils and groundwater: an evaluation. Eng Geol 60(1–4):193–207
Nicolaou SA, Gaida SM, Papoutsakis ET (2010) A comparative view of metabolite and substrate stress and tolerance in microbial bioprocessing: from biofuels and chemicals, to biocatalysis and bioremediation. Metab Eng 12(4):307–331
Paul D, Pandey G, Jain RK (2005) Suicidal genetically engineered microorganisms for bioremediation: need and perspectives. BioEssays 27(5):563–573
Perpetuo EA, Souza CB, Nascimento CAO (2011) Engineering bacteria for bioremediation. In: Carpi A (ed) Progress in molecular and environmental bioengineering – from analysis and modeling to technology applications. InTech, Rijeka, pp 605–632
Qu Y, Zhang X, Xu J, Zhang W, Guo Y (2014) Removal of hexavalent chromium from wastewater using magnetotactic bacteria. Sep Purif Technol 136:10–17
Ramasamy RK, Congeevaram S, Thamaraiselvi K (2011) Evaluation of isolated fungal strain from e-waste recycling facility for effective sorption of toxic heavy metal Pb (II) ions and fungal

protein molecular characterization-a Mycoremediation approach. Asian J Exp Biol Sci 2 (2):342–347
Riggle PJ, Kumamoto CA (2000) Role of a Candida albicans P1-type ATPase in resistance to copper and silver ion toxicity. J Bacteriol 182:4899–4905
Roane TM, Josephson KL, Pepper IL (2001) Dual-bioaugmentation strategy to enhance remediation of contaminated soil. Appl Environ Microbiol 67(7):3208–3215
Say R, Yimaz N, Denizli A (2003) Removal of heavy metal ions using the fungus *Penicillium canescens*. Adsorpt Sci Technol 21(7):643–650
Sharma S (2012) Bioremediation: features, strategies and applications. Asian J Pharm Life Sci 2 (2):202–213
Singh R, Singh P, Sharma R (2014) Microorganism as a tool of bioremediation technology for cleaning environment: a review. Proc Int Acad Ecol Environ Sci 4(1):1–6
Tastan BE, Ertugrul S, Donmez G (2010) Effective bioremoval of reactive dye and heavy metals by *Aspergillus versicolor*. Bioresour Technol 101(3):870–876
Thapa B, Kumar A, Ghimire A (2012) A review on bioremediation of petroleum hydro-carbon contaminants in soil. Kathmandu University. J Sci Eng Technol 8(1):164–170
Varma R, Turner A, Brown MT (2011) Bioaccumulation of metals by *Fucus ceranoides* in estuaries of South West England. Mar Pollut Bull 62(11):2557–2562
Volesky B (2004) Sorption and biosorption. BV-Sorbex, Inc., Montreal
Vullo DL, Ceretti HM, Daniel MA, Ramírez SA, Zalts A (2008) Cadmium, zinc and copper biosorption mediated by *Pseudomonas veronii* 2E. Bioresour Technol 99(13):5574–5581
Wasilkowski D, Swedziol Ż, Mrozik A (2012) The applicability of genetically modified microorganisms in bioremediation of contaminated environments. Chemik 66(8):822–826
Zabochnicka-Swiatek M, Krzywonos M (2014) Potentials of biosorption and bioaccumulation processes for heavy metal removal. Pol J Environ Stud 23(2):551–561

Microbial Clean-Up Strategy for Eating Garbage

14

Sheetal Rana, Abha Sharma, and Ranbir Singh Rana

Abstract

Bioremediation is the deliberate use of biological mechanisms to clean up pollutants, viz. hydrocarbons, oil, heavy metal, pesticides and dyes, by letting the microbes eat and digest toxic contaminants and consequently transform them into gases, water and other less toxic components. The indispensable habit of using products made out of plastic has led to the pollution havoc in the present day. The physical and chemical degradation methods do not provide an eco-friendly solution to disposal of garbage. Here, the usage of microorganisms has emerged as a key alternative offering solution to the challenges of reifying environment-friendly garbage clean-up. The resiliency of microorganisms to survive even the harshest of environmental conditions and the extreme diversity in microbial communities which comprise as many as 10,000 distinct microbial species per gram of soil make them highly effective in bioremediation of almost all environmental pollutants. As such, bioremediation is a highly promising solution for the degradation, eradication, immobilization and detoxification of chemical and physical waste materials. In addition, the process is also cheaper in equipment and labour costs in comparison to the physical and chemical treatment solutions. So bioremediation has a great contributory role to play in solving many existing and future environmental problems.

Keywords

Bioremediation · Microorganisms · Environment · Pollutants

S. Rana (✉)
Department of Organic Agriculture, CSKHPKV, Palampur, India

A. Sharma
Department of Agronomy, Forages and Grassland Management CSKHPKV, Palampur, India

R. S. Rana
Centre for Geoinformatics CSKHPKV, Palampur, India

S. G. Sharma et al. (eds.), *Microbial Diversity, Interventions and Scope*,
https://doi.org/10.1007/978-981-15-4099-8_14

14.1 Introduction

The expansion in urbanization, industrialization and population has rapidly increased the human contribution to environmental contamination in our recent century. Today the contamination is dangerous enough to potentially lead to extinction of rare and endangered species and also put our own existence in peril with the contamination of air, water and food sources. Existing chemical and physical treatment solutions for waste products lead to further contamination of environment because of the harmful residuals left after the treatment.

As such, the biotreatment of contaminated areas is increasingly finding leading applications in recent years. The traditional physical method of doing remediation is to dig up the contaminated soil and isolate and remove the areas covering the contaminated sites. These methods introduce risks during the excavation, handling and transportation of hazardous waste materials. Moreover, the cost of performing incineration, excavation, landfilling and storage makes these methods expensive and inefficient. Finding landfill sites for final disposal of contaminated waste materials is not trivial and requires several considerations to be made with security of individuals residing in neighbouring areas (Bollag and Bollag 1995).

Bioremediation offers an eco-friendly waste disposal alternative in order to solve the disadvantages of traditional waste disposal methods. Bioremediation is the method that relies on different microorganisms to degrade the pollutants into less poisonous types. Compared with conventional treatment, the main differentiating advantages of bioremediation are its relatively low cost and high efficiency with no additional nutrient requirement and minimal site disruption, thereby leading to greater public acceptance. However, the application of bioremediation is limited by the availability of microorganisms and also the presence of some chemicals with low water solubility and those which are not amenable to biodegradation (such as heavy metals, some chlorinated compounds, radionuclides).

The process of transforming natural assets accessible in the environment into types suitable for our usage is central to contemporary human activities. This method produces dangerous by products and therefore leads to the current issue of unnecessary air, water and land pollution. As such, any sustainable solution to the disposal of generated wastes in an environmentally safe way must conveniently integrate them back into the environment by reducing the concentration of harmful contaminants. Microorganisms (generally yeasts, bacteria or fungi) must be used. This is achieved by integrating the microorganisms or their products into the substrates to derive industrial products such as bioleaching (biomining), biodetergent, biotreatment of pulp, biotreatment of wastes (bioremediation), biofiltrations, aquaculture treatments, biotreatment of textiles, biocatalysts, biomass fuel production, biomonitoring and so on.

Microbes already play a significant role in safe disposal and reuse of garbage as the natural recycling of all living materials is performed by the microbes. All naturally produced materials are biodegradable. Microbes are nature's supreme waste dump. It is a testimony to their effectiveness that we, humans, have coerced them to clear up our natural messes for centuries. They have become invaluable and vital in discovering alternatives to several challenging human issues to guarantee the value of our surroundings. They had a beneficial impact on human and pet safety, DNA technology, economic safety and corporate and agricultural disposal methods. Microorganisms have been the catalyst in implementing feasible and cost-effective waste disposal solutions which otherwise would have been impossible via chemical or physical engineering methods.

Microorganisms have amazing metabolism and can develop easily in severe climate circumstances. This dietary adaptability of microorganisms to transform, alter and use poisonous pollutants to acquire power is utilized by bioremediation to biodegrade pollutants.

In contrast to naively gathering the pollutant and increasing the danger by transporting it, the method of bioremediation involves establishing a microbiological well to promote organized microbial growth to break down contaminants and convert damaging chemicals into less poisonous or non-toxic ionic and compound types. Bioremediators are these biological substances used to remediate contaminated locations. Therefore, the bioremediation method can be defined as the biotechnological method incorporating the implementation of microorganisms as bioremediators to remove many pollutant hazards. Typical primary bioremediators are bacteria, archaea and fungi, and their use is recommended to restore the initial natural environment and prevent further destruction (Vidali 2001; McKinney 1957).

14.1.1 Principles of Bioremediation

Bioremediation is the most efficient instrument for managing the polluted atmosphere, where organic waste is biodegraded to an innocuous state under regulated circumstances. Microorganisms, in fact, are able to destroy, degrade and even absorb damaging organic and nitrogen compounds (Jain and Bajpai 2012). Higher vegetation was also recorded to prevent such pollutants, mainly through the capacity to collect them in their bodies (Kumar et al. 2011). Bioremediation aims to encourage microorganisms to operate by providing optimum concentrations of oxygen and other chemicals necessary for their development to detoxify materials that are harmful to the atmosphere and human objects (Rathore 2017). Enzyme mediated all cellular responses. These relate to oxidoreductase, hydrolase, lyase, transferase, isomerase and ligase classes. Because of their non-specific and particular surface binding, many enzymes have extremely broad degradation ability. Bioremediation can also only be efficient if environmental circumstances permit microbial development; its implementation often includes manipulating environmental parameters to enable microbial development and degradation to continue at a quicker pace (Vidali 2001).

Bioremediation is based on natural attenuation, identified as natural procedures (mostly microbes but may also include fungi, algae and higher plants) in the environment that operate without human interference to decrease the quantity or toxicity of contaminants (Sharma 2011–2012).

Environmental biotechnology has been used to create effective solutions to solve difficult problems faced by humans in recent century. The steep advancement of molecular biology and biotechnology has allowed us to exploit the biological processes more efficiently to clean up polluted water and land areas through biological process. Biotechnological methods to handle garbage before or after it is carried into the surroundings are elements of biotechnological management techniques. Biotechnology can also be used industrially to develop goods and procedures that produce less cost, use less resources and require less power.

Recombinant DNA engineering has enhanced pollution reduction options and claims to further develop bioremediation (Vandevivere et al. 1998). Bioremediation utilizes real environment organisms, fungi and vegetation where the enzymes generated by these microbes target the contaminants and then transform these into non-toxic materials. To be efficient, environment conditions should promote microbial development (Mrozik et al. 2003). Bioremediation effectiveness relies on many variables, including the chemical type and intensity of pollutants, the physicochemical properties of the environment and their accessibility to microorganisms (El Fantroussi and Agathos 2005). Besides, microbes and pollutants do not distribute evenly in the surroundings. Many variables make monitoring and optimizing bioremediation procedures a complicated process.

Since this method is a challenging job, it includes three primary components – the environment, contaminants and microorganisms. Appropriate climate, humidity composition, pH, water availability and ion acceptors such as oxygen for aerobic microorganisms promote biological behaviour in the ecosystem (Fritsche and Hofrichter 2008).

14.1.2 Use of Microorganisms in Waste Management

Bacteria, fungi, algae and other microorganisms can be used for industrial waste treatment. Their use and efficiency relies on the sort of chemical pollution, environmental circumstances and microbial development supporting variables like resource accessibility and temperature, pH, humidity and availability of electron acceptors such as oxygen. These microorganisms are categorized as follows.

14.1.3 Use of Bacteria

Bacteria are the most used microorganism in any disposal scheme. They have different biochemical characteristics and metabolize most organic material in any industrial waste. Many microorganisms even eat exclusively on hydrocarbons (Yakimov et al. 2007). Biodegradation of hydrocarbons can happen under aerobic and anaerobic circumstances, as is the situation with *Pseudomonas* sp. and *Brevibacillus* sp. (Grishchenkov et al. 2000). Wiedemeier et al. (1995), however, indicate that anaerobic biodegradation may be much more essential. Among the hydrocarbon-degrading organisms isolated from the coastal environment (Floodgate

1984), the organisms corresponding to the ten species: *Bacillus*, *Corynebacterium*, *Staphylococcus*, *Streptococcus*, *Shigella*, *Alcaligens*, *Acinetobacter*, *Escherichia*, *Klebsiella* and *Enterobacter* were selected (Kafilzadeh et al. 2011) and *Bacillus* species was found to be the highest degrading organisms among all. Bacterial species capable of degrading organic hydrocarbons have been consistently extracted, primarily from land. These are generally Gram-negative bacteria, mostly belonging to the *Pseudomonas* genus. Biodegrading processes were also reported in the species *Mycobacterium*, *Corynebacterium*, *Aeromonas*, *Rhodococcus* and *Bacillus* (Mrozik et al. 2003). Although many bacteria can metabolize natural pollutants, one bacterium doesn't have the enzymatic capacity to deteriorate all or even most toxic chemicals in a polluted land. Mixed microbial populations have the most important biodegrading ability because more than one organism's genome data is crucial to degrade complicated mixtures of natural compounds in contaminated fields (Fritsche and Hofrichter 2005). Aerobic and anaerobic fungi can biotransform PCBs. Higher chlorinated PCBs are dehalogenated by anaerobic microorganisms, and lower chlorinated biphenyls are oxidized by aerobic bacteria (Seeger et al. 2001). Research on indigenous aerobic fungi has concentrated primarily on Gram-negative species of the species *Pseudomonas*, *Burkholderia*, *Ralstonia*, *Achromobacter*, *Sphingomonas* and *Comamonas*. However, several studies on PCB-degrading behaviour and gene identification of microbes involved in PCB degradation also suggested PCB-degrading ability of some Gram-positive species (genera *Rhodococcus*, *Janibacter*, *Bacillus*, *Paenibacillus and Microbacterium*) (Petric et al. 2007). Aerobic catabolic pathway for PCB degradation seems to be very similar for most of the bacteria and comprises four steps catalysed by the enzymes biphenyl dioxygenase (BphA), dihydrodiol dehydrogenase (BphB), 2, 3-dihydroxybihenyl dioxygenase (DHBD) (BphC) and hydrolase (BphD) (Taguchi et al. 2001).

Effective removal of pesticides by adding bacterium was noted earlier for many compounds, including atrazine (Struthers et al. 1998). Recent findings concerning pesticide-degrading bacteria embrace the chlorpyrifos-degrading bacterium *Providencia stuartii* isolated from agricultural soil (Surekha et al. 2008) and isolates *Bacillus*, *Staphylococcus* and *Stenotrophomonas* from cultivated and uncultivated soil able to degrade dichlorodiphenyltrichloroethane (DDT) (Kanade et al. 2012) (Table 14.1).

14.1.4 Fungi

In the waste management scheme, fungi that can metabolize several organic compounds may also be employed. Except under uncommon environmental circumstances, they dominate the bacteria. *Phanerochaete chrysosporium* fungi can degrade a wide variety of permanent or toxic environmental pollutants. Straws, sawdust or cobs are commonly used substrates. Fungi are most effective, particularly when the natural polymeric compounds break down, and are equipped with extracellular enzyme systems. They can also quickly colonize and enter substrates and transport and redistribute nutrients into their mycelium through their hyphal systems

Table 14.1 Examples of microorganisms involved in bioremediation

Contaminants	Microorganisms
Atrazine	*Pseudomonas* sp. (ADP)
2,4,6-Trinitrotoluene	*Methanococcus* sp.
Chlorpyrifos	*Enterobacter* sp.
Dibenzothiophene (DBT)	*Rhizobium meliloti*
Hexahydro-1,3,5-trinitro-1,3,5-triazine (RDX)	*Acetobacterium paludosum, Clostridium acetobutylicum*
PAHs	*Pseudomonas* sp., *Pycnoporus sanguineus, Coriolus versicolor, Pleurotus ostreatus, Fomitopsis palustris, Daedalea elegans*
Phenanthrene, PAH	*Agrobacterium, Bacillus, Burkholderia, Pseudomonas* and *Sphingomonas*
Polychlorinated biphenyl (PCB)	*Rhodococcus erythropolis, Rhizobium* sp.
Polycyclic aromatic hydrocarbon (PAH)	Fungi

Source: Kang (2014)

(Matavuly and Molitoris 2009). Mycorrhiza is a symbiosis between a fungus and vascular crop roots. The fungus colonizes the roots of the host plant in a mycorrhizal association, both intracellularly such as in arbuscular mycorrhizal (AMF) mushrooms and extracellularly, for instance, in ectomycorrhizal fungi. They are also a key element in soil life and chemistry. Mycorrhizal biological remedial treatment is called mycorrhizal remediation (Khan, 2006). Fungi have significant degrading functions that affect the recycling of recalcitrant materials (e.g. lignin) and the disposal of environmentally hazardous waste (Fritsche and Hofrichter 2005).

14.1.5 Algae

Algae that use sunlight as an energy source may also be used to manage inorganic waste such as ammonia, carbon dioxide, magnesium, potassium, iron, calcium, sulphate, phosphate and sodium. Algae and bacteria can be used together for the same waste components. Organic waste components are metabolized into inorganic components by bacteria, which are then used by the algae.

Species of *Chlorella*, *Anabaena inaequalis*, *Westiellopsis prolifica*, *Stigeoclonium lenue* and *Synechococcus* withstand heavy metals. A number of species of *Chlorella*, *Anabaena* and marine algae have been employed in the suppression of heavy metals, but operational circumstances restrict their practical implementation (Dwivedi 2012). Metals are absorbed by adsorption of algae. Unicellular algae have recorded metal chelation. Biosorption of heavy metals by brown algae is known for a long period, including the extraction of heavy metals by a variety of cell wall elements like alginate and fucoidane of heavy metals. Most study has been conducted in this field on marine and land algae (Davis et al. 2003). It has been indicated that the microalga *S. incrassatulus* can be used to remove Cr(VI), Cd

(II) and Cu(II) in continuous cultures (Pena-Castro et al. 2004). Heavy metal bioremediation was also recorded for green algae, for instance, *C. sorokiniana* for Cr(III) removal (Akhtar et al. 2008).

14.1.6 Protozoa

Protozoa are the most basic types of microorganisms to be used in waste disposal processes. The protozoa may assist in the decrease of the population of useless bacteria instead of being the primary part of treatment system. They therefore help to produce a clear effluent. The protozoans are the primary grazer for degrading bio-pollutants, which affects the relation of the protozoa with degrading bacteria. In order to examine the impact of protozoan flagellate Heteromita in the biodegradation of benzene and methylbenzene, Mattison and Harayama (2015) have developed a model for the food chain. The study discovered that the degrading rates of benzene and methylbenzene were 8.5 times better in bacteria than before during the logarithmic increasing population of the flagella population. The protozoa infusors clearly can speed the biodegradation of heterogeneous materials such as PAH. For instance, the naphthalene degradation speed can be four times higher than before.

The protozoa process that accelerates the biodegradation of organic contaminants can be expected to include several feasible hypotheses, primarily six of which are:

1. Mineralisation of nutrients that increases nutrient turnover
2. Activation of bacteria that regulates the amount, destroys old cells and excretes an active ingredient
3. Selective grazing, which decreases resource and space competition and is therefore useful for the development of degrading bacteria
4. Physical disruption that can boost the level of oxygen and the degrading matter surface
5. Direct degradation that can excrete specific degradation enzymes
6. Sym-metabolism that provides bacteria with energy and carbon resources during degradation (Chen et al. 2007)

14.1.7 Factors of Bioremediation

The management and optimization of procedures of bioremediation have many complicated variables: the presence of a pollutant-degrading microbial population, the accessibility of contaminants and the environmental variables (type of soil, temperature, pH, the presence of oxygen or other electron acceptors and nutrients) (Table 14.2).

14.1.7.1 Biological Factors

The biotic variables influence organic degradation through the competition between microorganisms for restricted carbon supplies, antagonistic relationships or

Table 14.2 Showing factors of bioremediation

Factors of bioremediation	Condition required
Microorganisms	Aerobic or anaerobic
Natural biological processes of microorganism	Catabolism and anabolism
Environmental factors	Temperature, pH, oxygen content, electron acceptor/donor
Nutrients	Carbon, nitrogen, oxygen, etc.
Soil moisture	25–28% of water holding capacity
Type of soil	Low clay or silt content

protozoan predation of microorganisms. Contaminant degradation speed often depends on contaminant density and the quantity of "catalyst" available. In that regard, both the number of organisms able to metabolize contaminants and the quantity of enzymes created within each cell are represented as the "catalyst" amount (Madhavi and Mohini 2012). Expression of particular enzymes can boost or reduce contaminant degradation. The magnitude of the contaminant metabolism should also involve particular enzymes, and the contaminant's "affinity" to them as well as their accessibility is essential. The major biological factors are as follows: mutation, horizontal gene transfer, enzyme activity, interaction (competition, succession and predation), its own growth until critical biomass is reached, population size and composition (Boopathy 2000).

14.1.7.2 Environmental Factors

The metabolic characteristics of the microorganisms and physicochemical properties of the targeted contaminants determine possible interaction during the process. The actual effective interaction between the two, however, relies on the site's environmental conditions. Microorganism reproduction and activities are influenced by pH, heat, humidity, land composition, water solubility, nutrients, location features, redox ability and oxygen material, absence of qualified human capital in this sector and pollutant bioavailability (contaminant quantity, form, solubility, chemical composition and poisoning). These above variables determine kinetics of degradation (Adams et al. 2015). Biodegradation can happen under a wide pH range; however, in most marine and land environments, pH 6.5–8.5 is usually ideal for biodegradation. Moisture affects contaminant consumption frequency because it affects the type and quantity of soluble products, as well as the osmotic stress and pH of natural and freshwater environments (Cases and De Lorenzo 2005). Most environmental variables are discussed below.

14.1.7.3 Availability of Nutrients

Adding oxygen adjusts the vital nutritional equilibrium for microbial development, as well as affects biodegradation frequency and efficiency. Nutrient mixing, particularly by supplying vital proteins such as N and P, can enhance the biodegradation effectiveness by optimizing the C:N:P proportion. Microorganisms need several elements such as carbon, oxygen and phosphorus to sustain and maintain their microbial operations. Adding a suitable amount of nutrients is a good approach to increase the cellular function of microorganisms and consequently the biodegradation speed in harsh settings (Couto et al. 2014; Phulia et al. 2013). Aquatic biodegradation is similarly restricted by nutrient accessibility (Thayasi et al. 2011). Like other species' dietary requirements, oil-eating microbes also demand adequate nutrients for their growth and development (Macaulay 2015).

14.1.7.4 Temperature

Among the most significant physical variables to determine microorganism longevity is heat (Das and Chandran 2011). In harsh freezing settings such as the Arctic, oil degradation through normal produces is very low, putting more stress on the microbes to clear up accumulated petroleum. The under-zero temperature of the water is responsible for the suppression or even freezing of the full cytoplasm of transportation canals in microbial bodies, thereby making most inactive (Yang et al. 2009). Moreover, the degradation of different compounds requires different temperatures. Temperature also speeds or slows down bioremediation processes as microbial physiological characteristics are extremely influenced. Microbial activity speed rises with heat, reaching its highest amount at optimum heat. It abruptly declines with higher or lower heat and ultimately stops after approaching a particular heat.

14.1.7.5 Concentration of Oxygen

Some bacteria need oxygen to improve their biodegradation level. Other species however require no oxygen for biodegradation. Biological degradation occurs aerobically and anaerobically, as for most living organisms, air is necessary for survival. In most cases, the presence of oxygen could improve the metabolism of hydrocarbons (Macaulay 2015).

14.1.7.6 Moisture Content

To achieve their growth, microorganisms necessitate sufficient oxygen. Biodegradation inhibitors and the soil humidity content have the most negative impact.

14.1.7.7 pH

pH for a substance, defined mainly as acidity, basicity and alkalinity, has its own effect on metabolic activity as well as influences removal process. Soil pH measurement could show microbial growth potential (Enim 2013). Lower or higher pH values have shown less metabolic processes (Wang et al. 2011).

14.1.7.8 Site Characterization and Selection

Enough remediation work should be done in order to properly characterize the magnitude and extent of contamination before a bioremediation solution is proposed. The following factors should at least be covered in this job. The horizontal and vertical extent of contamination should be completely determined, the sites must be specified and the justification for choosing them should be listed and the techniques for sampling and analysis should be well described.

14.1.7.9 Metal Ions

In tiny quantities, metals are essential for bacteria and fungi, but they prevent the metabolic activity of cells in large quantities. Metal compounds can affect the degradation rate directly and indirectly.

14.1.7.10 Toxic Compounds

The decontamination by microorganisms if slowed down when the concentration of toxic contaminants is high. The toxicity level and processes depend on the particular toxicants, concentration and microorganisms being exposed. Some organic and inorganic substances can be poisonous to specific types of microorganisms (Madhavi and Mohini 2012).

14.1.7.11 Prospects for Increasing the Effectiveness of Bioremediation

The efficacy and variety of the microorganisms used in the method directly affect bioremediation. The variety of microorganism communities enables the overall use of bioremediation techniques to restore sites contaminated with various pollutants. As a result, better knowledge of the role of catabolic paths and microbial metabolic functions facilitates improving the efficiency of bioremediation procedures. The restricted supply of nutrients is another significant factor that affects the efficacy of bioremediation techniques. The addition of nutrients is known as biostimulation and is a strategy to improve the efficiency of microbial operations. The indigenous microbes reveal a rise in their activity when biostimulation is applied and can resist the degrading impacts of extremely concentrated pollutants. However, it is also suggested that nutrients not be overused (Wang et al. 2012). The promotion of microbial population growth and diversity is an orthogonal strategy to increase the accessibility of nutrients. This strategy is known as "bioaugmentation" and benefits from the synergy among the various metabolic microbial procedures (Silva-Castro et al. 2012; Bhattacharya et al. 2015).

14.1.8 Bioremediation Strategies

14.1.8.1 Ex Situ Bioremediation

The pollutants are excavated and transferred to another site for processing in this method. The use of this method depends, for instance, on the circumstances at the contaminated site: is the pollutant deeply contaminating the soil, is it extremely concentrated, is the pollution costly to treat on the ground (Philp and Atlas 2005).

14.1.8.2 Biopile

This method excavates the pollutant over the floor into stacks and then adds nutrients to boost the development and activity of microbials. Bioremediation is also improved by aeration that increases the microbial activity. This is an economical method for supporting efficient biodegradation when sufficient amounts of nutrients and aerating exist (Whelan et al. 2015). It also avoids the gassing into the environment of low molecular weight (LMW) pollutants. The technology can be used also in cold areas (Whelan et al. 2015; Dias et al. 2015; Gomez and Sartaj 2014).

14.1.8.3 Windrows

The piled polluting soil is regularly shifted to improve bioremediation in this method. This ensures uniformity of the microbial population, pollutant concentration, water and aeration and nutrient accessibility throughout the polluted site (Barr 2002). Windrow is more efficient in the removal of hydrocarbons compared to biopiles. It is used, however, until the soil type is more friable (Coulon et al. 2010). The drawback is that greenhouse gases are released when anaerobic areas begin to grow in stacked areas (Hobson et al. 2005).

14.1.8.4 Bioreactor

This vessel is designed to provide microbial biological responses against contaminants that are fed as slurry. This technique offers closer control over parameters such as temperature, pH and aeration other than ex situ procedures, thereby efficiently improving biological responses to decrease bioremediation time. Furthermore, bioreactor also can be used to treat volatile organic compound contamination such as benzene.

14.1.8.5 Land Farming

The method can be considered either as ex situ or in situ depending on the pollutant concentration at the treatment location and is a low-price and extremely economical method. The treatment is in situ if excavated soil is processed at the same site; otherwise it is ex situ. If it is in situ and if the contaminant does not reach the depths of the soil, then it is not necessary to excavate polluted soil (Nikolopoulou et al. 2013). The groundwater pollute is stored after excavation over the floor to enable indigenous microbes to aerobically biodegrade (Philp and Atlas 2005; Paudyn et al. 2008; Volpe et al. 2012; Silva-Castro et al. 2015). Tillage and irrigation are the main activities of agriculture. Tilling enhances the accessibility of aeration and nutrients. This stimulates the operations and increases bioremediation of autochthonous microorganisms.

14.1.8.6 In Situ Bioremediation Technique

The polluted soil is not dug in this method; instead, contaminants are handled at the site with very less soil distortion. It is a cost-effective method because of reduced excavation costs. However, some construction and on-site machinery costs are involved for its efficient implementation. It is applied for the treatment of chlorinated solvents, dyes, heavy metals and polluted hydrocarbon locations (Folch et al. 2013; Kim et al. 2014; Frascari et al. 2015).

14.1.8.7 Bioventing

Bioventure takes place through regulated manipulation of the airflow to deliver oxygen to the unsaturated (vadosic) area and to boost indigenous microbial processes. In order to enhance bioremediation, nutrients and humidity are introduced (Philp and Atlas 2005). It is usually supplemented by other on-site techniques of bioremediation. It may be used to recover petroleum-polluted locations (Hohener and Ponsin 2014).

14.1.8.8 Bioslurping

This is a mixture of bioventing and vapour processing. It supplies an indirect source of oxygen (Gidarakos and Aivalioti 2007) and draws liquid in a stream by means of a pumping system. The biological procedures in this treatment refer to the aerobic biological degradation of hydrocarbons in the unsaturated area when air is introduced (Kim et al. 2014).

14.1.8.9 Biosparging

This technique is like bioventure. By better aeration, microbial activity is increased. Unlike bioventure, more ventilation is carried out in the saturated zone to allow those compounds that are volatile to move into the unsaturated area, promoting biodegradation (Philip and Atlas 2005).

14.1.8.10 Special Features of Bioremediation

- It is an environmentally friendly method with low labour, time and facility costs. It is secure. It is widely accepted by the public for treatment of waste at contaminated soils. It is risk-free as the amount of microbes decreases by itself when contaminant is removed. The residuals of the therapy are therefore completely harmless.
- The benefit of low cost and applicability over most polluted sites results in minimal disturbance to ordinary operations around the polluted site. There is no need to carry waste off-site, so any health and environmental risks that could occur from carrying toxic contaminants can be minimized.
- It guarantees that the pollutants are completely destroyed and that dangerous compounds are converted into harmless substances without any potential burden for processing and disposal of contaminated material.
- No dangerous chemicals need be introduced on the polluted site. The nutrients added to aid the development of microbes are only fertilizers commonly used in lawns and gardens that are not harmful.

14.1.9 Limitations of Bioremediation

- Only those compounds that are biodegradable are restricted to the applicability of bioremediation. Not all compounds can be degraded quickly and completely.
- Highly site-specific factors affecting the efficiency of biological restoration include the existence of metabolically competent microbial communities,

appropriate circumstances for microbial growth and suitable concentrations of nutrients and pollutants.

- Further study is necessary to bioremediate complicated contaminant mixtures as biodegradation variables in this situation are not yet been well known.
- Finally, bioremediation is limited by the longer processing time than standard alternatives, such as soil excavation and extraction or incineration.

References

Adams GO, Fufeyin PT, Okoro SE (2015) Bioremediation, biostimulation and bioaugmentation: a review. Int J Environ Bioremed Biodegrad 3(1):28–39

Akhtar N, Iqbal M, Iqbal ZS, Iqbal J (2008) Biosorption characteristics of unicellular green alga *Chlorella sorokiniana* immobilized in loofa sponge for removal of Cr(III). J Environ Sci 20:231–239

Barr D (2002) Biological methods for assessment and remediation of contaminated land: case studies. Construction Industry Research and Information Association, London

Bhattacharya M, Guchhait S, Biswas D, Datta S (2015) Waste lubricating oil removal in a batch reactor by mixed bacterial consortium: a kinetic study. Bioprocess Biosyst Eng 38:2095–2106

Bollag JM, Bollag WB (1995) Soil contamination and the feasibility of biological remediation. In: Skipper HD, Turco RF (eds) Bioremediation: science and applications: Soil Science Society of America. American Society of Agronomy, and Crop Science Society of America, Madison, p 1

Boopathy R (2000) Factors limiting bioremediation technologies. Bioresour Technol 74:63–67

Cases I, De Lorenzo V (2005) Genetically modified organisms for the environment: stories of success and failure and what we have learned from them. Int Microbiol 8:213–222

Chen X, Liu M, Hu F, Mao X, Li H (2007) Contributions of soil micro-fauna (protozoa and nematodes) to rhizosphere ecological functions. Acta Ecol Sin 27(8):3132–3143

Coulon F, Al Awadi M, Cowie W, Mardlin D, Pollard S, Cunningham C, Risdon G, Arthur P, Semple KT, Paton GI (2010) When is a soil remediated? Comparison of biopiled and windrowed soils contaminated with bunker-fuel in a full-scale trial. Environ Pollut 158:3032–3040

Couto N, Fritt-Rasmussen J, Jensen PE (2014) Suitability of oil bioremediation in an Artic soil using surplus heating from an incineration facility. Environ Sci Pollut Res 21(9):6221–6227

Das N, Chandran P (2011) Microbial degradation of petroleum hydrocarbon contaminants: an overview. Biotechnol Res Int:1–13

Davis TA, Volesky B, Mucci A (2003) A review of the biochemistry of heavy metal biosorption brown algae. Water Res 37:4311–4330

Dias RL, Ruberto L, Calabró A, Balbo AL, Del Panno MT, Mac Cormack WP (2015) Hydrocarbon removal and bacterial community structure in on-site biostimulated biopile systems designed for bioremediation of diesel-contaminated Antarctic soil. Polar Biol 38: 677–687

Dwivedi S (2012) Bioremediation of heavy metal by algae: current and future perspective. J Adv Lab Res Biol 3(3)

El Fantroussi S, Agathos SN (2005) Is bioaugmentation a feasible strategy for pollutant removal and site remediation? Curr Opin Microbiol 8:268–275

Enim AE (2013) Factors that determine bioremediation of organic compounds in the soil. Acad J Interdiscipl Stud 2(13):125–128

Floodgate G (1984) The fate of petroleum in marine ecosystems. In: Atlas RM (ed) Petroleum microbiology. Macmillan, New York, pp 355–398

Folch A, Vilaplana M, Amado L, Vicent R, Caminal G (2013) Fungal permeable reactive barrier to remediate groundwater in an artificial aquifer. J Hazard Mater 262:554–560

Frascari D, Zanaroli G, Danko AS (2015) In situ aerobic cometabolism of chlorinated solvents: a review. J Hazard Mater 283:382–399

Fritsche W, Hofrichter M (2005) Aerobic degradation of recalcitrant organic compounds by microorganisms. In: Jördening H-J, Winter J (eds) Environmental biotechnology: concepts and applications. Wiley-VCH Verlag GmbH & Co. KGaA, Weinheim. https://doi.org/10.1002/3527604286.ch7

Fritsche W, Hofrichter M (2008) Aerobic degradation by microorganisms. In: Rehm H-J, Reed G (eds) Biotechnology set, 2nd edn. Wiley-VCH VerlagGmbH, Weinheim. https://doi.org/10.1002/9783527620999.ch6m

Gidarakos E, Aivalioti M (2007) Large scale and long term application of bioslurping: the case of a Greek petroleum refinery site. J Hazard Mater 149:574–581

Gomez F, Sartaj M (2014) Optimization of field scale biopiles for bioremediation of petroleum hydrocarbon contaminated soil at low temperature conditions by response surface methodology (RSM). Int Biodeterior Biodegradation 89:103–109

Grishchenkov VG, Townsend RT, McDonald TJ, Autenrieth RL, Bonner JS, Boronin AM (2000) Degradation of petroleum hydrocarbons by facultative anaerobic bacteria under aerobic and anaerobic conditions. Process Biochem 35(9):889–896

Hobson AM, Frederickson J, Dise NB (2005) CH4 and N2O from mechanically turned windrow and vermin composting systems following in-vessel pre-treatment. Waste Manag 25:345–352

Hohener P, Ponsin V (2014) In situ vadose zone bioremediation. Curr Opin Biotechnol 27:1–7

Jain PK, Bajpai V (2012) Biotechnology of bioremediation – a review. Int J Environ Sci 3 (1):535–549

Kafilzadeh F, Sahragard P, Jamali H, Tahery Y (2011) Isolation and identification of hydrocarbons degrading bacteria in soil around Shiraz Refinery. Afr J Microbiol Res 4(19):3084–3089

Kanade SN, Ade AB, Khilare VC (2012) Malathion degradation by *Azospirillum lipoferum* Beijerinck. Sci Res Rep 2(1):94–103

Kang JW (2014) Removing environmental organic pollutants with bioremediation and phytoremediation. Biotechnol Lett 36(6):1129–1139

Khan AG (2006) Mycorrhizoremediation an enhanced form of phytoremediation. J Zhejiang Univ (Sci) 7(7):503–514

Kim S, Krajmalnik-Brown R, Kim J-O, Chung J (2014) Remediation of petroleum hydrocarbon-contaminated sites by DNA diagnosis-based bioslurping technology. Sci Total Environ 497:250–259

Kumar A, Bisht BS, Joshi VD (2011) Review on bioremediation of polluted environment: a management tool. Int J Environ Sci 1(6):1079–1093

Macaulay BM (2015) Understanding the behavior of oil-degrading micro-organisms to enhance the microbial remediation of spilled petroleum. Appl Ecol Environ Res 13(1):247–262

Madhavi GN, Mohini DD (2012) Review paper on – parameters affecting bioremediation. Int J Life Sci Pharma Res 2(3):77–80

Matavuly MN, Molitoris HP (2009) Marine fungi degraders of poly-3-hydroxyalkanoate based plastic materials. Proc Nat Sci Publ Matica Srpska 116:253–265

McKinney RE (1957) Activity of microorganisms in organic waste disposal. Appl Microbiol 5:00311–00033

Mrozik A, Piotrowska-Seget Z, Labuzek S (2003) Bacterial degradation and bioremediation of polycyclic aromatic hydrocarbons. Pol J Environ Stud 12(1):15–25

Nikolopoulou M, Pasadakis N, Norf H, Kalogerakis N (2013) Enhanced ex situ bioremediation of crude oil contaminated beach sand by supplementation with nutrients and rhamnolipids. Mar Pollut Bull 77:37–44

Paudyn K, Rutter A, Rowe RK, Poland JS (2008) Remediation of hydrocarbon contaminated soils in the Canadian Arctic by landfarming. Cold Reg Sci Technol 53:102–114

Pena-Castro JM, Martínez-Jeronimo F, Esparza-Garcia F, Canizares-Villanueva RO (2004) Heavy metals removal by the microalga *Scenedesmus incrassatulus* in continuous cultures. Bioresour Technol 94:219–222

Petric I, Hrak D, Fingler S, Vonina E, Cetkovic H, Kolar BA, UdikoviKoli N (2007) Enrichment and characterization of PCB-degrading bacteria as potential seed cultures for bioremediation of contaminated soil. Food Technol Biotechnol 45(1):11–20

Philp JC, Atlas RM (2005) Bioremediation of contaminated soils and aquifers. In: Atlas RM, Philp JC (eds) Bioremediation: applied microbial solutions for real-world environmental cleanup. American Society for Microbiology (ASM) Press, Washington, DC, pp 139–236

Phulia V, Jamwal A, Saxena N (2013) Technologies in aquatic bioremediation. In: Kumar P, Zaki BMSA, Chauhan A (eds) Freshwater ecosystem and xenobiotics. Discovery Publishing House PVT. Ltd., New Delhi, pp 65–91

Rathore AK (2017) Bioremediation: current research and application. IK International Publisher, New Delhi. ISBN:978-93-85909-60-3

Seeger M, Camara B, Hofer B (2001) Dehalogenation, denitration, dehydroxylation, and angular attack on substituted biphenyls and related compounds by a biphenyl dioxygenase. J Bacteriol 183:3548–3555

Sharma PD (2011–12) Ecology and environment, 11th edn. Rastogi Publications, Meerut, pp 346–374

Silva-Castro GA, Uad I, Gónzalez-López J, Fandiño CG, Toledo FL, Calvo C (2012) Application of selected microbial consortia combined with inorganic and oleophilic fertilizers to recuperate oil-polluted soil using land farming technology. Clean Techn Environ Policy 14:719–726

Silva-Castro GA, Uad I, Rodríguez-Calvo A, Gónzalez-López J, Calvo C (2015) Response of autochthonous microbiota of diesel polluted soils to land-farming treatments. Environ Res 137:49–58

Struthers JK, Jayachandran K, Moorman TB (1998) Biodegradation of atrazine by *Agrobacterium radiobacter* J14a and use of this strain in bioremediation of contaminated soil. Appl Environ Microbiol 64:3368–3375

Surekha Rani M, Lakshmi V, Suvarnalatha Devi KP, Jaya Madhuri R, Aruna S, Jyothi K, Narasimha G, Venkateswarlu K (2008) Isolation and characterization of a chlorpyrifos degrading bacterium from agricultural soil and its growth response. Afr J Microbiol Res 2:026–031

Taguchi K, Motoyama M, Kudo T (2001) PCB/biphenyl degradation gene cluster in *Rhodococcus rhodochrous*K37, is different from the well-known *bph* gene clusters in *Rhodococcus* sp. P6, RHA1, and TA42. RIKEN review. 42. Focused on ecomolecular science research

Vandevivere PC, Bianchi R, Verstraete W (1998) Treatment and reuse of wastewater from the textile wet-processing industry: review of emerging technologies. J Chem Technol Biotechnol 72:289–302

Vidali M (2001) Bioremediation – an overview. Pure Appl Chem 73(7):1163–1172

Volpe A, D'Arpa S, Del Moro G, Rossetti S, Tandoi V, Uricchio VF (2012) Fingerprinting hydrocarbons in a contaminated soil from an Italian natural reserve and assessment of the performance of a low-impact bioremediation approach. Water Air Soil Pollut 223:1773–1782

Wang Q, Zhang S, Li Y (2011) Potential approaches to improving biodegradation of hydrocarbons for bioremediation of crude oil pollution. Environ Prot J 2:47–55

Wang X, Wang Q, Wang S, Li F, Guo G (2012) Effect of biostimulation on community level physiological profiles of microorganisms in field-scale biopiles composed of aged oil sludge. Bioresour Technol 111:308–315

Whelan MJ, Coulon F, Hince G, Rayner J, McWatters R, Spedding T, Snape I (2015) Fate and transport of petroleum hydrocarbons in engineered biopiles in polar regions. Chemosphere 131:232–240

Wiedemeier TH, Miller RN, Wilson JT (1995) Significance of anaerobic processes for the intrinsic bioremediation of fuel hydrocarbons. In: Proceedings of the petroleum hydrocarbons and organic chemicals in groundwater – prevention, detection, and remediation conference, November 29–December 1, 1995, Houston, TX

Yakimov MM, Timmis KN, Golyshin PN (2007) Obligate oil-degrading marine bacteria. Curr Opin Biotechnol 18(3):257–266

Yang SZ, Jin HJ, Wei Z (2009) Bioremediation of oil spills in cold environments: a review. Pedosphere 19(3):371–381

Part VI

Microbes in Industries

Microbes in Food and Beverage Industry 15

Neegam Nain, Katoch Gunjan Kumari, Hridya Haridasan, and Shiwani Guleria Sharma

Abstract

Microorganisms play a crucial role in food spoilage and degradation. The unpleasant taste, odor and texture are the characteristics of spoiled food. However, despite being pathogenic, microorganisms play an important role in the production of many fermented food and beverages in household and food industries. Microbes are used in the fermentation of dairy products and development of alcoholic beverages. Yogurt, curd, sour cream, buttermilk, bread, and cheese are the basic examples of dairy products where microbes act as indispensable material for their production. Vegetables are also fermented to increase their shelf life and flavor. Pickles (cucumbers), sauerkraut (cabbage), soy sauce (soybeans), kimchi (Chinese cabbage), olives (green olives), etc. are examples of fermented vegetables. However, species of *Saccharomyces* play an inevitable role in the production of wine, beer, brews, champagne, etc. Today the demand for fermented food, probiotics and alcoholic beverages is increasing due to their taste and health benefits. This chapter reviews the different microorganisms which are involved in the production of food and beverages at industrial scale, and it also highlights the advantages of using the following microorganisms in food and beverage sector.

Keywords

Microbes · Food industry · Beverage industry · Yeast · Bacteria · Mold

---

N. Nain · K. G. Kumari
Department of Food Science and Technology, School of Agriculture, Lovely Professional University, Phagwara, Punjab, India

H. Haridasan · S. G. Sharma (✉)
Department of Microbiology, School of Bioengineering and Biosciences, Lovely Professional University, Phagwara, Punjab, India

S. G. Sharma et al. (eds.), *Microbial Diversity, Interventions and Scope*,
https://doi.org/10.1007/978-981-15-4099-8_15

15.1 Introduction

Microorganisms in food vary from bacteria to fungi. They survive by consuming the food constituents, i.e., carbohydrates, and utilize them in their metabolic pathways. These organisms can be divided as those that can cause spoilage and decrease food and beverage quality and those that aid in the processing of the foods and beverages and aid in preservation. Spoilage microorganisms are introduced into food from raw materials or from the industrial environment (Rawat 2015). These microorganisms contribute to major food loss in various food industries. These microorganisms are present in the environment and also on the surface of various vegetables and food. The major reason of the loss is mishandling during post-harvesting period. The growth of these microorganisms is favored by intrinsic and extrinsic factors. The intrinsic factors that favor their growth include water activity of the food material, pH, moisture, and nutrient content, while the extrinsic factors include temperature, relative humidity, and presence of gases or oxygen (Rawat 2015). The microbes also play a vital role in the assessment and maintenance of food safety and its quality in the industries (Martorell 2005).

Microbes play the most important role in the fermentation process, where they convert simple sugars to ethanol, acid, and carbon dioxide, such as in beer and wine industry or in yogurt and cheese industry. They are also used in the bakery industry to produce bread (Beresford et al. 2001). Microorganisms play an essential role in the processing of a range of foodstuffs in the industrial food products sector.

For the prevention of diseases in humans, antibiotics play a vital role. *Penicillium notatum*, a bacterium which produces penicillin, is such an example for the industries which use bacteria to develop antibiotic. *Saccharomyces cerevisiae* produce and retain beverages such as brandy, rum, whiskey, and beer. Enzymes such as lipase are also manufactured by microorganisms. *Saccharomyces cerevisiae* manufactures ethanol as one of the most essential industrial chemicals. The fungus *Trichoderma* prepares immunosuppressive agents such as cyclosporine. To preserve packed foods in food processing technology, a few microorganisms are used. The food usually consist lots of bacteria that may be useful to some including that which preserve food via fermentation products and others that cause human disease or spoilage of food.

The most important groups of microorganisms in food production include lactic acid bacteria, largely those of the genera *Pediococcus*, *Vagococcus*, *Leuconostoc*, *Oenococcus*, *Carnobacterium*, *Lactobacillus*, *Enterococcus*, *Lactococcus*, and *Streptococcus*. Carbohydrate fermentation combined with substrate-level phosphorylation is an important characteristic of lactic acid bacteria. The LAB is a source of growth inhibitors (bacteriocins) and large amounts of lactic acid, helping the flavor and texture of fermented products and inhibiting food spoilage. These bacteriocins are peptides which are ribosomically formed and extracellularly set free by bacteria. A variety of LAB are beneficial to people and animal, while others ruin beer, wine, and meats. For thousands of years, fermented milk products have been produced, but the microbiological basis of these fermentations has only been clarified last year (Beresford et al. 2001). Food microorganisms such as lactic acid bacteria are

Table 15.1 List of main microbes involved in food and beverage industries

S. no.	Microorganisms	Types of microorganisms	Food or beverage
1.	*Acetobacter aceti*	Bacteria	Chocolate and vinegar
2.	*Acetobacter cerevisiae*	Bacterium	Beer
3.	*Candida colliculosa*	Fungus	Cheese, kefir
4.	*Streptococcus, Leuconostoc, Lactobacillus*	Bacteria	Buttermilk
5.	*Streptococcus thermophiles, Lactobacillus bulgaricus*	Bacteria	Yogurt
6.	*Penicillium camemberti*	Fungus	Camembert cheese
7.	*Leuconostoc, Lactobacillus* species	Bacteria	Sauerkraut
8.	*Saccharomyces cerevisiae*	Yeast	Bread
9.	*Aspergillus niger*	Fungus	Citric acid
10.	*Acetobacter acute*	Bacteria	Acetic acid
11.	*Saccharomyces cerevisiae*	Yeast	Ethanol (green petrol)
12.	*Aspergillus oryzae*	Fungus	Soy sauce
13.	*Lactobacillus kimchii*	Bacteria	Kimchi
14.	*Mucor hiemalis*	Fungus	Soybean curd
15.	*Zymomonas mobilis*	Bacteria	Palm wine

predominant starter cultures that are used in processing fermented foodstuffs to get texture, flavor, and appearance. Current examples of fermented foods are milk products such as cheese, sour cream, and yogurt, meat products, and vegetable products (pickles, sauerkraut, olives) for which the commercial starter cultures are used. In order to be effective during the fermentation process, starting cultures should be controlled by microflora that is normal and produces the desired fermentation end products. A wide variety of activities such as lactose metabolism, proteinase activity, transport of oligopeptides, bacteriophage resistance mechanisms, production and immunity of bacteriocin, resistance of bacteria, manufacture of exopolysacchrides, and use of citrate are carried by plasmids containing a bacterial content of lactic acid. Progress in molecular technology has allowed superior strains of starter culture to be constructed with respect to food fermentation. Such strains have improved their immunity to bacteriophage and genetic stability and decreased variability, and performance is unpredictable. The incorporation of probiotic microorganisms to provide consumers with health benefits is yet another use of beneficial microorganisms in foods. The use of fermented foodstuffs has risen dramatically over the past two decades. Fermenting microorganisms (starter culture) used for food processing have been produced and available to meet this demand. This involves the creation of new and improved genetic engineering strains (Ali 2010) (Table 15.1).

15.2 Microorganisms Involved in Food and Beverage Industries

15.2.1 Molds

Mold generally refers to fungus that grows in the form of multicellular filaments called hyphae, and whole mass of these hyphae is known as mycelium. Fungi are eukaryotic and heterotrophs, who get nutrition from external environment or sources through absorption. However, the single-celled fungi are referred to as yeasts. The fungi are quite fuzzy or cottony in appearance. Molds are easily distinguishable from bacteria. They require less moisture as compared to bacteria and yeast. The minimum water activity for germination of spores has been deduced to be in between 0.62 and 0.93. The reduction in water activity can result in decrease in growth rate. Majority of the molds are considered to be mesophilic. The optimum temperature for molds is 25–30 °C. Although some molds are capable to grow at freezing temperature too. The reason why molds grow on the surface of spoiled food is that they require or need free oxygen for growth. Molds are able to grow over a wide range of pH. Molds have significant use in industries in manufacture of different foods. Molds majorly play part in cheese production (e.g., Roquefort, Camembert). They also play important part in the production of bread, Quorn, soy sauce, etc. Molds are widely used in pharmaceutical industry (Barnett 2003; Dagnas and Membre 2013).

Molds usually found on meat and poultry are *Aspergillus*, *Botrytis*, *Cladosporium*, *Fusarium*, *Geotrichum*, *Monilia*, *Mortierella*, *Mucor*, *Neurospora*, *Oidium*, *Oospora*, *Penicillium*, *Rhizopus*, *and Thamnidium*. These molds are easily detectable on other food components also. Molds being a spoilage microorganism also aid food and beverage industries at various levels (Dagnas and Membre 2013).

Molds have remarkable use at industrial scale in the production of different kinds of cheese. Blue-veined cheese is produced by the introduction of *Penicillium roqueforti*. Fungi participate in ripening and in enhancement of flavor and color of food. *P. nalgiovense* is used as starter culture for cured and fermented meat products. This species of mold promotes notably flavor of product, controls moisture loss, and arrests the development of mycotoxigenic fungal species. However, *P. camemberti* produces enzymes, some of which are associated with casein hydrolysis during cheese ripening, which aids in the development of flavor and texture of Camembert and Brie cheeses (Lasztity 2011; Smit et al. 2005). *Aspergillus flavus* is used in the production of various cheeses and soy sauce. *Rhizopus* is able to metabolize starch into glucose and then directly converts it into alcohol. *Mucor indicus* has a great possibility to be used as a nutritional rich source, e.g., fish feed. *Aspergillus niger* aids in the production of citric acid at industrial scale (Saxena 2015). Many fermentations also include species from *Neurospora*, *Monascus*, and *Actinomucor*. Typical products from temperate areas (cheese or soy sauce) are species of *Aspergillus* and *Penicillium*, while *Rhizopus, Amylomyces*, and *Mucor* are typical products from mainly tropical areas (tempe, tape). The foods developed by mold fermentation range from mold-rich cheeses and meats; delicious staples, such as tempe; flavoring foods including soy sauce; and sweet foods like candy and brem cake is important (Smit et al. 2005).

15.2.2 Yeasts

Yeasts are unicellular fungi that dwell in wide habitats. They are mostly found on the leaves and flowers of plants, soil, and water. They are also found on the skin of animals and sometimes in the intestinal tract as parasitic organisms or symbiotic organisms (Slávikovă and Vadkertiová 2003). These organisms divide asexually or sexually, asexually by budding, i.e., in *Saccharomyces*; direct division (fission), i.e., in *Schizosaccharomyces*; or growing as filaments (mycelium). Sexual reproduction is in the form of asci that contain eight haploid ascospores. The ascospores fuse with neighboring nuclei and divide by vegetative division or some fuse with other ascospores (Heslot and Gaillardin 1992).

Yeast was used in 6000 BC to make beer, wine, and bread by the Egyptians and later adopted by the Romans. This is how yeast is being currently used in the industry for alcoholic fermentation and baking. Their limitless capacity to break down simple sugars to produce alcohol, fats, enzymes, and heterologous proteins allows its application in the industry (Halasz and Lásztity 1991).

In baking, the most common yeast used is *Saccharomyces cerevisiae.* It is used as a leaving or raising agent. This occurs when the organisms break down the sugars present in the dough producing carbon dioxide gas that causes bubble formation in the dough causing it to rise. In brewing, several yeast species are used to brew beer. It is a very good protein source and a nutrient source rich in vitamin B. The products of brewer's yeast are normally found in liquid form, in powders, or in tablets. This is made from mainly cereals, i.e., barley, and they ferment the carbohydrates (sugar) to produce alcohol. In wine making, it is formed by the action of the yeast by fermentation on the sugars present in grape juice producing carbon dioxide as its by-product. Mostly, the yeast can be found on the grape skin, and it is sufficient to allow for the fermentation to occur. Other fermentative processes where yeast can be used are in soy sauce production (Otero et al. 1998; Corran 1975).

In dairy products, yeast is mainly used to enhance flavor and texture in manufacture of fermented milks, i.e., kefir and koumiss, and also in cheese production. They are being applied as secondary starter cultures in enriching the growth of lactic acid bacteria and boosting aroma of food substances. The mostly used species are *S. cerevisiae*, *D. hansenii*, and *K. marxianus* among others (Tofalo and Suzzi 2016). Starter culture includes yeasts for the production of certain kinds of fermented products such as cheese, bread, vegetable products, vinegar, etc. As alternative protein sources that meet requirements in a world of low production and rapidly developing populations, the importance of yeasts in food technology and human nutrition renders the development of food yeasts. There are also remarkable advantages of single cell protein (SCP) microorganisms relative to traditional protein sources (soya or meat). The advantages are well established. The high levels of protein in microorganisms are low contributing to the quick development of biomass, which can continue and is independent of environmental conditions. Yeasts can grow with or without oxygen as an optional anaerobe. Industrial alcohol and spirit such as brandy, rum, and tequila are processed by distillery yeast. Probiotic properties of yeasts have been documented and shown in the way of survival by gastrointestinal pathogens including *E. coli*, *Salmonella*, and *Shigella*. More precisely *S. boulardii* is a thermophilic nonpathogenic yeast used as a probiotic supplement for a wide range of gut disorders, such as diarrhea for over 50 years. The torula yeast is produced in sugar and mineral blends generally comprising molasses, celluloses, or brewing by-products. In the food industry, table olive is a major fermenting product. It cannot be ingested directly and must be processed depending on some special features such as the oleuropein bitter component, low sugar content, or high oil levels. Microorganisms such as lactic acid bacteria and yeast play significant roles in the production of table olives.

15.2.3 Bacteria

Bacteria are prokaryotic microorganisms present widely in natural environment. Bacteria divide at very high rate. Some of the bacteria can sustain themselves in extreme conditions. The minimum water activity for multiplication of bacteria is 0, and the optimal temperature for bacterial growth is 37 °C. Some species of bacteria are pathogenic or disease causing in nature, whereas some are involved in food spoilage. *Clostridium botulinum*, *C. perfringens*, and *Bacillus cereus* are known to cause food intoxications (Kirkland and Fierer 1996).

Bacteria play a vital role in the production of food and beverages in food industries as well as in household. Lactic acid bacteria play chief role in a variety of fermented drinks and products. This group of bacteria is able to break down carbohydrates and produce lactic acid, is involved in the degradation of proteins and lipids and manufacturing alcoholic beverages, and helps in the development of curd, yogurt, and fermented milk and is also involved in the processing of fermented fish, meat, vegetables, etc. Besides, they also help in enhancing the flavor, texture, and nutritional value of the fermented food (sauerkraut, kimchi). *Lactobacillus, Leuconostoc, Pediococcus, and Streptococcus* are the major LAB genera that are involved in food and beverage production.

However, in the production of wine (malolactic fermentation) and rice wine, *O. oeni* and *Lb. sakei* bacteria are involved (Saxena 2015). Many microorganisms have characteristics that can support food production and transformation. A variety of fermenting foods from crude animals and plants are produced by many food microorganisms. The fermentation of these microorganisms is the result of the acidic and partially organoleptic properties of fermentation products. The foods such as ripened cheeses, fermented sausage, sauerkraut, and pickles have not only a considerably longer life span than their raw materials but also the flavor and aroma characteristics that fermented organisms contribute directly or indirectly. LAB are the main entities engaged in dairy fermentation. Milk fermentations depended naturally on the LAB in raw milk before the availability of starter cultures.

15.3 Health Benefits of Microbes in Food

Benefits of microbial fermentations:

- **Microbial fermentation**

Various strains of bacteria and fungi are being used during food fermentation that yields various sorts of cultured products with enhanced flavor, taste, and aroma.

The following are examples of fermented products:

1. In dairy products, fermented milk products. During milk coagulation, various types of cheese are produced, e.g., soft unripened and ripened and various hard types. Mostly lactic acid–fermenting bacteria are used.
2. Vegetable, e.g., sauerkraut, pickles, and olive production.
3. Meat products, e.g., fermented sausage preparation.
4. Bread and alcoholic beverages, e.g., wine, beer, cider, and vinegar. During alcoholic beverage production, different yeast strains are used by fermentation of cereals and grains. Wines can be produced by action of molds on rotting grapes.

These microorganisms are used differently among various food industries, i.e., various fermentation industries, for example, breweries and vinegar manufacturers culture their own strains and inocula. In the dairy industry, meat industry, and bakeries, cultures are mostly obtained from suppliers that produce high-quality food ingredients (Hansen, 2004; Mogensen et al. 2002).

- **Food additives from microorganisms**

  Some microorganisms are used for the production of processing aids that are used in the food industry, i.e., lactase prepared from strains of *Aspergillus niger*, *Aspergillus oryzae*, and *Kluyveromyces lactis*. It is mainly used in the preparation of low-lactose or lactose-free foods for lactose-intolerant individuals. In lactase-treated milk, its advantage is that there is increased sweetness of the milk and hence sugar addition can be avoided in flavored milk manufacture. Lactase can also be used in ice cream, yogurt, and frozen dessert industries to enhance sweetness, creaminess, and digestibility. Studies also revealed that cheese prepared from hydrolyzed milk ripens faster than that prepared from normal milk (Neelakantan et al. 1999). Other food additives include alkaline phosphates, microbial lipases, and flavor enhancement, i.e., after fermentation.
- **Preservation**

Microorganisms can also be used in the preservation of the food products, i.e., enhancing the shelf life of food and beverage products (Chen and Hoover 2003). The production of antimicrobial products from microorganisms, i.e., lactic acid-producing bacteria, enhances the shelf-life such as bacteriocins, organic acids, hydrogen peroxides, carbon dioxide, and diacetyls.

– **Probiotic and functional food production**

These types of products allow the consumer to eat or drink healthy foods or drinks without bringing a change in their diet. Probiotics have been defined as "live microorganisms which when administered in adequate amounts confer a health benefit on the host" (FAO/WHO 2002). These organisms should have the ability to survive the gut passage, i.e., acid tolerant and resistance to bile. These microorganisms must be safe and effective and also maintain effectiveness and potency for the time period of the product's shelf-life (FAO/WHO 2002; Saad et al. 2013).

Various genera of microorganisms are utilized as probiotics. The frequently used strains are lactic acid bacteria, e.g., *Lactobacillus* and *Enterococcus* and from the genus of *Bifidobacterium* (Ouwehand et al. 2002; Saad et al. 2013; Bintsis 2018). Currently, probiotics main applications are in dairy product production, e.g., cheese, yogurt, and ice cream. Probiotics oldest source is from fermented dairy products and has strains of *Lactobacillus* and currently has been added to cooked pork meat, fruit juices, chocolate, and chewing gum (Bernardeau et al. 2006; Ouwehand et al. 2002; Ranadheera et al. 2010).

15.4 Drawbacks of Using Microbes in Food and Beverage Industries

Although microorganisms play a vital role in the production of many edible products in food and beverage industries, some of the researches have concluded the harmful impacts by consumption of fermented products. During fermentation of some non-alcoholic food, the microorganisms produce alcohol as by-product. Soy sauce is a fermented food which contains alcohol in small quantity. Lactic acid and ammonia are other products of fermentation, which are considered to affect human health. The processed food possesses fewer nutrients as compared to the food which is consumed in their natural, fresh, and unprocessed state. According to the research, the fermented food usually contains less vitamin B12 (Kharobe 2018).

Few microorganisms are capable of releasing toxins such as biogenic amines and aflatoxins. Dairy products are perishable commodities and highly vulnerable to be contaminated by pathogens. Sometimes proliferation occurs during manufacturing which in turn causes sporadic cases or outbreaks of disease (Fernández et al. 2015).

Bifidobacteria is a vital group of nonstarter microorganisms that are included in manufacturing of some dairy products, such as fermented milks; their proliferation will add up to increase in the level of lactate and acetate in final products (Cremonesi et al. 2012).

Microorganisms majorly produce beneficial compounds during dairy fermentation, but in some cases metabolic activities result in the discharge of toxic substances. Mycotoxins produced by some fungi and biogenic amines mainly due to the metabolic activity of some *Lactobacillus* bacteria are the two types of toxic compounds that have been found in dairy products. *Aspergillus*, *Fusarium*, and

Penicillium are the three genera of filamentous fungi involved in releasing mycotoxins (Delavenne et al. 2011).

15.5 Conclusion

Microbes in the food and beverage industry can be monitored to produce a modified product, i.e., fermented food and beverages, or if not checked, then they can be a leading cause of food deterioration after processing. Raw food materials before processing should be decontaminated to reduce risk of spoilage and contamination. During processing, the microbes used for production of food should be monitored and given optimum conditions, and contamination from outer sources should be reduced. There are various methods that can be used to get rid of spoilage-causing microorganisms, i.e., application of high temperature, pH, or aseptic conditions (Martorell 2005). Different strains of useful microbes are used for various purposes, and these microbes should be handled separately.

References

Ali AA (2010) Beneficial role of lactic acid bacteria in food preservation and human health: a review. Res J Microbiol 5(12):1213–1221

Barnett JA (2003) Beginnings of microbiology and biochemistry: the contribution of yeast research. Microbiology 149(3):557–567

Beresford TP, Fitzsimons NA, Brennan NL, Cogan TM (2001) Recent advances in cheese microbiology. Int Dairy J 11:259–274

Bernardeau M, Guguen M, Vernoux JP (2006) Beneficial *Lactobacilli* in food and feed: long- term use, biodiversity and proposals for specific and realistic safety assessments. FEMS Microbiol Rev 30:487–513

Bintsis T (2018) Lactic acid bacteria: their applications in foods. J Microbiol Mycol 6(2):89–94

Chen H, Hoover DG (2003) Bacteriocins and their food applications. Compr Rev Food Sci Food Saf 2:82–99

Corran HS (ed) (1975) A history of brewing. Davis & Charles, Newton Abbot

Cremonesi P, Vanoni L, Silvetti T, Morandi S, Brasca M (2012) Identification of *Clostridium beijerinckii, Cl. butyricum, Cl. sporogenes, Cl. tyrobutyricum* isolated from silage, raw milk and hard cheese by a multiplex PCR assay. J Dairy Res 79(3):318–323

Dagnas S, Membre JM (2013) Predicting and preventing mold spoilage of food products. J Food Prot 76(3):538–551

Delavenne E, Mounier J, Asmani K, Jany JL, Barbier G, Blay GL (2011) Fungal diversity in cow, goat and ewe milk. Int J Food Microbiol 151(2):247–251

FAO/WHO (2002) Joint FAO/WHO (Food and Agriculture Organization/World Health Organization) Working Group report on drafting guidelines for the evaluation of probiotics in food, London, Ontario, Canada

Fernández M, Hudson JA, Korpela R, de los Reyes-Gavilán CG (2015) Impact on human health of microorganisms present in fermented dairy products. Biomed Res Int 2015:13

Halasz A, Lásztity R (1991) Use of yeast biomass in food production. CRC Press, Boca Raton. 352p

Hansen EB (2004) Micro-organisms. In: Hui YH, Meunier-Goddik L, Hansen ÅS, Josephsen J, Nip WK, Stanfield PS, Toldrá F (eds) Handbook of food and beverage fermentation technology, vol 2. Marcel Dekker, Inc., New York, pp 9–21

Heslot H, Gaillardin C (1992) Molecular biology and genetic engineering of yeast. CRC Press, Boca Raton

Kharobe VB (2018) Advantages and disadvantages of traditional fermentation of dairy products. Int J Recent Trend Sci Technol 2018:52–55

Kirkland TN, Fierer J (1996) Coccidiomycosis: a reemerging infectious disease. Emerg Infect Dis 2 (3):192–199

Lasztity R (2011) Microorganisms important in food microbiology. Food quality and standards. Vol 3

Martorell MT (2005) Fernandez-Espinar and A. Querol. Int J Food Microbiol 101:293–302

Mogensen G, Salminen S, O'Brien J, Ouwehand A, Holzapfel W, Shortt C, Fonden R, Miller GD, Donohue D, Playne M, Crittenden R, Salvadori B, Zink R (2002) Inventory of micro-organisms with a documented history of use in food. Bull IDF 377:10

Neelakantan S, Mohanty A, Kaushik JK (1999) Production and use of microbial enzymes for dairy processing. Curr Sci 77:143–148

Otero MA, Cabello AJ, Vasallo MC, Garcia L, Lopez JC (1998) Preparation of an imitation soy sauce from hydrolyzed dried yeast *Candida utilis*. J Food Process Preserv 22:419–432

Ouwehand AC, Salminen S, Isolauri E (2002) Probiotics: an overview of beneficial effects. Antonie Van Leeuwenhoek 82:279–289

Ranadheera R, Baines S, Adams M (2010) Importance of food in probiotic efficacy. Food Res Int 43:1–7

Rawat S (2015) Food spoilage: microorganisms and their prevention. Asian J Plant Sci Res 5 (4):47–56

Saad N, Delattre C, Urdaci M, Schmitter JM, Bressollier P (2013) An overview of the last advances in probiotic and prebiotic field. LWT Food Sci Technol 50:1–16

Saxena R (2015) Beneficial role of microorganisms in food industry. Microbiology

Sláviková E, Vadkertiová R (2003) The diversity of yeasts in the agricultural soil. J Basic Microbiol 43(5):430–436

Smit G, Smit BA, Engels WJ (2005) Flavour formation by lactic acid bacteria and biochemical flavour profiling of cheese products. FEMS Microbiol Rev 29(3):591–610

Tofalo R, Suzzi G (2016) Yeasts. In: Encyclopedia of food and health. Academic, Oxford, pp 593–599. https://doi.org/10.1016/b978-0-12-384730-0.00292-5

Microbes in Pharmaceutical Industry 16

Divya Kapoor, Pankaj Sharma, Mayur Mukut Murlidhar Sharma, Anju Kumari, and Rakesh Kumar

Abstract

Microbes are ubiquitous in nature, and the tremendous potential of microbes is an unquestionable and unhidden phenomenon. The pharmaceutical aspects of microbial diversity can be judiciously explored for patronizing and safeguarding human health standards. Several studies have unveiled the remarkable wonders which target to prolong and ease the human health by treating pathogenic diseases and by curing different metabolic disorders, thereby sustaining human regime. Generally, the aim of pharmaceutical microbiology is to offer acquaintance and consider the importance of the occurrence of bacteria, yeasts, moulds, viruses and toxins in diverse pharmacological raw materials, products, intermediates and the environs advocating therapeutic construction as well as the microbial regulator of medicinal harvests, manufacturing surroundings and people. Meanwhile, the outline of this functional theme area of microbiology, above the ages, pharmacological microbiology, has advanced and stretched expressively to comprehend numerous other sides, e.g. examination and expansion of novel anti-infective representatives, the use of microbes to perceive mutagenic and oncogenic prospective in medications and the usage of microbes in the production of insulin and various other human growth hormone. An array of bioactive composites sequestered using different approaches have not only exposed prominence in diverse pharmaceutical and biotechnological solicitations but have also

D. Kapoor · P. Sharma · R. Kumar (✉)
Department of Microbiology, CCS Haryana Agricultural University, Hisar, Haryana, India
e-mail: sehrawatrk@hau.ac.in

M. M. M. Sharma
Department of Agriculture and Life Industry, Kangwon National University, Chuncheon, Republic of Korea

A. Kumari
Center of Food Science & Technology, CCS Haryana Agricultural University, Hisar, Haryana, India

S. G. Sharma et al. (eds.), *Microbial Diversity, Interventions and Scope*,
https://doi.org/10.1007/978-981-15-4099-8_16

augmented the indulgence of mankind in exploring variety of microbiota and targeting their diverse functions and the credulous biology behind their production. The defensible and pecuniary stream of the dynamic pharmaceutical elements is often calmer to accomplish for composites fashioned through microbial fermentation attitudes versus the gardening of slower developing macroorganisms. This article stresses on microbial fabricators and their potential to engender innovative biologically active compounds and their starring role in the simplification of human life.

Keywords

Therapeutic · Antibiotics · Bacteriocins · Biosurfactants · Nutraceuticals

16.1 Introduction

The pharmaceutical industry has suffered from a key change in the previous years. The financial side of the industry requires certain new and epic drugs annually for the company's lucrativeness. Accordingly, this has resulted into a wide-ranging succession of unions among the hulks of the industrial sectors which was further accompanied by their acquirements of biopharmaceutical corporations running at a smaller scale. However, the above-mentioned approach was also not able to resolve the pecuniary problem as the marketable productivity of the pooled corporations was not even found to be equivalent to the entire quantity of products commercialized by the establishments prevailing earlier than the merger obsession seized over (Demain 2002). A wide variety of products, like anticancer cytotoxic drugs and vaccines, anti-infectious disease antibiotics and vaccines, to hormonal disorder therapy and many other indications, which owe their origin to different microbes, find applications in the pharmaceutical industry. Microbes are often used for producing a passel of different primary and secondary products to benefit mankind for many years (Demain 2004). The pharmaceutical industry is primarily inclined on microbial systems for several different types of molecules, for instance, small peptides and some small molecular mass organic molecules, greater molecules comprising proteins and nucleic acids as well as diverse macromolecules, for instance, lipids and carbohydrate polymers, plus numerous amalgamations of product types, for instance, lipopeptides, lipopolysaccharides and peptidoglycan. The active pharmaceutical ingredient of any drug is represented by some of these product types. The secondary metabolites finding their root from microbial as well as plant systems have significantly enhanced the life term of human beings all through the twentieth century. They have also contributed a lot in plummeting pain and sorrows and thus have transfigured the medicines. A major proportion of the permitted medications are either natural harvests or anyhow interrelated to them, even exclusive of biologicals, for instance, vaccines and monoclonal antibodies. (https://www.manufacturingchemist.com/news/article_page/Production_of_pharmaceutical_compounds_through_microbial_fermentation/61614). The three billion years of bacterial residence on the earth has resulted in the creation of nature's own kind of combinatorial chemistry, which appears to be much more striking and mysterious than its chemical counterpart, and thereby has allowed the foundation of millions of unusual and unique secondary metabolites. The progression of combinatorial biosynthesis has also opened different and novel ways for synthesizing new products by any technique of biological origin corresponding to combinatorial chemistry. Moreover, these natural products, although endowed with structures having higher levels of spatial complexity as equalled to the synthetic and artificial chemicals, have been prodigiously advantageous to mankind. The discriminatory as well as novel approach of targeting pathogenic bacteria and fungi by secondary metabolites of microbial origin ushered in the antibiotic age, and for over five decades, the humanity has been promoted from this extraordinary possession of 'wonder drugs'. The marketplace for such brilliant composites is over \$30 billion per annum (Demain 2002). This incredible pharmaceutical prospective of microorganisms is being exploited even before the advent of fermentation industry, for instance, cow pox vaccine served as the first proteinaceous vaccine that was developed by Jenner in 1796 even much before the birth of fermentation industry which happened in the early 1900s, landmarked by the development of the first large-scale anaerobic fermentations meant for manufacturing different metabolites. The major milestone achieved in the pharmaceutical industry was the discovery of penicillin in 1927 which proved to be a major milestone for advocating role of microbes in pharmaceutical industry. However, its expansion did not ascend awaiting the beginning of the 1940s, erstwhile the time of unearthing of another major antibiotic streptomycin. The opening protein pharmaceutical manufactured was the hormone insulin by Banting and Best in 1922 (Demain and Vaishnav 2009). The beginning of genetic engineering made available the recombinant proteins into the marketplace that completely transformed the state of the pharma industries (Demain 2004). Later, the modern era of biotechnology commenced in 1971 with the founding of the Cetus Corporation in California prior to the innovation of recombinant DNA by Berg, Cohen and Boyer in California (Demain and Vaishnav 2009). At present, there are above 200 permitted peptide as well as protein pharmaceutical products that appear on the FDA list. The major recombinant protein pharmaceuticals that are manufactured are human growth hormone (HGH), human insulin, albumin, factor VIII and numerous others. These biopharmaceuticals have proved to be influential in profoundly refining human health (Swartz 1996). The advancement in molecular biology is the most important dynamic strength in fuelling biopharmaceutical investigation studies for the fabrication of different kinds of products. The biopharma industry is highly polygonal, dealing with proteomics, metabolomics, antisense molecules, ribozymes, monoclonal antibodies, genomics, pharmacogenomics, combinatorial chemistry and biosynthesis, high-throughput screening, bioinformatics, nanobiotechnology, gene therapy, tissue engineering and countless additional stuffs. The evolution of genetic engineering has also largely transformed the aspects of medicine, pharmacology as well as industry. The sound understanding of microbial systems has also provided a unique and irreplaceable platform for an ever-increasing number of different microbial metabolites which offer a pharmaceutical aptitude for enhancing the livelihood of mankind. Although the acknowledgement given to microbes for their highly beneficial attributes can never be accounted totally, in this chapter an effort has been made to crease the evidence on different roles of microbes for their unique and lucrative pharmaceutical aspects.

16.2 Therapeutic Products

The sustainable and financially viable streaming of the active pharmaceutical ingredient appears frequently calmer to attain for several composites whose production is basically grounded on microbial fermentation attitudes as paralleled with the rearing of sluggishly growing macroorganisms. A number of developments in the methodologies employed in fermentation technologies, biosynthesis and synthesis offer various prospects meant for both creation and delivering drug leads that seem either unavailable or non-realistic by any single method unconventionally (Waters et al. 2010). The act of drug innovation from sources of natural origin has been reliant on the high-throughput screens and selections of extracts primed on plants, microorganisms and a few higher organisms from a very long time. Among microorganisms, predominantly bacteria belonging to the order *Actinomycetales*, along with various filamentous fungi, have ascertained to be exceptionally rich bases of novel antibiotics, cholesterol-lowering drugs, anticancer therapeutics and so on. However, with the lapse of time, the finding of unique composites experienced a dramatic drop from these sources. This significant drop is primarily governed by the frequent and repeated isolation of the previously acknowledged complexes. This propensity provoked a general perception that the potential of microbes as foundations for novel therapeutics compounds has been shattered (Newman and Cragg 2007; Bentley et al. 2002). However, the genetic analysis of several microbes has exposed the unparalleled capability of these microbes to synthesize several novel complexes hitherto unnoticed due to the employment of orthodox approaches and methodologies meant for cultivating, extracting and bioactivity testing. Superficially, the genes coding the biosynthesis of such novel complexes are supposed to express at a very truncated level in the laboratory environments which makes the recognition process an enormously challenging job (Zazopoulos et al. 2003). Therefore, microbes are a source of several kinds of therapeutic compounds which are produced as either primary or secondary metabolites.

16.2.1 Antibiotics

The attribute of utmost significance of microbial systems is that microbes can afford an ecologically viable supply of preparatory ingredients meant for the assembly of exceptional classes of drugs along with the multifarious and exclusive chemical space shielded by the natural products they yield (Cuevas and Francesch 2009). Microbial systems are greatly acknowledged for producing numerous compounds that exhibit biocidal activities. Since their clinical employment eight decades ago, antibiotics have become the foundation of modern medicine (Bush et al. 2011). Although a milestone has been travelled if the history of antibiotic production is traced, the prevalence of infectious diseases still accounts for the second major cause of the global deaths, where the bacterial infections are responsible for around 17 million deaths on an annual basis which primarily affects children and the old ones. The term *antibiotic* was first introduced by Selman Waksman in 1941 for designating any small molecule made by a microbe that antagonizes the growth of other microbes. Although the term was used very later after the discovery by Alexander Fleming in 1928, which marked the beginning of the microbial drug era when he discovered in a Petri dish seeded with *Staphylococcus aureus* that a compound produced by a mould killed the bacteria. The mould, branded as *Penicillium notatum*, was producing an active agent that was later named as penicillin. Later, this penicillin was isolated as a yellow powder and was used as a potent antibacterial compound during World War II. By using Fleming's method, other naturally occurring substances, such as chloramphenicol and streptomycin, were isolated (Demain and Sanchez 2009). From 1945–1955, the development of penicillin, along with streptomycin, chloramphenicol and tetracycline, which are produced by soil bacteria, ushered in the antibiotic age (Clardy et al. 2009). The microbial systems are solely exploited by the antibiotic industry where the genus *Streptomyces* solely accounts for approximately 80% of the antibiotics (Watve et al. 2001) including the representative antibiotics like streptomycin, cephalosporin, chloramphenicol, kanamycin, etc. According to an analysis, it is appraised that the screening of around 10,000 actinomycetes would yield almost 2500 antibiotic producers. In addition to it, 2250 of these would synthesize streptothricin, 125 streptomycin and 40 tetracycline. The antibiotic vancomycin is anticipated to be produced by one in a hundred thousand; erythromycin, by one in a million; and daptomycin, the newest antibiotic, by one in ten million. The soil bacteria that synthesize so many antibiotics live in unusually complex multispecies surroundings, and tracing both neighbours and ancestors will be a daunting task (Clardy et al. 2009). The industrial production of naturally produced antibiotics is carried out by the process of microbial fermentation.

There is always a quest to cultivate the operative next-generation antimicrobial rehabilitations, therefore a domineering appreciative of bacterial response towards antibiotics and further leveraging this considerate for developing treatments that grow the drug effectiveness beyond the contemporary practices (Stokes et al. 2019). The current collection of antibiotics has principally ensued from screens premeditated for identification of molecules constraining microbial progression in vitro (Brown and Wright 2016). Although an extensive gamut of distinct bioactive composites has been exposed by targeting this approach, only a few cellular progressions are targeted (Kohanski et al. 2010). Although exceptions can never be circumvented, the major cellular processes that are often targeted are (1) biogenesis of cell envelope (2) DNA replication, (3) transcription and (4) protein biosynthesis. These processes present the principal roles in expediting cell growth accompanied with the division. These practices are also found to consume a chief segment of the energy reserve of the cell, where around 70% of the total ATP consumption is solely accounted by the protein biosynthesis (Stouthamer 1973). Therefore, the disconcertion of such energy-overwhelming processes is not astonishing as antibiotics persuade substantial, yet recurrently disregarded, trepidations to metabolic homeostasis (Belenky et al. 2015; Zampieri et al. 2017). The efficacy of such biotechnologically important molecules can be acknowledged by the fact that these molecules amend the metabolic state of the bacterial cells thereby pushing them towards a stage of death or stasis. Furthermore, the vulnerability of the bacterial cells towards any particular antibiotic compound depends on the metabolic state of the cell. Therefore, the efficacy of any particular antibiotic compound can be largely enhanced by altering the metabolic state of the cell (Stokes et al. 2019). So, by considering these facts, the processes of antibiotic development against any pathogen burden the investigators to put the bacterial metabolism at the forefront.

The universal market share of antibiotic industry was found to be equivalent to USD 45.31 billion in 2018, and it is also speculated that it will reach USD 62.06 billion by 2026 and would expand at a compound annual growth rate of 4.0% over the estimated period. The cumulative upsurge in the cases of infectious diseases accompanied by imbalanced supply of antibiotics is the chief driving force for the proliferation of this market. Furthermore, the accumulative exertions commenced by major companies in the quest to develop advanced products will also support market growth. According to data published by the Pew Charitable Trust, in March 2016, about 37 promising molecules were being investigated within the US market. Majority of these are in phase II clinical trials and are anticipated to hit the market between 2018 and 2020 (Anonymous 2019). Diarrhoea is designated to be the foremost reason accounting for global death among children; therefore, it also necessitates antibiotic involvement in order to avoid the morbidity. The other infectious diseases acknowledged to pose high burden are HIV/AIDS, pneumonia, malaria and tuberculosis. Furthermore, the emergence of new infections, such as the Zika and Ebola virus, is also supporting the development and uptake of antibiotics. Further, the increasing phenomenon of antibiotic resistance also upsurges the research in this field.

16.2.2 Bacteriocins

Bacteriocins are peptides having low-molecular weight which are released by producer killer bacterial cell in order to destroy the other sensitive and closely related cells inhabiting the similar environment and competing for foodstuff as well as additional nutrients (Juturu and Wu 2018). The bacteriocins are classified as of narrow spectrum if the bacteriocins are found to be inhibitory to the members fitting to the same species, whereas they are pronounced as broad spectrum if these are found to be killer against different bacteria finding their origin from another groups. However, the producer microbe develops resistance towards their own killer peptidal molecules by producing explicit immunity proteins (Cotter et al. 2005). Bacteriocins are ribosomally made antimicrobial peptides and are conventionally being employed as food additives, which are either supplemented or shaped by the starter cultures throughout the fermentation process (Chikindas et al. 2018). Unusually, a few bacteriocins are also reported which along with the possession of their innate antibacterial property also display supplementary antiviral as well as antifungal activities. Although a vast array of Gram-positive and Gram-negative microbes along with *Archaea* are reported to produce bacteriocins, the bacteriocins from Gram-positive bacteria, exclusively from lactic acid bacteria, have comprehensively been inspected, bearing in mind their excessive biosafety as well as broad industrial applications.

The foremost description on the innovation of such antibacterial element fashioned by *Lactococcus lactis* subsp. *lactis* was in the year 1933 by Whitehead (Whitehead 1933; Juturu and Wu 2018) which was later entitled nisin (Mattick and Hirsch 1947). Nisin is also the sole bacteriocin that had got successful in getting FDA endorsement for its solicitation as a preservative in food industry. Its antimicrobial properties are well explained by a clear understanding of its crystal structure. It causes the death of pathogens, explained by constraining the synthesis of peptidoglycan layer, which is coupled with pore formation in cellular membrane, which ultimately clues the cell lysis (Juturu and Wu 2018). Some bacteriocins have different mechanism of action. They kill the sensitive cells by interfering with the progression of cell division by averting the directed cell from the normal processes of cell cycle (Hasper et al. 2006). The bacteriocins fashioned by numerous bacteria are found to be dynamic against various human as well as animal microbial pathogens, comprising vancomycin-resistant enterococci (VRE) and methicillin-resistant *Staphylococcus aureus* (MRSA) without displaying toxicity, for instance, the bacteriocins lacticin 3147 and nisin A are reported to be operative against MRSA and VRE. Their usage as antimicrobials is further advocated by their physical stability (Morgan et al. 2005; Piper et al. 2009; Ahmad et al. 2017). Such properties of these bioactive compounds make them potent molecules for treating hospital acquired infections. Bacteriocins have also been widely found to the pathogens of respiratory tract which further suggests their role in treating deadly infections of respiratory column (De Kwaadsteniet et al. 2009; Ahmad et al. 2017). There are several other properties of bacteriocins which make them superior molecules and extend their applications from food to human health. There is greater opportunity of mounting these antimicrobial peptides into next-generation antibiotics which is also escorted with the quick expansion of genetics along with nanotechnology. This development has also cemented the way to even additional charming uses, for instance, unique transporter molecules and the usage as anticancer compound. Furthermore, a few bacteriocins have also been reported to control quorum sensing process which further advocates innovative applications for these substances (Hong et al. 2012; Chikindas et al. 2018). Other bacteriocins, for example, subtilosin A manufactured by *Bacillus subtilis*, are also known to possess antiviral (Quintana et al. 2014) and spermicidal accomplishments (Sutyak et al. 2008) which extend their usage in diverse fields. The spermicidal activity possessed by subtilin and lacticin 3147 projects their use as safe and natural contraceptive in humans along with animals. Another bacteriocin, fermenticin, also displays striking immobilization of sperm as well as spermicidal activity, thus suggesting it as a striking agent for articulating antibacterial vaginosis as well as in the formulation of contraceptive products (Kaewnopparat et al. 2013). The antiviral activity of bacteriocin is attributable to its ability to interfere with the later phases of viral replication (Reddy et al. 2004). In addition to the possession of antiviral property, the bacteriocins might possibly be employed in the postsurgical control of pathogens (van Staden et al. 2012). The application of bacteriocins as food additive and food preservative has been well explored. The novel and emergent possible character for bacteriocins is to be expected at the functional foods where these bacteriocin-producing microbes will be taken directly either as an ingredient of the food or as a health stimulating over-the-counter preparation which aims to positively modulate the gastrointestinal microbiome (Galvez et al. 2011; Murphy et al. 2013). Besides these properties, several reports have claimed the possible use of these compounds as probiotics in seafood industry, as intestinal protectors, in treating dermal infections, in veterinary medicines which highlight the importance of these peptidal molecules (Table 16.1).

16.2.3 Biosurfactants

The term 'surfactant' represents a class of molecules which are decisive towards the operation of both cleaning along with formulated products and also represents a monetary value of $32–36 billion yearly market. A major proportion of surfactants currently dominating the market finds its origin either from oil or from natural produces via chemical reactions (Sharma et al. 2019). A number of surfactants finding usage in diverse fields have got restricted usage due to their toxicity to humans and aquatic life (Dolman et al. 2019). Therefore, biosurfactants appear to be pertinent molecules which could address several problems caused by chemical surfactants along with the possession of different novel traits which extend their application in diverse fields. Biosurfactants are surface-active mixtures finding their origin from biological sources and are often produced as an extracellular product or as a constituent of cell membrane by different microorganisms such as bacteria, yeast or fungi (Cameotra and Makkar 1998; Mulligan 2005; Kaur et al. 2017). The solicitation of biosurfactants in enhanced oil recovery, bioremediation, agriculture, food production, chemistry along with cosmetics and pharmaceuticals (Pacwa-Plociniczak et al. 2011) is on an increase, every year. Biosurfactants encompass an extensive collection of chemical assemblies, for instance, lipopeptides, glycolipids, polysaccharide-protein complexes, phospholipids, fatty acids and neutral lipids. The ability of possession of miscellaneous belongings and different physiological roles for different classes of biosurfactants extends their applications in diverse fields (Rodrigues et al. 2006). Furthermore, these biotechnologically important particles can be tailor-prepared to suit diverse uses by altering the growth substrate or growth environments. These are deliberated as secondary metabolites which possibly play indispensable roles meant for the survival of producer by aiding transport of nutrients or by assisting microbe–host interactions or by acting as biocide agents.

Table 16.1 Bacteriocin and their pharmaceutical applications

Sr. No.	Bacteriocin	Producer microbe	Properties	Possible use	References
1	Plantaricin LPL-1	*Lactobacillus plantarum LPL-1*	Bactericidal activity against food-borne bacteria	Food preservation	Wang et al. (2018)
2	Enterocin CRL35	*Enterococcus mundtii CRL35*	Antimicrobial activity	Antibiotic	Farizano et al. (2019)
3	Bacteriocin AS-48	*Enterococcus faecalis UGRA10*	Kills *Trypanosoma cruzi*	Antiprotozoal drug	Martín-Escolano et al. (2019)
4	Class III bacteriocin	*Lysinibacillus JX402121*	Antimicrobial activity against food-borne pathogen *Bacillus pumilus*	Food preservation	Ahmad et al. (2019)
5	Bacteriocin BL and SL	*B. amyloliquefaciens* ELI149	Antimicrobial	Broad spectrum antibiotic	Salazar et al. (2017)
6	Bacteriocin ST16Pa	*Lactobacillus plantarum* ST16Pa	Antimicrobial	Antimicrobial drug formulation and food preservation	Sabo et al. (2019)
7	Nisin-Z	*Lactococcus lactis* subsp. *lactis*	Antimicrobial	Broad spectrum antimicrobial drug	Kaktcham et al. (2019)
8	DF01 bacteriocin	*Lactobacillus brevis* DF01	Anti-biofilm	Antimicrobial agent for food preservation	Kim et al. (2019)
9	Bacteriocin ABC transporter	*Enterococcus casseliflavus*	Antimicrobial and antioxidant	Clinical applications	Indira et al. (2018)
10	Plantaricin JK	*Lactococcus lactis*	Antimicrobial	Clinical applications	Xu et al. (2019)
11	GEn09, GEn12, GEn14 and GEn17	*Enterococcus durans*	Antiviral	Clinical applications	Cavicchioli et al. (2018)
12	Pediocin	*Enterococcus faecium*	Antiviral	Antiviral drug	Todorov et al. (2010)
13	HY 449	*Lactococcus sp. HY 449*	Antimicrobial	Cosmetic formulations	Oh et al. (2006)
14	Nisin	*Lactococcus lactis*	Anticancer	Anticancer drugs	Ahmadi et al. (2017)
15	Fermenticin HV6b	*Lactobacillus fermentum* HV6b MTCC10770	Sperm immobilization and spermicidal activity	Contraceptive	Kaur et al. (2013)

The other roles of biosurfactants include amassing the surface area and bioavailability of hydrophobic water-insoluble substrates, heavy metal binding, bacterial pathogenesis, quorum sensing and biofilm formation (Singh and Cameotra 2004). Their antiviral, antifungal and antibacterial actions make them pertinent molecules, thus extending their applications in battling diseases. Additionally, they also behave as anti-adhesive agents against numerous pathogens, thereby promoting their efficacy as appropriate anti-adhesive coatings for insertional materials used in medical procedures (Rodrigues et al. 2006). Such properties of these naturally produced amphipathic molecules have increased their usage and commercial potential in the medical field. Biosurfactant production among bacteria is well explored nominating *Pseudomonas aeruginosa*, *P. putida*, *P. stutzeri* (Joshi and Shekhawat 2014), *Bacillus subtilis*, *B. pumilus*, *B. licheniformis*, *Lactococcus lactis* (Rodrigues et al. 2006), *Lactobacillus* spp., *Streptococcus thermophilus* and *Nocardioides* spp. (Khopade et al. 2011) as some of the promising candidates known for producing biosurfactants mainly rhamnolipid in nature (Rahman et al. 2002). Contrary to this, relatively fewer fungi such as *Aspergillus ustus* (Kiran et al. 2009), *Ustilago maydis* (Alejandro et al. 2011), *Trichosporon ashii* (Chandran and Das 2010) *Candida bombicola* (Felse et al. 2007), *C. lipolytica* (Sarubbo et al. 2007), *C. ishiwadae* (Thanomsub et al. 2004), *C. batistae* (Konishi et al. 2008), *C. antarctica* (Hua et al. 2003), etc. are known to produce biosurfactants. The possession of several traits outspreads their application in several fields, but the altitudinous cost of biosurfactant production usually restricts their usage. Sophorolipid, for instance, the most economic and broadly obtainable microbial biosurfactant, has newly been published at $ 34/kg active matter (Roelants et al. 2019). This cost is many times greater than the amount of representative specialty surfactants, which highlights an important requisite for reducing the production price with the intention of making their usage in an extensive array of bulk applications feasible and to overcome the industry challenges associated with the standard chemical surfactants (Dolman et al. 2019). Although all the pharmaceutical aspects of these bioactive compounds of microbial origin seem impossible to know, however, the increasing research in this area is perpetually unveiling different novel properties of these compounds. In this section, an attempt has been made to cover the different activities of biosurfactants which have assigned several pharmaceutical roles to these molecules.

16.2.3.1 Antimicrobial and Anti-adhesive Activity

The possession of antibacterial, antifungal, antiviral and anti-adhesive properties has projected biosurfactants as a potential substitute to the traditional antibiotics beside numerous food-borne pathogens (Banat et al. 2010; Sharma and Saharan 2016). The most extensively studied bacterium for biosurfactant production is *Bacillus* which produces cyclic lipopeptides and lipoproteins which are further comprised by lichenysins, surfactins, fengicins and bacillomycin as chief kinds of biosurfactants. Surfactin is usually known for the possession of antibacterial and antifungal activity. However, its applications are not just restricted to its antibacterial properties; it also possesses antiviral and antitumor properties and also induces formation of ion channels in lipid bilayer membranes (Kim et al. 1998). Lipopeptides along with their homologues represent a collection of active peptides having alike arrangements and similar molecular mass (Pathak et al. 2014). Fengycin, which is one among the most vital and encouraging lipopeptides (Ongena et al. 2005), is a capable antifungal agent against pathogenic fungi (Wang et al. 2004). It occurs as a mixture of isoforms which vary in fatty acid chain lengths and in the amino acid configuration of the peptide ring (Vanittanakom et al. 1986). Biosurfactants act on the pathogen by disrupting its cytoplasmic membranes, which ultimately leads to cell lysis and outflow of metabolites, and disorder the protein conformations, which eventually modifies imperative membrane functions (Tahmourespour et al. 2011; Sharma and Saharan 2016). Their mechanism of action grants them the ability to be active against a number of Gram-positive and Gram-negative pathogenic and semi-pathogenic microbial strains including multidrug-resistant strains (Das et al. 2009). Biosurfactants have also been found to constrain the linkage of pathogens to different solid substrates or to different infection sites (Das et al. 2009); thus, the prior bonding of biosurfactants to such solid surfaces may establish a novel and operative means of battling establishment by pathogenic microorganisms (Rivardo et al. 2009). For instance, the pre-coating of vinyl urethral catheters by the biosurfactant surfactin solution before inoculating with media caused a significant reduction in the total biofilm formed by different pathogens like, *Salmonella enterica*, *Salmonella typhimurium*, *E. coli* and *Proteus mirabilis* (Rodrigues et al. 2004).

16.2.3.2 Active Ingredient in Cosmetic Formulations

There are various challenges faced by the pharmaceutical industries, and one among those challenges is the solubilization as well as preservation of the active ingredients of the cosmetics. Consequently, many of these preparations encompass irritant chemical additives which enhance the shelf-life along with the solubility of different hydrophobic constituents. The employment of biosurfactant for solubilization and preservatives due to the possession of their surfactant properties and configuration makes them more biocompatible than their chemical equivalents (Rodríguez-López et al. 2019). Their role as active ingredients of cosmetics is further advocated by their antimicrobial activities. They are also largely used as the chief vital constituent in different products like soap, moisturizers, shampoo, hair conditioners, toothpastes, cleansers, shower gel, creams and many additional skin care and healthcare products (Akbari et al. 2018). It has also been found that biosurfactants play a very key role in maintaining skin moisturizing ability like ceramides (Kitagawa et al. 2011) which are epidermal lipids vital for dermal barrier and dryness. The amalgamation of biosurfactants in different cosmetics products such as skin care lotions and moisturizing creams can advance the worth of the final produce and also supports in roughness improvement (Kitagawa et al. 2011; Vecino et al. 2017). The cosmetic preparations have to deal with emulsions (Masmoudi et al. 2005). The use of plant-based essential oils as an imperative constituent for conditioning and anti-ageing commitments in cosmetics (Ferreira et al. 2017) requires an emulsifier for stabilizing these emulsions; therefore, natural and greener surfactants appear safe owing to their ability of curing in a natural way (Vijayakuma and Saravanan 2015). The supreme thought-provoking complications, which are stimulated by the use of chemical surfactants in different cosmetics preparations, are the appearance of dermal irritations as well as allergic reactions. The artificial surfactants interact with proteins, eliminate lipids from the epidermal surface by disorganizing the intercellular structure of such lipids and also affect the living cells in the skin. The employment of biosurfactants in these formulations can effectively overcome these harmful effects (Vecino et al. 2017). Additionally, several biosurfactants contain fatty acid chains which prevent the skin from harmful effects of ultraviolet light of sun by preventing the generation of free radicals which would otherwise hurt and damage the cross-linkage of several elastic proteins, incapacitate the antioxidant enzymes and also ground the peroxidation of the lipids of cell membrane, thereby leading to the appearance of wrinkles. Therefore, biosurfactants act as natural antioxidants and prevent the skin from harmful effects of ultraviolet light.

The efficiency of any cosmetic products depends on its capability to infiltrate the barrier possessed by the skin. Therefore, the permeation of desired product through skin seems to be an important and essential requirement for the relevant transport of bioactive composites. The employment of biosurfactants in different cosmetic preparations could be supportive to support the penetration of the active principles of cosmetic into the skin (Alonso et al. 2015). Moreover, the increasing awareness for environmental sustainability has also largely accounted for shifting the lifestyles of consumers seeking for green alternatives. This green trend could have largely sparked the interest of consumers in products containing 'natural' components in cosmetic and personal care products. Since biosurfactants are very encouraging and fascinating substances as they are solely grounded on renewable and sustainable resources, they seem to be more biocompatible as compared to the synthetic surfactants (Rodrigues et al. 2006; Gudiña et al. 2013).

16.2.3.3 Anticancer Properties

The second highest number of global deaths is still accounted by cancer which has added to 8.8 million deaths in the year 2015 (WHO 2017), and the most surprising and thought-provoking thing is that this number could possibly reach 17 million by the year 2020 (Obtel et al. 2015). The typical treatment for cancer is highly based on chemotherapy (Demain and Sanchez 2009). The major limitation restricting the chemotherapy is the appearance of chemoresistance (Alfarouk et al. 2015). Additionally, a large number of studies have advocated that most of the chemotherapeutic drugs are extremely cytotoxic and are also reported to target very proliferative cells non-specifically, which ultimately results in only an insignificant advance in the survival of patient (Sak 2012; Dy and Adjei 2013). Although there are numerous foundations for anticancer drugs, microbes have engrossed considerable attention owing to their ease in production management and their capability of fabricating different bioactive metabolites, like biosurfactants (Chiewpattanakul et al. 2010). Indeed, several reports have revealed that biosurfactants are among the microbial metabolites that show promising biological activities (Dey et al. 2015). Biosurfactants are often reported to show cytotoxic activity against cancer cells. For instance, surfactin can be used for numerous biomedical solicitations owing to its anticancer properties (Liu et al. 2012). The amphiphilic nature of biosurfactants makes possible the effective hydrophobic interactions amid the fatty acid and the phospholipids which seems to be a characteristic feature responsible for imparting the anticancer activity (Wu et al. 2017). The biosurfactant, surfactin, displays potential cytotoxicity on human breast cancer cells through encouraging apoptosis as a result of increase in ROS generation and cutting the mitochondrial membrane potential. The peptide moiety of the surfactin molecule binds with the cancer cells through the polar heads of membrane lipids and thus controls the proliferation. It is also found that the cytotoxic activity of biosurfactant keeps on increasing with a perpetual increase in the biosurfactant concentration and the increasing concentration of biosurfactant does not affect any toxicity against human normal cell lines. Various cell abnormalities, for example, cell rounding up, cell shrinkage, membrane blebbing and a greater number of floating dead cells, are observed in the presence of biosurfactant, whereas the control cells lack any such abnormality and also show adherence along with the maintenance of a well-organized structure (Ramalingam et al. 2019) (Table 16.2).

16.2.4 Immunobiotics

The microorganisms are of utmost importance to human health attributable to their contribution to food digestion and the expansion and ideal operation of the immune system (Hooper and Gordon 2001). The growing attention in the positive roles of microbes has ensued in the assortment of species endowed with reputed health-endorsing capabilities for the treatment of conditions in which microbiota—or their optimal functioning—have been disrupted. The term ‘probiotic’ was coined from the food industry in order to designate the live microbial food constituents which are valuable to human health. However, the benefits given by these microbial components of food were not largely elaborated, and the mechanisms behind the health promotion activities also seemed to be quite unclear. It has been established that probiotics support the human health by triggering the mutual mucosal system by stimulating the gut antigen-presenting cells in order to encourage protection as well as to switch the regulatory mechanisms (Hessle et al. 2000; Clancy 2003). Therefore, it urges the requirement of a new term in order to ascertain bacteria have health promotion activities by driving mucosal immune mechanisms, as equated to those with severely local effects. Therefore, the term ‘immunobiotics’ comes into play as an appropriate term for fulfilling this requirement (Clancy 2003). There are several strains of lactic acid bacteria which have the ability to influence human as well as animal health by controlling the systemic along with mucosal immune systems (Villena et al. 2016). Such immune regulatory and probiotic LAB immunobiotics also confer defence against different viral infections by modifying adaptive as well as innate antiviral immunity. Numerous reports have also claimed that the immunobiotic LAB also curtail the extent of diarrhoea along with a decline in the number of occurrences, stabilize the permeability of gut and upsurge the fabrication of viral specific antibodies (Basu et al. 2009; Liu et al. 2010; Maragkoudakis et al.

Table 16.2 Biosurfactants and their pharmaceutical applications

Sr. No.	Microorganism	Biosurfactant	Class	Application	References
1	*Bacillus subtilis*	Surfactin	Lipopeptide	Antibacterial	Yuliani et al. (2018)
2	*Pseudomonas aeruginosa*	Rhamnolipids	Glycolipids	Antibacterial	de Freitas et al. (2019)
3	*Serratia rubidaea* SNAU02	Rhamnolipids	Glycolipid	Biocontrol against Fusarium wilt	Nalini and Parthasarathi (2018)
4	*Lactobacillus pentosus*	Sophorolipids	Glycolipid	Antibacterial	Rodríguez-López et al. (2019)
5	*Bacillus subtilis*	Fengycin	Lipopeptide	Antifungal	Sa et al. (2018)
6	*Micromonospora marina*	Surfactin	Lipopeptide	Anticancer	Ramalingam et al. (2019)
7	*Bacillus stratosphericus* FLU5	Surfactin and pumilacidin	Lipopeptide	Useful as biological material	Hentati et al. (2019)
8	*Acinetobacter junii B6*	–	Lipopeptide	Pharmaceutical applications	Ohadi et al. (2018)
9	*Acinetobacter indicus M6*	–	Glycolipoprotein	Antimicrobial, anti-biofilm, antitumour	Karlapudi et al. (2018)
10	*Bacillus sp.* BS3	–	Lipopeptide	Antibacterial, antiviral, antitumor	Donio et al. (2013)
11	*Cyberlindnera saturnus*	Cybersan	Glycolipid	Antimicrobial, therapeutic	Balan et al. (2019)
12	*Lactobacillus paracasei*	–	Glycolipopeptide	Stabilizing agent in cosmetics	Ferreira et al. (2017)
13	*Wickerhamomyces anomalus* CCMA 0358		Glycolipid	Antimicrobial	Souza et al. (2017)
14	*Candida albicans* SC5314	Sophorolipid	Glycolipid	Food emulsifier, antimicrobial	Gaur et al. (2019)
15	*Candida glabrata* CBS138	Sophorolipid	Glycolipid	Food emulsifier, antimicrobial	Gaur et al. (2019)
16	*Candida* sp. AH62	Sophorolipid	Glycolipid	Antimicrobial	Archana et al. (2019)
17	*Candida bombicola* URM 3718	CB	–	Toothpaste formulations, antimicrobial	Resende et al. (2019)
18	*Pseudomonas aeruginosa* UCP 0992	PB	–	Toothpaste formulations, antimicrobial	Resende et al. (2019)
19	*Bacillus methylotrophicus* UCP 1616	BB	–	Toothpaste formulations, antimicrobial	Resende et al. (2019)
20	*Pseudomonas aeruginosa* UCP 0992	PB	–	Mouthwash formulations	Farias et al. (2019)

2010). In a study by Majamaa et al. (1995), the children receiving *L. rhamnosus* GG were found to have shorter durations of diarrhoea. The shielding outcome as a result of intake was found to be associated with increased intestinal as well as serum IgA concentration and a greater quantity of cells secreting antibody specific to rotaviruses. The developing nations usually encounter the problems of viral mucosal infections. Therefore, the employment of immunobiotics for improving the consequence of such viral infections has been anticipated by several researchers (Bardach et al. 2011; Edmond et al. 2012; Gentile et al. 2012). There are several instances which have regularly advocated the importance and efficacy of the administration of immunobiotic bacteria on human health. The immunobiotic bacterium *Lactobacillus rhamnosus* CRL1505 has been reported to improve the mucosal immunity along with a reduction in the occurrence and harshness of the intestinal as well as respiratory infection in children. The frequency of such infectious happenings received a major and significant decline from 66% in the placebo group to 34% in the group that received the probiotic yogurt containing immunobiotic formulation. Moreover, there was also a significant reduction in the manifestation of pointers of disease rigorousness such as fever and the need for antibiotic treatment in children receiving the immunobiotic yogurt (Villena et al. 2012). The fact that the immunomodulatory efficiencies of different immunobiotic bacteria are highly strain specific should be taken into consideration while supervising the patient suffering from infection. Additionally, the interactions between the intestinal cells and the immunobiotic bacteria trigger some molecular signals, which in turn control the expression of several genes governing immune response, which further plays a central role in any kind of immune response elicited by the bacteria.

The beneficial effects of immunobiotics are not only restricted to humans. Various studies have also found their beneficial effects like antiviral and anti-inflammatory accomplishments in animals, for instance, the administration of LAB downregulated the employment of viral-activated monocytes/macrophages into the intestinal tract, thereby limiting the inflammation induced by the virus (Zhang et al. 2008). The immunobiotic bacteria have also shown elevated levels of cytokines in response to viral infections. Furthermore, they are also reported to promote the improvement of intestinal epithelial tight connections that is supposed to contribute towards the conservancy as well as re-establishment of the gut homeostasis following a viral infection (Liu et al. 2010; Azevedo et al. 2012). The increasing research has also proved that even nonviable beneficial bacteria could deliver various health promotion properties (Kataria et al. 2009; Lahtinen and Endo 2011). Comparatively, nonviable cells seem to be largely beneficial for the food industry as they provide novel product solicitations, augment the shelf life of foodstuffs and also condense the prices for storage as well as delivery (Miyazawa et al. 2011) (Table 16.3).

16.2.5 Psychobiotics

The occurrence of different psychological disorders along with depression is on a perpetual increase, and it has been assessed that about 4.4% (322 million) of the

Table 16.3 Different health benefits conferred by immunobiotic bacteria

Sr. No.	Microorganism	Health benefit	Reference
1	*Lactobacillus rhamnosus* GG	Protective against intestinal viral infections	Villena et al. (2016)
2	*Lactobacillus rhamnosus* CRL1505 *L. plantarum* CRL1506	Recovering intestinal injury by regulation of proinflammatory cytokines production	Tada et al. (2016)
3	*Lactobacillus acidophilus and Lactobacillus reuteri*	Modulation of cytokine responses on infection with human rotavirus	Azevedo et al. (2012)
4	*Lactobacillus paracasei*	Control meningococcal infection	Belkacem et al. (2018)
5	*Lactobacillus rhamnosus* CRL1505 and *Lactobacillus plantarum* CRL1506	Immunomodulatory response	Albarracin et al. (2017)
6	*S. thermophilus* CRL807	Immunomodulatory response along with the reduction in oxidative stress	Del Carmen et al. (2014)

universal population are engulfed by depression and about 3.6% (264 million) are grieved from anxiety syndromes (WHO 2017). The presently accessible pharmacological and psychosomatic treatments are not found to be much effective in terms of modest short-term reimbursements, austere ill effects and age restrictions, which has put a dire need to develop new alternatives (Warda et al. 2019). The term 'psychobiotics' has been newly devised to designate another evolving class of probiotics which find their significance towards psychiatry (Dinan et al. 2013). These microbes are endowed with 'mind-altering' traits, and they have got the unique ability of producing numerous biologically active composites, for instance, peptides and mediators which are usually allied with neurotransmission in mammals. Such neuroactive compounds grant these microbes with the unique ability of imparting health benefit to patients experiencing psychiatric illness (Wall et al. 2014). These microbes have the capability of fabricating and transporting different neuroactive compounds such as gamma-aminobutyric acid, serotonin, catecholamines and acetylcholine, which predominantly act on the brain-gut axis. A vast array of such compounds has been isolated from the bacteria residing in human gut. The neurotransmitters released by the bacteria in the intestine are supposed to encourage epithelial cells to secrete compounds that sequentially moderate the process of neural signalling within the enteric nervous system and subsequently signal brain function and behaviour of the host. The effects are supposed to be mediated via the vagus nerve, spinal cord or neuroendocrine systems. The possession of antidepressant or anxiolytic activity by certain psychobiotics has also been reported and confirmed in various preclinical appraisals (Dinan et al. 2013).

There are increasing number of evidences which at various stages have proved the ability of different microorganisms to alter the cognitive and emotional processes by acting through the brain-gut axis (Heijtz et al. 2011). The brain-gut axis is a platform for providing bidirectional communiqué between the gut and the brain along with the inclusion of the highly complex intestinal microbiome (Collins et al. 2012; Cryan and Dinan 2012). This unique communication platform assimilates neural, hormonal as well as immunological signalling amid the gut and the brain which is considered to be critical to uphold homeostasis (Collins et al. 2012). In recent times, though, this axis notion was further extended to the 'microbiota-gut-brain axis', when it seemed quite clear that it is not only the intestinal tract that can shake the working of the central nervous system but also its 100 trillion microbial populations affect it and accordingly mood and behaviour (Heijtz et al. 2011; Neufeld et al. 2011). This fact is strongly advocated by the evidence that any changes in the configuration of the gut microbes deteriorates the gastrointestinal, neuroendocrine or immune pathways and relationships, which sequentially might alter the brain-gut communications and accordingly result in any kind of disease (Cryan and O'Mahony 2011). It is also believed that microbes are pharmacological agents which work as drug delivery vehicles owing to their aptitude of synthesizing several neuroactive complexes (Lyte 2011). They either transport the pre-synthesized neurochemicals present in bacterium in adequate amounts at time of consumption or actively produce it when they are inside the gastrointestinal tract.

A large number of bacteria residing the gastrointestinal tract have the inherent capability of producing numerous neuromodulators as well as neurotransmitters. For instance, GABA productions have largely been reported by the members of *Bifidobacterium* spp. and *Lactobacillus* spp., whereas *Bacillus* spp. and *Escherichia* spp. are greatly acknowledged for producing norepinephrine. Furthermore, *Enterococcus* spp., *Streptococcus* spp. and *Escherichia* spp. are reportedly known for producing serotonin, whereas *Bacillus* spp. and *Lactobacillus* spp. are accredited with the production of dopamine, acetylcholine and histamine, respectively (Wall et al. 2014). Other metabolites of bacterial origin such as short-chain fatty acids and long-chain fatty acids are also supposed to have neuroactive functions. The bacteria that are known for producing short-chain fatty acids are *Clostridium*, *Bifidobacterium*, *Bacteroides*, *Eubacterium*, *Propionibacterium*, *Lactobacillus*, etc. (Macfarlane and Macfarlane 2012), whereas *Bifidobacterium* (Wall et al. 2012) is mainly known for producing long-chain fatty acids.

Since the brain-gut axis is a bidirectional platform, it can also be taken into consideration that the brain strongly affects the microbial configuration of the gut. Several reports have also claimed that the microbial community of the gut experiences substantial changes during the exposure to stress. It has been studied in mice systems that *Bacteroides* spp. often decrease as compared to their levels in control mice, while there is a relative increase in the number of *Clostridium* spp. (Bailey et al. 2011). The mechanisms behind the correlation between stress and alteration of the gut microbial configuration are quite indistinct but may comprise an altered microbial habitat attributable to stress-persuaded deviations in the intestinal motility and mucin exudation (Collins and Bercik 2009). The relative level of noradrenaline increases in the gut lumen during the state of stress, and this may also account for the alterations in the microbial composition. Such chemical substances are acknowledged to modify the gene expression in some bacteria, which ultimately ends up in favoured progression of definite microbial communities (Collins et al. 2012). The major development of brain takes place in utero and endures after birth also. The infants born through vaginal delivery get their gastrointestinal tract inhabited by the bacteria in the lower birth canal and perineum of mother; consequently, the microbial inhabitants of the newborns born by caesarian segment vary from those that are delivered through the genital tract (Juárez et al. 2008). The design of electrical movement in the brain is found to be less compound in children born by caesarian section as compared to the neonates of the same age group born by vaginal delivery. Therefore, it is limpid clear that altered establishment arrangements of microbiota strongly affect the early post-delivery brain growth which pose longer-term significances (Kim et al. 2003). Thus, the effect of different microbes on the development of the brain up to such an extent greatly unveils their significant and irreplaceable role which further proposes their role as psychobiotic agents for curing different kinds of mental illness((Table 16.4).

16.3 Hormones

Microbes provide incredible stands for synthesizing a vast array of molecules (Rani et al. 2019; Singh et al. 2019), and hormones fitting to diverse chemical families are one among those molecules of biopharmaceutical importance (Jeandet et al. 2013). These proteins are biological molecules of great importance as they play an imperative part in catalysing diverse biochemical reactions fundamental to metabolism, behave as organizational constituents of biological clusters and are also accountable for different inter- as well as intracellular connections and cell signalling happenings which are found to be critical for sustaining life. Although the human cell systems synthesize a great deal of proteins which further assimilate into an enormously multifaceted physiologic grid and accomplish specific activities as catalysers, signalling agents or structural machineries, the malfunctioning of these proteins often marks the engulfment by stark pathologies such as diabetes (Vajo et al. 2001), dwarfism (Takeda et al. 2010), cystic fibrosis (Cutting 2005), thalassaemia or impaired blood clotting, among many others (Sanchez-Garcia et al. 2016). As a result, it can be clearly taken into consideration that the paucities in the fabrication of any explicit polypeptide or the production of any mutant and worthless kind protein of biological worth ordinarily lead to pathologies oscillating from slight to austere one. In human beings, such kind of ailments are cured by the medical injection of that particular omitted protein via exterior bases, to touch usual concentrations at systemic or tissular levels (Manning et al. 1989; Ferrer-Miralles et al. 2009). So, a number of human proteins are of great pharmaceutical significance, but getting hold of them from their ordinary bases seems to be a difficult task. The perpetual developments in the recombinant DNA technologies, from the late 1970s by employing *Escherichia coli* as a biological framework, propose a highly persuasive platform for the controlled as well as scalable manufacture of different polypeptides of human concern in a way that otherwise seems very expensive. Within 4 years of the development of recombinant DNA technology, genetically engineered bacteria

Table 16.4 Psychobiotic bacteria encompassing different health benefits

Sr. No	Microorganism	Role	Possible mechanism	References
1.	*Lactobacillus rhamnosus* JB-1	Regulation of emotional behaviour	Transformed expression of inhibitory GABA receptors Improved levels of N-acetyl aspartate and glutamate	Bravo et al. (2011), Janik et al. (2016)
2	*Mycobacterium vaccae*	Reduced anxiety	Induction of short-term physiological changes which affect behaviour	Matthews and Jenks (2013)
3	*Bifidobacterium longum* 1714	Stress reduction, memory improvement	Lower cortisol output and improved frontal midline electroencephalographic mobility	Allen et al. (2016)
4	*L. helveticus* R0052 and *B. longum* R0175	Reduction in anxiety and alleviated psychological distress	Decreased cortisol value Alteration of gut microbial community	Messaoudi et al. (2011)
5	*Lactobacillus plantarum* PS128	Reduction in anxiety and depression like behaviours of mice	Elevated levels of dopamine and serotonin	Liu et al. (2015); Liu et al. (2016)
6	*Lactobacillus helveticus* NS8	Abridged anxiety, depression and cognitive dysfunction	Improved levels of serotonin, norepinephrine and brain-derived neurotrophic factors in the hippocampus	Liang et al. (2015)
7	*Lactobacillus fermentum* PS150	Reduced anxiety and depression	Lowered interferon-γ gene expression and corticosterone levels in brain and plasma	Liu et al. (2019)
8	*Lactobacillus plantarum* 299v	Reduced depression and improvement in cognitive tasks	Reduced kynurenine concentration	Rudzki et al. (2019)
9	*Lactobacillus gasseri* CP2305	Stress reduction Improved sleep quality	Prevention of increases in basal salivary cortisol release Expression of stress-responsive microRNAs Enhancement in parasympathetic nerve activity	Nishida et al. (2017)
10	Heat killed *Lactobacillus fermentum* and *Lactobacillus delbrueckii*	Increased sociability Decreased stress	Lower levels of corticosterone and altered microbiota	Warda et al. (2019)
11	*Lactobacillus paracasei* PS23	Antidepressant and anxiolytic effects	Upsurge in hippocampal glucocorticoid receptor, mineralocorticoid receptor and brain-derived neurotrophic factor proteins, escalation in serotonergic and dopaminergic activities in the hippocampus, prefrontal cortex and striatum Improvement of the gut microbiota	Wei et al. (2019)

were making human insulin and human growth hormone (Lancini and Demain 2013). The microbial cells of high expediency, for instance, yeast and bacteria, which are easy to nurture are often employed for the production of such compounds. The recombinant protein insulin from recombinant *E. coli* was approved for clinical use in the early 1980s, for the treatment of diabetes (Ferrer-Miralles et al. 2009). The approval of recombinant insulin for clinical purposes has laid a foundation stone for the development as well as improvement of numerous heterologous protein fabrication machineries. This has also engendered precise microbial strains, which are strongly amended for protein fabrication, and has also endorsed the advanced amalgamation of yeasts and other eukaryotic machineries for this purpose. The microbes *Escherichia coli* and *Saccharomyces cerevisiae* are mainly exploited for the production of such recombinant proteins. However, other viable production systems such as cold-adapted bacteria, filamentous fungi and alternate yeast species can also be indentured for producing hormones of pharmaceutical importance.

E. coli has also been genetically modified for the expression of another recombinant protein gonadotropin-releasing hormone which has the potential to be employed as a successful agent for controlling fertility as well as hormone-reliant diseases (Xu et al. 2006). *E. coli* has always been the first choice of researchers for the manufacture of recombinant proteins and has been largely exploited for primarily cloning, genetic alteration as well as small-scale production for the research drives (Jenkins 2007). However, the recombinant proteins produced by *E. coli* are found to be lacking the post-translational modifications (Walsh and Jefferis 2006). Conversely, among the eukaryotic microbes especially yeast is customarily targeted when the protein of interest is not fashioned by the prokaryotic systems in the required forms or the protein is devoid of any particular post-translational modification which are typical for the biological activity of that certain protein and cannot be induced in any artificial way on the purified end product (Jenkins 2007). The permitted protein yields especially hormones manufactured using yeast systems are acquired absolutely from *Saccharomyces cerevisiae*, for instance, insulin, insulin analogues, non-glycosylated human growth hormone somatotropin, glucagon, etc. (Gerngross 2004).

The increased understanding of microbiome research has also unveiled numerous hidden aspects of human microbial ecology, i.e. learning the microbial communities residing within human bodies as well as the genes they comprise (Neuman et al. 2015). The ever-increasing research in this area has also revealed the bacterial modulation of human hormonal secretion. The microbial establishment in the human intestine has a critical role to play in the development immunity (Elahi et al. 2013) along with the development of human endocrine system (Clarke et al. 2013). The commensal microbiota colonizing the gut has the capability of producing and secreting hormones. This crosstalk is supposed to affect the host metabolism, immunity as well as behaviour. Furthermore, such an interaction is bidirectional, as the microbial inhabitants are also supposed to be both affected by and to affect host hormones (Neuman et al. 2015). It has also been proved that the involvement of enzymes in metabolizing the host hormones (comprising dopamine, melatonin, epinephrine, norepinephrine, serotonin, etc.) would have advanced from the mechanisms of horizontal gene transfer from bacteria (Iyer et al. 2004). Therefore, the judicial indenturing of gut microflora for the production of different hormones could prove to be a successful step for boosting the research accompanied with the synthesis of either novel or some unidentified hormones as well as hormonal systems.

16.4 Enzymes

Enzymes are deliberated to be the potential biocatalyst of utmost importance catalysing an enormous number of reactions, and microbes represent the principal and most convenient foundations of many enzymes (Demain and Adrio 2008). A greater proportion of the novel enzymes have been premeditated with the input of protein engineering along with the metagenomics. Furthermore, the application of diverse molecular techniques has also improved the eminence as well as performance of microbial enzymes, thereby extending their applications in numerous industries (Chirumamilla et al. 2001; Nigam 2013). The microbes which are usually targeted globally for synthesizing economically viable groundings of numerous enzymes for marketable applications are largely represented by bacteria, actinomycetes and fungi (Pandey et al. 1999). Predominantly, the enzymes of microbial origin find extensive usage in industries and medicine. The enzymes of microbial origin have expanded the appreciation comprehensively attributable to their pervasive uses in numerous industrial sectors, e.g. food, agriculture, energy, medicine and chemicals (Li et al. 2012; Choi et al. 2015). The total industrial production of enzymes is majorly dominated by the enzymes of microbial origin as 50% are produced by fungi and yeast and 35% from bacteria whereas the lingering 15% are from plants (Saranraj and Naidu 2014; Liu and Kokare 2017). The microbial enzymes are also more active and stable than plant and animal enzymes. In addition, the microorganisms represent an alternative source of enzymes because they can be cultured in large quantities in a short time by fermentation and owing to their biochemical diversity and susceptibility to gene manipulation (Anbu et al. 2013). Industries are looking for new microbial strains in order to produce different enzymes to fulfil the current enzyme requirements. The enzymes which are produced extracellularly by numerous filamentous fungi enzymes are usually explored much, for instance, the enzyme β-d-galactosidase which is accountable for catalysing the conversion of lactose sugar to glucose and galactose. Prominently, the universal marketplace for lactase is on perpetual and significant hike attributable to its prominence in the therapy of lactose intolerance disorder. It is usually marketed in the form of capsules which are further supposed to be employed as food supplement for people intolerant towards lactose prior to the consumption of milk and other dairy produces (O'Connell and Walsh 2008; Oliveira et al. 2011). Additionally, these enzymes are also active participants for the production of galactooligosaccharides which find explicit uses in diverse functional foods, for instance, low-energy diets, and are also employed as an imperative additive in different fermented milk and yields, breads and beverages (Ruiz-Matute et al. 2012; Zabian Bassetto et al. 2014).

These oligosaccharides are, however, indigestible but advantageous for humans as well as livestock. Their ingestion allows the population of bifidobacteria, the inhabitants of colon, to subdue the action of different putrefactive microflora, thereby plummeting the fabrication of different noxious products of microbial fermentations, thus circumventing the problem of intestinal constipation, and also to upsurge the fabrication of vitamins B complex (Jozala et al. 2016).

Several enzymes produced by microbial systems are of medicinal and therapeutic importance also. The enzyme asparaginase finds application in the treatment of some particular hematopoietic diseases such as acute lymphoblastic leukaemia and non-Hodgkin lymphoma. The unceasing growth of tumour cells requires an exogenous source of the amino acid asparagine for their propagation. The presence of this enzyme diminishes the asparagine level from the bloodstream which ultimately leads to the cell death. Nevertheless, the enzyme retrieved from the bacterial sources such as *E. coli* and *Erwinia chrysanthemi* brings some side effects along with it such as it leads to some austere immunological reactions. Consequently, the enzyme from fungal resources appears to be an effective alternative to that of bacterial sources as an antitumoural agent attributable to its steadiness and optimum pH near physiological environments. The enzymes belonging to the class collagenolytic proteases also found direct usage in clinical therapy, which also encompass wound remedy, sciatica in herniated intervertebral discs and retained placenta, and as a pretreatment for enhancing adenovirus-mediated cancer gene therapy (Watanabe 2004). These are reportedly produced by the fungus *Aspergillus niger* (Kumar and Takagi 1999). Presently, the most protuberant employment of microbial enzymes for medicinal purposes is for the exclusion of burns and dead skin by indenturing proteolytic enzymes and clot rupturing by fibrinolytic enzymes, for instance, nattokinase is a persuasive fibrinolytic enzyme and is often employed as an auspicious representative for thrombosis remedy (Sumi et al. 1987; Cho et al. 2010). Acid protease, rhodanese and dextranase can also be employed for handling alimentary dyspepsia, cyanide poisoning and tooth decay, respectively (Okafor and Okeke 2007). The microbial enzyme polyphenol oxidase is involved in the synthesis of 3, 4-dihydroxylphenylalanine which is used for the cure of Parkinson's disease (Faber 1997). Tyrosinase, an imperative oxidase enzyme, is tangled in melanogenesis as well as in the fabrication of l-dihydroxy phenylalanine (L-DOPA). L-DOPA finds usage as a precursor molecule for the fabrication of dopamine which is a potent drug for the treatment of Parkinson's disease and to control the myocardium neurogenic injury (Ikram-ul-Haq et al. 2002; Zaidi et al. 2014). Chitosanase finds usage as a catalyst for hydrolysis of chitosan to biologically active chitosan oligosaccharides, which is used as antimicrobial, antioxidant, lowering of blood cholesterol and high blood pressure, controlling arthritis, protective effects against infections and improving antitumor properties (Singh et al. 2016). Additionally, the microbial enzymes also find usage in clinical diagnosis meant for the quantifiable resolution of diabetes as well as other health disarrays, for instance, glucose oxidase for glucose; urease and glutamate dehydrogenase for urea; lipase, carboxyl esterase and glycerol kinase for triglycerides; urate oxidase for uric acid; and creatinase and sarcosine oxidases for creatinine (Dordick 2013; Le Roes-Hill and Prins 2016). Cholesterol oxidase has also been reported for useful biotechnological applications in the detection and conversion of cholesterol. Putrescine oxidase is used to detect biogenic amines, such as putrescine, a marker for food spoilage (Le Roes-Hill and Prins 2016).

Although the employment of enzymes displays a vast array of advantages as compared to other orthodox approaches, however, they also possess several limitations for their employment in healthcare and other industries. For instance, the employment of enzymes in countless mammalian systems requires 37 °C and a pH of 7.4 as optimal parameters, and in addition to it, the activity of enzyme is exceedingly sensitive to any shift in such parameters. The elevated temperatures and a major deviation from the optimal pH lead to the enzyme denaturation, thereby restricting the usage in non-physiological environments. Furthermore, they are also prone to substrate or product inhibition and their harvests can also elicit allergic responses. Also, the economy underlying the isolation and purification processes and their challenging retrieval for succeeding reuse also daunt their use (Johannes et al. 2006; Singh et al. 2016).

16.5 Nutraceuticals

The term 'nutraceutical' evolved as a result of hybridization of the terms 'nutrition' and 'pharmaceutical'. It was proposed by Stephen L. DeFelice in 1989 and was defined as 'any substance that is a food or a part of food and provides medical or health benefits, including the prevention and treatment of disease' (DeFelice 1995). However, later this notion was amended as 'a product isolated or purified from foods that is generally sold in medicinal forms not usually associated with food' (Pandey et al. 2010). The high organizational as well as functional specificity of nutraceuticals towards the bioactive composites which are accountable for providing medicinal as well as physiological assistances for a longer time as compared to pure nutritional or pharmaceutical effects makes them different from functional foods and drugs. They find their derivation from plants, animals, microbes and marine bases. Nutraceuticals are extensively employed in lieu of their health-endorsing or disease-averting assets particularly in precluding aging-allied illness comprising arthritis, diabetes, oxidative stress, osteoporosis, depression, gastrointestinal diseases, inflammation, cardiovascular diseases and cancer (Jain and Ramawat 2013). There has been a perpetual uplift in the global nutraceutical market. The international nutraceutical marketplace was projected to surpass \$171.8 billion in 2014 and grasp \$241.1 billion by 2019, while the US market alone was about \$75.9 billion. Microbes prove to be a very imperative tool for producing nutraceuticals. The approach of metabolic engineering in microbial systems seems to be an alluring attitude for the accomplishment production of value-added nutraceuticals. The prompt exposition of different biosynthetic pathways for natural products as well as the ease of genetic manipulation of microbes have greatly permitted the enlargement of microbes for the fabrication of innumerable nutraceuticals. In various approaches like harmonization of the process parameters of fermentation methods, metabolic engineering largely adds towards scaling up the nutraceutical production (Wang et al. 2016).

The microorganisms mainly *E. coli* and *S. cerevisiae* have commonly been established as reference organisms for the de novo or semi-de novo synthesis of different types of polyphenolic compounds. The principal gains of indenturing *E. coli* take account of its reckless growth coupled with the ease of genetic manipulation, whereas for *S. cerevisiae* its generally recognized as safe (GRAS) eminence as well as its capability of functional expression of plant metabolic enzymes makes it the platform organism (Leonard and Koffas 2007). Microbes produce a large number of compounds, such as alkaloids, polyphenolic compounds, polysaccharides, terpenoids, amino acids, etc., which are known for imparting potential health benefits. Several nutrients imparting health benefits such as polyunsaturated fatty acids (PUFAs) cannot be synthesized by the human body; therefore, these has to be supplied externally through the diet and these are known as 'essential fatty acids' (Escott-Stump and Mahan 2000). These are fundamental for the maintenance of body functioning and belong to the classes ω-3 and ω-6. The conventional sources for fatty acids are represented by porcine liver and fish oil. Conversely, microbial lipids, which are fashioned by oleaginous microbes, for instance, bacteria, algae and fungi, are also an auspicious foundation. Several fungi belonging to the different genera, for instance, *Aspergillus*, *Candida*, *Yarrowia*, *Zygosaccharomyces*, *Rhodotorula, Cryptococcus, Mortierella, Mucor*, etc. are largely acknowledged as eminent producer of polyunsaturated fatty acids (Ochsenreither et al. 2016).

Prebiotics and probiotics are also considered as another important category of nutraceuticals. Probiotics are living microbes which are administered in suitable quantities and deliberate an advantageous health outcome on the host. Characteristically, it is demarcated as a viable dietary supplement of microbial origin that constructively affects the host by its possessions in the intestinal tract. However, it was primarily proposed for usage with animal feed products. Nevertheless, the subsequent definition has been anticipated for human beings: 'a live microbial food ingredient that is beneficial to health'(Salminen et al. 1998). The historical background of probiotics can be traced back to the first ingestion of fermented milks, over 2000 years ago. The attention of scientific community in this field heightened from the efforts of Metchnikoff (1907), which were meant for the transformation of the toxic microbiota of the large intestine into a host-pleasant cluster of *Bacillus bulgaricus* (Hord 2008). There is employment of definite in order to cure several gastrointestinal (GI) complications, for instance, acute diarrhoea, lactose intolerance and antibiotic-allied GI side effects (Doron et al. 2005). These agents are acknowledged for the possession of several properties, such as non-pathogenic and non-toxic nature, resilient to gastric acid, adherence to gut epithelial tissues manufacturing antibacterial elements (Suvarna and Boby 2005). Several reports also claim that the intake of probiotics cuts the risk of systemic environments, such as allergy, asthma, cancer and several other infections of the ear and urinary tract (Lenoir-Wijnkoop et al. 2007). Probiotics when ingested in the form of fermented milk products have also exhibited cholesterol depressing possessions, whereas the supplementation of non-digestible and fermentable carbohydrate prebiotics has presented to diminish triacylglycerol intensities. Several studies have revealed that the bacteria *Bifidobacteria bifidum, Lactobacillus acidophilus* and *Lactobacillus bulgaricus* subordinate the cholesterol level in a noteworthy manner when its level is quite high (Ramaa et al. 2006). There are numerous classes of nutraceuticals which are also known for the possession of anticancer activity such as the marine-derived products fucoidans, chitosan, astaxanthin, β-carotene and phlorotannins which are eminent anticancer molecules. The induction of apoptosis through the intake of nutraceuticals play a major role in destruction of cancer cells. Other nutraceuticals of microbial origin are also known for displaying the anti-inflammatory activity. The impending anti-inflammatory nutraceuticals, for instance, fucoidan, chitin, chitosan and glucosamine, play a substantial role in the avoidance of inflammatory response either by constraining the fabrication of nitric oxide (NO), prostaglandin E2, inducible NO synthase, cyclooxygenase-2 or pro-inflammatory cytokines (Vidanarachchi et al. 2012). The possession of antioxidant activity is also supposed to be associated with the anti-inflammatory activity of such compounds (Guerin et al. 2003). The main products showing antioxidant activity are β-carotene, astaxanthin, fucoxanthin, polyphenols and fucoidans. Therefore, the intake of nutraceuticals confers a great deal of benefits to the human beings such as the anti-obesity activity, anticancer activity and prevention of cardiovascular disease, and microbes behave as the superior agents for producing these class of compounds.

16.6 Microbial Toxins for Clinical Applications

Microbial toxins represent another class of microbial metabolites which possess pharmaceutical potential and therefore can be commercialized and targeted for rearing various benefits to human beings. The microorganisms are able to produce such naturally toxic compound as an outcome of the long-term evolutionary process. Such complex evolutionary pathways have also awarded these toxic molecules with the ability to target the indispensable mechanisms underneath the processes of utmost importance to different living entities. These molecules are capable of attacking constituents of protein formation apparatus, signal transduction paths, actin polymerization and vesicle transport across cell along with the elicitation of immune as well as inflammatory reactions (Fabbri et al. 2008). Thereby, these toxins of microbial origin are progressively finding usage as appreciated tools for critically examining diverse physiological phenomena at cellular level, and recently, a proportion of these bioactive compounds has found medicinal usage for battling different human diseases. More importantly, the proteinous toxins fabricated largely by bacteria are mainly found to be actively engaged in affecting central metabolic and molecular pathways governing the physiology of eukaryotic organisms, thereby suggesting their possible employment in medical industries (Alfano et al. 2005; Fabbri et al. 2008). Thereby, such traits render them the ability of behaving as specific chemical daggers meant mainly for the segmentation of physiological progressions (Schlessinger 1993). In the previous decades, the employment of these molecules has allowed a clear understanding as well as characterization of numerous intracellular signalling pathways. Furthermore, they have also been employed and established as efficacious apparatuses serving in medical industries either amalgamated to the segments of monoclonal antibodies with the intention of engendering immunotoxins or to other ligands with the purpose of readdressing their toxic prospective in the direction of anticipated cellular objectives. Such immunotoxins have hence been established with the intention of destroying cancer cells, with recognized effectiveness (Alfano et al. 2005). Some toxins such as those secreted by *Clostridium botulinum* which is a Gram-positive and spore-making bacterium are known to cause a continual blockage of acetylcholine discharge which subsequently induces flaccid paralysis. After entering into the host, the neurotoxin undergoes several steps of processing and thereby ultimately affects the phenomenon of neurotransmission and ultimately causes muscle paralysis. The transformation of this life frightening toxin to a medical miracle has taken a long time. Its therapeutic potential was firstly acknowledged by Justinus Kerner in 1817 (Erbguth and Naumann 1999). It was used primarily as an agent meant for muscle-waning in order to correct investigational strabismus in monkeys. That particular use of this neurotoxin marked its further employment for diverse commercial formulations. This toxin now finds diverse applications in a number of different neurological human problems. The different neurological problems target by botulinum neurotoxin range from hemi-facial spasms, blepharospasm, cervical dystonia, oromandibular dystonia to laryngeal dystonia. Its injections are also acknowledged for treating Parkinson's disease (Cordivari et al. 2001). Moreover, it is also encouragingly used as an important cosmetic agent which specifically targets the reduction of facial wrinkles (Truong and Jost 2006; Hackett and Kam 2007). *Bacillus anthracis* is another bacterium known for producing deadly toxins that are vital for establishing pathogenesis. The toxin is supposed to act on cytosolic objectives, thereby resulting in the impairment of immune system. It also inactivates several kinases and the incongruous inactivation grounds the principle behind its application in treating cancer (Bodart et al. 2002). Another microbe, *Bordetella pertussis*, is known for producing a deadly toxin known as adenylate cyclase toxin which acts as virulent factor of this bacterium. The toxin is a peptidal molecule which on binding to the target promotes uncontrollable conversion of ATP to cAMP. It is reportedly known for enhancing antibody levels, thereby suggesting its potential for using as immunological adjuvant (Adkins et al. 2012). It has also been employed as a vehicle for delivering antigens. The toxin is also known for providing immunity against tumours and melanoma in mice. The toxins of *Vibrio cholera* and *E. coli* are heat-labile toxins which are proteinaceous in nature and are hexameric in nature. These molecules are found to be actively engaged in the activation of numerous signalling cascades. These molecules are potent enough to trigger the anti-inflammatory cytokines, thereby suggesting their possible fabrication as anti-inflammatory agents. Moreover, their potential for being used as vehicles for cellular admission as well as downregulation of inflammatory responses is infuriating substantial curiosity for promoting their use as new drug delivery vehicles as well as therapeutics for controlling human immune syndromes (de Haan and Hirst 2009). Innumerable autoimmune disorders, for instance, rheumatoid arthritis, diabetes and multiple sclerosis, are solely reliant on the stimulation of inflammatory response in addition to tissue damage. These toxins have been found to be highly acknowledged by several researchers for precluding stimulation of pro-inflammatory autoimmune diseases in different animal models (Haan and Hirst 2000). Another toxin of pharmaceutical importance is the toxin of *Bacillus thuringiensis* which is popularly known as cry1ac toxin. It is produced by this entomopathogenic bacteria during the sporulation phase. This protein shows a high cytotoxic activity against different insects, invertebrates and nematodes. Upon consumption, they are cleaved by the alkaline pH of the gut yielding activated toxin which forms pores in the epithelial cells of the gut, thereby causing death of the insect. Furthermore, these proteins are found to be exceedingly precise to their target insect whereas inoffensive to the human beings (Aronson and Shai 2001; Zhang et al. 2006).

The World Health Organization's warning of depletion of antibiotics due to increased phenomenon of antibiotic resistance in microbes has also sparked new interest in microbial toxins for their antibacterial properties. Microcin B17 is an antibacterial peptide that is known for interfering with the function of bacterial topoisomerase. It is supposed to constrain the DNA replication, thereby leading to the stimulation of the SOS system, DNA damage which ultimately leads to cell death (Herrero and Moreno 1986; Collin and Maxwell 2019). Therefore, the increasing roles of microbial toxins make them important molecules for the pharmaceutical industry, and the increasing phenomenon of antibiotic resistance projects their usage as novel antimicrobial molecules for promoting human health (Table 16.5).

16.7 Conclusion and Future Prospects

The contribution of microbial systems to the health and well-being of human beings across the world can never be ignored. In addition to manufacturing numerous primary metabolites, for instance, vitamins, amino acids and nucleotides, they are evolved with a propitious capability of producing secondary metabolites, which represent a major proportion of pharmaceuticals occupying the present-day market. They have escorted the mankind from a pre-antibiotic era where saving life was considered to be the thing of utmost importance to such a stage where life was enhanced up to a level which marked the human exploration of different products targeting the beauty enhancement. The diverse array of secondary metabolites of pharmaceutical importance secreted by microbes and the ability of a single microorganism to release different kinds of secondary metabolites under different environments have largely attracted the attention of mankind for their non-controversial role in holding the backbone of pharmaceutical industries. Microbial systems have allowed mankind to prevent from a deadly infectious disease by the advent of antibiotics, and now with the increasing phenomenon of antibiotic resistance, only microbes seem to be promising candidates attributable to the multifarious mechanisms of action of the secondary metabolites secreted by them. The other metabolites like bacteriocins and biosurfactants also possess a gamut of assets which rewards them the ability for their employment as anticancer molecules.

Table 16.5 Microbial toxins and their therapeutic possessions

Sr. No.	Microbial toxin	Microorganism	Role	Possible mechanism	References
1	Pseudomonas aeruginosa exotoxin A (PEA)	*Pseudomonas aeruginosa*	Anti-HIV activity	Selective killing of HIV infected cells	Alfano et al. (2005)
2	Botulinum neurotoxin (BoNt)	*Clostridium botulinum*	Muscle tone disorders	Block the release of acetylcholine	Fabbri et al. (2008)
3	Anthrax toxin	*Bacillus anthracis*	Anticancer drug	Inactivation of kinases	Bodart et al. (2002)
4	Adenylate cyclase toxin	*Bordetella pertussis*	Adjuvant potential	Unchecked transformation of cellular ATP to cAMP	Adkins et al. (2012)
5	Microcin B17	*Escherichia coli*	Antibacterial activity	Targets bacterial gyrase	Collin and Maxwell (2019)
6	VacA	*Helicobacter pylori*	Immunomodulatory activities	Interact with CD18 on human CD4+ T-cells	Donaldson and Williams (2009)
7	TcdA and TcdB	*Clostridium difficile*	Immunomodulatory activities	Glucosyltransferase activity	Sewald et al. (2008)
8	Listeriolysin O	*Listeria monocytogenes*	Immunomodulatory activities	Induction of apoptosis of lymphocytes	Carrero et al. (2004)
9	Cereulide	*Bacillus cereus*	Immunomodulatory activities	Inhibits NK cell cytotoxicity and cytokine fabrication	Paananen et al. (2002)
10	Shiga toxin	*Shigella dysenteriae*	Immunomodulatory activities	Induce cytokine secretion by macrophages	Tesh et al. (1994)

Moreover, these molecules also appear to be a ray of hope for targeting multidrug-resistant microbial pathogens. Several primary metabolites like enzymes have also contributed a lot in easing the process of secondary metabolite synthesis. The possession of modulation of immune systems has allowed their usage as potent agents for treating mental disorders as well as for boosting immune systems. The intellectual scientific attitude of human beings has put the deadly microbial toxins to various beneficial uses, thus allowing the exploitation of deadly pathogens for benefiting mankind. Although a lot has been achieved in this journey of the microbial world, a large is yet to explore. Since a major proportion of the microbial world is yet to culture, their metabolic patterns are yet to explore which could unveil a large number of hidden phenomena of microbial systems. The cultured microbes also represent a great deal of their belongings, but the detection systems that are prominently used are meant for detecting only a limited number of compounds. Some compounds are secreted in a very low level that is beyond the detection level of sensing machineries. Therefore, the human approach of exploring microbial systems seems to be a dwarf one. Thus, the proper understanding of metabolic patterns of microbes along with their behaviours in different environments can also make a difference. So, taken all together, it can be concluded that microbes are the smallest creatures with the largest possessions which on judicial exploration can help to sustain the mankind for a longer time.

References

Adkins I, Holubova J, Kosova M, Sadilkova L (2012) Bacteria and their toxins tamed for immunotherapy. Curr Pharm Biotechnol 13(8):1446–1473. https://doi.org/10.2174/138920112800784835

Ahmad V, Khan MS, Jamal QMS, Alzohairy MA, Al Karaawi MA, Siddiqui MU (2017) Antimicrobial potential of bacteriocins: in therapy, agriculture and food preservation. Int J Antimicrob Agents 49(1):1–11. https://doi.org/10.1016/j.ijantimicag.2016.08.016

Ahmad V, Ahmad K, Baig MH, AL-Shwaiman HA, Al Khulaifi MM, Elgorban AM, Khan MS (2019) Efficacy of a novel bacteriocin isolated from Lysinibacillus sp. against Bacillus pumilus. LWT 102:260–267

Ahmadi S, Ghollasi M, Hosseini HM (2017) The apoptotic impact of nisin as a potent bacteriocin on the colon cancer cells. Microb Pathog 111:193–197. https://doi.org/10.1016/j.micpath.2017.08.037

Akbari S, Abdurahman NH, Yunus RM, Fayaz F, Alara OR (2018) Biosurfactants—a new frontier for social and environmental safety: a mini review. Biotechnol Res Innov 2:81–90

Albarracin L, Kobayashi H, Iida H, Sato N, Nochi T, Aso H, Salva S, Alvarez S, Kitazawa H, Villena J (2017) Transcriptomic analysis of the innate antiviral immune response in porcine intestinal epithelial cells: influence of immunobiotic lactobacilli. Front Immunol 8:57

Alejandro CS, Humberto HS, María JF (2011) Production of glycolipids with antimicrobial activity by Ustilago maydis FBD12 in submerged culture. Afr J Microbiol Res 5:2512–2523

Alfano M, Rizzi C, Corti D, Adduce L, Poli G (2005) Bacterial toxins: potential weapons against HIV infection. Curr Pharm Des 11(22):2909–2926

Alfarouk KO, Stock CM, Taylor S, Walsh M, Muddathir AK, Verduzco D, Bashir AH, Mohammed OY, Elhassan GO, Harguindey S, Reshkin SJ (2015) Resistance to cancer chemotherapy: failure in drug response from ADME to P-gp. Cancer Cell Int 15(1):71

Allen AP, Hutch W, Borre YE, Kennedy PJ, Temko A, Boylan G, Murphy E, Cryan JF, Dinan TG, Clarke G (2016) Bifidobacterium longum 1714 as a translational psychobiotic: modulation of stress, electrophysiology and neurocognition in healthy volunteers. Transl Psychiatry 11:e939

Alonso C, Lucas R, Barba C, Marti M, Rubio L, Comelles F, Morales JC, Coderch L, Parra JL (2015) Skin delivery of antioxidant surfactants based on gallic acid and hydroxytyrosol. J Pharm Pharmacol 67(7):900–908

Anbu P, Gopinath SC, Cihan AC, Chaulagain BP (2013) Microbial enzymes and their applications in industries and medicine. Biomed Res Int 2013:1–2

Anonymous (2019). https://www.grandviewresearch.com/industry-analysis/antibiotic-market

Archana K, Reddy KS, Parameshwar J, Bee H (2019) Isolation and characterization of sophorolipid producing yeast from fruit waste for application as antibacterial agent. Environ Sustain 2:1–9

Aronson AI, Shai Y (2001) Why Bacillus thuringiensis insecticidal toxins are so effective: unique features of their mode of action. FEMS Microbiol Lett 195(1):1–8

Azevedo MS, Zhang W, Wen K, Gonzalez AM, Saif LJ, Yousef AE (2012) Lactobacillus acidophilus and lactobacillus reuteri modulate cytokine responses in gnotobiotic pigs infected with human rotavirus. Benefic Microbes 3(1):33–42. https://doi.org/10.3920/BM2011.0041

Bailey MT, Dowd SE, Galley JD, Hufnagle AR, Allen RG, Lyte M (2011) Exposure to a social stressor alters the structure of the intestinal microbiota: implications for stressor-induced immunomodulation. Brain Behav Immun 25(3):397–407

Balan SS, Kumar CG, Jayalakshmi S (2019) Physicochemical, structural and biological evaluation of Cybersan (trigalactomargarate), a new glycolipid biosurfactant produced by a marine yeast, Cyberlindnera saturnus strain SBPN-27. Process Biochem 80:171–180

Banat I, Franzetti A, Gandolfi I, Bestetti G, Martinotti M, Fracchia L, Smyth T, Marchant R (2010) Microbial biosurfactants production, applications and future potential. Appl Microbiol Biotechnol 87:427–444

Bardach A, Ciapponi A, Garcia-Marti S, Glujovsky D, Mazzoni A, Fayad A (2011) Epidemiology of acute otitis media in children of Latin America and the Caribbean: a systematic review and meta-analysis. Int J Pediatr Otorhinolaryngol 75:1062–1070. https://doi.org/10.1016/j.ijporl.2011.05.014

Basu S, Paul DK, Ganguly S, Chatterjee M, Chandra PK (2009) Efficacy of high dose lactobacillus rhamnosus GG in controlling acute watery diarrhea in Indian children: a randomized controlled trial. J Clin Gastroenterol 43(3):208–213. https://doi.org/10.1097/MCG.0b013e31815a5780

Belenky P, Jonathan DY, Porter CB, Cohen NR, Lobritz MA, Ferrante T, Jain S, Korry BJ, Schwarz EG, Walker GC, Collins JJ (2015) Bactericidal antibiotics induce toxic metabolic perturbations that lead to cellular damage. Cell Rep 13(5):968–980

Belkacem N, Bourdet-Sicard R, Taha MK (2018) Lactobacillus paracasei feeding improves the control of secondary experimental meningococcal infection in flu-infected mice. BMC Infect Dis 18(1):167

Bentley SD, Chater KF, Cerdeño-Tarraga AM, Challis GL, Thomson NR, James KD, Harris DE, Quail MA, Kieser H, Harper D et al (2002) Complete genome sequence of the model actinomycete Streptomyces coelicolor A3(2). Nature 417:141–147

Bodart JF, Chopra A, Liang X, Duesbery N (2002) Anthrax, MEK and cancer. Cell Cycle 1(1):7–12

Bravo JA, Forsythe P, Chew MV, Escaravage E, Savignac HM, Dinan TG, Bienenstock J, Cryan JF (2011) Ingestion of lactobacillus strain regulates emotional behavior and central GABA receptor expression in a mouse via the vagus nerve. Proc Natl Acad Sci 108(38):16050–16055

Brown ED, Wright GD (2016) Antibacterial drug discovery in the resistance era. Nature 529:336–343

Bush K, Courvalin P, Dantas G, Davies J, Eisenstein B, Huovinen P, Jacoby GA, Kishony R, Kreiswirth BN, Kutter E, Lerner SA (2011) Tackling antibiotic resistance. Nat Rev Microbiol 9 (12):894

Cameotra SS, Makkar RS (1998) Synthesis of biosurfactants in extreme conditions. Appl Microbiol Biotechnol 50:520–529

Carrero JA, Calderon B, Unanue ER (2004) Listeriolysin 0 from listeria monocytogenes is a lymphocyte apoptogenic molecule. J Immunol 172(8):4866–4874

Cavicchioli VQ, de Carvalho OV, de Paiva JC, Todorov SD, Júnior AS, Nero LA (2018) Inhibition of herpes simplex virus 1 (HSV-1) and poliovirus (PV-1) by bacteriocins from Lactococcus lactis subsp. lactis and Enterococcus durans strains isolated from goat milk. Int J Antimicrob Agents 51(1):33–37

Chandran P, Das N (2010) Biosurfactant production and diesel oil degradation by yeast species Trichosporon asahii isolated from petroleum hydrocarbon contaminated soil. Int J Eng Sci Technol 2:6942–6953

Chiewpattanakul P, Phonnok S, Durand A, Marie E, Thanomsub BW (2010) Bioproduction and anticancer activity of biosurfactant produced by the dematiaceous fungus Exophiala dermatitidis SK80. J Microbiol Biotechnol 20:1664–1671. https://doi.org/10.4014/jmb.1007.07052

Chikindas ML, Weeks R, Drider D, Chistyakov VA, Dicks LMT (2018) Functions and emerging applications of bacteriocins. Curr Opin Biotechnol 49:23–28

Chirumamilla RR, Muralidhar R, Marchant R, Nigam P (2001) Improving the quality of industrially important enzymes by directed evolution. Mol Cell Biochem 224(1/2):159–168

Cho YH, Song JY, Kim KM, Kim MK, Lee IY, Kim SB, Kim HS, Han NS, Lee BH, Kim BS (2010) Production of nattokinase by batch and fed-batch culture of Bacillus subtilis. New Biotechnol 27(4):341–346

Choi JM, Han SS, Kim HS (2015) Industrial applications of enzyme biocatalysis: current status and future aspect. Biotechnol Adv 33:1443–1454. https://doi.org/10.1016/j.biotechadv.2015.02.014

Clancy R (2003) Immunobiotics and the probiotic evolution. FEMS Immunol Med Microbiol 38 (1):9–12. https://doi.org/10.1016/s0928-8244(03)00147-0

Clardy J, Fischbach MA, Currie CR (2009) The natural history of antibiotics. Curr Biol 19(11): R437–R441

Clarke G, Grenham S, Scully P, Fitzgerald P, Moloney RD, Shanahan F, Dinan TG, Cryan JF (2013) The microbiome-gut-brain axis during early life regulates the hippocampal serotonergic system in a sex-dependent manner. Mol Psychiatry 18(6):666

Collin F, Maxwell A (2019) The microbial toxin microcin B17: prospects for the development of new antibacterial agents. J Mol Biol 431:3400–3426

Collins SM, Bercik P (2009) The relationship between intestinal microbiota and the central nervous system in normal gastrointestinal function and disease. Gastroenterology 136:2003–2014

Collins SM, Surette M, Bercik P (2012) The interplay between the intestinal microbiota and the brain. Nat Rev Microbiol 10:735–742

Cordivari C, Misra VP, Catania S, Lees AJ (2001) Treatment of dystonic clenched fist with botulinum toxin. Mov Disord 16(5):907–913

Cotter PD, Hill C, Ross RP (2005) Bacteriocins: developing innate immunity for food. Nat Rev Microbiol 3:777–788

Cryan JF, Dinan TG (2012) Mind-altering microorganisms: the impact of the gut microbiota on brain and behaviour. Nat Rev Neurosci 13:701–712

Cryan JF, O'Mahony SM (2011) The microbiome-gut-brain axis: from bowel to behaviour. Neurogastroenterol Motil 23:187–192

Cuevas C, Francesch A (2009) Development of Yondelis (trabectedin, ET-743): a semisynthetic process solves the supply problem. Nat Prod Rep 26:322–337

Cutting GR (2005) Modifier genetics: cystic fibrosis. Annu Rev Genomics Hum Genet 6:237–260

Das P, Mukherjee S, Sen R (2009) Antiadhesive action of a marine microbial surfactant. Colloids Surf B: Biointerfaces 71(2):183–186

de Freitas FJ, Vieira EA, Nitschke M (2019) The antibacterial activity of rhamnolipid biosurfactant is pH dependent. Food Res Int 116:737–744

de Haan L, Hirst TR (2009) Cholera toxin: a paradigm for multi-functional engagement of cellular mechanisms (review). Mol Membr Biol 21(2):77–92

De Kwaadsteniet M, Doeschate KT, Dicks LMT (2009) Nisin F in the treatment of respiratory tract infections caused by Staphylococcus aureus. Lett Appl Microbiol 48:65–70

DeFelice SL (1995) The nutraceutical revolution: its impact on food industry R&D. Trends Food Sci Technol 6(2):59–61
Del Carmen S, de LeBlanc AD, Martin R, Chain F, Langella P, Bermúdez-Humarán LG, LeBlanc JG (2014) Genetically engineered immunomodulatory Streptococcus thermophilus strains producing antioxidant enzymes exhibit enhanced anti-inflammatory activities. Appl Environ Microbiol 80(3):869–877
Demain AL (2002) Prescription for an ailing pharmaceutical industry. Nat Biotechnol 20 (4):331–331. https://doi.org/10.1038/nbt0402-331
Demain AL (2004) The biopharmaceutical revolution. Chem Today (Chim Oggi) 22:11–12
Demain AL, Adrio JL (2008) Contributions of microorganisms to industrial biology. Mol Biotechnol 38(1):41–55
Demain AL, Sanchez S (2009) Microbial drug discovery: 80 years of progress. J Antibiot 62(1):5
Demain AL, Vaishnav P (2009) Production of recombinant proteins by microbes and higher organisms. Biotechnol Adv 27(3):297–306
Dey G, Bharti R, Sen R, Mandal M (2015) Microbial amphiphiles: a class of promising new-generation anticancer agents. Drug Discov Today 20:136–146. https://doi.org/10.1016/j.drudis.2014.09.006
Dinan TG, Stanton C, Cryan JF (2013) Psychobiotics: a novel class of psychotropic. Biol Psychiatry 74(10):720–726
Dolman BM, Wang F, Winterburn JB (2019) Integrated production and separation of biosurfactants. Process Biochem 83:1–8. https://doi.org/10.1016/j.procbio.2019.05.002
Donaldson DS, Williams NA (2009) Bacterial toxins as immunomodulators. In: Pathogen-derived immunomodulatory molecules. Springer, New York, pp 1–18
Donio MB, Ronica SF, Viji VT, Velmurugan S, Jenifer JA, Michaelbabu M, Citarasu T (2013) Isolation and characterization of halophilic Bacillus sp. BS3 able to produce pharmacologically important biosurfactants. Asian Pac J Trop Med 6(11):876–883
Dordick JS (2013) Biocatalysts for industry. ISBN 9781475745979
Doron S, Snydman DR, Gorbach SL (2005) Lactobacillus GG: bacteriology and clinical applications. Gastroenterol Clin N Am 34:483–498. https://doi.org/10.1016/j.gtc.2005.05.011
Dy GK, Adjei AA (2013) Understanding, recognizing, and managing toxicities of targeted anticancer therapies. CA Cancer J Clin 63(4):249–279
Edmond K, Scott S, Korczak V, Ward C, Sanderson C, Theodoratou E (2012) Long term sequelae from childhood pneumonia; systematic review and meta-analysis. PLoS One 7:e31239. https://doi.org/10.1371/journal.pone.0031239
Elahi S, Ertelt JM, Kinder JM, Jiang TT, Zhang X, Xin L, Chaturvedi V, Strong BS, Qualls JE, Steinbrecher KA, Kalfa TA (2013) Immunosuppressive CD71+ erythroid cells compromise neonatal host defence against infection. Nature 504(7478):158
Erbguth FJ, Naumann M (1999) Historical aspects of botulinum toxin: Justinus Kerner (1786–1862) and the "sausage poison". Neurology 53(8):1850–1853
Escott-Stump E, Mahan LK (2000) Krause's food, nutrition and diet therapy, 10th edn. WB Saunders Company, Philadelphia, pp 553–559
Fabbri A, Travaglione S, Falzano L, Fiorentini C (2008) Bacterial protein toxins: current and potential clinical use. Curr Med Chem 15(11):1116–1125. https://doi.org/10.2174/092986708784221430
Faber K (1997) Biotransformations in organic chemistry: a textbook. Springer, Berlin
Farias JM, Stamford TC, Resende AH, Aguiar JS, Rufino RD, Luna JM, Sarubbo LA (2019) Mouthwash containing a biosurfactant and chitosan: an eco-sustainable option for the control of cariogenic microorganisms. Int J Biol Macromol 129:853–860
Farizano JV, Masías E, Hsu FF, Salomón RA, Freitag NE, Hebert EM, Minahk C, Saavedra L (2019) PrfA activation in listeria monocytogenes increases the sensitivity to class IIa bacteriocins despite impaired expression of the bacteriocin receptor. Biochim Biophys Acta (BBA) Gen Subj 1863(8):1283–1291

Felse PA, Shah V, Chan J, Rao KJ, Gross RA (2007) Sophorolipid biosynthesis by the yeast Candida bombicola from industrial fatty acid residues. Enzym Microb Technol 40(2):499–504

Ferreira A, Vecino X, Ferreira D, Cruz JM, Moldes AB, Rodrigues LR (2017) Novel cosmetic formulations containing a biosurfactant from lactobacillus paracasei. Colloids Surf B: Biointerfaces 155(4):522–529. https://doi.org/10.1016/j.colsurfb.2017.04.026

Ferrer-Miralles N, Domingo-Espín J, Corchero J, Vázquez E, Villaverde A (2009) Microbial factories for recombinant pharmaceuticals. Microb Cell Factories 8(1):17. https://doi.org/10.1186/1475-2859-8-17

Galvez A, Abriouel H, Ben Omar N, Lucas R (2011) Food applications and regulations. In: Drider D, Rebuffat S (eds) Prokaryotic antimicrobial peptides. Springer, New York, pp 353–390

Gaur VK, Regar RK, Dhiman N, Gautam K, Srivastava JK, Patnaik S, Kamthan M, Manickam N (2019) Biosynthesis and characterization of sophorolipid biosurfactant by Candida spp.: application as food emulsifier and antibacterial agent. Bioresour Technol 285:121314

Gentile A, Bardach A, Ciapponi A, Garcia-Marti S, Aruj P, Glujovsky D (2012) Epidemiology of community-acquired pneumonia in children of Latin America and the Caribbean: a systematic review and meta-analysis. Int J Infect Dis 16:e5–e15. https://doi.org/10.1016/j.ijid.2011.09.013

Gerngross TU (2004) Advances in the production of human therapeutic proteins in yeasts and filamentous fungi. Nat Biotechnol 22:1409–1414

Gudiña EJ, Rangarajan V, Sen R, Rodrigues LR (2013) Potential therapeutic applications of biosurfactants. Trends Pharmacol Sci 34(12):667–675

Guerin M, Huntley ML, Olaizola M (2003) Haematococcus astaxanthin: applications for human health and nutrition. Trends Biotechnol 21:210–216

Haan L, Hirst TR (2000) Cholera toxin and related enterotoxins: a cell biological and immunological perspective. J Nat Toxins 9(3):281–297

Hackett R, Kam PC (2007) Botulinum toxin: pharmacology and clinical developments: a literature review. Med Chem 3(4):333–345

Hasper HE, Kramer NE, Smith JL, Hillman JD, Zachariah C, Kuipers OP, de Kruijff B, Breukink E (2006) An alternative bactericidal mechanism of action for lantibiotic peptides that target lipid II. Science 313:1636–1637

Heijtz RD, Wang S, Anuar F, Qian Y, Björkholm B, Samuelsson A, Hibberd ML, Forssberg H, Pettersson S (2011) Normal gut microbiota modulates brain development and behavior. Proc Natl Acad Sci 108(7):3047–3052

Hentati D, Chebbi A, Hadrich F, Frikha I, Rabanal F, Sayadi S, Manresa A, Chamkha M (2019) Production, characterization and biotechnological potential of lipopeptide biosurfactants from a novel marine Bacillus stratosphericus strain FLU5. Ecotoxicol Environ Saf 167:441–449

Herrero M, Moreno F (1986) Microcin B17 blocks DNA replication and induces the SOS system in Escherichia coli. Microbiology 132(2):393–402

Hessle C, Andersson B, Wold AE (2000) Gram-positive bacteria are potent inducers of monocytic interleukin-12 (IL-12) while gram-negative bacteria preferentially stimulate IL-10 production. Infect Immun 68:3581

Hong K-W, Koh C-L, Sam C-K, Yin W-F, Chan K-G (2012) Quorum quenching revisited—from signal decays to signalling confusion. Sensors (Basel) 12:4661–4696

Hooper LV, Gordon JI (2001) Commensal host-bacterial relationships in the gut. Science 292 (5519):1115–1118

Hord NG (2008) Eukaryotic microbiotic crosstalk: potential mechanisms for health benefits of prebiotics and probiotics. Annu Rev Nutr 28:215–231. https://doi.org/10.1146/annurev.nutr.28.061807.155402

https://www.manufacturingchemist.com/news/article_page/Production_of_pharmaceutical_compounds_through_microbial_fermentation/61614

Hua Z, Chen J, Lun S, Wang X (2003) Influence of biosurfactants produced by Candida antarctica on surface properties of microorganism and biodegradation of n-alkanes. Water Res 3:4143–4150

Ikram-ul-Haq, Ali S, Qadeer MA (2002) Biosynthesis of L-DOPA by Aspergillus oryzae. Bioresour Technol 85(1):25–29

Indira M, Venkateswarulu TC, Vidya Prabhakar K, Abraham Peele K, Krupanidhi S (2018) Isolation and characterization of bacteriocin producing Enterococcus casseliflavus and its antagonistic effect on Pseudomonas aeruginosa. Karbala Int J Mod Sci 4:361–368. https://doi.org/10.1016/j.kijoms.2018.09.002

Iyer LM, Aravind L, Coon SL, Klein DC, Koonin EV (2004) Evolution of cell–cell signaling in animals: did late horizontal gene transfer from bacteria have a role? Trends Genet 20 (7):292–299

Jain N, Ramawat KG (2013) Nutraceuticals and antioxidants in prevention of diseases. In: Natural products: phytochemistry, botany and metabolism of alkaloids, phenolics and terpenes. Springer, Heidelberg, pp 2559–2580

Janik R, Thomason LA, Stanisz AM, Forsythe P, Bienenstock J, Stanisz GJ (2016) Magnetic resonance spectroscopy reveals oral lactobacillus promotion of increases in brain GABA, N-acetyl aspartate and glutamate. NeuroImage 125:988–995

Jeandet P, Vasserot Y, Chastang T, Courot E (2013) Engineering microbial cells for the biosynthesis of natural compounds of pharmaceutical significance. Biomed Res Int 2013:780145

Jenkins N (2007) Modifications of therapeutic proteins: challenges and prospects. Cytotechnology 53:121–125

Johannes T, Simurdiak MR, Zhao H (2006) Biocatalysis. In: Lee S (ed) Encyclopedia of chemical processing. Taylor & Francis, New York, pp 101–110

Joshi PA, Shekhawat DB (2014) Screening and isolation of biosurfactant producing bacteria from petroleum contaminated soil. Eur J Exp Biol 4(4):164–169

Jozala AF, Geraldes DC, Tundisi LL, Feitosa VD, Breyer CA, Cardoso SL, Mazzola PG, Oliveira-Nascimento LD, Rangel-Yagui CD, Magalhães PD, Oliveira MA (2016) Biopharmaceuticals from microorganisms: from production to purification. Braz J Microbiol 47:51–63

Juárez I, Gratton A, Flores G (2008) Ontogeny of altered dendritic morphology in the rat prefrontal cortex, hippocampus, and nucleus accumbens following cesarean delivery and birth anoxia. J Comp Neurol 507:1734–1747

Juturu V, Wu JC (2018) Microbial production of bacteriocins: latest research development and applications. Biotechnol Adv 36:2187–2200. https://doi.org/10.1016/j.biotechadv.2018.10.007

Kaewnopparat S, Dangmanee N, Kaewnopparat N, Srichana T, Chulasiri M, Settharaksa S (2013) In vitro probiotic properties of lactobacillus fermentum SK5 isolated from vagina of a healthy woman. Anaerobe 22:6–13

Kaktcham PM, Kouam EM, Tientcheu ML, Temgoua JB, Wacher C, Ngoufack FZ, de Lourdes P-CM (2019) Nisin-producing Lactococcus lactis subsp. lactis 2MT isolated from freshwater Nile tilapia in Cameroon: bacteriocin screening, characterization, and optimization in a low-cost medium. LWT 107:272–279

Karlapudi AP, Venkateswarulu TC, Srirama K, Kota RK, Mikkili I, Kodali VP (2018) Evaluation of anti-cancer, anti-microbial and anti-biofilm potential of biosurfactant extracted from an Acinetobacter M6 strain. J King Saud Univ-Sci 32(1):223–227

Kataria J, Li N, Wynn JL, Neu J (2009) Probiotic microbes: do they need to be alive to be beneficial? Nutr Rev 67(9):546–550

Kaur B, Balgir PP, Mittu B, Kumar B, Garg N (2013) Biomedical applications of fermenticin HV6b isolated from lactobacillus fermentum HV6b MTCC10770. Biomed Res Int 2013:1–8

Kaur K, Sangwan S, Kaur H (2017) Biosurfactant production by yeasts isolated from hydrocarbon polluted environments. Environ Monit Assess 189(12):603

Khopade A, Biao R, Liu X, Mahadik K, Zhang L, Kokare C (2011) Production and stability studies of the biosurfactant isolated from marine Nocardiopsis spp. B4. Desalination 285:198–204

Kim K, Jung SY, Lee DK, Jung JK, Park JK, Kim DK, Lee CH (1998) Suppression of inflammatory responses by surfactin, a selective inhibitor of platelet cytosolic phospholipase A2. Biochem Pharmacol 55(7):975–985. https://doi.org/10.1016/s0006-2952(97)00613-8

Kim HR, Jung KY, Kim SY, Ko, Lee YM, Kim JM (2003) Delivery modes and neonatal EEG: spatial pattern analysis. Early Hum Dev 75(1–2):35–53

Kim N-N, Kim WJ, Kang S-S (2019) Anti-biofilm effect of crude bacteriocin derived from Lactobacillus brevis DF01 on Escherichia coli and Salmonella Typhimurium. Food Control 98:274–280

Kiran GS, Hema TA, Gandhimathi R, Selvin J, Thomas TA, Ravji TR, Natarajaseenivasan K (2009) Optimization and production of a biosurfactant from the sponge-associated marine fungus Aspergillus ustus MSF3. Colloids Surf B: Biointerfaces 73(2):250–256

Kitagawa M, Suzuki M, Yamamoto S, Sogabe A, Kitamoto D, Imura T, Morita T (2011). U.S. patent application no. 13/170,432

Kohanski MA, Dwye DJ, Collins JJ (2010) How antibiotics kill bacteria: from targets to networks. Nat Rev Microbiol 8:423–435

Konishi M, Fukuoka T, Morita T, Imura T, Kitamoto D (2008) Production of new types of sophorolipids by Candida batistae. J Oleo Sci 57:359–369

Kumar CG, Takagi H (1999) Microbial alkaline proteases: from a bioindustrial viewpoint. Biotechnol Adv 17(7):561–594

Lahtinen SJ, Endo A (2011) Health effects of nonviable probiotics. In: Lahtinen S, Ouwehand AC, Salminen S, von Wright A (eds) Lactic acid bacteria: microbiological and functional aspects, 4th edn. CRC Press, Boca Raton, pp 670–685

Lancini G, Demain AL (2013) Bacterial pharmaceutical products. In: The prokaryotes: applied bacteriology and biotechnology. Springer, Heidelberg, pp 257–280. https://doi.org/10.1007/978-3-642-31331-8_28

Le Roes-Hill M, Prins A (2016) Biotechnological potential of oxidative enzymes from Actinobacteria. In: Actinobacteria – basics and biotechnological applications. https://doi.org/10.5772/61321

Lenoir-Wijnkoop I, Sanders ME, Cabana MD, Caglar E, Corthier G, Rayes N, Sherman PM, Timmerman HM (2007) Probiotic and prebiotic influence beyond the intestinal tract. Nutr Rev 65:469–489. https://doi.org/10.1111/j.1753-4887.2007.tb00272.x

Leonard E, Koffas MA (2007) Engineering of artificial plant cytochrome P450 enzymes for synthesis of isoflavones by Escherichia coli. Appl Environ Microbiol 73(22):7246–7251

Li S, Yang X, Yang S, Zhu M, Wang X (2012) Technology prospecting on enzymes: application, marketing and engineering. Comput Struct Biotechnol J 2(3):e201209017

Liang S, Wang T, Hu X, Luo J, Li W, Wu X, Duan Y, Jin F (2015) Administration of lactobacillus helveticus NS8 improves behavioral, cognitive, and biochemical aberrations caused by chronic restraint stress. Neuroscience 310:561–577

Liu X, Kokare C (2017) Microbial enzymes of use in industry. In: Biotechnology of microbial enzymes. Elsevier, London/New York, pp 267–298. https://doi.org/10.1016/b978-0-12-803725-6.00011-x

Liu F, Li G, Wen K, Bui T, Cao D, Zhang Y (2010) Porcine small intestinal epithelial cell line (IPEC-J2) of rotavirus infection as a new model for the study of innate immune responses to rotaviruses and probiotics. Viral Immunol 23(2):135–149. https://doi.org/10.1089/vim.2009.0088

Liu X, Ren B, Gao H, Liu M, Dai H, Song F, Zhang L (2012) Optimization for the production of surfactin with a new synergistic antifungal activity. PLoS One 7(5):e34430

Liu WH, Yang CH, Lin CT, Li SW, Cheng WS, Jiang YP, Wu CC, Chang CH, Tsai YC (2015) Genome architecture of lactobacillus plantarum PS128, a probiotic strain with potential immunomodulatory activity. Gut Pathogens 7(1):22

Liu YW, Liu WH, Wu CC, Juan YC, Wu YC, Tsai HP, Wang S, Tsai YC (2016) Psychotropic effects of lactobacillus plantarum PS128 in early life-stressed and naïve adult mice. Brain Res 1631:1–2

Liu YW, Ong JS, Gan CY, Khoo BY, Yahaya S, Choi SB, Low WY, Tsai YC, Liong MT (2019) Lactobacillus fermentum PS150 showed psychotropic properties by altering serotonergic pathway during stress. J Funct Foods 59:352–361

Lyte M (2011) Probiotics function mechanistically as delivery vehicles of neuroactive compounds: microbial endocrinology in the design and use of probiotics. BioEssays 33(8):574–581
Macfarlane GT, Macfarlane S (2012) Bacteria, colonic fermentation, and gastrointestinal health. J AOAC Int 95(1):50–60
Majamaa H, Isolauri E, Saxelin M, Vesikari T (1995) Lactic acid bacteria in the treatment of acute rotavirus gastroenteritis. J Pediatr Gastroenterol Nutr 20(3):333–338. https://doi.org/10.1097/00005176-199504000-00012
Manning MC, Patel K, Borchardt RT (1989) Stability of protein pharmaceuticals. Pharm Res 6:903–918
Maragkoudakis PA, Chingwaru W, Gradisnik L, Tsakalidou E, Cencic A (2010) Lactic acid bacteria efficiently protect human and animal intestinal epithelial and immune cells from enteric virus infection. Int J Food Microbiol 141(Suppl 1):S91–S97. https://doi.org/10.1016/j.ijfoodmicro.2009.12.024
Martín-Escolano R, Cebrián R, Martín-Escolano J, Rosales MJ, Maqueda M, Sánchez-Moreno M, Marín C (2019) Insights into Chagas treatment based on the potential of bacteriocin AS-48. Int J Parasitol Drugs Drug Resist 10:1–8
Masmoudi H, Le Y, Piccerelle P, Kister J (2005) The evaluation of cosmetic and pharmaceutical emulsions aging process using classical techniques and a new method: FTIR. Int J Pharm 289:117–131. https://doi.org/10.1016/j.ijpharm.2004.10.020
Matthews DM, Jenks SM (2013) Ingestion of Mycobacterium vaccae decreases anxiety-related behavior and improves learning in mice. Behav Process 96:27–35. https://doi.org/10.1016/j.beproc.2013.02.007
Mattick ATR, Hirsch A (1947) Further observations on an inhibitory substance (nisin) from lactic streptococci. Lancet 2:5–7
Messaoudi M, Lalonde R, Violle N, Javelot H, Desor D, Nejdi A, Bisson JF, Rougeot C, Pichelin M, Cazaubiel M, Cazaubiel JM (2011) Assessment of psychotropic-like properties of a probiotic formulation (lactobacillus helveticus R0052 and Bifidobacterium longum R0175) in rats and human subjects. Br J Nutr 105(5):755–764
Metchnikoff E (1907) The prolongation of life. Putnam & Sons, New York
Miyazawa K, He F, Kawase M, Kubota A, Yoda K, Hiramatsu M (2011) Enhancement of immunoregulatory effects of lactobacillus gasseri TMC0356 by heat treatment and culture medium. Lett Appl Microbiol 53(2):210–216
Morgan SM, O'Connor PM, Cotter PD, Ross RP, Hill C (2005) Sequential actions of the two component peptides of the lantibiotic lacticin 3147 explain its antimicrobial activity at nanomolar concentrations. Antimicrob Agents Chemother 49:2606–2611
Mulligan CN (2005) Environmental applications for biosurfactants. Environ Pollut 133:183–198
Murphy EF, Clarke CF, Marques TM, Hill C, Stanton C, Ross RP, O'Doherty RM, Shanahan F, Cotter PD (2013) Antimicrobials: strategies for targeting obesity and metabolic health? Gut Microbes 4:48–53
Nalini S, Parthasarathi R (2018) Optimization of rhamnolipid biosurfactant production from Serratia rubidaea SNAU02 under solid-state fermentation and its biocontrol efficacy against Fusarium wilt of eggplant. Ann Agrarian Sci 16(2):108–115
Neufeld K, Kang N, Bienenstock J, Foster J (2011) Reduced anxiety-like behavior and central neurochemical change in germ-free mice. Neurogastroenterol Motil 23:255–264
Neuman H, Debelius JW, Knight R, Koren O (2015) Microbial endocrinology: the interplay between the microbiota and the endocrine system. FEMS Microbiol Rev 39(4):509–521
Newman DJ, Cragg GM (2007) Natural products as sources of new drugs over the last 25 years. J Nat Prod 70:461–477
Nigam P (2013) Microbial enzymes with special characteristics for biotechnological applications. Biomol Ther 3(4):597–611
Nishida K, Sawada D, Kuwano Y, Tanaka H, Sugawara T, Aoki Y, Fujiwara S, Rokutan K (2017) Daily administration of paraprobiotic lactobacillus gasseri CP2305 ameliorates chronic stress-associated symptoms in Japanese medical students. J Funct Foods 36:112–121

O'Connell S, Walsh G (2008) Application relevant studies of fungal beta-galactosidases with potential application in the alleviation of lactose intolerance. Appl Biochem Biotechnol 149 (2):129–138

Obtel M, Lyoussi B, Tachfouti N, Pelissier SM, Nejjari C (2015) Using surveillance data to understand cancer trends: an overview in Morocco. Arch Public Health 73(1):45

Ochsenreither K, Glück C, Stressler T, Fischer L, Syldatk C (2016) Production strategies and applications of microbial single cell oils. Front Microbiol 7:1539

Oh S, Kim SH, Ko Y, Sim JH, Kim KS, Lee SH, Park S, Kim YJ (2006) Effect of bacteriocin produced by Lactococcus sp. HY 449 on skin-inflammatory bacteria. Food Chem Toxicol 44 (4):552–559

Ohadi M, Dehghannoudeh G, Forootanfar H, Shakibaie M, Rajaee M (2018) Investigation of the structural, physicochemical properties, and aggregation behavior of lipopeptide biosurfactant produced by Acinetobacter junii B6. Int J Biol Macromol 112:712–719. https://doi.org/10.1016/j.ijbiomac.2018.01.209

Okafor N, Okeke B (2007) Biocatalysis: immobilized enzymes and immobilized cells. In: Modern industrial microbiology and biotechnology. Science Publishers, Enfield, pp 398–420

Oliveira C, Guimarães PMR, Domingues L (2011) Recombinant microbial systems for improved β-galactosidase production and biotechnological applications. Biotechnol Adv 29(6):600–609

Ongena M, Jacques P, Toure Y, Destain J, Jabrane A, Thonart P (2005) Involvement of fengycin-type lipopeptides in the multifaceted biocontrol potential of Bacillus subtilis. Appl Microbiol Biotechnol 69:29–38

Paananen A, Mikkola R, Sareneva T et al (2002) Inhibition of human natural killer cell activity by cereulide, an emetic toxin from bacillus cereus. Clin Exp Immunol 129(3):420–428

Pacwa-Plociniczak M, Ptaza GA, Piotrowska-Seget Z, Cameotra SS (2011) Environmental applications of biosurfactants: recent advances. Int J Mol Sci 12:633–654

Pandey A, Selvakumar P, Soccol CR, Nigam P (1999) Solid-state fermentation for the production of industrial enzymes. Curr Sci 77:149–162

Pandey M, Verma RK, Saraf SA (2010) Nutraceuticals: new era of medicine and health. Asian J Pharm Clin Res 3(1):11–15

Pathak KV, Bose A, Keharia H (2014) Characterization of novel Lipopeptides produced by Bacillus tequilensis P15 using liquid chromatography coupled electron spray ionization tandem mass spectrometry (LC–ESI–MS/MS). Int J Pept Res Ther 20:133–143

Piper C, Draper LA, Cotter PD, Ross RP, Hill C (2009) A comparison of the activities of lacticin 3147 and nisin against drug-resistant Staphylococcus aureus and Enterococcus species. J Antimicrob Chemother 64:546–551

Quintana VM, Torres NI, Wachsman MB, Sinko PJ, Castilla V, Chikindas M (2014) Antiherpes simplex virus type 2 activity of the antimicrobial peptide subtilosin. J Appl Microbiol 117:1253–1259

Rahman KSM, Banat IM, Thahira J, Thayumanavan T, Lakshmanaperumalsamy P (2002) Bioremediation of gasoline contaminated soil by a bacterial consortium amended with poultry litter, coir pith, and rhamnolipid biosurfactant. Bioresour Technol 81(1):25–32

Ramaa C, Shirode A, Mundada A, Kadam V (2006) Nutraceuticals – an emerging era in the treatment and prevention of cardiovascular diseases. Curr Pharm Biotechnol 7(1):15–23. https://doi.org/10.2174/138920106775789647

Ramalingam V, Varunkumar K, Ravikumar V, Rajaram R (2019) Production and structure elucidation of anticancer potential surfactin from marine actinomycete Micromonospora marina. Process Biochem 78:169–177

Rani K, Sharma P, Kumar S, Wati L, Kumar R, Gurjar DS, Kumar D (2019) Legumes for sustainable soil and crop management. In: Sustainable management of soil and environment. Springer, Singapore, pp 193–215

Reddy KV, Aranha C, Gupta SM, Yedery RD (2004) Evaluation of antimicrobial peptide nisin as a safe vaginal contraceptive agent in rabbits: in vitro and in vivo studies. Reproduction 128:117–126

Resende AH, Farias JM, Silva DD, Rufino RD, Luna JM, Stamford TC, Sarubbo LA (2019) Application of biosurfactants and chitosan in toothpaste formulation. Colloids Surf B: Biointerfaces 181:77–84

Rivardo F, Turner RJ, Allegrone G, Ceri H, Martinotti MG (2009) Anti-adhesion activity of two biosurfactants produced by Bacillus spp. prevents biofilm formation of human bacterial pathogens. Appl Microbiol Biotechnol 83(3):541–553

Rodrigues L, Van der Mei HC, Teixeira J, Oliveira R (2004) Influence of biosurfactants from probiotic bacteria on formation of biofilms on voice prostheses. Appl Environ Microbiol 70 (7):4408–4410

Rodrigues L, Banat IM, Teixeira J, Oliveira R (2006) Biosurfactants: potential applications in medicine. J Antimicrob Chemother 57:609–618

Rodríguez-López L, Rincón-Fontán M, Vecino X, Cruz JM, Moldes AB (2019) Preservative and irritant capacity of biosurfactants from different sources: a comparative study. J Pharm Sci 108:2296–2304. https://doi.org/10.1016/j.xphs.2019.02.010

Roelants S, Van Renterghem L, Maes K, Everaert B, Vanlerberghe B, Demaeseneire S, Soetaert W (2019) Microbial biosurfactants: from lab to market. In: C. Press (ed) Microbial biosurfactants and their environmental and industrial applications. CRC Press, Boca Raton/London/New York

Rudzki L, Ostrowska L, Pawlak D, Małus A, Pawlak K, Waszkiewicz N, Szulc A (2019) Probiotic lactobacillus plantarum 299v decreases kynurenine concentration and improves cognitive functions in patients with major depression: a double-blind, randomized, placebo controlled study. Psychoneuroendocrinology 100:213–222

Ruiz-Matute AI, Corzo-Martínez M, Montilla A, Olano A, Copovi P, Corzo N (2012) Presence of mono-, di- and galactooligosaccharides in commercial lactose-free UHT dairy products. J Food Compos Anal 28(2):164–169

Sa RB, An X, Sui JK, Wang XH, Ji C, Wang CQ, Li Q, Hu YR, Liu X (2018) Purification and structural characterization of fengycin homologues produced by Bacillus subtilis from poplar wood bark. Australas Plant Pathol 47(3):259–268

Sabo SS, Converti A, Ichiwaki S, Oliveira RPS (2019) Bacteriocin production by lactobacillus plantarum ST16Pa in supplemented whey powder formulations. J Dairy Sci 1:87–99. https://doi.org/10.3168/jds.2018-14881

Sak K (2012) Chemotherapy and dietary phytochemical agents. Chemother Res Pract 2012:1–11

Salazar F, Ortiz A, Sansinenea E (2017) Characterisation of two novel bacteriocin-like substances produced by *Bacillus amyloliquefaciens* ELI149 with broad-spectrum antimicrobial activity. J Glob Antimicrob Resist 11:177–182. https://doi.org/10.1016/j.jgar.2017.08.008

Salminen S, Bouley C, Boutron-Ruault MC (1998) Br J Nutr 80(Suppl):S147–S171

Sanchez-Garcia L, Martín L, Mangues R, Ferrer-Miralles N, Vázquez E, Villaverde A (2016) Recombinant pharmaceuticals from microbial cells: a 2015 update. Microb Cell Factories 15 (1):33

Saranraj P, Naidu MA (2014) Microbial pectinases: a review. Glob J Tradit Med Syst 3:1–9

Sarubbo LA, Farias CB, Campos-Takaki GM (2007) Co-utilization of canola oil and glucose on the production of a surfactant by Candida lipolytica. Curr Microbiol 54(1):68–73

Schlessinger D (1993) Bacterial toxins. In: Schaechter M, Medoff G, Einstein BI (eds) Mechanism of microbial disease, 2nd edn. Williams and Wilkins, Baltimore, pp 162–175

Sewald X, Gebert-Vogl B, Prassl S et al (2008) Integrin subunit CD 18 is the T-lymphocyte receptor for the helicobacter pylori vacuolating cytotoxin. Cell Host Microbe 3(1):20–29

Sharma D, Saharan BS (2016) Functional characterization of biomedical potential of biosurfactant produced by lactobacillus helveticus. Biotechnol Rep 11:27–35

Sharma P, Sangwan S, Kaur H (2019) Process parameters for biosurfactant production using yeast Meyerozyma guilliermondii YK32. Environ Monit Assess 191(9):531

Singh P, Cameotra SS (2004) Potential applications of microbial surfactants in biomedical sciences. Trends Biotechnol 22:142–146

Singh R, Kumar M, Mittal A, Mehta PK (2016) Microbial enzymes: industrial progress in 21st century. 3 Biotech 6(2):174

Singh A, Sharma P, Kumari A, Kumar R, Pathak DV (2019) Management of root-knot nematode in different crops using microorganisms. In: Varma A, Tripathi S, Prasad R (eds) Plant biotic interactions. Springer, Cham

Souza KST, Gudiña EJ, Azevedo Z, de Freitas V, Schwan RF, Rodrigues LR, Teixeira JA (2017) New glycolipid biosurfactants produced by the yeast strain Wickerhamomyces anomalus CCMA 0358. Colloids Surf B: Biointerfaces 154:373–382. https://doi.org/10.1016/j.colsurfb.2017.03.041

Stokes JM, Lopatkin AJ, Lobritz MA, Collins JJ (2019) Bacterial metabolism and antibiotic efficacy. Cell Metab 30(2):251–259

Stouthamer AH (1973) A theoretical study on the amount of ATP required for synthesis of microbial cell material. Antonie Van Leeuwenhoek 39:545–565

Sumi H, Hamada H, Tsushima H, Mihara H, Muraki H (1987) A novel fibrinolytic enzyme (nattokinase) in the vegetable cheese Natto; a typical and popular soybean food in the Japanese diet. Experientia 43(10):1110–1111

Sutyak KE, Anderson RA, Dover SE, Feathergill KA, Aroutcheva AA, Faro S, Chikindas ML (2008) Spermicidal activity of the safe natural antimicrobial peptide subtilosin. Infect Dis Obstet Gynecol 2008:540758. https://doi.org/10.1155/2008/540758

Suvarna VC, Boby VU (2005) Probiotics in human health: a current assessment. Curr Sci 88:1744–1748

Swartz JR (1996) Escherichia coli recombinant DNA technology. In: Neidhardt FC (ed) Escherichia coli and Salmonella: cellular and molecular biology, 2nd edn. American Society of Microbiology Press, Washington, DC, pp 1693–1711

Tada A, Zelaya H, Clua P, Salva S, Alvarez S, Kitazawa H (2016) Immunobiotic lactobacillus strains reduce small intestinal injury induced by intraepithelial lymphocytes after toll-like receptor 3 activation. Inflamm Res 65(10):771–783. https://doi.org/10.1007/s00011-016-0957-7

Tahmourespour A, Salehi R, Kermanshahi RK, Eslami G (2011) The anti-biofouling effect of lactobacillus fermentum-derived biosurfactant against Streptococcus mutans. Biofouling 27:385–392

Takeda A, Cooper K, Bird A, Baxter L, Frampton GK, Gospodarevskaya E, Welch K, Bryant J (2010) Recombinant human growth hormone for the treatment of growth disorders in children: a systematic review and economic evaluation. Health Technol Assess (Winch Eng) 14(42):1–209

Tesh VL, Ramegowda B, Samuel JE (1994) Purified Shiga-like toxins induce expression of proinflammatory cytokines from murine peritoneal macrophages. Infect Immun 62 (11):5085–5094

Thanomsub B, Watcharachaipong T, Chotelersak K, Arunrattiyakorn P, Nitoda T, Kanzaki H (2004) Monoacylglycerols: glycolipid biosurfactants produced by a thermotolerant yeast, Candida ishiwadae. J Appl Microbiol 96:588–592

Todorov SD, Wachsman M, Tomé E, Dousset X, Destro MT, Dicks LM, de Melo Franco BD, Vaz-Velho M, Drider D (2010) Characterisation of an antiviral pediocin-like bacteriocin produced by Enterococcus faecium. Food Microbiol 27(7):869–879

Truong DD, Jost WH (2006) Botulinum toxin: clinical use. Parkinsonism Relat Disord 12 (6):331–355

Vajo Z, Fawcett J, Duckworth WC (2001) Recombinant DNA technology in the treatment of diabetes: insulin analogs. Endocr Rev 22:706–717

van Staden AD, Brand AM, Dicks LM (2012) Nisin F-loaded brushite bone cement prevented the growth of Staphylococcus aureus in vivo. J Appl Microbiol 112:831–840

Vanittanakom N, Loeffler W, Koch U, Jung G (1986) Fengycin-a novel antifungal lipopeptide antibiotic produced by Bacillus subtilis F-29- 3. J Antibiot 39:888–901

Vecino X, Cruz JM, Moldes AB, Rodrigues LR (2017) Biosurfactants in cosmetic formulations: trends and challenges. Crit Rev Biotechnol 37(7):911–923. https://doi.org/10.1080/07388551.2016.1269053

Vidanarachchi JK, Kurukulasuriya MS, Malshani Samaraweera A, Silva KFST (2012) Applications of marine nutraceuticals in dairy products. Adv Food Nutr Res 65:457–478. https://doi.org/10.1016/b978-0-12-416003-3.00030-5

Vijayakuma S, Saravanan V (2015) Biosurfactants—types sources and applications. Res J Microbiol 10(5):181–192. https://doi.org/10.3923/jm.2015.181.192

Villena J, Salva S, Nuñez M, Corzo J, Tolaba R, Faedda J (2012) Probiotics for everyone! The novel immunobiotic *Lactobacillus rhamnosus* CRL1505 and the beginning of social probiotic programs in Argentina. Int J Biotechnol Wellness Ind 1:189–198. https://doi.org/10.6000/1927-3037/2012.01.03.05

Villena J, Vizoso-Pinto MG, Kitazawa H (2016) Intestinal innate antiviral immunity and immunobiotics: beneficial effects against rotavirus infection. Front Immunol 7:563. https://doi.org/10.3389/fimmu.2016.00563

Wall R, Marques TM, O'Sullivan O, Ross RP, Shanahan F, Quigley EM, Dinan TG, Kiely B, Fitzgerald GF, Cotter PD, Fouhy F, Stanton C (2012) Contrasting effects of Bifidobacterium breve NCIMB 702258 and Bifidobacterium breve DPC 6330 on the composition of murine brain fatty acids and gut microbiota. Am J Clin Nutr 95(5):1278–1287

Wall R, Cryan JF, Ross RP, Fitzgerald GF, Dinan TG, Stanton C (2014) Bacterial neuroactive compounds produced by psychobiotics. In: Microbial endocrinology: the microbiota-gut-brain axis in health and disease. Springer, New York, pp 221–239. https://doi.org/10.1007/978-1-4939-0897-4_10

Walsh G, Jefferis R (2006) Post-translational modifications in the context of therapeutic proteins. Nat Biotechnol 24:1241–1252

Wang J, Liu J, Wang X, Yao J, Yu Z (2004) Application of electrospray ionization mass spectrometry in rapid typing of fengycin homologues produced by Bacillus subtilis. Lett Appl Microbiol 39:98–102

Wang J, Guleria S, Koffas MA, Yan Y (2016) Microbial production of value-added nutraceuticals. Curr Opin Biotechnol 37:97–104

Wang Y, Qin Y, Zhang Y, Wu R, Li P (2018) Antibacterial mechanism of plantaricin LPL-1, a novel class IIa bacteriocin against listeria monocytogenes. Food Control 121:888–899. https://doi.org/10.1016/j.foodcont.2018.10.025

Warda AK, Rea K, Fitzgerald P, Hueston C, Gonzalez-Tortuero E, Dinan TG, Hill C (2019) Heat-killed lactobacilli alter both microbiota composition and behaviour. Behav Brain Res 362:213–223

Watanabe K (2004) Collagenolytic proteases from bacteria. Appl Microbiol Biotechnol 63 (5):520–526

Waters AL, Hill RT, Place AR, Hamann MT (2010) The expanding role of marine microbes in pharmaceutical development. Curr Opin Biotechnol 21(6):780–786

Watve MG, Tickoo R, Jog MM, Bhole BD (2001) How many antibiotics are produced by the genus Streptomyces? Arch Microbiol 176(5):386–390

Wei CL, Wang S, Yen JT, Cheng YF, Liao CL, Hsu CC, Wu CC, Tsai YC (2019) Antidepressant-like activities of live and heat-killed lactobacillus paracasei PS23 in chronic corticosterone-treated mice and possible mechanisms. Brain Res 1711:202–213

Whitehead HR (1933) A substance inhibiting bacterial growth, produced by certain strains of lactic streptococci. Biochem J 27:1793–1800

WHO (2017) Cancer: key facts [Online]. World Health Organization. Available online at: http://www.who.int/mediacentre/factsheets/fs297/en/

Wu YS, Ngai SC, Goh BH, Chan KG, Lee LH, Chuah LH (2017) Anticancer activities of surfactin and potential application of nanotechnology assisted surfactin delivery. Front Pharmacol 8:761

Xu J, Li W, Wu J, Zhang Y, Zhu Z, Liu J, Hu Z (2006) Stability of plasmid and expression of a recombinant gonadotropin-releasing hormone (GnRH) vaccine in Escherichia coli. Appl Microbiol Biotechnol 73(4):780–788

Xu Y, Yang L, Li P, Gu Q (2019) Heterologous expression of class IIb bacteriocin Plantaricin JK in Lactococcus Lactis. Protein Expr Purif 159:10–16

Yuliani H, Perdani MS, Savitri I, Manurung M, Sahlan M, Wijanarko A, Hermansyah H (2018) Antimicrobial activity of biosurfactant derived from Bacillus subtilis C19. Energy Procedia 153:274–278

Zabian Bassetto R, Rodrigues MCC, Almeida MM, de Chiquetto NC (2014) Caracterização da produção de galactooligossacarideos por fermentação sequencial. Evidência Ciência e Biotecnogia 14:57–68

Zaidi KU, Ali AS, Ali SA, Naaz I (2014) Microbial tyrosinases: promising enzymes for pharmaceutical, food bioprocessing, and environmental industry. Biochem Res Int 2014:1–16

Zampieri M, Zimmermann M, Claassen M, Sauer U (2017) Nontargeted metabolomics reveals the multilevel response to antibiotic perturbations. Cell Rep 19(6):1214–1228

Zazopoulos E, Huang K, Staffa A, Liu W, Bachmann BO, Nonaka K, Ahlert J, Thorson JS, Shen B, Farnet CM (2003) A genomics-guided approach for discovering and expressing cryptic metabolic pathways. Nat Biotechnol 21:187–190

Zhang X, Candas M, Griko NB, Taussig R, Bulla LA Jr (2006) A mechanism of cell death involving an adenylyl cyclase/PKA signaling pathway is induced by the Cry1Ab toxin of Bacillus thuringiensis. Proc Natl Acad Sci U S A 103(26):9897–9902

Zhang W, Wen K, Azevedo MS, Gonzalez A, Saif LJ, Li G (2008) Lactic acid bacterial colonization and human rotavirus infection influence distribution and frequencies of monocytes/macrophages and dendritic cells in neonatal gnotobiotic pigs. Vet Immunol Immunopathol 121(3–4):222–231. https://doi.org/10.1016/j.vetimm.2007.10.001

Industrial Potential of Microbial Enzymes 17

Sonali and Richa Arora

Abstract

Enzymes, specifically those with microbial origin, are considered as the potential biocatalysts for many reactions and have extensive uses in industries. Microbes are alternative source of enzymes which can be grown in large quantities in a less time period by fermentation process and are more efficient as well as active. For the fulfillment of the recent necessities of different enzymes, industries are searching for new microbial strains. Lipases and other polymer-hydrolyzing enzymes such as cellulases, chitinases, and amylases are nowadays used for industrial applications. Some microbes are responsible for the production of thermostable enzymes such as *Bacillus subtilis*, *Bacillus amyloliquefaciens*, and *Bacillus licheniformis* reported to produce thermostable amylase. Members of the *Bacillus* sp., *Streptomyces* sp., *Thermoascus aurantiacus*, and *Fusarium proliferatum* have been responsible to produce xylanases. For chitinases *Bacillus licheniformis*, *Bacillus* sp., and *Streptomyces thermoviolaceus* were known to be the chief sources. Amylases, proteases or lipases have great marketable potential in many textile industries.

Keywords

Microbial enzymes · Enzyme market · Industrial application · Biocatalysts

Sonali · R. Arora (✉)
School of Bioengineering and Biosciences, Lovely Professional University, Phagwara, Punjab, India
e-mail: richa.22440@lpu.co.in

S. G. Sharma et al. (eds.), *Microbial Diversity, Interventions and Scope*,
https://doi.org/10.1007/978-981-15-4099-8_17

Abbreviations

BOD	Biological oxygen demand
COD	Chemical oxygen demand
COS	Chitosan oligosaccharides
DSM	Dutch State Mines
ELISA	Enzyme-linked immunosorbent assay
IPR	Intellectual property rights
SOD	Superoxide dismutase

17.1 Introduction

The microbial enzymes are essential entities acting as metabolic catalysts which are accommodated in different industries and applications. The commercial use of these enzymes is universal and holds specific importance (Adrio and Demain 2014). These enzymes are overpowering the chemical catalysts. A huge number of industries are involving the use of microbial enzymes to manufacture a good quality product (Kumar and Singh 2013). The categorization of the enzymes based upon the different sources of microbial production is 50%, 35%, and 20% by fungi, yeast, and bacteria, respectively (Saranraj and Naidu 2014). There is a huge demand for enzymes in the industrial sector for sustained solutions. These potent enzymes are present in the environment which are promising and will last as long as the eternities will (Liu and Kokare 2017). A large group of microorganisms such as actinomycetes, bacteria, yeast, and fungi are susceptible to produce intracellular, intercellular, and extracellular enzymes. A number of microbial enzymes, such as amylases, proteases, cellulases, and laccases, are secreted extracellularly. *Saccharomyces cerevisiae* and *Aspergillus niger* produce catalase enzyme which is intracellular in nature (Fiedurek and Gromada 2000). The use of chemical production process for the manufacture of pharmaceutical and chemicals has certain limitations such as low catalytic effectiveness, deficiently in enantiomeric specificity for chiral formation; requirement for elevated temperature, pressure and low pH. Apart from this, the excessive usage of organic solvents results in the overproduction of organic wastes and pollutants. Microbial enzymes are practically useful as their applications makes them viable because of their function without requirement of substrate functional groups protection, have long half-life, a high stereo-selectivity, and additionally they work magnificent on substrates which are not natural (Johnson 2013). These could be monitored and altered genetically for the enhancement of major assets such as stability, substrate specificity, and specific activity (Zhang et al. 2015). The advantages are simultaneously balanced by disadvantages such as the requirement of specific cofactors. Worldwide industrial enzyme market commercializes Novozymes for acting as the leading competitor in the industry, followed by Dutch State Mines (DSM), DuPont, and among others. On the basis of product excellence, performance of company, innovation capability, and usage of IPR (intellectual property rights), the companies aim to come up in the global market. Mainly, North America and Europe are superpower countries holding a large number of consumers for industrial enzymes. Asia-Pacific region is competing against a vigorous hike in the demand for enzymes especially in countries such as Japan, China, and India, which is adversely affecting economies of these countries (Adrio and Demain 2014). Regardless, the enzyme potential and its industrial application have been hindered in uncertain ways of disturbing its stability, catalytic efficiency, and specificity. These limitations have been overcome by the isolation of enzymes from natural sources, immobilization, and abrupt mutations (Elleuche et al. 2014). Massive increase in the oil prices has led the manufacturers to look for a cheap resource; therefore, biomass is being considered as a sustainable option. Urgency in compensating with the undue advantage of environmental resources has put on a huge pressure on green technologies to resonate the chemical processes with safer, cleaner, and more environment-friendly biocatalysts. Therefore, these green technologies have substituted the chemical processes after the advancement in enzyme engineering (Choi et al. 2015). *Lactococcus lactis*, *Lactobacillus*, *Streptococcus*, and *Propionibacterium* are used in industries for supply of casein-degrading enzymes which convert methionine via new pathways into flavor formatting aromatic products used for cheese ripening. Also, to improve flavor formation of several cheeses, glutamate dehydrogenase enzyme is useful (Budnik et al. 2017). This review is mainly focusing on the use of microbial enzymes in different processes which are employed by industries and the Indian market of enzymes.

17.2 Industries Demanding Enzymes: Biotechnological Tools

A large number of industrial processes involves the use of enzymes in several aspects such as brewing, food processing, making of detergents, fermented foods, textiles, and pharmaceuticals. Some of the enzyme classes, their microbial sources, and industrial applications are mentioned in Table 17.1.

17.2.1 Food Industries

With the increase in global population, potential activities of enzymes can be invested in compensating the growing demand for food quality. The total production of food is maintained by the involvement of biomolecules which correspond in food flavor, aroma, color, texture, and nutritive values (Singh et al. 2016a). The participation of enzymes in food manufacturing and ingredient industry has affected the market quality of product in terms of its safety. Furthermore, the growing awareness in fields, for example, fat alteration and sweetener technology, has led to the proliferation of enzyme usage (Li et al. 2012). In food industries, the application of enzymes is bifurcated to various areas, for example, dairy, baking, juice making, and breweries. All around the world, bakery invests in the effective working of microbial enzymes where they play a major role to enhance dough constancy, smoothness of crumb and its structure, and product's shelf life (Singh et al.

Table 17.1 Reaction catalyzed by different classes of enzymes produced by microbes and their industrial applications

Classes of enzymes	EC no.	Reaction catalyzed	Enzyme involved	Microbial sources	Applications	References
Oxidoreductases	EC1	Transfer of electrons from one molecule to another	Laccase	*Trametes hirsuta*	Non-chlorine bleaching, delignification	Moiseenko et al. (2018),Upadhyay et al. (2016), Ghorai et al. (2009), and Barad et al. (2012)
			Catalase	*Aspergillus niger*	Cheese processing	
			Glucose oxidase	*Aspergillus niger, Penicillium chrysogenum*	Oxygen removal from beer Dough strengthening	
Transferases	EC2	Transfer of groups of atoms from one molecule to another	Transglutaminase	*Streptomyces* sp., *Streptoverticillium* sp.	Protein cross-linking Laminated dough strength	Kuraishi et al. (2001), Weijers et al. (2008), and Maiorano et al. (2008)
			Glycosyltransferase	*Bacillus* sp.	Synthesis of oligosaccharides	
			Fructosyltransferase	*Penicillium rugulosum, Aspergillus niger, Aspergillus oryzae*	Used as sweeteners, reduce cholesterol	
Hydrolases	EC3	Breakdown of substrate by water, and the reaction is called as hydrolysis. In common, larger molecules are cleaved to smaller fragments with the help of hydrolases enzyme	Lipases	*Aspergillus niger, A. oryzae, Aspergillus flavus*	Faster cheese ripening, flavor-customized cheese, dough stability and conditioning, fat stain elimination, synthesis of pharmaceuticals, polymers, biodiesels, biosurfactants, skin care	Contesini et al. (2010), Kuhad et al. (2011), Beg et al. (2001), and de Souza et al. (2010)
			Cellulase	*Trichoderma reesei, Penicillium funiculosum, Aspergillus niger, Bacillus* sp.	Fruit liquefaction, color clarification, deinking, drainage improvement	

			Xylanase	*Streptomyces thermoviolaceus*, *Aspergillus* sp., *Bacillus* sp.	Dough conditioning, enhanced digestibility of starch	
			Amylase	*Bacillus licheniformis*, *Aspergillus* sp., *Bacillus* sp.	Flour adjustment, bread softness, starch hydrolysis, carbohydrate stain removal	
Lyases	EC4	This enzyme catalyzes the formation of double bonds through the removal of groups	α-Acetolactate decarboxylases	*Brevibacillus brevis*, *Saccharomyces cerevisiae*	Beer maturation	Cejnar et al. (2016) and Wong et al. (2000)
			Alginate lyase	*Alteromonas sp.*, *Pseudomonas aeruginosa*	Protoplasting of seaweed	
Isomerases	EC5	Transfer of groups from one position to another in same molecule. By this, these enzymes rearrange the atoms of substrate and change its structure	Glucose isomerase	*Bacillus coagulans*, *Actinomyces missouriensis*, *Lactobacillus brevis*	High-fructose corn syrup	Dordick (2013)
Ligases	EC6	Joining of molecules with covalent bonds. In this new bonds are formed	Synthetases	*Streptomyces albulus*, *Bacillus* sp., *Acinetobacter* sp.	Water softener, drug delivery and cosmetics, feed preservative	Rehm (2010)

2016a). Another important part of food enzyme industry is dairy enzyme which enhances organoleptic components such as aroma, color, flavor, etc. corresponding to high milk product yield. The optimum use of dairy enzymes depends upon the renounced dairy market varying from coagulant to bio-protective enzymes in order to increase the shelf life and dairy product safety. For the manufacturing of yogurt, cheese, and many more milk products, dairy enzymes are utilized (Qureshi et al. 2015). In other industries such as fruit juice industry, enzymes are applied in processes to enhance operation effectiveness such as in shedding, juicing, clearing up, and removal for improving the product value (Kumar 2015). The major functions of enzymes such as cellulases and pectinases throughout fruit juice processing, for maceration, liquefaction, and elucidation, develop yield and cost-efficacy (Garg et al. 2016). Some of the examples of these enzymes used in food industries are explained below.

17.2.1.1 Transglutaminase

Transglutaminase enzyme produced by microbes catalyzes the cross-coupling of the amine side chain of lysine with the amide side chain of glutamine, resulting in the formation of an isopeptide linkage in a mature protein (Zhang et al. 2018). This enzyme is utilized extensively in the food industry called as "meat glue" for the modification of protein for large-scale purposes for the improvement of texture and protein binding. Such processes involve the coating of transglutaminase on the blocks of food item and left for cross-linking (Dorr and Fuerst 2018). Transglutaminase is an enzyme which is indulged in the quality improvement of product manufactured in baking industry, for example, flour, bread, and the texture of cooked pasta (Kieliszek and Misiewicz 2014). Similarly the same enzyme participates in the milk protein polymerization and functional property improvement of dairy products (Rossa et al. 2011) (Fig. 17.1).

17.2.1.2 Lipases

Aspergillus sp. (Contesini et al. 2010), *Bacillus laterosporus* (Djafar et al. 2010), *Candida* sp. (Salihu et al. 2012), and *Hansenula anomala* (Longo and Sanromán 2006) are some of the sources of lipase enzyme. Lipases are involved in improving flavors of bakery products and also release short chains of fatty acids which help in increasing the shelf life of bakery products (Fernandes 2010). Lipases when used in combination with emulsifiers such as insulin have beneficiary effects such as rheological modifications including cake crumb modification with high homogenous cell structure (Rodríguez-García et al. 2014). Lipases are also used for enhancement of flavors, quicker cheese preparation, manufacture of modified milk products, and milk fat lipolysis (Sharma et al. 2001).

17.2.1.3 Amylases

Industries such as baking, brewing, digestion cure products, syrups, fruit juices, etc. use amylases (Couto and Sanromán 2006). Majorly, α-amylase is used in the baking industry. When amylase is added into the dough before kneading it, then it degrades the starch present in flour. Amylase, when added to dough, enhances the pace of

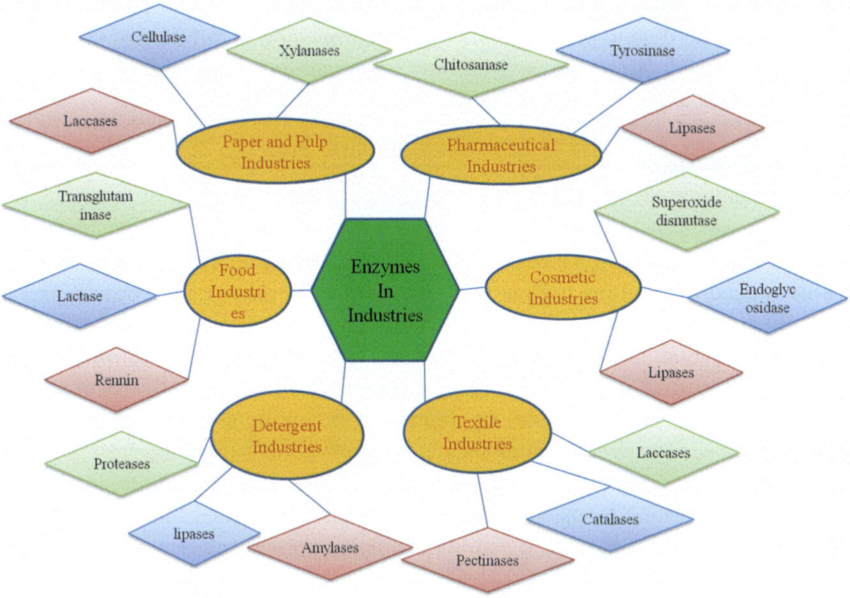

Fig. 17.1 Industrial enzymes and their different applications (self-made)

fermentation and dough viscosity reduction, which improves amount and consistency of the product (Wang et al. 2018). Alongside, it produces sugar in the dough to enhance the taste, coating color, and toasting qualities of the bread. Apart from fermentable compound generation, α-amylase is efficient in anti-staling effect in bread baking along with enhancing softness of baked products by increasing their shelf life (Souza 2010). *Bacillus stearothermophilus*, a thermostable maltogenic amylase producer, is commercially involved in the bakery industry (Park et al. 2018). Amylases participate in beer or fruit juice clarification for the pretreatment of animal feed to increase the digestibility of the fiber (Ghorai et al. 2009).

17.2.2 Paper and Pulp Industries

A microbial enzyme in paper and pulp is helpful in sustainability issues where industries have shown adverse effect on ecosystem. Efficient enzyme utilization has reduced time consumption, energy consumption, and chemical usage. These are useful in improved deinking, bleaching in paper and pulp industry, and waste treatment, resulting in the increase of biological oxygen demand (BOD) and chemical oxygen demand (COD) (Srivastava and Singh 2015). Presently, the pre-bleaching of kraft pulp is the major application of enzymes. Enzymes involved in increase pulp fibrillation and water preservation decrease time consumption in virgin pulps. Enzymes having recycled fibers have been used for deinking for the purpose of restoration of their bonding and increase in its freeness (Saxena and Singh

Chauhan 2017). Some major applications nowadays include vessel picking reduction in tropical hardwood pulps and the careful elimination of xylan from dissolving pulp (Bajpai 1999). The following are few important enzymes used on paper and pulp industries.

17.2.2.1 Xylanases

There are many microbes which are responsible for the production of xylanase, for example, *Aspergillus terreus* (Lakshmi et al. 2009), *Trichoderma harzianum* (Seyis and Aksoz 2005), and *Cryptococcus albidus* (Harris and Ramalingam 2010). Papermaking procedure requires high amount of chemicals endangering waste removal problem (Verma and Satyanarayana 2013). The pulp and paper industry is looking for innovative and new methods to replace harmful chemicals utilized in the processes of papermaking. Biopulping is the preferential method for the pretreatment of woody or nonwoody material by lignin-degrading fungi previous to regular pulping process. However, limitations are the pretreatment time and loss in yield, since the organisms will attack at the same time on both polysaccharides and lignin. To eradicate such limitations, xylanase pretreatment is provided for distribution of sodium hydroxide to enhance the traditional pulping process in both hardwoods and softwoods (Woldesenbet et al. 2012). For the purpose of biobleaching, the enzyme should be functional at higher temperature and should be thermostable, alkalophilic, and cellulase-free xylanase (Walia et al. 2015). The lignin associated with hemicellulosic fraction is removed by enzymatic activity by providing minute harm to the pulp as it has poor cellulolytic activity. Apart from the xylanase utilization in bleaching, it is exceptionally helpful in increasing pulp fibrillation by decreasing thrashing times of pulp to prolonged freeness in reused fibers (Walia et al. 2017).

17.2.2.2 Cellulases

Some of the applications indulged into paper industries such as biopulping, dewatering, enzymatic deinking and pulp fiber bio-characterization, improvement in handsheet strength of fibers, decreasing problems of drainage uses cellulase enzyme. Moreover, these enzymes are useful in manufacturing of recyclable cardboard and soft paper such as towel and sanitary paper and removal of adhered paper (Singh et al. 2016b). Cellulases are needed for complete breakdown of cellulose in the biomass for biofuel production unlike the paper industries and textile industries. They utilize entire and discrete cellulose components so that proper cellulose breakdown is achieved for improved and best quality paper and fabric properties. Some novel products are obtained by the biotechnological techniques where cellulosic biomass is considered to be the potential source of sustainable approach. Product viability is more in the case of cellulose-mediated pumping than in mechanical pulping because cellulose-mediated pumping is considered to be energy saving. Moreover, it enhances the physical properties of product such as interbonding of fibers and mechanical strength (Chen et al. 2012). Solid waste production and deforestation are reduced by the recycling of paper waste for new product manufacturing. Collectively cellulase and hemicellulase participate in deinking of waste papers for the improvement of the recycled paper quality (Ibarra et al. 2012). Enzymatic deinking produces minimal or negligible use of alkali, enhanced brightness, and strength properties. Likewise, alkaline-yellowing effects of the environment is prevented by enzyme-mediated deinking. The cellulase treatment is opted to eradicate fibrils and colloidal substance to fix drainage-related problems in paper mills. Besides, manufacturing of soft papers and making of biodegradable cardboards are major applications of cellulase (Kuhad et al. 2011). Some of the microbes which are involved in cellulase production are *Aspergillus oryzae* (Sher et al. 2017), *Fusarium solani* (Bhatti et al. 2013), *Trichoderma atroviride* (Grigorevski-Lima et al. 2013), *Bacillus subtilis* (Shabeb et al. 2010), etc.

17.2.2.3 Laccases

In paper production, lignin has to be separated and degraded from wood pulp. Traditionally, the separation was done by polluting chlorine-containing reagents (Virk et al. 2012). Laccase is accounted for the chlorophenol degradation and decolorization, chlorolignin containing black liquor and pulp paper mill wastewater (Chandra et al. 2015). Effluent of pulp paper mill contains pigments from primary mixture along with plant extracts having high amount of cellulose and heavy metals. The major component of phenolic compound is lignin, which is solubilized and eliminated while in pulping process. Filamentous fungi can deteriorate wastewater with the help of various classes of the enzymes. The area of interest has been increased in biopulping as laccases play a significant function in it. In the process of lignin degradation, laccase acts upon minute phenolic lignin portions where the substrate functions on lignin polymer to cause degradation. Conventionally delignifying or decolorizing of paper pulp is done either by chlorine- or oxygen-based chemical oxidant (such as ClO_2 and O_2) (Bajpai and Bajpai 1992). In the pulp and paper industry, chemical bleaching is prioritized, although it lacks the environmental safety as they generate hazardous chemical and chlorinated by-products. Hence, laccase delignification systems can be adopted only if they come up with proper solution to their drawbacks. Thus, these can be employed to current pulp production line (Chandra and Chowdhary 2015). Here are a few examples of microbes which are responsible for laccase production: these are *Phlebia radiata* (Kantelinen et al. 1989), *Azospirillum lipoferum* (Gupta et al. 2017), *Streptomyces coelicolor* (Gupta et al. 2012), *Coprinus cinereus* (Lin et al. 2013), *Lentinus tigrinus* (Hsu et al. 2012), etc.

17.2.3 Cosmetic Industries

Enzyme application in cosmetics industry has increased drastically. In commercial products such as creams, mouthwashes, hair spray, hair dyeing, and toothpastes, some of the enzymes are utilized as the free radical eliminator and also have some other applications (Li et al. 2012). Here are few enzymes used in cosmetics.

17.2.3.1 Superoxide Dismutase

There are many microbial sources of superoxide dismutase (SOD) which is used by the industries, few of them are *Thermomyces lanuginosus* which is known to be thermophilic fungus (Li et al. 2005), *Thermoascus aurantiacus* (Guo et al. 2007), and *Anoxybacillus gonensis* (Bhatia et al. 2018). SOD is a fine example of cosmetic enzyme so as to capture free radicals for the prevention of skin damage which occurs due to environmental pollution, toxic waste, pathogenic bacteria, and other dangerous factors. The mixture of SOD and peroxidase acting up as free radical eradicator reduces UV-induced erythema, i.e., redness of the skin in sunscreen cream (Babizhayev 2006).

17.2.3.2 Endoglycosidase

Endoglycosidase and papain are the enzymes that are found in toothpaste and mouthwash for whitening of teeth and removing patches, odor, deposits on teeth, and gum tissues (Vashist et al. 2019). These enzymes also remove vitamin precursor, fatty alcohol, and a few enzymes attached with polymeric molecules. In addition to all of the above, these enzymes are also utilized in eye contact lens cleaners to get rid of protein films (Li et al. 2012). *Mucor hiemalis* (Katoh et al. 2016) and *Rhodococcus* sp. (Ito and Yamagata 1989) are the examples of microorganisms responsible for production of endoglycosidases.

17.2.3.3 Lipases

Lipase shows actions in surfactants and in perfume manufacturing; therefore, they are customized for perfumes and cosmetics. By the esterification of glycerols, two surfactants, monoacylglycerols and diacylglycerols, are produced and are used in cosmetics and perfume industries (Sharma et al. 2011). *Candida cylindracea* (Krishnan et al. 2017) and *Rhizomucor miehei* (Montiel et al. 2019) are major lipase-producing microbes used in cosmetics.

17.2.4 Detergent Industries

Detergent industries are easy suppliers of the enzymes. Enzymes increase detergent's functioning in order to increase the activity to remove stains and make its nature environment friendly. These enzymes can remove residues of starchy foods such as custard, any gravy, potato, and chocolate and some other minor oligosaccharides. Thus, they are indulged in the manufacturing of dishwashing detergents (Mitidieri et al. 2006). Most of the detergent companies used proteolytic enzymes produced by *Bacillus* sp. Examples of such companies are Tide and Dynamo (Kumar et al. 2008). *Bacillus* spp. which are known to produce proteolytic enzymes are *Bacillus cereus* (Banik and Prakash 2004), *Bacillus brevis* (Banerjee et al. 1999), etc.

17.2.4.1 Lipases

Lipases rely onto fabric exterior to construct a complex called fabric-lipase which will break chemical bonds whenever water is added. There is a growing application of lipases in detergent industries, and they are thought to be appropriate for washing powders and liquid detergents if they are thermophilic, alkalophilic, water soluble, low substrate specific, and tolerant to detergent proteases and other surfactants (Chauhan et al. 2013; Sarmah et al. 2018). In 1994 *Humicola lanuginosa* was used to produce the first industrial lipase by, a Denmark-based company, Novozymes, and the lipase was known as lipolase (Priji et al. 2016).

17.2.4.2 Amylases

The enzyme alpha-amylase is crucial in detergent industries which is prominent in the improvement of laundry bleach composition and bleaching without the color-darkening effect (Saini et al. 2017). Several detergents contain such enzymes (Mitidieri et al. 2006). These enzymes remove the stains of starchy food such as potatoes, etc. Termamyl is produced by α-amylase which is used in detergent industries released from *Bacillus licheniformis*. Apart from these applications, α-amylase is predominantly helpful in dishwashing and de-starching detergents (Liu and Kokare 2017). There are also other microbial sources which produce α-amylases: *Pseudoalteromonas arctica* (Lu et al. 2010), *Bacillus stearothermophilus* (Srivastava and Baruah 1986), *Clostridium perfringens* (Vester et al. 2015), and *Aspergillus niger* (Varalakshmi et al. 2009).

17.2.4.3 Proteases

Microbes involved in protease production are *Bacillus pumilus* (Baweja et al. 2016), *Caldicoprobacter guelmensis* (Bouacem et al. 2015), *Bacillus amyloliquefaciens* (Guleria et al. 2016), *Bacillus circulans* (Patil et al. 2016), etc. These enzymes are the area interest for detergent industry because of its potential to remove proteinaceous stains and to obtain miraculous results that cannot be obtained by conventional detergent technologies. Apart from laundry detergents, they are also used as a boom for dishwashing detergents in both industrial and domestic sector (Gupta et al. 2002).

17.2.5 Pharmaceutical Industries

Manufacturing of chiral medicines involving the formation of chiral pharmaceutical intermediates, sustainably as well as financially, is considered to play a vital role in pharmaceutical industries. Many enzymes such as lipases, proteases, and ketoreductases are extensively useful in the production of chiral alcohols, carboxylic acids, and amines (Zheng and Xu 2011). These enzymes are widely used as beneficial drugs in the context of physical condition issues related to enzymatic insufficiency and digestive disorders and also in diagnostic methods such as enzyme-linked immunosorbent assay (ELISA) and diabetes testing kits (Mane and Tale 2015). Nowadays, maximum use of the microbial enzymes is involved in the elimination of burnt skin by proteolytic enzymes and clot busting by fibrinolytic enzymes (Zaidi et al. 2014). Pharmaceutical industries use some microbes, for example, *Escherichia* sp. (Swartz 2001), *Bacillus* sp. (Mondal et al. 2000), and *Leuconostoc* (Sanchez and Demain 2011), for enzyme production.

17.2.5.1 Chitosanase

Chitosanase is produced by microbes such as *Streptomyces* sp. (Sinha et al. 2012), *Bacillus* sp. (Su et al. 2006), *Trichoderma koningii* sp. (da Silva et al. 2012), and *Pseudomonas* sp. (Wang et al. 2007). Chitosanase enzyme hydrolyzes the chitosan to biologically working chitosan oligosaccharides (COS). COS is used as antimicrobial and antioxidant and in lowering of blood pressure, lessening the cholesterol, regulating arthritis, protection against infections, and bettering antitumor properties (Thadathil and Velappan 2014).

17.2.5.2 Tyrosinase

Tyrosinases are abundant in nature which support several biological functions and play an important role in defense mechanism (especially in melanogenesis) (Nunes and Vogel 2018). Tyrosine-related melanogenesis causes hair pigmentation, skin pigmentation, and eye pigmentation in mammals because it is the most important part of the skin which has to be protected by UV radiation (Ray et al. 2007).

17.2.5.3 Lipases

Lipases are extensively utilized in medical and pharmaceutical productions. Lipases secreted by *Candida rugosa* helps in the production of lovastatin, a common drug that helps in reducing the intensity of serum cholesterol. *Serratia marcescens* lipase has been broadly utilized in the production of diltiazem hydrochloride for the asymmetric hydrolysis of key intermediate, i.e., 3-phenylglycidic acid ester (Andualema and Gessesse 2012). For the formulation of various enantiopure molecules, several lipases are appropriate to be used such as alcohols, amides, esters, and carboxylic acids. These molecules are used in anti-inflammatory, anticancer, antiviral, antihypertensive, anticholesterol, and anti-Alzheimer disease drugs (Houde et al. 2004).

17.3 Trends of Enzymatic Market in India

Worldwide, the beverage, food, detergent, or drug industries produce such kind of enzymes which are directly or indirectly related to human consumption. The growing population demands safety in food and drugs which has turned up as a challenge for the manufacturers nowadays. The agriculture-based economy of India has been estimated to exceed 7.9% by 2018 (Singh et al. 2016a) and may call for an open market to worldwide enzyme-based companies for their valuable investments. Biotech division of India accounts 2% among the global biotech marketplace as it is the center of attraction globally because of its investment schemes (Binod et al. 2013). Indian bio-industrial area which contains enzyme companies has reached up to Rs. 3950 million in a year from 2006 to 2007 and enrolled 5.33% growth in these years. The import activity of India is 70% of the overall enzymes which is a rough idea about its consumption in the nation, the major part of which is inclined toward detergent production and textile, starch, and pharmaceutical industries (Chandel et al. 2007). The global shares of import in India are from USA, Europe, and China which are 40%, 25%, and 15%, respectively. Due to a large number of opportunities outside India, a lot of companies from India are growing their brand name outside the country, in competition with countries including China (Binod et al. 2013). In the year 2012–2013, Advanced Enzyme Technologies Ltd., an Indian company having high demand globally, invested 30% of the share in the enzyme industry, and the Denmark-based company Novozymes invested 44% of the shares. Other well-known manufacturers were Rossari Biotech, Lumis Biotech, Maps Enzymes, Zytex, and others (www.crisil.com).

17.4 Conclusions

Enzyme industries are the leading industry globally, and many investors are interested in the profit earned by the export and import scheme of such enzymes. By the advancement in biotechnological techniques, many shares have been sold on the beneficiary of enzymes as they are working tremendously in detergent, food, agriculture, and pharmaceutical industries. A remarkable quantity of enzymes is provided by microorganisms, and these enzymes also have a broad use in many industries, such as food industries, animal feed, paper and pulp industry, textile industries, detergent industries, and pharmaceutical industries. The high specificity, fast action, and biodegradability are some of the exclusive properties of enzymes which allow enzyme-assisted processes in industry to work under milder reaction conditions, with better yields and a decrease in waste production. The naturally occurring biocatalysts are cultured and designed properly to increase their metabolic yield which will end up in benefitting the industry with improved product. By the advent of enzyme technology, the recombinant technology will take up its lead in a way that their combination will make cumulative benefit.

Acknowledgments One of the authors (Sonali) is very thankful to Lovely Professional University, Phagwara, for providing Master's registration (Reg. No. 11400774).

References

Adrio J, Demain A (2014) Microbial enzymes: tools for biotechnological processes. Biomol Ther 4:117–139

Andualema B, Gessesse A (2012) Microbial lipases and their industrial applications. Biotechnology 11(3):100–118

Babizhayev MA (2006) Biological activities of the natural imidazole containing peptidomimetics n-acetylcarnosine, carcinine and L carnosine in ophthalmic and skin care products. Life Sci 78:2343–2357

Bajpai P (1999) Application of enzymes in the pulp and paper industry. Biotechnol Prog 15 (2):147–157

Bajpai P, Bajpai PK (1992) Biobleaching of Kraft pulp. PRO 27(6):319–325

Banerjee UC, Sani RK, Azmi W, Soni R (1999) Thermostable alkaline protease from Bacillus brevis and its characterization as a laundry detergent additive. Process Biochem 35 (1–2):213–219

Banik RM, Prakash M (2004) Laundry detergent compatibility of the alkaline protease from *Bacillus cereus*. Microbiol Res 159(2):135–140

Barad S, Horowitz SB, Moscovitz O, Lichter A, Sherman A, Prusky D (2012) A Penicillium expansum glucose oxidase–encoding gene, GOX2, is essential for gluconic acid production and acidification during colonization of deciduous fruit. Mol Plant-Microbe Interact 25(6):779–788

Baweja M, Tiwari R, Singh PK, Nain L, Shukla P (2016) An alkaline protease from *Bacillus pumilus* MP 27: functional analysis of its binding model toward its applications as detergent additive. Front Microbiol 7:1195

Beg Q, Kapoor M, Mahajan L, Hoondal GS (2001) Microbial xylanases and their industrial applications: a review. Appl Microbiol Biotechnol 56(3–4):326–338

Bhatia K, Mal G, Bhar R, Attri C, Seth A (2018) Purification and characterization of thermostable superoxide dismutase from Anoxybacillus gonensis KA 55 MTCC 12684. Int J Biol Macromol 117:1133–1139

Bhatti HN, Batool S, Afzal N (2013) Production and characterization of a novel (beta)-Glucosidase from Fusarium solani. Int J Agric Biol 15(1):140–144

Binod P, Palkhiwala P, Gaikaiwari R, Nampoothiri KM, Duggal A, Dey K, Pandey A (2013) Industrial enzymes-present status and future perspectives for India. J Sci Ind Res 72:271–286

Bouacem K, Bouanane-Darenfed A, Laribi-Habchi H, Elhoul MB, Hmida-Sayari A, Hacene H, Bejar S (2015) Biochemical characterization of a detergent-stable serine alkaline protease from Caldicoprobacter guelmensis. Int J Biol Macromol 81:299–307

Budnik LT, Scheer E, Burge PS, Baur X (2017) Sensitising effects of genetically modified enzymes used in flavour, fragrance, detergence and pharmaceutical production: cross-sectional study. Occup Environ Med 74:39–45

Cejnar R, Hložková K, Kotrba P, Dostálek P (2016) Surface-engineered Saccharomyces cerevisiae displaying α-acetolactate decarboxylase from Acetobacter aceti ssp xylinum. Biotechnol Lett 38 (12):2145–2151

Chandel AK, Rudravaram R, Rao LV, Ravindra P, Narasu ML (2007) Industrial enzymes in bioindustrial sector development: an Indian perspective. J Commer Biotechnol 13(4):283–291

Chandra R, Chowdhary P (2015) Properties of bacterial laccases and their application in bioremediation of industrial wastes. Environ Sci Processes Impacts 17(2):326–342

Chandra R, Yadav S, Kumar V (2015) Microbial degradation of lignocellulosic waste and its metabolic products. Environmental waste management. CRC Press, Boca Raton, pp 250–293

Chauhan M, Chauhan RS, Garlapati VK (2013) Evaluation of a new Lipase from Staphylococcus sp. for detergent additive capability. BioMed Res Int 2013:374967

Chen Y, Wan J, Zhang X et al (2012) Effect of beating on recycled properties of unbleached eucalyptus cellulose fiber. Carbohydr Polym 87:730–736

Choi JM, Han SS, Kim HS (2015) Industrial applications of enzyme biocatalysis: current status and future aspects. Biotechnol Adv 33(7):1443–1454

Contesini FJ, Lopes DB, Macedo GA, da Graça Nascimento M, de Oliveira Carvalho P (2010) *Aspergillus* sp. lipase: potential biocatalyst for industrial use. J Mol Catal B Enzym 67 (3–4):163–171

Couto SR, Sanromán MA (2006) Application of solid-state fermentation to food industry- a review. J Food Eng 76:291–302

da Silva LC, Honorato TL, Franco TT, Rodrigues S (2012) Optimization of chitosanase production by Trichoderma koningii sp. under solid-state fermentation. Food Bioprocess Technol 5 (5):1564–1572

de Souza RL, Barbosa JMP, Zanin GM, Lobão MWN, Soares CMF, Lima ÁS (2010) Partitioning of porcine pancreatic lipase in a two-phase systems of polyethylene glycol/potassium phosphate aqueous. Appl Biochem Biotechnol 161(1-8):288–300
Djafar F, Purwadaria T, Sinurat AP (2010) Isolation of endophytic bacteria from palm oil fruits and characterization of their lipases. Microbiol Indones 4(2):4
Dordick JS (Ed) (2013). Biocatalysts for industry. Springer, Boston
Dorr BM, Fuerst DE (2018) Enzymatic amidation for industrial applications. Curr Opin Chem Biol 43:127–133
Elleuche S, Schroder C, Sahm K, Antranikian G (2014) Extremozymes-biocatalysts with unique properties from extremophilic microorganisms. Curr Opin Biotechnol 29:116–123
Fernandes P (2010) Enzymes in food processing: a condensed overview on strategies for better biocatalysis. Enzym Res 2010:1. https://doi.org/10.4061/2010/862537
Fiedurek J, Gromada A (2000) Production of catalase and glucose oxidase by *Aspergillus niger* using unconventional oxygenation of culture. J Appl Microbiol 89:85–89
Garg G, Singh A, Kaur A et al (2016) Microbial pectinases: an ecofriendly tool of nature for industries. 3 Biotech 6(1):47–59
Ghorai S, Banik SP, Verma D, Chowdhury S, Mukherjee S, Khowala S (2009) Fungal biotechnology in food and feed processing. Food Res Int 42(5–6):577–587
Grigorevski-Lima AL, de Oliveira MMQ, do Nascimento RP, da Silva Bon EP, RRR C (2013) Production and partial characterization of cellulases and xylanases from Trichoderma atroviride 676 using lignocellulosic residual biomass. Appl Biochem Biotechnol 169(4):1373–1385
Guleria S, Walia A, Chauhan A, Shirkot CK (2016) Purification and characterization of detergent stable alkaline protease from Bacillus amyloliquefaciens SP1 isolated from apple rhizosphere. J Basic Microbiol 56(2):138–152
Guo F, Liu S, Chen J, Wang Y, Li D (2007) Purification, characterization, and molecular cloning of a thermostable superoxide dismutase from Thermoascus aurantiacus. Biosci Biotechnol Biochem 71(1):1090–1093
Gupta R, Beg Q, Lorenz P (2002) Bacterial alkaline proteases: molecular approaches and industrial applications. Appl Microbiol Biotechnol 59(1):15–32
Gupta A, Nederlof I, Sottini S, Tepper AW, Groenen EJ, Thomassen EA, Canters GW (2012) Involvement of Tyr108 in the enzyme mechanism of the small laccase from Streptomyces coelicolor. J Am Chem Soc 134(44):18213–18216
Gupta V, Capalash N, Gupta N, Sharma P (2017) Bio-prospecting laccases in the bacterial diversity of activated sludge from pulp and paper industry. Indian J Microbiol 57(1):75–82
Harris AD, Ramalingam C (2010) Xylanases and its application in food industry: a review. J Exp Sci 1:1–11
Houde A, Kademi A, Leblanc D (2004) Lipases and their industrial applications. Appl Biochem Biotechnol 118:155–170
Hsu CA, Wen TN, Su YC, Jiang ZB, Chen CW, Shyur LF (2012) Biological degradation of anthroquinone and azo dyes by a novel laccase from Lentinus sp. Environ Sci Technol 46 (9):5109–5117
Ibarra D, Monte MC, Blanco A et al (2012) Enzymatic deinking of secondary fibers: cellulases/ hemicellulases versus laccase-mediator system. J Ind Microbiol Biotechnol 39:1–9
Ito M, Yamagata T (1989) Purification and characterization of glycosphingolipid-specific endoglycosidases (endoglycoceramidases) from a mutant strain of *Rhodococcus* sp. Evidence for three molecular species of endoglycoceramidase with different specificities. J Biol Chem 264 (16):9510–9519
Johnson EA (2013) Biotechnology of non-Saccharomyces yeasts- the ascomycetes. Appl Microbiol Biotechnol 97:503–517
Kantelinen A, Hatakka A, Viikari L (1989) Production of lignin peroxidase and laccase by Phlebia radiata. Appl Microbiol Biotechnol 31(3):234–239

Katoh T, Katayama T, Tomabechi Y, Nishikawa Y, Kumada J, Matsuzaki Y, Yamamoto K (2016) Generation of a mutant Mucorhiemalisendoglycosidase that acts on core-fucosylated N-glycans. J Biol Chem 291(44):23305–23317
Kieliszek M, Misiewicz A (2014) Microbial transglutaminase and its application in the food industry. A review. Folia Microbiol 59:241–250
Krishnan S, Madhumitha SR, Bharathi P, Ravi TK (2017) Effect of immobilization on the enantioselective esterification of ibuprofen using lipase from Candida cylindracea. Manipal J Pharm Sci 3(1):15–23
Kuhad RC, Gupta R, Singh A (2011) Microbial cellulases and their industrial applications. Enzyme Res 2011:10
Kumar S (2015) Role of enzymes in fruit juice processing and its quality enhancement. Adv Appl Sci Res 6(6):114–124
Kumar A, Singh S (2013) Directed evolution: tailoring biocatalysis for industrial application. Crit Rev Biotechnol 33:365–378
Kumar D, Savitri TN, Verma R, Bhalla TC (2008) Microbial proteases and application as laundry detergent additive. Res J Microbiol 3(12):661–672
Kuraishi C, Yamazaki K, Susa Y (2001) Transglutaminase: its utilization in the food industry. Food Rev Intl 17(2):221–246
Lakshmi GS, Rao CS, Rao RS, Hobbs PJ, Prakasham RS (2009) Enhanced production of xylanase by a newly isolated Aspergillus terreus under solid state fermentation using palm industrial waste: a statistical optimization. Biochem Eng J 48(1):51–57
Li DC, Gao J, Li YL, Lu J (2005) A thermostable manganese-containing superoxide dismutase from the thermophilic fungus Thermomyces lanuginosus. Extremophiles 9(1):1–6
Li S, Yang X, Yang S et al (2012) Technology prospecting on enzymes: application, marketing and engineering. Comput Struct Biotechnol J 2:1–11
Lin Y, Zhang Z, Tian Y, Zhao W, Zhu B, Xu Z, Yao Q (2013) Purification and characterization of a novel laccase from Coprinus cinereus and decolorization of different chemically dyes. Mol Biol Report 40(2):1487–1494
Liu X, Kokare C (2017) Microbial enzymes of use in industry. In: Biotechnology of microbial Enzymes. Academic, Amsterdam, pp 267–298
Longo MA, Sanromán MA (2006) Production of food aroma compounds: microbial and enzymatic methodologies. Food Technol Biotechnol 44(3):335–353
Lu M, Wang S, Fang Y, Li H, Liu S, Liu H (2010) Cloning, expression, purification, and characterization of cold-adapted α-amylase from Pseudoalteromonas arctica GS230. Protein J 29(8):591–597
Maiorano AE, Piccoli RM, Da Silva ES, de Andrade Rodrigues MF (2008) Microbial production of fructosyltransferases for synthesis of pre-biotics. Biotechnol Lett 30(11):1867
Mane P, Tale V (2015) Overview of microbial therapeutic enzymes. Int J Curr Microbiol App Sci 4 (4):17–26
Mitidieri S, Martinelli AHS, Schrank A, Vainstein MH (2006) Enzymatic detergent formulation containing amylase from Aspergillus niger: a comparative study with commercial detergent formulations. Bioresour Technol 97(10):1217–1224
Moiseenko KV, Savinova OS, Vasina DV, Kononikhin AS, Tyazhelova TV, Fedorova TV (2018) Laccase Isoenzymes of Trametes hirsuta LE-BIN072: degradation of industrial dyes and secretion under the different induction conditions. Appl Biochem Microbiol 54(9):834–841
Mondal KC, Banerjee R, Pati BR (2000) Tannase production by Bacillus licheniformis. Biotechnol Lett 22(9):767–769
Montiel MC, Máximo F, Serrano-Arnaldos M, Ortega-Requena S, Murcia MD, Bastida J (2019) Biocatalytic solutions to cyclomethicones problem in cosmetics. Eng Life Sci 19(5):370–388
Nunes CS, Vogel K (2018) Tyrosinases—physiology, pathophysiology, and applications. In: Enzymes in human and animal nutrition. Academic, London, pp 403–412
Park SH, Na Y, Kim J, Dal Kang S, Park KH (2018) Properties and applications of starch modifying enzymes for use in the baking industry. Food Sci Biotechnol 27:1–14

Patil U, Mokashe N, Chaudhari A (2016) Detergent-compatible, organic solvent-tolerant alkaline protease from Bacillus circulans MTCC 7942: purification and characterization. Prep Biochem Biotechnol 46(1):56–64

Priji P, Sajith S, Faisal PA, Benjamin S (2016) Microbial lipases– properties and applications. J Microbiol Biotechnol Food Sci 6(2):799

Qureshi MA, Khare AK, Pervez A (2015) Enzymes used in dairy industries. Int J Appl Res 1 (10):523–527

Ray K, Chaki M, Sengupta M (2007) Tyrosinase and ocular diseases: some novel thoughts on the molecular basis of oculocutaneous albinism type 1. Prog Retin Eye Res 26(4):323–358

Rehm BH (2010) Bacterial polymers: biosynthesis, modifications and applications. Nat Rev Microbiol 8(8):578

Rodríguez-García J, Sahi SS, Hernando I (2014) Functionality of lipase and emulsifiers in low-fat cakes with inulin. LWT-Food Sci Technol 58(1):173–182

Rossa PN, De Sa EMF, Burin VM et al (2011) Optimization of microbial transglutaminase activity in ice cream using response surface methodology. LWT-Food Sci Technol 44:29–34

Saini R, Saini HS, Dahiya A (2017) Amylases: characteristics and industrial applications. J Pharmacogn Phytochemist 6(4):1865–1871

Salihu A, Alam MZ, Abdul Karim MI, Salleh HM (2012) Lipase production: an insight in the utilization of renewable agricultural residues. Resour Conserv Recycl 58:36–44

Sanchez S, Demain AL (2011) Enzymes and bioconversions of industrial, pharmaceutical, and biotechnological significance. Organic Process Research & Development 15(1):224–230

Saranraj P, Naidu MA (2014) Microbial pectinases: a review. Glob J Traditional Med Sys 3:1–9

Sarmah N, Revathi D, Sheelu G, Yamuna Rani K, Sridhar S, Mehtab V, Sumana C (2018) Recent advances on sources and industrial applications of lipases. Biotechnol Prog 34(1):5–28

Saxena A, Singh Chauhan P (2017) Role of various enzymes for deinking paper: a review. Crit Rev Biotechnol 37(5):598–612

Seyis I, Aksoz N (2005) Xylanase production from Trichoderma harzianum 1073 D 3 with alternative carbon and nitrogen sources. Food Technol Biotechnol 43(1):37–40

Shabeb MS, Younis MA, Hezayen FF, Nour-Eldein MA (2010) Production of cellulase in low-cost medium by Bacillus subtilis KO strain. World Appl Sci J 8(1):35–42

Sharma R, Chisti Y, Banerjee UC (2001) Production, purification, characterization and applications of lipases. Biotechnol Adv 19:627–662

Sharma D, Sharma B, Shukla AK (2011) Biotechnological approach of microbial lipase: a review. Biotechnology 10(1):23–40

Sher H, Faheem M, Ghani A, Mehmood R, Rehman H, Bokhari SA (2017) Optimization of cellulase enzyme production from Aspergillus oryzae for industrial applications. World J Biol Biotechnol 2(2):155–158

Singh R, Kumar M, Mittal A, Mehta PK (2016a) Microbial enzymes: industrial progress in 21st century. 3 Biotech 6:174

Singh S, Singh VK, Aamir M, Dubey MK, Patel JS, Upadhyay RS, Gupta VK (2016b) Cellulase in pulp and paper industry. In New and Future Developments in Microbial Biotechnology and Bioengineering. Elsevier, pp 152–162

Sinha S, Tripathi P, Chand S (2012) A new bifunctional chitosanase enzyme from Streptomyces sp. and its application in production of antioxidant chitooligosaccharides. Appl Biochem Biotechnol 167(5):1029–1039

Souza PMD (2010) Application of microbial α-amylase in industry-a review. Braz J Microbiol 41 (4):850–861

Srivastava RAK, Baruah JN (1986) Culture conditions for production of thermostable amylase by Bacillus stearothermophilus. Appl Environ Microbiol 52(1):179–184

Srivastava N, Singh P (2015) Degradation of toxic pollutants from pulp & paper mill effluent. Discovery 40(183):221–227

Su C, Wang D, Yao L, Yu Z (2006) Purification, characterization, and gene cloning of a chitosanase from Bacillus species strain S65. J Agric Food Chem 54(12):4208–4214

Swartz JR (2001) Advances in Escherichia coli production of therapeutic proteins. Curr Opin Biotechnol 12(2):195–201

Thadathil N, Velappan SP (2014) Recent developments in chitosanase research and its biotechnological applications: a review. Food Chem 150:392–399. https://doi.org/10.1016/j.foodchem.2013.10.083

Upadhyay P, Shrivastava R, Agrawal PK (2016) Bioprospecting and biotechnological applications of fungal laccase. 3 Biotech 6(1):15

Varalakshmi KN, Kumudini BS, Nandini BN, Solomon J, Suhas R, Mahesh B, Kavitha AP (2009) Production and characterization of a-Amylase from Aspergillus niger JGI 24 isolated in Bangalore. Pol J Microbiol 58(1):29–36

Vashist P, Kanchana R, Devasia VLA, Shirodkar PV, Muraleedharan UD (2019) Biotechnological implications of hydrolytic enzymes from marine microbes. In: Advances in biological science research. Academic, London, pp 103–118

Verma D, Satyanarayana T (2013) Production of cellulase-free xylanase by the recombinant Bacillus subtilis and its applicability in paper pulp bleaching. Biotechnol Prog 29:1441. https://doi.org/10.1002/btpr.1826

Vester JK, Glaring MA, Stougaard P (2015) An exceptionally cold-adapted alpha-amylase from a metagenomic library of a cold and alkaline environment. Appl Microbiol Biotechnol 99 (2):717–727

Virk AP, Sharma P, Capalash N (2012) Use of laccase in pulp and paper industry. Biotechnol Prog 28:21–32

Walia A, Mehta P, Guleria S, Shirkot CK (2015) Modification in the properties of paper by using cellulase-free xylanase produced from alkalophilic Cellulosimicrobium cellulans CKMX1 in biobleaching of wheat straw pulp. Can J Microbiol 61:1–11

Walia A, Guleria S, Mehta P, Chauhan A, Parkash J (2017) Microbial xylanases and their industrial application in pulp and paper biobleaching: a review. 3 Biotech 7(1):11

Wang Y, Zhou PG, Yu JX, Pan XR, Wang PP, Lan WQ, Tao SD (2007) Antimicrobial effect of chitooligosaccharides produced by chitosanase from pseudomonas CUY8. Asia Pac J Clin Nutr 16(S1):174–177

Wang X, Pei D, Teng Y, Liang J (2018) Effects of enzymes to improve sensory quality of frozen dough bread and analysis on its mechanism. J Food Sci Technol 55(1):389–398

Weijers CA, Franssen MC, Visser GM (2008) Glycosyltransferase-catalyzed synthesis of bioactive oligosaccharides. Biotechnol Adv 26(5):436–456

Woldesenbet F, Virk AP, Gupta N, Sharma P (2012) Effect of microwave irradiation on xylanase production from wheat bran and biobleaching of eucalyptus Kraft pulp. Appl Biochem Biotechnol 167:100–108

Wong TY, Preston LA, Schiller NL (2000) Alginate lyase: review of major sources and enzyme characteristics, structure-function analysis, biological roles, and applications. Annu Rev Microbiol 54(1):289–340

www.crisil.com

Zaidi KU, Ali AS, Ali SA et al (2014) Microbial tyrosinases: promising enzymes for pharmaceutical, food bioprocessing, and environmental industry. Biochem Res Int 2014:1. https://doi.org/10.1155/2014/854687

Zhang Y, Ge J, Liu Z (2015) Enhanced activity of immobilized or chemically modified enzymes. ACS Catal 5:4503–4513

Zhang Y, Park KY, Suazo KF, Distefano MD (2018) Recent progress in enzymatic protein labelling techniques and their applications. Chem Soc Rev 47(24):9106–9136

Zheng GW, Xu JH (2011) New opportunities for biocatalysis: driving the synthesis of chiral chemicals. Curr Opin Biotechnol 22:784–792

Part VII

Microbes in Medicine

Living Medicines for Health and Disease Management 18

Surender Jangra and Ramesh Pothuraju

Abstract

Consumption of probiotics in the form of fermented products has a long history. Since the last two decades, probiotics has gained the attention of the scientific community because of their health beneficial effects. Positive effects of probiotics on metabolic disorders such as nonalcoholic fatty liver disease, immune diseases, obesity, diabetes, insulin resistance, cardiovascular disease, irritable bowel syndrome, and inflammatory bowel disease have been reported, but exact mechanism of action of probiotics in amelioration of these disorders is yet to be elucidated. Generally, genera *Lactobacillus* and *Bifidobacterium* are employed as probiotics. Different probiotics act differently in conferring health beneficial effects. Moreover, health-promoting effects of probiotics are dependent on the strain. Furthermore, probiotic dosages, feeding schedule, mechanism of action, and long-term effects on health are yet to be elucidated. Therefore, further studies are required to explain the health beneficial effects of probiotics before it can be rationally prescribed to patients. In this chapter, we will discuss about role of probiotics in the prevention and treatment of various metabolic disorders.

Keywords

Living medicine · Probiotics · Metabolic disorders · Gut health · Gut microbiota · Low-grade inflammation

Authors Surender Jangra and Ramesh Pothuraju have been equally contributed to this chapter.

S. Jangra (✉)
School of Bioengineering and Biosciences, Lovely Professional University, Phagwara, Punjab, India

R. Pothuraju (✉)
Department of Biochemistry and Molecular Biology, University of Nebraska Medicine Center, Omaha, NE, USA

S. G. Sharma et al. (eds.), *Microbial Diversity, Interventions and Scope*,
https://doi.org/10.1007/978-981-15-4099-8_18

18.1 Introduction

Once considered as problem of developed countries, metabolic disorders such as nonalcoholic fatty liver disease, obesity, diabetes, insulin resistance, and cardiovascular diseases are getting prevalent in developing countries too. Two major factors for the startling rise of these disorders are sedentary lifestyle and consumption of inadequate diets. Therefore, active lifestyle and proper energy intake are recommended for the treatment of these disorders, but success rate is extremely less and highly disappointing. Therefore, various methods like pharmacological and surgical methods have been developed for their treatments. Pharmacological medicines, viz., orlistat (inhibits pancreatic lipase) and sibutramine, have been approved for the obesity management. Some drugs like rimonabant have been withdrawn from the market due to their side effects (Zanella and Ribeiro Filho 2009). Surgical methods like liposuction, bariatric surgery, and Roux-en-Y are used to treat obesity in humans, but both surgical and pharmacological methods are associated with certain side effects and are not much effective in treating obesity in humans (Pothuraju et al. 2014; Zanella and Ribeiro Filho 2009). Antidiabetic drugs act in many ways. Sulfonylureas increase insulin secretion, thiazolidinediones decrease insulin resistance in peripheral tissues, α-glucosidase inhibitors such as acarbose decrease intestinal glucose uptake, and biguanides such as metformin inhibit gluconeogenesis (Abuissa et al. 2005; Marchetti 2005). But these pharmacological approaches seem to be insufficient in maintaining the normal blood glucose level. Moreover, treatment of diabetic patients for long term causes higher costs to the society. To treat the complications associated with cardiovascular diseases, statins are widely used, but long-term use of statins is also associated with certain side effects (Levine et al. 1995). Therefore, due to the lack of significant methods in the management of metabolic disorders, development of safe and effective method/s has become the priority of the scientific community all over the globe.

Today living medicines (probiotics) are gaining the attention of the researchers' all over the globe for the prevention/treatment of disorders discussed above. Literatures explaining the positive effects of probiotics on the metabolic disorders are increasing exponentially. Probiotics have been defined as live microorganisms that confer health benefits upon the host when administered in adequate amount (FAO/WHO 2002). When consumed orally, probiotics reaches to the colon, colonizes there, and confers health benefits to the host. Genera *Lactobacillus and Bifidobacterium* are generally employed as probiotics. Beneficial effects of probiotics on hypercholesterolemia (Shin et al. 2010), cancer (Kumar et al. 2010), intestinal integrity (Mennigen and Bruewer 2009), irritable bowel syndrome (Aragon et al. 2010), nonalcoholic fatty liver disease (Ma et al. 2013), immune diseases (Borchers et al. 2009; Gill and Prasad 2008), obesity (Jangra et al. 2019; Kang et al. 2013; Park et al. 2013; Rather et al. 2014), and insulin resistance

(Ma et al. 2008; Okubo et al. 2013) have been described elsewhere. In this chapter, we have discussed about the role and mechanism of action of probiotics in management of metabolic disorders.

18.2 Living Medicines (Probiotics) for Health and Disease Management

18.2.1 Probiotics, Gut Permeability, and Low-Grade Inflammation

Trillions of microorganisms are living in the colon of humans. This gut microbiota has been reported to be associated with various functions such as metabolism of nutrients, production of metabolites (short-chain fatty acids), vitamins production (vitamin B12 and vitamin K), xenobiotics and drugs metabolism, improvement of gut integrity, immune system modulation, and removal of pathogens from the intestine by colonization resistance (Jandhyala et al. 2015). Structure of the gut flora is determined by various factors such as stress, diet, medications, environmental factors, and infections. Literatures have confirmed that dysbiosis in gut microbiota leads to the development of various disorders such as irritable bowel syndrome (Principi et al. 2018), inflammatory bowel diseases (Nell et al. 2010; Sokol et al. 2008), colorectal cancer (Arthur et al. 2012; Scanlan et al. 2008), allergic diseases (McLoughlin and Mills 2011), nonalcoholic fatty liver diseases (Abu-Shanab and Quigley 2010; Henao-Mejia et al. 2012), arteriosclerotic diseases (Koeth et al. 2013; Wang et al. 2011), and metabolic syndromes such as obesity and diabetes (Turnbaugh et al. 2008; Turnbaugh et al. 2006).

Consumption of energy-rich diets has been known to cause dysbiosis of the gut microbiota (Firmicutes/Bacteroidetes ratio increases) in both humans and mice (Ley et al. 2005). Furthermore, intake of energy-rich diets triggers the production of lipopolysaccharides (LPS) by the altered gut microbiota (Cani et al. 2008; Kim et al. 2012; Moreira et al. 2012). LPS produced in the intestinal lumen decreases the expression of tight junction proteins that leads into increased gut permeability. This results into increased absorption of LPS into blood stream from the intestinal lumen (metabolic endotoxemia). LPS in the circulation reaches to adipose tissue, liver, and muscles, and induces oxidative stress as well as the production of pro-inflammatory cytokines such as TNFα and IL6. These cytokines trigger low-grade inflammation that results into high-fat-diet-induced metabolic disorders such as nonalcoholic fatty liver disease, diabetes, obesity, and insulin resistance (Fig. 18.1) (Cani et al. 2008). Additionally, intake of a high fat diet (HFD) has been reported to induce the inflammation in intestine (Ding et al. 2010). It is surmised that intestinal inflammation, which occurs in the early stages of obesity development (Ding et al. 2010), induces the inflammation in the mesenteric adipose tissue through the release of pro-inflammatory cytokines (Fig. 18.1) (Li et al. 2008). Therefore, it can be thought that improvement in the intestinal inflammation could improve adipose tissue inflammation, which results in the reduction of body fat content.

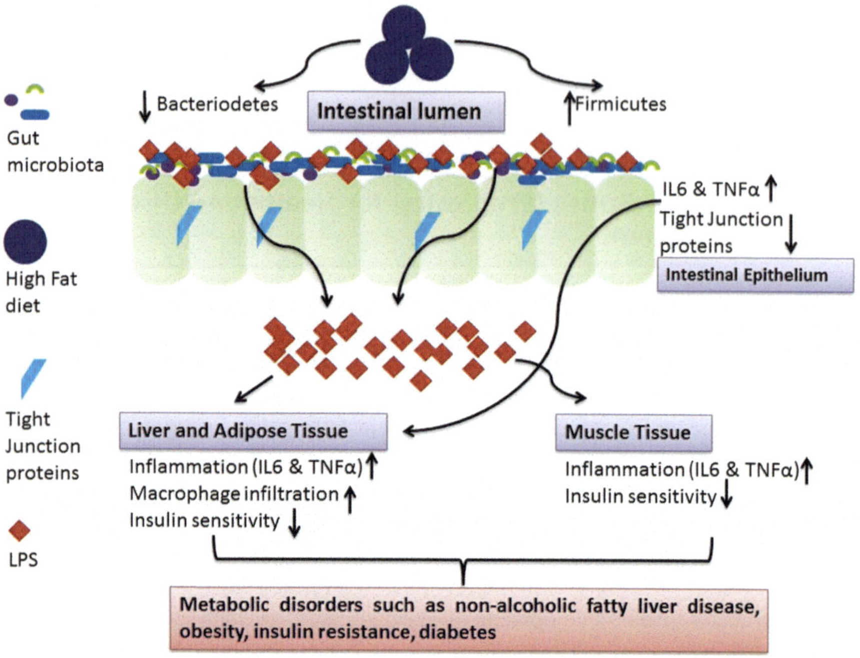

Fig. 18.1 Effect of feeding of high fat diet on gut microbiota structure, gut permeability, intestinal & adipose tissue inflammation, and metabolic disorders

Lactobacillus sakei OK67 has been reported to attenuate high fat diet (HFD)-induced obesity and blood glucose intolerance by decreasing metabolic endotoxemia (low LPS in colon and blood) and inflammation in the colon of mice with concomitantly enhancing the levels of tight junction proteins (Lim et al. 2016). Similarly, *L. brevis* OK56 ameliorated HFD-induced obesity by reducing the metabolic endotoxemia and NF-kB signaling in the colon of mice. Also, the levels of tight junction proteins were higher in OK56-fed group (Kim et al. 2015). In another study, a negative correlation between abundance of bifidobacteria and metabolic endotoxemia was reported. Therefore, increasing the bifidobacterial counts in the colon could be considered as a significant method in reducing the high-fat-diet induced metabolic diseases (Cani et al. 2007). Administration of *L. curvature* HY7601 and *L. plantarum* KY1032 altered the structure of gut microbiota significantly. Furthermore, metabolism and inflammation linked genes in the liver and adipose tissue of mice were also modulated. This results in the lowering of body fat content in mice (Park et al. 2013). Therefore, normalizing the gut microbiota through intake of probiotics could be considered as an ideal way of ameliorating the complications associated with metabolic disorders.

18.2.2 Probiotics and Weight Management

In obesity a large amount of fat is accumulated in the body that alters health, which leads into cascade of other diseases like insulin resistance, diabetes, cardiovascular diseases, and even cancer. In recent years, incidences of childhood obesity are increasing gradually, which could lead to higher obese adult population in the near future. In obesity preadipocytes are differentiated into mature adipocytes with concomitantly accumulation of fat droplets, which leads to increased adipocyte size and number (Sears et al. 1996). Furthermore, a large amount of triglycerides are also accumulated in the adipose tissue, which results in increased levels of free fatty acids in the circulation. Thereafter, free fatty acids are accumulated in liver and muscles as a lipid-burden hypothesis (Nadler and Attie 2001). Adipokines like TNFα, IL-6, IL-1, and leptin are secreted by adipose tissue under the influence of free fatty acids, which results into decreased vascularization, with the development of hypoxia and inflammation (Sears et al. 1996). Literatures have suggested that occurrence of obesity and its comorbidities are not solely due to the changes in the host genome or sedentary lifestyle or consumption of more energy than expended. Changes in gut microbiota (dysbiosis) are also considered as one of the reasons for the development of obesity (Cani and Delzenne 2009). However, the role of gut microbiota in metabolic disorders is still controversial and debatable.

Many research groups from all over the globe have reported the positive effects of probiotics like lactobacilli and bifidobacteria on obesity and gut microbiota in both rodents and humans. Different probiotics act differently in counteracting the obesity. Some probiotics like *L. rhamnosus* PL60 and *L. plantarum* PL62 exhibited the beneficial effects on obesity by producing the conjugated linoleic acid (Lee et al. 2006; Lee et al. 2007). Lyophilized mixture of lactobacilli, bifidobacteria, and *S. thermophiles* (VSL#3) exerted the beneficial effects on obesity and its related comorbidities through increasing the natural killer T cells in liver of mice (Ma et al. 2008). *L. gasseri* SBT2055 inhibited the absorption of fat in the intestine of rats (Hamad et al. 2008), and reduced the levels of genes linked with inflammation in epididymal adipose tissue of mice (Miyoshi et al. 2014). This led to decreased adipocyte size and body weight. *L. paracasei* F19 decreased the body fat by enhancing the expression of fasting-induced adipose factor (FIAF), an inhibitor of lipoprotein lipase (LPL) (Aronsson et al. 2010). LPL is a rate-limiting enzyme for the deposition of triglycerides in the adipose tissue (Serra et al. 2017). Therefore, low deposition of triglycerides in the adipose tissue could be expected under influence of higher FIAF expression. *L. gasseri* BNR17 ameliorated the obesity by enhancing the levels of genes linked with oxidation of fat (Kang et al. 2013). *L. sakei* OK67 ameliorated the obesity and its related comorbidities by enhancing the abundance of tight junction proteins. This led to improved metabolic endotoxemia and inflammation (Lim et al. 2016). *L. casei* NCDC19 exhibited anti-obesity effects by increasing bifidobacterial number in the colon (Rather et al. 2014), and enhancing the expression of genes related to energy expenditure (Jangra et al. 2019). *L. plantarum* LMT1–48 inhibited the pathogenic bacteria (*Enterobacter cloacae*) as well as increased the microbiota diversity in the intestine (Jin et al. 2019). However, despite the large amount of data, effects of probiotics on obesity are still controversial and debatable. Moreover, studies have confirmed that positive effects of probiotics on metabolic disorders are strain specific (Fåk and Bäckhed 2012; Qiao et al. 2015; Yin et al. 2010).

Only a few species/strains of *Lactobacillus* and *Bifidobacterium* genera are effective in the management of body weight. Furthermore, few strains included in *Lactobacillus* genera such as *L. acidophilus* NCDC13 have no significant effect on the body weight (Arora et al. 2012), whereas few strains such as *L. acidophilus* and *L. reuteri* L6798 increased the body weight perhaps because of their pro-inflammatory effects (Fåk and Bäckhed 2012; Million et al. 2012). Moreover, probiotic dose, feeding schedule, mechanism of action and long-term effects on obesity and its related disorders are yet to be elucidated. Therefore, further studies are required to explain the health beneficial effects of probiotics on obesity before it can be rationally prescribed to the obese individuals.

18.2.3 Probiotics and Diabetes

Obesity also contributes to the onset of insulin resistance, a situation in which body don't respond properly to the insulin and resulted into higher blood glucose and insulin levels. It is also widely accepted that the incidences of diabetes decreases as body weight decreases (Taylor 2008; Tuomilehto et al. 2001). Low-grade inflammation (characterized by high levels of pro-inflammatory cytokines such as TNFα, IL6 etc.) is considered as one of the causes for the onset of insulin resistance and type 2 diabetes as well as other harmful effects associated with obesity (Greenberg and Obin 2006; Lee et al. 2009). Toll-like receptor 4 (TLR4), present on insulin target tissues, is thought to mediate the development of low-grade inflammation and insulin resistance (Kim and Sears 2010). It gets activated by binding with exogenous ligands (LPS & dietary fatty acids) as well as endogenous ligands (circulating free fatty acids). Activation of TLR4 causes the disruption of the insulin signaling through inhibitory phosphorylation of insulin receptor substrate, and also leads to the higher expression of pro-inflammatory cytokines genes (Kim and Sears 2010). Higher levels of these cytokines has been reported to mediate insulin resistance in adipose tissues through disruption of insulin signaling (Hotamisligil et al. 1995), but the exact mechanisms associated with low-grade inflammation are yet to be elucidated.

Bifidobacteria and lactobacilli strains have been reported to prevent insulin resistance and diabetes, and various mechanisms have been reported in literatures explaining the positive effects of probiotics on insulin resistance and diabetes. *L. gasseri* SBT2055 improved the insulin resistance in mice by reducing the levels of genes linked with low-grade inflammation in epididymal adipose tissue of mice (Miyoshi et al. 2014). *L. casei* Shirota ameliorated the insulin resistance state in obese mice through reduction of LPS levels in the blood (Naito et al. 2011). *L. reuteri* GMNL-263 ameliorated the insulin resistance through enhancement of expression of glut4 (insulin-dependent glucose transporter) and pparγ (activates the glut4 expression) in adipose tissue (Hsieh et al. 2013). *L. gasseri* BNR17 exhibited the antidiabetic effects by increasing glut4 and decreasing insulin levels (Kang et al. 2013). *L. casei* NCDC19 significantly blunted the rise of high-fat- and sucrose-diet-induced blood glucose, insulin, and HOMA-IR score in mice, which could be linked with the higher expression of genes (*adiponectin*, *cpt1*, *ppar alpha*, *pgc1beta*, and *foxa2*) involved in energy expenditure (Jangra et al. 2019).

18.2.4 Probiotics and Hypercholesterolemia

Hypercholesterolemia is a major risk factor for hypertension, cardiovascular diseases and coronary heart diseases. In hypercholesterolemia, total cholesterol, LDL-cholesterol, and triglycerides levels increase abnormally while levels of HDL-cholesterol decrease. Due to accumulation of lipids in the body, levels of pro-inflammatory cytokines (TNF alpha and IL6) increase (Coppack 2001), and activity of antioxidant enzymes gets adversely affected. Many studies have confirmed the positive effects of probiotics on hypercholesterolemia and its associated disorders. *L. plantarum* 9-41-A and *L. fermentum* M1-16 increased the excretion of cholesterol and bile salts through feces, and consequently, cholesterol level decreases (Xie et al. 2011). Feeding of *L. plantarum* Biocenol LP96 resulted into significantly reduced serum triglycerides and VLDL-cholesterol levels while the feeding of *L. plantarum* LS/07 resulted into significantly reduced serum total cholesterol only (Salaj et al. 2013). Feeding of *L. casei* NCDC19 probiotic dahi to high-fat-diet fed C57BL/6 mice resulted into significantly decreased levels of total and LDL-cholesterol in blood plasma (Rather et al. 2014). *L. rhamnosus* MTCC5957 and *L. rhamnosus* MTCC5897 reduced hepatic and blood lipids in Wistar rats fed high cholesterol diet by increasing cholesterol excretion through feces and decreasing the inflammatory markers in liver (Yadav et al. 2019). Similarly, feeding of *L. casei* NCDC19 decreased the levels of serum triglycerides and VLDL-cholesterol significantly in C57BL/6 mice (Jangra et al. 2019).

It has not been established as to how the responses on lipid parameters are different as a consequence of administration of different probiotic organisms. The differences in results on administration of different probiotic strains were ascribed to the confounding variables such as different sources and property of lactobacilli strains. More than one mechanism has been proposed by different research groups for lowering cholesterol by probiotics. Decreased absorption of cholesterol from the intestine into blood due to adhesion and incorporation of cholesterol on the bacterial cell surface/cell membrane leads to increased excretion of cholesterol from the body through feces (Anila et al. 2016; Tabuchi et al. 2004). Deconjugation of bile salts by probiotic due to its bile salt hydrolase activity is another possible mechanism for lowering cholesterol (Nguyen et al. 2007). Higher expression of low-density lipoprotein receptor due to administration of probiotics (Kumar et al. 2013; Park et al. 2007) has also been reported, but the exact mechanism for lowering cholesterol is still uncertain and controversial.

18.2.5 Probiotics and Cancer

Human gastrointestinal tract contains more than 10^{14} microorganisms and these microorganisms plays an important role in treating several metabolic diseases along with cancers (Belkaid and Hand 2014; Pothuraju et al. 2018). Recent studies also suggest the role of gut microbiota in improving food absorption, resistance to infection, host immune system and metabolism (Eslami et al. 2019). Nowadays, researchers are attracted toward beneficial microorganisms which includes probiotics which when administered adequately confers health benefits to the host. The alteration in the gut microbiome is responsible for initiation and progression of inflammation, genetic and epigenetic alterations and, finally, leads to tumor formation (Eslami et al. 2019). The use of probiotics in the management of cancer has no clear evidence; however, some researchers claimed that administration of probiotics maintains proper rebalance of gut microbiome and is also effective against radio- and chemotherapies (Drago 2019). Cancer cells show uncontrolled proliferation and resistance to apoptosis and chemotherapeutic drugs. Incubation of *Lactobacillus rhamnosus*, *Bifidobacterium lactis*, *Escherichia coli* K12, and *Atopobium minutum* with intestinal Caco-2 cell line showed apoptosis (Altonsy et al. 2010). Another study showed activation of apoptotic death in gastric as well as colon cancer cell lines, which might be due to the production of short-chain fatty acids (SCFAs) in the culture medium containing *Propionibacterium freudenreichii*. The activation of apoptotic process is due to upregulation of caspase 3, free oxygen radicals' production, and mitochondrial permeability of the cancer cells (Kumar et al. 2015). Gut microbiota manipulation by probiotics might be able to reduce the colon cancer risk and improve the safety (Hendler and Zhang 2018). Furthermore, role of probiotics in the prevention of lung cancer is underscanty (Sharma et al. 2018). In a clinical trial, patients who received chemotherapy alone had severe constipation and showed decreased levels of *Lactobacillus*, *Bifidobacterium,* and *Bacteroides*; however, these organisms were reverted upon probiotic administration in lung cancer patients. Overall, understanding the role of probiotics is important in randomized clinical trials and also their use in potential adjuvant therapy for treating several cancers in near future.

18.3 Conclusion

Probiotics are attracting researchers' attention as important functional food ingredients in modulation of complications associated with metabolic disorders. However, the role of probiotics in the management of metabolic disorders is still controversial and debatable. Therefore, it can be exciting to explore the indigenous strains of probiotic bacteria for their efficacy in the regulation of metabolic disorders in both animal models and humans before it can be rationally prescribed to patients.

References

Abuissa H, Bell DSH, O'Keefe JH (2005) Strategies to prevent type 2 diabetes. Curr Med Res Opin 21(7):1107–1114

Abu-Shanab A, Quigley EMM (2010) The role of the gut microbiota in non-alcoholic fatty liver disease. Nat Rev Gastroenterol Hepatol 7(12):691–701

Altonsy MO, Andrews SC, Tuohy KM (2010) Differential induction of apoptosis in human colonic carcinoma cells (Caco-2) by Atopobium, and commensal, probiotic and enteropathogenic bacteria: mediation by the mitochondrial pathway. Int J Food Microbiol 137(2–3):190–203

Anila K, Kunzes A, Bhalla T (2016) *In vitro* cholesterol assimilation and functional enzymatic activities of putative probiotic lactobacillus sp isolated from fermented foods/beverages of north west india. J Nutr Food Sci 6(467):1–5

Aragon G, Graham DB, Borum M, Doman DB (2010) Probiotic therapy for irritable bowel syndrome. Gastroenterol Hepatol 6(1):39–44

Aronsson L, Huang Y, Parini P, Korach-André M, Håkansson J, Gustafsson J-Å, Pettersson S, Arulampalam V, Rafter J (2010) Decreased fat storage by *Lactobacillus Paracasei* is associated with increased levels of angiopoietin-like 4 protein (ANGPTL4). PLoS One 5(9):e13087

Arora T, Anastasovska J, Gibson G, Tuohy K, Sharma RK, Bell J, Frost G (2012) Effect of *Lactobacillus acidophilus* NCDC 13 supplementation on the progression of obesity in diet-induced obese mice. Br J Nutr 108(8):1382–1389

Arthur JC, Perez-Chanona E, Mühlbauer M, Tomkovich S, Uronis JM, Fan T-J, Campbell BJ, Abujamel T, Dogan B, Rogers AB, Rhodes JM, Stintzi A, Simpson KW, Hansen JJ, Keku TO, Fodor AA, Jobin C (2012) Intestinal inflammation targets cancer-inducing activity of the microbiota. Science 338(6103):120–123

Belkaid Y, Hand TW (2014) Role of the microbiota in immunity and inflammation. Cell 27;157 (1):121–141

Borchers AT, Selmi C, Meyers FJ, Keen CL, Gershwin ME (2009) Probiotics and immunity. J Gasteroenterol 44(1):26–46

Cani PD, Delzenne NM (2009) Interplay between obesity and associated metabolic disorders: new insights into the gut microbiota. Curr Opin Pharmacol 9(6):737–743

Cani PD, Neyrinck AM, Fava F, Knauf C, Burcelin RG, Tuohy KM, Gibson GR, Delzenne NMJD (2007) Selective increases of bifidobacteria in gut microflora improve high-fat-diet-induced diabetes in mice through a mechanism associated with endotoxaemia. Diabetologia 50 (11):2374–2383

Cani PD, Bibiloni R, Knauf C, Waget A, Neyrinck AM, Delzenne NM, Burcelin R (2008) Changes in gut microbiota control metabolic endotoxemia-induced inflammation in high-fat diet–induced obesity and diabetes in mice. Diabetes 57(6):1470–1481

Coppack SW (2001) Pro-inflammatory cytokines and adipose tissue. Proc Nutr Soc 60(3):349–356

Ding S, Chi M, Scull B, Rigby R, Schwerbrock N, Magness S, Jobin C, Lund P (2010) High-fat diet: bacteria interactions promote intestinal inflammation which precedes and correlates with obesity and insulin resistance in mouse. PLoS One 5(8):e12191

Drago L (2019) Probiotics and Colon Cancer. Microorganisms 7(3):66

Eslami M, Yousefi B, Kokhaei P, Hemati M, Nejad ZR, Arabkari V, Namdar A (2019) Importance of probiotics in the prevention and treatment of colorectal cancer. J Cell Physiol 234:17127. https://doi.org/10.1002/jcp.28473

Fåk F, Bäckhed F (2012) *Lactobacillus reuteri* prevents diet-induced obesity, but not atherosclerosis, in a strain dependent fashion in Apoe−/− mice. PLoS One 7(10):e46837

FAO/WHO (2002) Joint FAO/WHO working group on drafting guidelines for the evaluation of probiotics in food. London, Ontario, Canada. (2002). April, May. Guidelines for evaluation of probiotics in food. FAO/WHO, Rome

Gill H, Prasad J (2008) Probiotics, immunomodulation, and health benefits. In: Bösze Z (ed) Bioactive components of milk. Springer New York, New York, pp 423–454

Greenberg AS, Obin MS (2006) Obesity and the role of adipose tissue in inflammation and metabolism. Am J Clin Nutr 83(2):461S–465S

Hamad EM, Sato M, Uzu K, Yoshida T, Higashi S, Kawakami H, Kadooka Y, Matsuyama H, El-Gawad IAA, Imaizumi K (2008) Milk fermented by *Lactobacillus gasseri* SBT2055 influences adipocyte size via inhibition of dietary fat absorption in Zucker rats. Br J Nutr 101 (5):716–724

Henao Mejia J, Elinav E, Jin C, Hao L, Mehal WZ, Strowig T, Thaiss CA, Kau AL, Eisenbarth SC, Jurczak MJ, Camporez J-P, Shulman GI, Gordon JI, Hoffman HM, Flavell RA (2012) Inflammasome-mediated dysbiosis regulates progression of NAFLD and obesity. Nature 482:179

Hendler R, Zhang Y (2018) Probiotics in the treatment of colorectal cancer. Medicines 5(3):101

Hotamisligil GS, Arner P, Caro JF, Atkinson RL, Spiegelman BM (1995) Increased adipose tissue expression of tumor necrosis factor-α in human obesity and insulin resistance. J Clin Invest 95:2409–2415

Hsieh FC, Lee CL, Chai CY, Chen WT, Lu YC, Wu CS (2013) Oral administration of *Lactobacillus reuteri* GMNL-263 improves insulin resistance and ameliorates hepatic steatosis in high fructose-fed rats. Nutr Metab 10(1):35

Jandhyala SM, Talukdar R, Subramanyam C, Vuyyuru H, Sasikala M, Nageshwar Reddy D (2015) Role of the normal gut microbiota. World J Gastroenterol 21(29):8787–8803

Jangra S, Sharma RK, Pothuraju R, Bhakri G (2019) Milk fermented with *Lactobacillus casei* NCDC19 improves high fat and sucrose diet alters gene expression in obese mice. Int Dairy J 90:15–22

Jin CW, Jin DH, Uk JH, Ho JH, Yeung-Hyen K, Hoon KT (2019) Antiobesity effects of *Lactobacillus plantarum* LMT1-48 accompanied by inhibition of enterobacter cloacae in the intestine of diet-induced obese mice. J Med Food 22(6):560–566. https://doi.org/10.1089/jmf.2018.4329

Kang JH, Yun SI, Park MH, Park JH, Jeong SY, Park HO (2013) Anti-obesity effect of *Lactobacillus gasseri* BNR17 in high-sucrose diet-induced obese mice. PLoS One 8(1):e54617

Kim JJ, Sears DD (2010) TLR4 and insulin resistance. Gastroenterol Res Pract 2010:212563., 11 pages. https://doi.org/10.1155/2010/212563

Kim KA, Gu W, Lee IA, Joh EH, Kim DH (2012) High fat diet-induced gut microbiota exacerbates inflammation and obesity in mice via the TLR4 signaling pathway. PLoS One 7(10):e47713

Kim KA, Jeong JJ, Kim DH (2015) *Lactobacillus brevis* OK56 ameliorates high-fat diet-induced obesity in mice by inhibiting NF-κB activation and gut microbial LPS production. J Funct Foods 13:183–191

Koeth RA, Wang Z, Levison BS, Buffa JA, Org E, Sheehy BT, Britt EB, Fu X, Wu Y, Li L, Smith JD, DiDonato JA, Chen J, Li H, Wu GD, Lewis JD, Warrier M, Brown JM, Krauss RM, Tang WHW, Bushman FD, Lusis AJ, Hazen SL (2013) Intestinal microbiota metabolism of L-carnitine, a nutrient in red meat, promotes atherosclerosis. Nat Med 19(5):576–585

Kumar M, Kumar A, Nagpal R, Mohania D, Behare P, Verma V, Kumar P, Poddar D, Aggarwal PK, Henry CJK, Jain S, Yadav H (2010) Cancer-preventing attributes of probiotics: an update. Int J Food Sci Nutr 61(5):473–496

Kumar M, Rakesh S, Nagpal R, Hemalatha R, Ramakrishna A, Sudarshan V et al (2013) Probiotic *Lactobacillus rhamnosus* GG and aloe vera gel improve lipid profiles in hypercholesterolemic rats. Nutrition 29:574–579

Kumar KS, Sastry N, Polaki H, Mishra V (2015) Colon cancer prevention through probiotics: an overview. J Can Sci Ther 7(2):081–092

Lee HY, Park JH, Seok SH, Baek MW, Kim DJ, Lee KE, Paek KS, Lee Y, Park JH (2006) Human originated bacteria, *Lactobacillus rhamnosus* PL60, produce conjugated linoleic acid and show anti-obesity effects in diet-induced obese mice. Biochim Biophys Acta (BBA) – Mol Cell Biol Lipids 1761(7):736–744

Lee K, Paek K, Lee HY, Park JH, Lee Y (2007) Antiobesity effect of trans-10,cis-12-conjugated linoleic acid-producing *Lactobacillus plantarum* PL62 on diet-induced obese mice. J Appl Microbiol 103(4):1140–1146

Lee DE, Kehlenbrink S, Lee H, Hawkins M, Yudkin JS (2009) Getting the message across: mechanisms of physiological cross talk by adipose tissue. Am J Physiol Endocrinol Metab 296(6):E1210–E1229

Levine GN, Keaney JF, Vita JA (1995) Cholesterol reduction in cardiovascular disease – clinical benefits and possible mechanisms. N Engl J Med 332(8):512–521

Ley RE, Bäckhed F, Turnbaugh P, Lozupone CA, Knight RD, Gordon JI (2005) Obesity alters gut microbial ecology. Proc Natl Acad Sci USA 102(31):11070–11075

Li H, Lelliott C, Håkansson P, Ploj K, Tuneld A, Verolin-Johansson M, Benthem L, Carlsson B, Storlien L, Michaëlsson E (2008) Intestinal, adipose, and liver inflammation in diet-induced obese mice. Metabolism 57(12):1704–1710

Lim S-M, Jeong J-J, Woo KH, Han MJ, Kim D-H (2016) *Lactobacillus sakei* OK67 ameliorates high-fat diet–induced blood glucose intolerance and obesity in mice by inhibiting gut microbiota lipopolysaccharide production and inducing colon tight junction protein expression. Nutr Res 36(4):337–348

Ma X, Hua J, Li Z (2008) Probiotics improve high fat diet-induced hepatic steatosis and insulin resistance by increasing hepatic NKT cells. J Hepatol 49(5):821–830

Ma Y-Y, Li L, Yu C-H, Shen Z, Chen L-H, Li Y-M (2013) Effects of probiotics on nonalcoholic fatty liver disease: a meta-analysis. World J Gastroenterol 19(40):6911–6918

Marchetti P (2005) New-onset diabetes after liver transplantation: from pathogenesis to management. Liver Transpl 11(6):612–620

McLoughlin RM, Mills KHG (2011) Influence of gastrointestinal commensal bacteria on the immune responses that mediate allergy and asthma. J Allergy Clin Immunol 127(5):1097–1107

Mennigen R, Bruewer M (2009) Effect of probiotics on intestinal barrier function. Ann NY Acad Sci 1165(1):183–189

Million M, Angelakis E, Paul M, Armougom F, Leibovici L, Raoult D (2012) Comparative meta-analysis of the effect of *Lactobacillus* species on weight gain in humans and animals. Microb Pathog 53(2):100–108

Miyoshi M, Ogawa A, Higurashi S, Kadooka Y (2014) Anti-obesity effect of *Lactobacillus gasseri* SBT2055 accompanied by inhibition of pro-inflammatory gene expression in the visceral adipose tissue in diet-induced obese mice. Eur J Nutr 53(2):599–606

Moreira APB, Texeira TFS, Ferreira AB, do Carmo Gouveia Peluzio M, de Cássia Gonçalves Alfenas R (2012) Influence of a high-fat diet on gut microbiota, intestinal permeability and metabolic endotoxaemia. Br J Nutr 108(5):801–809

Nadler ST, Attie AD (2001) Please pass the chips: genomic insights into obesity and diabetes. J Nutr 131(8):2078–2081

Naito E, Yoshida Y, Makino K, Kounoshi Y, Kunihiro S, Takahashi R, Matsuzaki T, Miyazaki K, Ishikawa F (2011) Beneficial effect of oral administration of *Lactobacillus casei* strain Shirota on insulin resistance in diet-induced obesity mice. J Appl Microbiol 110(3):650–657

Nell S, Suerbaum S, Josenhans C (2010) The impact of the microbiota on the pathogenesis of IBD: lessons from mouse infection models. Nat Rev Microbiol 8:564

Nguyen TDT, Kang JH, Lee MS (2007) Characterization of *Lactobacillus plantarum* PH04, a potential probiotic bacterium with cholesterol-lowering effects. Int J Food Microbiol 113 (3):358–361

Okubo T, Takemura N, Yoshida A, Sonoyama K (2013) KK/Ta mice administered *Lactobacillus plantarum* strain No. 14 have lower adiposity and higher insulin sensitivity. Biosci Microbiota, Food Health 32(3):93–100

Park Y, Kim J, Shin Y, Kim S, Whang K (2007) Effect of dietary inclusion of *Lactobacillus acidophilus* ATCC 43121 on cholesterol metabolism in rats. J Microbiol Biotechnol 17:655e662

Park D-Y, Ahn Y-T, Park S-H, Huh C-S, Yoo S-R, Yu R, Sung M-K, McGregor RA, Choi M-S (2013) Supplementation of *Lactobacillus curvatus* HY7601 and *Lactobacillus plantarum* KY1032 in diet-induced obese mice is associated with gut microbial changes and reduction in obesity. PLoS One 8(3):e59470

Pothuraju R, Sharma RK, Chagalamarri J, Jangra S, Kumar Kavadi P (2014) A systematic review of Gymnema sylvestre in obesity and diabetes management. J Sci Food Agric 94(5):834–840

Pothuraju R, Rachagani S, Junker WM, Chaudhary S, Saraswathi V, Kaur S, Batra SK (2018) Pancreatic cancer associated with obesity and diabetes: an alternative approach for its targeting. J Exp Clin Cancer Res 37(1):319

Principi N, Cozzali R, Farinelli E, Brusaferro A, Esposito S (2018) Gut dysbiosis and irritable bowel syndrome: the potential role of probiotics. J Infect 76(2):111–120

Qiao Y, Sun J, Xia S, Li L, Li Y, Wang P, Shi Y, Le G (2015) Effects of different *Lactobacillus reuteri* on inflammatory and fat storage in high-fat diet-induced obesity mice model. J Funct Foods 14:424–434

Rather SA, Pothuraju R, Sharma RK, De S, Mir NA, Jangra S (2014) Anti-obesity effect of feeding probiotic dahi containing *Lactobacillus casei* NCDC 19 in high fat diet-induced obese mice. Int J Dairy Technol 67(4):504–509

Salaj R, Stofilová J, Soltesová A, Hertelyová Z, Hijová E, Bertková I, Strojný L, Kružliak P, Bomba A (2013) The effects of two *Lactobacillus plantarum* strains on rat lipid metabolism receiving a high fat diet. Sci World J 2013:135142–135142

Scanlan PD, Shanahan F, Clune Y, Collins JK, O'Sullivan GC, O'Riordan M, Holmes E, Wang Y, Marchesi JR (2008) Culture-independent analysis of the gut microbiota in colorectal cancer and polyposis. Environ Microbiol 10(3):789–798

Sears IB, MacGinnitie MA, Kovacs LG, Graves RA (1996) Differentiation-dependent expression of the brown adipocyte uncoupling protein gene: regulation by peroxisome proliferator-activated receptor gamma. Mol Cel Biol 16(7):3410–3419

Serra MC, Ryan AS, Goldberg AP (2017) Reduced LPL and subcutaneous lipid storage capacity are associated with metabolic syndrome in postmenopausal women with obesity. Obes Sci Pract 3 (1):106–114

Sharma A, Viswanath B, Park YS (2018) Role of probiotics in the management of lung cancer and related diseases: an update. J Funct Foods 40:625–633

Shin HS, Park SY, Lee DK, Kim SA, An HM, Kim JR, Kim MJ, Cha MG, Lee SW, Kim KJ, Lee KO, Ha NJ (2010) Hypocholesterolemic effect of sonication-killed *Bifidobacterium longum* isolated from healthy adult Koreans in high cholesterol fed rats. Arch Pharm Res 33 (9):1425–1431

Sokol H, Pigneur B, Watterlot L, Lakhdari O, Bermúdez-Humarán LG, Gratadoux J-J, Blugeon S, Bridonneau C, Furet J-P, Corthier G, Grangette C, Vasquez N, Pochart P, Trugnan G, Thomas G, Blottière HM, Doré J, Marteau P, Seksik P, Langella P (2008) *Faecalibacterium prausnitzii* is an anti-inflammatory commensal bacterium identified by gut microbiota analysis of Crohn disease patients. Proc Natl Acad Sci USA 105(43):16731–16736

Tabuchi M, Tamura A, Yamada N, Ishida T, Hosoda M, Hosono A (2004) Hypocholesterolemic effects of viable and heat-sterilized cells of *Lactobacillus GG* in rats fed a high-cholesterol diet. Milchwissenschaft 59(5–6):249–253

Taylor RJD (2008) Pathogenesis of type 2 diabetes: tracing the reverse route from cure to cause. Diabetologia 51(10):1781–1789

Tuomilehto J, Lindström J, Eriksson JG, Valle TT, Hämäläinen H, Ilanne-Parikka P, Keinänen-Kiukaanniemi S, Laakso M, Louheranta A, Rastas M, Salminen V, Aunola S, Cepaitis Z, Moltchanov V, Hakumäki M, Mannelin M, Martikkala V, Sundvall J, Uusitupa M (2001) Prevention of type 2 diabetes mellitus by changes in lifestyle among subjects with impaired glucose tolerance. N Engl J Med 344(18):1343–1350

Turnbaugh PJ, Ley RE, Mahowald MA, Magrini V, Mardis ER, Gordon JI (2006) An obesity-associated gut microbiome with increased capacity for energy harvest. Nature 444 (7122):1027–1031

Turnbaugh PJ, Hamady M, Yatsunenko T, Cantarel BL, Duncan A, Ley RE, Sogin ML, Jones WJ, Roe BA, Affourtit JP, Egholm M, Henrissat B, Heath AC, Knight R, Gordon JI (2008) A core gut microbiome in obese and lean twins. Nature 457:480

Wang Z, Klipfell E, Bennett BJ, Koeth R, Levison BS, DuGar B, Feldstein AE, Britt EB, Fu X, Chung Y-M, Wu Y, Schauer P, Smith JD, Allayee H, Tang WHW, DiDonato JA, Lusis AJ, Hazen SL (2011) Gut flora metabolism of phosphatidylcholine promotes cardiovascular disease. Nature 472:57

Xie N, Cui Y, Yin Y-N, Zhao X, Yang J-W, Wang Z-G, Fu N, Tang Y, Wang X-H, Liu X-W, Wang C-L, Lu F-G (2011) Effects of two Lactobacillus strains on lipid metabolism and intestinal microflora in rats fed a high-cholesterol diet. BMC Complement Altern Med 11:53–53

Yadav R, Vij R, Kapila S, Khan SH, Kumar N, Meena S, Kapila R (2019) Milk fermented with probiotic strains *Lactobacillus rhamnosus* MTCC5957 and *Lactobacillus rhamnosus* MTCC5897 ameliorates the diet-induced hypercholesterolemia in rats. Ann Microbiol 69 (5):483–494

Yin Y-N, Yu Q-F, Fu N, Liu X-W, Lu F-G (2010) Effects of four bifidobacteria on obesity in high-fat diet induced rats. World J Gastroenterol 16(27):3394–3401

Zanella MT, Ribeiro Filho FF (2009) Emerging drugs for obesity therapy. Arq Bras Endocrinol Metabol 53:271–280

Fighting Microbes with Microbes 19

Boopathi Seenivasan, Chiranth M. Prakash, and Vani Janakiraman

Abstract

Microbes exhibit a strong association with human beings by colonizing different parts of the body. These microbes can be either beneficial or harmful. Pathogenic microbes are known to cause serious infections in humans and in other multicellular organisms which disturb the host physiology. These pathogenic microbes have intrinsic traits which contribute to their survival under hostile conditions, evasion of host immune responses and resistance to various therapeutic agents which in turn confers them with near invincibility. Therefore, exploration of novel agents which could specifically target and kill microbes is very much on the demand. Interestingly, one such agent could be microbes themselves. Utilizing microbial components and/or microbial whole cells either to target pathogens directly or at modulating the biological fitness of the host including boosting host immune responses. In this chapter, we discuss these various modes by which microbes and their products could be employed in combating microbial infections, eventually to improve healthcare.

Keywords

Bacteria · Gut flora · Commensal · Faecal microbial transplantation

B. Seenivasan · C. M. Prakash · V. Janakiraman (✉)
Department of Biotechnology, Bhupat and Jyoti Mehta School of Biosciences, Indian Institute of Technology Madras, Chennai, India
e-mail: vani@iitm.ac.in

S. G. Sharma et al. (eds.), *Microbial Diversity, Interventions and Scope*,
https://doi.org/10.1007/978-981-15-4099-8_19

19.1 Introduction

Diverse microbial communities establish an intimate association with the human body. This complex mixture of microbial communities encompasses high density of bacteria along with archaea, fungi, protozoa and viruses (Sommer and Bäckhed 2013). This human-microbiome interaction begins in the early stages of birth and lasts throughout life (Sommer and Bäckhed 2013). Also, upon environmental exposure microbes colonize in different parts of human body, including skin, oral cavity, gastrointestinal tract and vaginal sites (Dethlefsen et al. 2007).

Gastrointestinal tract of human harbours 10–100 trillion microbes that belong to Firmicutes and Bacteroidetes phyla (Turnbaugh et al. 2007). Gut microbes play a central role in protecting the body by contributing to intestinal homeostasis, metabolism and host immunity (Sommer and Bäckhed 2013). These gut microbes also protect the host from pathogenic microorganisms (Hooper and Macpherson 2010). It is important to note that health and disease conditions are critically driven by the balance between beneficial and pathogenic bacterial populations in addition to the host responses. Perturbations in the gut microbial community by pathogens or other factors result in dysbiosis that is associated with high risk of celiac disease, gastric cancer, autism, obesity, anorexia and Crohn's disease (Clemente et al. 2012). Broad-spectrum antibiotics kill commensal bacteria and thus collapse the microbial community structure in the body. Overuse of antibiotics for any microbial infection not only leads to the development of antibiotic resistant microbes but also hampers the gastrointestinal microbial homeostasis which in turn becomes an etiological factor for obesity, diabetes, asthma (O'Toole and Gautam 2018). It is estimated that death toll caused by antibiotic resistance microbes will escalate to 10 million in 2050 (De Kraker et al. 2016). Therefore, there is an urgent need to substitute the present intervention strategies for efficient control of infection-borne diseases. Employing commensal microbes or probiotics or molecules derived from them in controlling diseases or infection could be an attractive strategy. Further, the recent developments in genetic engineering such as CRISPR-Cas9 systems seem to offer promise for designing improved antimicrobials. In this chapter, we summarize how microbes could be used as direct tools to tackle pathogenic microbes or as indirect tools to modulate the host responses to mitigate pathogenesis.

19.2 Faecal Microbial Transplantation (FMT)

As mentioned earlier, gut microbial dysbiosis contributes to the development of different disorders. Faecal microbial transplantation (FMT) is one of the approaches that is being tried as an intervention strategy to restore the microbial balance. Clinical studies using FMT to restore microbial diversity for treating *Clostridium difficile*-associated colitis have been proven successful (Borody et al. 2003; De Leon et al. 2013). Human intestine has a complex microbial diversity in which most of the bacterial population has not been unveiled. In fact, composition of the microbial population in healthy donors varies from individual to individual. Therefore, it is imperative to design microbial therapy employing bacterial population with proven beneficial effects to precise out treatment (Skelly et al. 2019). Once native microbiota or probiotics arrive the destination in the host body, with the newly achieved balance, an array of mechanisms are executed by the bacteria, which could impact the pathogens by direct or indirect mechanisms (Pickard et al. 2017). Delivering of microbes through FMT effectively eliminate harmful microbes by competing for nutrients and producing antimicrobial molecules (bacteriocins).

19.3 Effects of Beneficial Microbes Against Pathogens

19.3.1 Direct Effects Against Pathogens

- In a polymicrobial community, commensal microbes compete with other microbes for niche and nutrients. Bacteria produce antibacterial molecules called bacteriocins, that are varied in structure and chemical modification, could be used as on alternate to treat antibiotic-resistant bacteria (Cotter et al. 2013). This proteinaceous bacteriocins kill the target organisms by affecting cell membrane or cell wall integrity or by modulating metabolic pathways (Hols et al. 2019). Bacteriocins are classified into class I, class II, class III and class IV based on physiochemical properties (Ahmed et al. 2017). Commensal bacteria are known to produce novel types of antibacterial molecules (Donia and Fischbach 2015) which in turn provides several clinical advantages. Donia et al. (2014) discovered 3118 biosynthetic gene clusters from the human microbial genomes, and they showed that genes encoding antibacterial molecules are found throughout the genomes and metagenomes of human microbiota. This gives an idea that commensal microbiome could be a treasure to explore novel therapeutic molecules. Rigorous studies on the natural products of commensal bacteria are being expedited due to the novel and versatile nature (Table.19.1). Recently, it was found that commensal bacteria can produce a novel class of antibiotics in which 7α-dehydroxylating bacteria utilize tryptophan as a precursor to produce 1-acetyl-β-carboline and turbomycin A antibiotics that inhibit the growth of *C. difficile* in the presence of secondary bile acid deoxycholic acid (Kang et al. 2019). Thus, antimicrobials of commensal bacteria seem to play a vital role in colonization resistance.

Bacteria compete with other microbes either by producing antibacterial molecules in the extracellular milieu or by delivering effector proteins into the competitors in a contact-dependent manner using type VI secretory system (T6SS) (Benz and Meinhart 2014). It is increasingly becoming apparent that T6SS is also crucially involved in the inter-bacterial competition and host-microbes interaction (Jani and Cotter 2010). Bacteria cause a detrimental effects on their competitor by delivering effector proteins such as peptidoglycan-degrading enzymes (viz. amidase/glycosidase), membrane-attacking enzymes (viz. phospholipase) and nucleic-acid-degrading enzymes (viz. DNase) through T6SS (Russell et al. 2014a). Previously, it was believed that only pathogenic microbes are attributed to have such specialized

Table 19.1 List of commensal bacteria-derived products against different kind of potential pathogens

Molecule name	Site of source in host	Source	Antimicrobial activity	References
Colicin V	Human faeces	Metagenomic library	*E. coli*	Cohen et al. (2017)
Lugdunin	Nasal	*Staphylococcus lugdunensis*	Methicillin-resistant *S. aureus*, glycopeptide intermediate resistant *S. aureus* and vancomycin-resistant *Enterococcus* isolates	Zipperer et al. (2016)
Salivaricin D	Faeces from healthy infant	*Streptococcus salivarius* 5M6c	*Streptococcus pyogenes* and *Streptococcus pneumoniae*	Birri et al. (2012)
Salivaricin A2 Salivaricin B	Oral	*Streptococcus salivarius* strain K12	*Enterococcus faecalis* ATCC	Hyink et al. (2007)
Ruminococcin C	Human faeces	*Ruminococcus gnavus* E1	*Clostridium perfringens*	Pujol et al. (2011)
Microcin 25,	Infant faeces	*E. coli* AY25	*Shigella flexneri*, *Salmonella enteritidis* and *E. coli*	Salomon and Farias (1992)
1-Acetyl-β-carboline	Human faeces	*Clostridium scindens*	*Clostridium difficile*	Kang et al. (2019)

T6SS system to deliver toxic proteins. A recent study postulates that even Bacteroidetes, the most prevalent of the gut microbes, might also utilize T6SS to exhibit colonization resistance and to maintain the community structure of the commensal microbes (Russell et al. 2014b).

- Well-known mechanisms by which commensal microbes eliminate the invading pathogens are by limiting the space and nutrients. Commensal bacteria sequester nutrients that are needed for the growth of pathogenic microorganisms, which consequently affect the growth and efficacy of the colonization of the pathogens (Kamada et al. 2013). For instance, commensal *E. coli* acquires the amino acid proline efficiently and impedes the growth of shiga toxin-producing *E. coli* O157: H7 that is well known as a food-borne pathogen (Momose et al. 2008). Maltby et al. (2013) have demonstrated that commensal *E. coli* HS and *E. coli* Nissle 1917 utilize the available sugars and restrict the access to the nutrients for pathogenic *E.coli* O157: H7, thereby they exhibit colonization resistance. Likewise, probiotic *E.coli* Nissle inhibits the colonization of *Salmonella* Typhimurium by acquiring iron and creates nutrient-limited conditions which ultimately impact the growth of the pathogen (Deriu et al. 2013). Similarly, commensal *Mucispirillum schaedleri* competes with *Salmonella enterica serovar* Typhimurium possibly for nitrate and formate, thereby obstructing pathogen invasion and virulence gene expression (Herp et al. 2019). Commensal-derived indole can inhibit the epithelial cell attachment of *Salmonella* Typhimurium (Kohli et al. 2018) and *Candida albicans* (Oh et al. 2012). Commensal Enterobacteriaceae and Clostridia create epithelial hypoxia by consuming available oxygen which in turn inhibit the growth of the enteric *Salmonella enterica* serovar Enteritidis (Litvak et al. 2019).

- Bacteria communicate with each other through signalling molecules in order to regulate gene expression upon population density, termed as quorum sensing (QS) (Waters and Bassler 2005). Gram-positive bacteria utilize pheromone-like signalling peptides as a QS signalling molecules, whereas Gram-negative bacteria produce acyl homoserine lactones. Though QS signals are species or genus specific, autoinducer-2 (AI-2) is recognized as an interspecies communicating signal that is being produced, detected and responded across the bacterial kingdom (Bassler 2002). Pathogenic bacteria employ QS mechanism for the production of virulence factors. Many reports evidently have shown that intestinal commensal bacteria produce AI-2 molecules (Lukáš et al. 2008; Thompson et al. 2015). Hsiao et al. (2014) have shown that AI-2 of commensal *Ruminococcus obeum* plays an imperative role in attenuating the virulence factors of *V. cholera*. Recent evidence suggests that probiotic *Lactobacillus sakei* NR28 exhibits quorum quenching activity against enterohaemorrhagic *E. coli* O157:H7 by inhibiting AI-2 activity and virulence factors (Park et al. 2014). Since most of the intestinal pathogens utilize AI-2 to regulate virulence genes expression, it widens the opportunity to utilize quorum quenching compounds for therapeutic interventions. Indeed, analogues of the QS signals, competitively binding with respective receptors of the pathogen, can be manipulated to attenuate the virulence factors of the pathogens (Boopathi et al. 2017). Paharik et al. (2017) have demonstrated that autoinducing peptide (AIP) from commensal *Staphylococcus caprae* competitively bind with the receptors of *Staphylococcus aureus* which in turn reduces pathogen burden and intradermal infection. It was found that AIP of *S. caprae* binds with AgrC, a ligand-binding receptor protein *S. aureus*, which in turn affects the capability of *S. aureus* to defend the host immune system.

19.3.2 Indirect Effects Against Pathogens

In addition to the direct mechanisms, commensal microbes restrict the colonization of pathogens by indirect mechanisms: (1) stimulate the production of antimicrobial peptides of the host, (2) obstruct the translocation of pathogen into epithelial cells by restoring the function of tight junction and (3) enhance innate and adaptive immune system.

In addition to the direct mechanisms, commensal microbes restrict the colonization of pathogens by indirect mechanisms, in which commensal bacteria stimulate the production of antimicrobial peptides of the host, obstruct the translocation of pathogen into epithelial cells by restoring the function of tight junction and enhance innate and adaptive immune system.

Clostridium difficile is a Gram-positive anaerobic spore-forming bacterium which causes severe diarrhoea in the patients who had antibiotic treatments. Commensal microbiota converts the taurocholate (primary bile salt) into deoxycholate (secondary bile salt) using 7α-dehydroxylase. These secondary bile salts inhibit the germination of *C. difficile* spores (Buffie et al. 2015), whereas primary bile salts promote spore germination of *C. difficile* (Sorg and Sonenshein, 2008). It implies that the balance between primary and secondary bile salts determine the colonization of *C. difficile*.

19.3.2.1 Immunomodulatory Effects

Regulation of immune homeostasis is important to restrict colonization of pathogens. Commensal microbes enhance host immune responses by regulating the functions of macrophages, neutrophils and T cells. Following are the examples of commensal microbes mediated immune modulatory effects against pathogens:

- Intestinal epithelial cells produce antimicrobial peptides that control intestinal homeostasis. *Bifidobacterium breve NCC2950* induces the expression of antimicrobial peptide RegIIIγ, a peptidoglycan-binding C-type lectin that inhibit Gram-positive bacteria (Natividad et al. 2013).
- Butyrate plays a key role in triggering the antimicrobial activity of the macrophages by modulating the metabolism, in which reduced mTOR kinase activity and increased LC3-mediated host defence lead to resistance against enteropathogens (Schulthess et al. 2019). Commensal-derived butyrate controls the growth of *Salmonella enterica serovar* Typhimurium by inducing the expression of antimicrobial calprotectin, ROS and autophagosome formation in macrophages (Schulthess et al. 2019).
- Uptake of molecules occurs in the gastrointestinal tract through the intestinal barrier, in which tight-junction components play a key role in the transport of ions and small molecules (Vancamelbeke and Vermeire 2017). Toxins such as TcdA and TCdB of *C. difficile* disrupt the connection between zonula occludens-1 and actin filaments, which consequently increase the paracellular permeability (Nusrat et al. 2001). Commensal-derived butyrate diminishes intestinal inflammation and promotes intestinal barrier function in *C. difficile*-infected mice by activating transcription factor HIF-1 in intestinal epithelial cells, which in turn decreases the epithelial permeability and translocation of the pathogen, thereby giving protection against *C. difficile* toxin-induced colitis (Fachi et al. 2019).
- Epithelial layer is covered with mucus layer that physically acts as a barrier hindering the microorganisms from accessing the cell directly (Cornick et al. 2015). Glycoproteins such as mucins are differentially expressed in the mucus layer. Pathogens such as *C. difficile* decrease the production of MUC2, which is an important component of the mucus layer whereas *Bifidobacterium* and *E. coli* protect the host by triggering the production of mucins (Libertucci and Young 2019).

19.3.2.2 Effect on Distal Organs

Recent studies have highlighted that gut microbiota do cross-talk with distantly located organs such as brain, liver, lung, bone and heart (Feng et al. 2018). Perturbations in the structure of the gut microbial community and function, caused by diet, stress or diseased condition, lead to dysbiosis that facilitates microbial-derived products to get into the circulatory system (Jacobs et al. 2017). Plethora of evidences have highlighted the protective role of gut microbiota from pulmonary diseases. For instance, the gut of the germ-free mice colonized with probiotic *Bifidobacterium longum* 5^{1A} protects the host from pulmonary infection by reducing the load of *Klebsiella pneumoniae* (Vieira et al. 2016). Similarly, commensal microbiota-administered mice have shown resistance against *Streptococcus pneumoniae* (Schuijt et al. 2016) and *Escherichia coli* K1 (Deshmukh et al. 2014). Additionally, certain bacteria also produce enzymes such as glycosidases, β-glucuronidase, azoreductases and nitroreductases that can transform procarcinogenic molecules into active carcinogens. Probiotic *Bifidobacterium adolescentis* SPM0212 can inhibit the functions of harmful faecal enzyme and colon cancer cell line proliferation (Kim et al. 2008).

19.4 Emerging Therapeutics Approaches

19.4.1 Phage Therapy

Phage therapy is the direct administration of bacteriophages into the patients in order to target a particular bacterial infection. The mode of using bacteriophages in therapy began way back in the year 1919, but the discovery of antibiotics as a potent therapeutic agent stalled the advancements of phage therapy (Altamirano and Barr 2019). However, with the rampant increase in emergence of antibiotic resistant strains, phage therapy has again gained limelight. Phages are non-living biological entities which consist of a protein capsid which encloses its nucleic acid. The strain-specific receptors help the phages determine its host to which they inject their nucleic acid, which then manipulates the bacterial machinery to produce more phage particles and eventually lyse and destroy the host cells (Lin et al. 2017). The presence of highly strain-specific receptors in the phage is one of the most important features in phage therapy.

Technological improvements have made way for the development of phage cocktails which precisely act upon an array of pathogens (Altamirano and Barr 2019). Engineering the phages to express biofilm-degrading enzymes, such as EPS depolymerase (Hughes et al. 1998) and Dispersin B (Lu and Collins 2007), has shown to help the phages to seep into the biofilm matrix and act on biofilm-based bacterial infections. Similarly bacteriophages are also engineered to overexpress endolysins which degrade the peptidoglycan layer of species-specific bacteria (Borysowski et al. 2006; Gervasi et al. 2014).

The combined use of recombinant phages with antibiotics has been experimentally shown to enhance the killing of resistant bacteria by many folds (Lu and Collins

2009). Human trials employing phage therapy conducted on clinically significant diseases like shigellosis, cholera, typhoid, chronic otitis and MDR *S. aureus*-induced diabetic foot ulcer have shown great success (Lin et al. 2017). Multiple studies have shown that phages do not disrupt host metabolism and microbiota and do not activate the host immune system (Chan et al. 2013). The development of phage cocktail which precisely act upon an array of pathogens has broadened phage therapy against various bacterial infections. In a recent report, a cocktail of three phages was used to completely treat a patient suffering from cystic fibrosis with disseminated *Mycobacterium abscessus* infection (Dedrick et al. 2019). As of date, few bacteriophages have been approved for clinical usage, while in the next few years one can witness a surge in phage-mediated therapies (Ghosh et al. 2019).

19.4.2 Predatory Bacteria

Bdellovibrio bacteriovorus and *Micavibrio aeruginosavorus* are two Gram-negative bacteria belonging to the phylum proteobacteria. These two species are gaining interest for their abilities to combat antimicrobial resistant Gram-negative bacteria (Dashiff et al. 2011) (Kadouri et al. 2013). Species *Bdellovibrio* attach, invade and lyse other Gram-negative bacteria, while the *Micavibrio* species attach, feed itself externally and divide without entering the prey bacteria. Interestingly both the predatory bacteria have shown their potential to prey on ESKAPE pathogens (*Enterobacter genus, Staphylococcus aureus, Klebsiella pneumoniae, Acinetobacter baumannii, Pseudomonas aeruginosa and Escherichia coli*) which are a threat to human health (Negus et al. 2017). Studies have shown that predatory bacteria do not induce pro-inflammatory response in human macrophage cell lines and epithelial cell lines and show no cytotoxic response (Gupta et al. 2016; Monnappa et al. 2016). In addition, the bacteria were found to show a modest increase in the levels of pro-inflammatory cytokines in mouse upon infection through intranasal and intravenous routes. Thus, predatory bacteria could be used as an ideal biological agent to fight drug-resistant bacterial infections (Negus et al. 2017).

19.4.3 Fighting Pathogens Through Engineered Bacteria

Several bacterial species have been engineered to overexpress protein(s) which in turn can aid in killing or inhibiting other microbes. A potent anti-HIV protein CV-N which is originally synthesized by *Nostoc ellipsosporum* was made to overexpress in a commensal *L. jensenii* (Bolmstedt et al. 2001). The engineered *L. jensenii* which colonizes and forms biofilms in the vaginal mucosa was found to continuously secrete CV-N and prevent HIV transmission in macaques (Lagenaur et al. 2011). *L. jensenii* engineered to overexpress and deliver RANTES and C1C5 RANTES has also been proposed to be an ideal system to block HIV-1 infection (Secchi et al. 2009).

The empty envelopes of Gram-negative bacteria also called as ghosts have been found to activate broad range of cell types in innate and adaptive immunity. Different types of antigens when loaded into the cytoplasmic lumen and periplasmic space of the ghosts have demonstrated the ability of ghosts as a vaccine candidate and also as an adjuvant. The potential of bacterial ghosts as vaccines and as potential adjuvants against various diseases is reviewed elsewhere (Lubitz et al. 2009; Hajam et al. 2017).

19.5 Conclusion

Surging evidence suggest that microbes could be efficiently put to use as therapeutic agents to control infectious diseases. Strong evidence about the gut-living commensal bacteria governing the fate of microbial infections at different parts of the body accelerate the hope for designing better therapeutic interventions using microbial agents. Further, genetically modified beneficial microbes could be used for targeted delivery of specific molecules to tackle pathogens. Applying the principle of quorum-sensing inhibition as a therapeutic intervention is also a promising strategy to attenuate pathogenesis of the antibiotic-resistant bacteria. Thus, we surmise that microbes are endowed with potentials and hence could be explored for precision medicine for infection.

References

Ahmad V, Khan MS, Jamal QMS, Alzohairy MA, Al Karaawi MA, Siddiqui MU (2017) Antimicrobial potential of bacteriocins: in therapy, agriculture and food preservation. Int J Antimicrob Agents 49(1):1–11

Altamirano FLG, Barr JJ (2019) Phage therapy in the postantibiotic era. Clin Microbiol Rev 32(2): e00066–e00018

Bassler BL (2002) Small talk: cell-to-cell communication in bacteria. Cell 109(4):421–424

Benz J, Meinhart A (2014) Antibacterial effector/immunity systems: it's just the tip of the iceberg. Curr Opin Microbiol 17:1–10

Birri DJ, Brede DA, Nes IF (2012) Salivaricin D, a novel intrinsically trypsin-resistant antibiotic from *Streptococcus salivarius* 5M6c isolated from a healthy infant. Appl Environ Microbiol 78 (2):402–410

Bolmstedt AJ, O'Keefe BR, Shenoy SR, McMahon JB, Boyd MR (2001) Cyanovirin-N defines a new class of antiviral agent targeting N-linked, high-mannose glycans in an oligosaccharide-specific manner. Mol Pharmacol 59(5):949–954

Boopathi S, Vashisth R, Manoharan P, Kandasamy R, Sivakumar N (2017) Stigmatellin Y–an anti-biofilm compound from *Bacillus subtilis* BR4 possibly interferes in PQS–PqsR mediated quorum sensing system in *Pseudomonas aeruginosa*. Bioorg Med Chem Lett 27 (10):2113–2118

Borody TJ, Warren EF, Leis S, Surace R, Ashman O (2003) Treatment of ulcerative colitis using fecal bacteriotherapy. J Clin Gastroenterol 37(1):42–47

Borysowski J, Weber-Dąbrowska B, Górski A (2006) Bacteriophage endolysins as a novel class of antibacterial agents. Exp Biol Med 231(4):366–377

Buffie CG, Bucci V, Stein RR, McKenney PT, Ling L, Gobourne A et al (2015) Precision microbiome reconstitution restores bile acid mediated resistance to *Clostridium difficile*. Nature 517(7533):205

Chan BK, Abedon ST, Loc-Carrillo C (2013) Phage cocktails and the future of phage therapy. Future Microbiol 8(6):769–783

Clemente JC, Ursell LK, Parfrey LW, Knight R (2012) The impact of the gut microbiota on human health: an integrative view. Cell 148(6):1258–1270

Cohen LJ, Han S, Huang YH, Brady SF (2017) Identification of the colicin V bacteriocin gene cluster by functional screening of a human microbiome metagenomic library. ACS infect Dis 4 (1):27–32

Cornick S, Tawiah A, Chadee K (2015) Roles and regulation of the mucus barrier in the gut. Tissue Barriers 3(1–2):e982426

Cotter PD, Ross RP, Hill C (2013) Bacteriocins—a viable alternative to antibiotics? Nat Rev Microbiol 11(2):95

Dashiff A, Junka RA, Libera M, Kadouri DE (2011) Predation of human pathogens by the predatory bacteria *Micavibrio aeruginosavorus* and *Bdellovibrio bacteriovorus*. J Appl Microbiol 110(2):431–444

De Kraker ME, Stewardson AJ, Harbarth S (2016) Will 10 million people die a year due to antimicrobial resistance by 2050? PLoS Med 13(11):e1002184

De Leon LM, Watson JB, Kelly CR (2013) Transient flare of ulcerative colitis after fecal microbiota transplantation for recurrent *Clostridium difficile* infection. Clin Gastroenterol Hepatol 11 (8):1036–1038

Dedrick RM, Guerrero-Bustamante CA, Garlena RA, Russell DA, Ford K, Harris K et al (2019) Engineered bacteriophages for treatment of a patient with a disseminated drug-resistant *Mycobacterium abscessus*. Nat Med 25(5):730

Deriu E, Liu JZ, Pezeshki M, Edwards RA, Ochoa RJ, Contreras H et al (2013) Probiotic bacteria reduce *Salmonella typhimurium* intestinal colonization by competing for iron. Cell Host Microbe 14(1):26–37

Deshmukh HS, Liu Y, Menkiti OR, Mei J, Dai N, O'leary CE et al (2014) The microbiota regulates neutrophil homeostasis and host resistance to *Escherichia coli* K1 sepsis in neonatal mice. Nat Med 20(5):524

Dethlefsen L, McFall-Ngai M, Relman DA (2007) An ecological and evolutionary perspective on human–microbe mutualism and disease. Nature 449(7164):811

Donia MS, Fischbach MA (2015) Small molecules from the human microbiota. Science 349 (6246):1254766

Donia MS, Cimermancic P, Schulze CJ, Brown LCW, Martin J, Mitreva M et al (2014) A systematic analysis of biosynthetic gene clusters in the human microbiome reveals a common family of antibiotics. Cell 158(6):1402–1414

Fachi JL, de Souza Felipe J, Pral LP, da Silva BK, Corrêa RO, de Andrade MCP et al (2019) Butyrate protects mice from *Clostridium difficile*-induced colitis through an HIF-1-dependent mechanism. Cell Rep 27(3):750–761

Feng Q, Chen WD, Wang YD (2018) Gut microbiota: an integral moderator in health and disease. Front Microbiol 9:151

Gervasi T, Horn N, Wegmann U, Dugo G, Narbad A, Mayer MJ (2014) Expression and delivery of an endolysin to combat *Clostridium perfringens*. Appl Microbiol Biotechnol 98(6):2495–2505

Ghosh C, Sarkar P, Issa R, Haldar J (2019) Alternatives to conventional antibiotics in the era of antimicrobial resistance. Trends Microbiol 27:323

Gupta S, Tang C, Tran M, Kadouri DE (2016) Effect of predatory bacteria on human cell lines. PLoS One 11(8):e0161242

Hajam IA, Dar PA, Won G, Lee JH (2017) Bacterial ghosts as adjuvants: mechanisms and potential. Vet Res 48(1):37

Herp S, Brugiroux S, Garzetti D, Ring D, Jochum LM, Beutler M et al (2019) Mucispirillum schaedleri antagonizes *Salmonella* virulence to protect mice against colitis. Cell Host Microbe 25(5):681–694

Hols P, Ledesma-García L, Gabant P, Mignolet J (2019) Mobilization of microbiota commensals and their bacteriocins for therapeutics. Trends Microbiol 27:690. https://doi.org/10.1016/j.tim.2019.03.007

Hooper LV, Macpherson AJ (2010) Immune adaptations that maintain homeostasis with the intestinal microbiota. Nat Rev Immunol 10(3):159

Hsiao A, Ahmed AS, Subramanian S, Griffin NW, Drewry LL, Petri WA et al (2014) Members of the human gut microbiota involved in recovery from *Vibrio cholerae* infection. Nature 515 (7527):423

Hughes KA, Sutherland IW, Jones MV (1998) Biofilm susceptibility to bacteriophage attack: the role of phage-borne polysaccharide depolymerase. Microbiology 144(11):3039–3047

Hyink O, Wescombe PA, Upton M, Ragland N, Burton JP, Tagg JR (2007) Salivaricin A2 and the novel lantibiotic salivaricin B are encoded at adjacent loci on a 190-kilobase transmissible megaplasmid in the oral probiotic strain *Streptococcus salivarius* K12. Appl Environ Microbiol 73(4):1107–1113

Jacobs MC, Haak BW, Hugenholtz F, Wiersinga WJ (2017) Gut microbiota and host defense in critical illness. Curr Opin Crit Care 23(4):257–263

Jani AJ, Cotter PA (2010) Type VI secretion: not just for pathogenesis anymore. Cell Host Microbe 8(1):2–6

Kadouri DE, To K, Shanks RM, Doi Y (2013) Predatory bacteria: a potential ally against multidrug-resistant Gram-negative pathogens. PLoS One 8(5):e63397

Kamada N, Chen GY, Inohara N, Núñez G (2013) Control of pathogens and pathobionts by the gut microbiota. Nat Immunol 14(7):685

Kang JD, Myers CJ, Harris SC, Kakiyama G, Lee IK, Yun BS et al (2019) Bile acid 7-α-dehydroxylating gut bacteria secrete antibiotics that inhibit *Clostridium difficile*: role of secondary bile acids. Cell Chem Biol 26(1):27–34

Kim Y, Lee D, Kim D, Cho J, Yang J, Chung M et al (2008) Inhibition of proliferation in colon cancer cell lines and harmful enzyme activity of colon bacteria by *Bifidobacterium adolescentis* SPM0212. Arch Pharm Res 31(4):468

Kohli N, Crisp Z, Riordan R, Li M, Alaniz RC, Jayaraman A (2018) The microbiota metabolite indole inhibits *Salmonella* virulence: involvement of the PhoPQ two-component system. PLoS One 13(1):e0190613

Lagenaur LA, Sanders-Beer BE, Brichacek B, Pal R, Liu X, Liu Y et al (2011) Prevention of vaginal SHIV transmission in macaques by a live recombinant *Lactobacillus*. Mucosal Immunol 4(6):648

Libertucci J, Young VB (2019) The role of the microbiota in infectious diseases. Nat Microbiol 4 (1):35

Lin DM, Koskella B, Lin HC (2017) Phage therapy: an alternative to antibiotics in the age of multi-drug resistance. World J Gastrointest Pharmacol Ther 8(3):162

Litvak Y, Mon KK, Nguyen H, Chanthavixay G, Liou M, Velazquez EM et al (2019) Commensal Enterobacteriaceae protect against *Salmonella* colonization through oxygen competition. Cell Host Microbe 25(1):128–139

Lu TK, Collins JJ (2007) Dispersing biofilms with engineered enzymatic bacteriophage. Proc Natl Acad Sci 104(27):11197–11202

Lu TK, Collins JJ (2009) Engineered bacteriophage targeting gene networks as adjuvants for antibiotic therapy. Proc Natl Acad Sci 106(12):4629–4634

Lubitz P, Mayr UB, Lubitz W (2009) Applications of bacterial ghosts in biomedicine, Pharmaceutical biotechnology. Springer, New York, pp 159–170

Lukáš F, Gorenc G, Kopečný J (2008) Detection of possible AI-2-mediated quorum sensing system in commensal intestinal bacteria. Folia Microbiol 53(3):221–224

Maltby R, Leatham-Jensen MP, Gibson T, Cohen PS, Conway T (2013) Nutritional basis for colonization resistance by human commensal *Escherichia coli* strains HS and Nissle 1917 against *E. coli* O157: H7 in the mouse intestine. PloS one 8(1):e53957

Momose Y, Hirayama K, Itoh K (2008) Competition for proline between indigenous *Escherichia coli* and *E. coli* O157: H7 in gnotobiotic mice associated with infant intestinal microbiota and its contribution to the colonization resistance against *E. coli* O157: H7. Antonie Van Leeuwenhoek 94(2):165–171

Monnappa AK, Bari W, Choi SY, Mitchell RJ (2016) Investigating the responses of human epithelial cells to predatory bacteria. Sci Rep 6:33485

Natividad JM, Hayes CL, Motta JP, Jury J, Galipeau HJ, Philip V et al (2013) Differential induction of antimicrobial REGIII by the intestinal microbiota and *Bifidobacterium breve* NCC2950. Appl Environ Microbiol 79(24):7745–7754

Negus D, Moore C, Baker M, Raghunathan D, Tyson J, Sockett RE (2017) Predator versus pathogen: how does predatory *Bdellovibrio bacteriovorus* interface with the challenges of killing gram-negative pathogens in a host setting? Annu Rev Microbiol 71:441–457

Nusrat A, von Eichel-Streiber C, Turner JR, Verkade P, Madara JL, Parkos CA (2001) *Clostridium difficile* toxins disrupt epithelial barrier function by altering membrane microdomain localization of tight junction proteins. Infect Immun 69(3):1329–1336

Oh S, Go GW, Mylonakis E, Kim Y (2012) The bacterial signalling molecule indole attenuates the virulence of the fungal pathogen *Candida albicans*. J Appl Microbiol 113(3):622–628

O'Toole RF, Gautam SS (2018) The host microbiome and impact of tuberculosis chemotherapy. Tuberculosis 113:26–29

Paharik AE, Parlet CP, Chung N, Todd DA, Rodriguez EI, Van Dyke MJ et al (2017) Coagulase-negative staphylococcal strain prevents *Staphylococcus aureus* colonization and skin infection by blocking quorum sensing. Cell Host Microbe 22(6):746–756

Park H, Yeo S, Ji Y, Lee J, Yang J, Park S et al (2014) Autoinducer-2 associated inhibition by *Lactobacillus sakei* NR28 reduces virulence of enterohaemorrhagic *Escherichia coli* O157: H7. Food Control 45:62–69

Pickard JM, Zeng MY, Caruso R, Núñez G (2017) Gut microbiota: role in pathogen colonization, immune responses, and inflammatory disease. Immunol Rev 279(1):70–89

Pujol A, Crost EH, Simon G, Barbe V, Vallenet D, Gomez A, Fons M (2011) Characterization and distribution of the gene cluster encoding RumC, an anti-*Clostridium perfringens* bacteriocin produced in the gut. FEMS Microbiol Ecol 78(2):405–415

Russell AB, Peterson SB, Mougous JD (2014a) Type VI secretion system effectors: poisons with a purpose. Nat Rev Microbiol 12(2):137

Russell AB, Wexler AG, Harding BN, Whitney JC, Bohn AJ, Goo YA et al (2014b) A type VI secretion-related pathway in Bacteroidetes mediates interbacterial antagonism. Cell Host Microbe 16(2):227–236

Salomon RA, Farías RN (1992) Microcin 25, a novel antimicrobial peptide produced by *Escherichia coli*. J Bacteriol 174(22):7428–7435

Schuijt TJ, Lankelma JM, Scicluna BP, de Sousa e Melo F, Roelofs JJ, de Boer JD et al (2016) The gut microbiota plays a protective role in the host defence against pneumococcal pneumonia. Gut 65(4):575–583

Schulthess J, Pandey S, Capitani M, Rue-Albrecht KC, Arnold I, Franchini F et al (2019) The short chain fatty acid butyrate imprints an antimicrobial program in macrophages. Immunity 50 (2):432–445

Secchi M, Xu Q, Lusso P, Vangelista L (2009) The superior folding of a RANTES analogue expressed in lactobacilli as compared to mammalian cells reveals a promising system to screen new RANTES mutants. Protein Expr Purif 68(1):34–41

Skelly AN, Sato Y, Kearney S, Honda K (2019) Mining the microbiota for microbial and metabolite-based immunotherapies. Nat Rev Immunol 19:1

Sommer F, Bäckhed F (2013) The gut microbiota—masters of host development and physiology. Nat Rev Microbiol 11(4):227

Sorg JA, Sonenshein AL (2008) Bile salts and glycine as co germinants for *Clostridium difficile* spores. J Bacteriol 190(7):2505–2512

Thompson JA, Oliveira RA, Djukovic A, Ubeda C, Xavier KB (2015) Manipulation of the quorum sensing signal AI-2 affects the antibiotic-treated gut microbiota. Cell Rep 10(11):1861–1871

Turnbaugh PJ, Ley RE, Hamady M, Fraser-Liggett CM, Knight R, Gordon JI (2007) The human microbiome project. Nature 449(7164):804

Vancamelbeke M, Vermeire S (2017) The intestinal barrier: a fundamental role in health and disease. Expert Rev Gastroenterol Hepatol 11(9):821–834

Vieira AT, Rocha VM, Tavares L, Garcia CC, Teixeira MM, Oliveira SC et al (2016) Control of *Klebsiella pneumoniae* pulmonary infection and immunomodulation by oral treatment with the commensal probiotic *Bifidobacterium longum* 51A. Microbes Infect 18(3):180–189

Waters CM, Bassler BL (2005) Quorum sensing: cell-to-cell communication in bacteria. Annu Rev Cell Dev Biol 21:319–346

Zipperer A, Konnerth MC, Laux C, Berscheid A, Janek D, Weidenmaier C et al (2016) Human commensals producing a novel antibiotic impair pathogen colonization. Nature 535(7613):511

Impact of Human Microbiome on Health 20

Sujata Das, Charu Khanna, Shalini Singh, Shilpa Nandi, and Reema Verma

Abstract

The human genome in the recent years, by the advent of technological advancements, has emerged as a major prolocutor for reciprocity between the human body and the food consumed. As known, microbiome comprises all the genetic materials within a microbiota and can thereby be also referred to as metagenome of the microbiota. Contemporary researches have revealed the influence of microbiome not only on human mind and health status, but also in wide range of disease switching, ranging from cardio-metabolic diseases, allergies and obesities to life-threatening diseases such as cancer. Though the complete mechanism of many diseases is yet unclear, research works have revealed that the metabolites, nutrients and microbes can be regarded as the key players for such physiological state. The major approach of this chapter is to enlighten the interrelationship of the microbiome on the human health either in a synergistic or in an antagonistic manner.

Keywords

Genome · Microbiome · Microbiota

20.1 Introduction

In the last few decades, immense initiatives have been taken in understanding the role of microbiome on human health. In the present day, either in the field of therapeutic development or medical treatment, the impact of microbiome is

S. Das (✉) · S. Singh · S. Nandi · R. Verma
School of Bioengineering and Biosciences, Lovely Professional University, Phagwara, Punjab, India

C. Khanna
School of Pharmaceutical Sciences, Lovely Professional University, Phagwara, Punjab, India

S. G. Sharma et al. (eds.), *Microbial Diversity, Interventions and Scope*,
https://doi.org/10.1007/978-981-15-4099-8_20

especially apparent in the studies of different microbial communities and the human microbiome. In the recent times, ‘omics’ has provided a significant impetus on all the facets of biological research which has ushered the field to gain in momentum. The study, which is contemporarily popularised as study of ‘human microbiome’, is an outcome of the advancement in the field of genomics and other fields of microbiology, which has given the classical microbiology a new outlook and perspective. Thereby, the attention was driven and directed towards the genomics, which was the major objective of the Human Microbiome Project (HMP) (Turnbough and Wilson 2007). The HMP was used for the characterisation of all the microbial communities living in the human body, eventually switching the interest towards, not only, the type of microbes existing in the human body but also to the role and activity they perform both in the case of a healthy and diseased individual. Since 2003, after the publication of the first human genome (Chial 2008), the biomedical research on microbiome has obtained a significant scientific attention. There has been an immense leap from the culture-based surveys of various tissues or organs, for example, of gut and oral cavity, to molecular profiling of the microbial communities and their biochemical products like enzymes, proteins, and amino acids in all the different ecological niches of the human body (Eckburg et al. 2005; Gill et al. 2006; Costello et al. 2009) for this subject. This chapter attempts to put an insight into the distribution and diversification of human microbiome, the behaviour of human microbiome on the human health and microbiome as a paradigm for the future nutritional and medicinal strategies for human benefits.

20.2 Systemic Microbiome: Its Distribution and Diversification

Over 100 trillion microbes are estimated to be residing both inside and over the surface of humans, possessing genome which is approximately 150 times to that of the entire human being (Wang et al. 2017). This includes microbes of different families such as bacteria, fungi, archaea and viruses, contributing about eight million unique protein-coding genes as compared to the human genome, which comprises only around 22,000 protein-coding genes (Tomayko et al. 2013). This variation and the nature of the organism, specific to the specific anatomic site of the body, is a consequence of their specific growth requirement. The other determinants for such growth include coevolution of microorganism, their extensive interaction amongst each other and with the human host, variation in the composition and function with respect to the population, human life span and inconsistency of the body sites, ecologic condition, difference in the oxygen tension, airway luminal temperature, mechanism of muco-ciliary clearance, sex, genetics and socio-economic status. Hence, there has been a development of the concept of interdependence, in variety of physiologic, immunologic and metabolic processes, which ultimately determines the microbiome community in a particular site of the human body system.

20.2.1 The Gut Microbiome

Amongst all the human systemic microbiome, the gut microbiome, composed of the genetic material of the microorganisms in the gut, occupies a very essential and special position. They play an important role in various physiological processes like metabolism, immunity development and nourishment supply. The genotype and the immune system of the host have been shown to contribute towards the development of gut microbiota (Thaiss et al. 2016). In response to environmental factors, such as diet, pathogens and xenobiotic substances, a crosstalk occurs between the human immune system and the microbiome. For instance, the myeloid cells, epithelial layer and the innate lymphoid cells, part of the immune system, crosstalks with the gut microbiota for which the microbiome composition, host physiology and disease susceptibility are the main consequences of such crosstalks and feedback loops between them. Along with the bacterial community, like Firmicutes and Bacteroidetes species (Table 20.1), these interactions are also contributed by the other microbiota like fungi (Pothoulakis 2009), archaea and viruses (Breitbart et al. 2003). Though the understanding of the immunological relationship between the fungi and archaea is limited currently, the trans-kingdom commensalism is expected to be formed from infancy (LaTuga et al. 2011).

The principal constituents are the bacteriome, virome and mycobiome, whose strong interdependence maintains the functionality of the gut microbiota, if imbalanced may also affect the other systems in various ways. Since the time of birth of an individual, when the sterile gut of the neonate gets exposed to the biota of mother's vagina during the vaginal delivery or hospital microbiota in case of caesarean section (which may even include the multidrug-resistant species), the microbes starts their colonisation with an eventual change by the age of 3–5 years, by when an individual starts resembling bacterial community to that of an adult both structurally and functionally (Bull and Plummer 2014). In adults, the composition of gut microbiota is uneven throughout the length of the gut. As compared to small intestine, which is rich in the species related to phylum *Firmicutes*, colon on the contrary exhibits the presence of members of phylum Bacteroidetes. The microbiome of lumen and that attached to the epithelial lining even show differences. The stool sample exhibited the presence of *Bacteroides, Streptococcus, Ruminococcus, Lactobacillus, Enterococcus, Bifidobacterium* and *Clostridium* which presented the lumen community while on the mucous layer detected the presence of *Enterococcus, Lactobacillus* and *Clostridium* (Swidsinski et al. 2005).

20.2.2 The Microbiome of the Lungs and the Airways

During the initiation of HMP, the airways and lungs were exempted from the study, believing these parts to be sterile in nature (Moffatt and Cookson 2017). This fact was always acceptable because of the negative results yielded by the various standard microbiological culture tests of the healthy individuals (Faner et al. 2017). Its study was also a challenge owing to the difficulty in assessing the lower

Table 20.1 Distribution and diversity of human microbiome

Site	Name of dominant species	Average genome size	References
Oral	*Streptococcus, Haemophilus, Actinomyces, Prevotella*	2.11 Mb	Nayfach and Pollard (2015) and Gao et al. (2018)
Gut	*Firmicutes, Bacteroidetes, Bacteroides, Streptococcus, Ruminococcus, Enterococcus, Bifidobacterium, Clostridium, Enterococcus, Lactobacillus* and *Clostridium*	2.5–5.8 Mb	Nayfach and Pollard (2015), Bull and Plummer (2014), and Bäckhed et al. (2012)
Skin	*Firmicutes, Proteobacteria, Bacteroides, Actinobacteria, Corynebacterium* spp., *Staphylococcus* spp., *Propionibacteriun* spp., *Malassezia* spp., *Cryptococcus* spp., *Epiciccum* spp., *Aspergillus* spp., and *Rhodotorula* spp. Gram negative organism abundance	2.23 Mb	Grice and Segre (2011), Ross et al. (2017), and Byrd et al. (2018)
Respiratory system	**URT:** *Staphylococcus* spp., *Corynebacterium* spp., *Propionibacterium* spp., *Moraxella* spp., *Streptococcus* spp. and *Dolosigranulum* spp. *Haemophilus* spp., *Rothia* spp., *Neisseria* spp., *Streptococcal* spp., *Veillonella* spp., *Leptotrichia* spp., *Prevotella* spp., *Penicillium* spp., *Candida* spp., *Aspergillus* spp., and *Alternaria* spp., human bocavirus, human adenovirus, human rhinovirus, human coronavirus, polyoma viruses and other *Anelloviridae* family		Nayfach and Pollard (2015), Man et al. (2017), Faner et al. (2017), Moffatt and Cookson (2017), Dickson et al. (2016), and Yatera et al. (2018)
	LRT: *Acinetobacter* spp., *Staphylococcus* spp. and *Ureaplasma* spp., *Haemophilus* spp., *Moraxella* spp., *Streptococcus* spp., *Staphylococcus* spp., *Tropheryma whipplei*, bacteriophages, *Anelloviridae, Systenosterma, Erethecium* and *Malassezia* genera		

(continued)

Table 20.1 (continued)

Site	Name of dominant species	Average genome size	References
Cardiovascular system	*H. pylori,* herpes simplex virus, *Cytomegalovirus, Chlamydia pneumoniae, Mycoplasma pneumoniae, Epstein Barr virus*	–	Clifford and Hoffman (2015)
Urinary system	*Corneybacterium, Escherichia, Ureaplasma, Mycoplasma, enterococcus, Aerococcus, Staphylococcus, Gemella, Anaerococcus, Prevotella, Finegoldia., Actinobaculum, Aerococcus, Anaerococcus, Gardnerella, Bulkholderia, Corneybacterium, Bifidobacterium, Actinobacteria, Bacteroidetes, Rhodobacter, Alloscardovia, Trueperella, Atopobium, Sneathia, Enterobacteriaceae, Shigella, Prevotella., Saccharofermentans, Proteiniphilum, Parvimonas* and *Jonquetella*	2.11 Mb	Nayfasch and Pollard (2015)

tract without the invasive techniques such as bronchoscopy. Hence, there has been a delay in the systemic microbiome assay until the first study indicating the similarity of bacterial density of this part with the upper small bowel of human body was reported (Man et al. 2017). This has been made possible due to the advances in molecular techniques independent of culture practices (Faner et al. 2017). Human respiratory system is divided into upper respiratory tract (URT) and lower respiratory tract (LRT) with alveoli, present in the LRT, acquiring the surface area nearly 70 m^2 (Man et al. 2017). This complete tract is occupied by the niche-specific microbiota with higher density dwelling in URT.

The development of microbiota has been thought to effect on the morphological genesis of this system (Man et al. 2017). During the first hours of a healthy neonate, non-specific microbes, presumed of maternal origin has been detected. Abundance of *Staphylococcus* spp. in the first week, in the URT due to niche specification, has also been detected, which is then occupied by the *Corynebacterium* spp., dominated by the *Dolosigranulum* spp. The *Moraxella* species has its dominance at the age of 4–6 months. The individuals with the possession of such microbiota have been found to possess stable microbiome community along with better airway health (Morris et al. 2013; Segal et al. 2013) This healthy development is prone to disturbance under certain conditions such as usage of antibiotics, oxygen tension, temperature and pH, presence of other siblings, seasonal variations, vaccinations, exposure to smoke and host genetics (Man et al. 2017). Attempt to observe the diversity of flora in URT and LRT is well observed (Table 20.1).

20.2.3 The Microbiome of the Cardiovascular System

Studies using species specific molecular techniques have declared that the disease-free arteries and veins are microbe free in nature (Jin et al. 2019; Sobol 2014). Very few studies do report the presence of bacterial and viral genome in some vessels of healthy subjects. Amongst the microbes, *Helicobacter pylori* and herpes simplex virus are the most prevalent followed by *Cytomegalovirus, Chlamydia pneumoniae, Mycoplasma pneumoniae* and *Epstein Barr virus* and were also detected in healthy aorta, saphenous veins and internal mammary arteries (Clifford and Hoffman 2015).

20.2.4 The Microbiome of the Cardiovascular System

Studies using species-specific molecular techniques have declared that the disease-free arteries and veins are microbe free in nature (Jin et al. 2019; Sobol 2014). Very few studies do report the presence of bacterial and viral genome in some vessels of healthy subjects. Amongst the microbes, *Helicobacter pylori* and herpes simplex virus are the most prevalent followed by *Cytomegalovirus, Chlamydia pneumoniae, Mycoplasma pneumoniae* and *Epstein Barr virus* and were also detected in healthy aorta, saphenous veins and internal mammary arteries (Clifford and Hoffman 2015).

20.2.5 Urogenital Microbiome

The human urinary tract along with urine was earlier considered to be ‘sterile’, until the modern-day research, which has confirmed the presence of microbes in this system, significantly in healthy individuals. The advances in techniques such as 16 s rRNA sequencing have considerably helped in revealing the normal microbiota of the human body system. After analysis of the urine, the most common genera reported are *Lactobacillus* (more in women) and *Streptococcus* (more in men), both of which, along with few other groups, deliver a protective role in this system against different pathogens. Various bacterial taxa, as illustrated (Table 20.1), had been recognised in healthy adults, but some specific genera such as *Saccharofermentans, Proteiniphilum, Parvimonas* and *Jonquetella* were found in persons with age more than 70 years. The variation of urinary microbiome related to age and sex may be due to differences in voiding habits, hygiene, urinary metabolites, anatomic structures, hormonal variation and histology. Even the vaginal microbiome at premenstrual phase, reproductive age and post-menopausal phase exhibits variation (Aragón et al. 2018).

20.2.6 The Microbiome of the Nervous System

Central nervous system is considered to be one of the most immune privileged systems because of its closed compartmentalisation. It is isolated by physical barriers like blood-brain barriers and blood-cerebrospinal fluid barriers, from the circulatory system (Obermeier et al. 2013; Ransohoff and Engelhardt 2012). Thus, lack of lymphatic drainage, expression of major histocompatibility complex by the parenchymal cells and anti-inflammatory environment of the central nervous system accounts for such privileged status of seclusion from the microbiota (Berer and Krishnamoorthy 2014).

Though, in our body, there exists a bidirectional communication system, involving hormonal, immunological and neural signalling pathways, between the brain and the gut, accessed by the microbial flora of the intestine and certain metabolites, also known as the gut-brain axis. It is estimated that nearly 90% of serotonin (5-HT), a neurotransmitter, is produced in the intestine under the influence of gut microbiota, and the activation of its receptors in the enteric nervous system is responsible for the neuroprotection and adult neurogenesis in the mouse model (De Vadder et al. 2018).

20.2.7 The Microbiome of the Skin

The largest organ forming the external interface of the human body to the environment is the skin. Nearly 1.5–2.0 m^2 of the skin covers an average human with 2–3 mm depth. Three tissue layers are found in the skin: epidermis, dermis and hypodermis. Epidermis is colonised by millions of the microbes such as bacteria, fungi, arthropods and even viruses. It acts as the physical, anatomical and immunological barrier to various pathogens extending protection of the body, unless this barrier is broken or there is an imbalance between commensal and pathogenic organisms resulting to cutaneous or systemic diseases. Their acidic pH, continuous shedding of epidermal cells, hydrophobic nature, salinity and association with the antimicrobial compounds make them an efficient barrier (Ross et al. 2017). Though microbes do exist on them in spite of the above characteristics, the number of microbes inhabiting the skin ranges from one million to about one billion in each cm^2. Although, human skin is inhabited by quite diverse microflora, but most commonly found bacterial phyla includes *Proteobacteria, Corynebacteria, Propionibacteria, Bacteroidetes, Firmicutes and Staphylococcus* spp. Amongst fungi, *Malassezia* spp., *Cryptococcus* spp., *Epiciccum* spp., *Aspergillus* spp. and *Rhodotorula* spp. are most commonly found. The factors that affect the prevalence and dominance of community on the skin include biological sex, skin depth, skin location (skin thickness, folds, density of hairs), age, health, geographical location, ethnicity, use of lotions, soaps, cosmetics, and antibiotics and hygiene practices.

20.3 Influence of Microbiome on Human Health

As humans are known to have a constant symbiosis with microorganisms, the human microbial community, inhabiting the various system exhibits their influence on them. There are nearly 100 trillion symbiotic microorganisms that exist on and within the human body and have shown to play very important roles both in human health and disease causation. The influence of this microbiome on various systems is as follows:

20.3.1 Influence of Microbiome on Maintenance of Human-Gut Environment

Amongst all the systems, the highest and heterogeneous microbial density resides in colon and is mostly codependent in nature and present along both longitudinal (proximal to distal) and axial (mucosal to lumen) gradients of the gastrointestinal (GI) tract. The microbiome has been reported to aid in food digestion, vitamin biosynthesis, bile acid biotransformation, building of innate immunity, maintenance of intestinal barrier (Valdes et al. 2018), etc. Thus, gut is an "essential organ", carrying approximately 150 times more genes than are found in the entire human genome (Wang et al. 2017) (Fig. 20.1).

20.3.1.1 Metabolism

Microbial inhabitants within the host often contribute to metabolism such as:

- Bile salt metabolism
- Synthesis of essential and non-essential amino acids
- Replication of virulence factors in enteric pathogen
- Pro-drug transformation into active drugs and
- Metabolism of xenobiotic compounds
- Antibiotics by chemical transformation (Sarkar et al. 2018)

Gut microbiota and host interaction result in the secretion of a series of metabolites including trimethylamine-N-oxide (TMAO); short-chain fatty acid (SCFA) such as acetate and butyrate; secondary bile acid; and indoxyl sulphates that activate numerous signalling pathways affecting host physiological processes (Jin et al. 2019).

20.3.1.2 Contribution Towards the Host Immune System

When a child is born, the immune system at birth is under a relative state of immaturity. The developing immune system is characterised by a skewed T- and B-cell development and a blunted inflammatory cytokine production with respect to the regulatory responses. The consequence of such an underdeveloped immune system is high susceptibility to infections. Thereby, the regulatory environment ensures the establishment of the microbiota which ultimately helps in immune

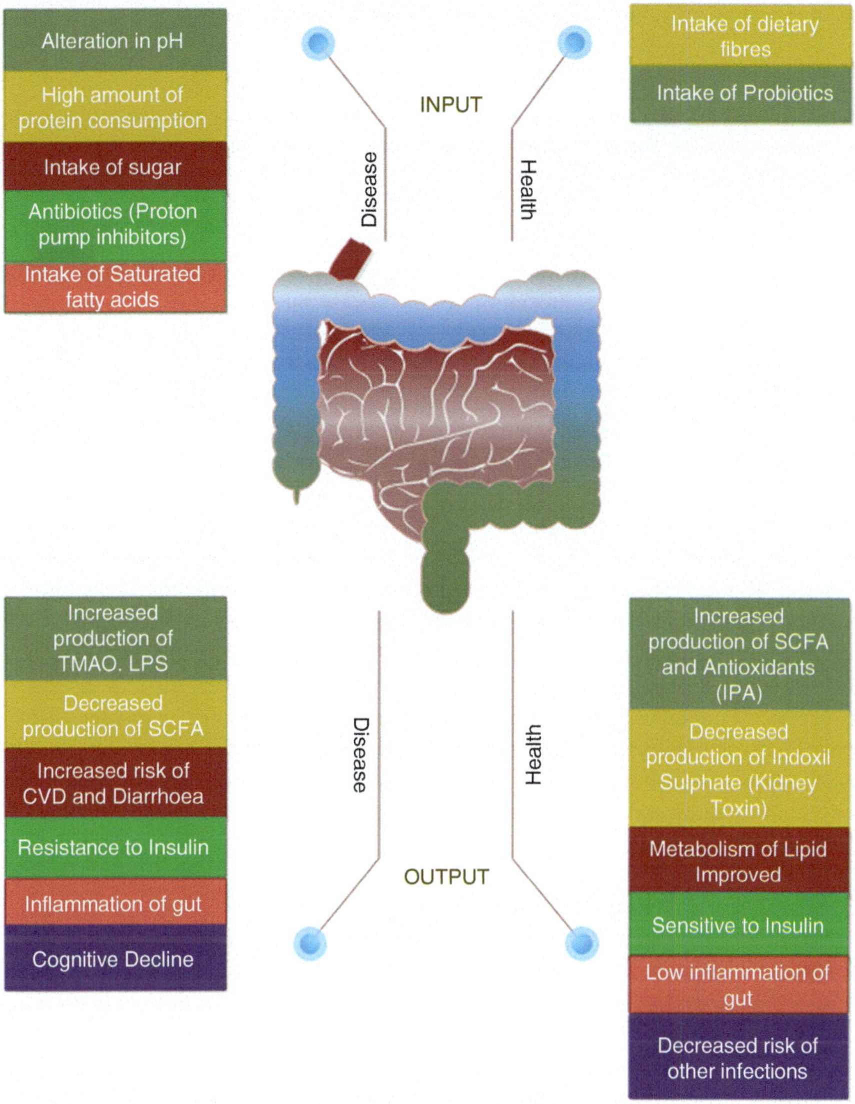

Fig. 20.1 Schematic diagram showing the role of the gut microbiota in health and disease with some inputs and outputs. *CVD* cardiovascular disease, *IPA* indole propionic acid, *LPS* lipopolysaccharide, *SCFA* short-chain fatty acids, *TMAO* trimethylamine-N-oxide

regulation and limits the mucosal inflammation following their colonisation (Elahi et al. 2013).

Many intestinal bacteria also prevent the colonisation of pathogen by producing antimicrobial substances such as bacteriocin (inhibit pathogen growth), by competing for nutrition and attachment sites. This act is termed as barrier/competitive-exclusion effect (Collado et al. 2010). Exposure to intestinal bacteria has also been found to prevent certain allergic responses in the hosts.

Thereby, the commensals, specifically the bacterial species and the products or metabolites derived from them, are considered as an intrinsic regulator of all the immune responses for the upliftment and restoration of human meta-organism's health (Belkaid and Hand 2014).

20.3.1.3 Gut Microbiota and Associated Diseases

Gut bacteria are intrinsically linked with the health of our entire body. However, a change in gut microbiota composition, termed as dysbiosis, can result in enhanced susceptibility of the host towards pathology. Dysbiosis can cause various diseases such as inflammatory bowel disease (IBD) and irritable bowel syndrome (IBS), chronic kidney disease (CKD), cardiovascular disease (CVD), atherosclerosis, obesity, autism, allergy, asthma, hypertension, coronary artery disease, and heart failure (Tang and Hazen 2017; Backhed, et al., 2012).

20.3.1.3.1 Irritable Bowel Syndrome (IBS) and Inflammatory Bowel Disease (IBD)

IBS and IBD are related to bowel disorder with signs and symptoms of abdominal cramping, pain and bloating associated with features of disordered defecation in IBS and ongoing inflammation of all part of GI tract in IBD. Although IBS is a multifactorial disease (Soares 2014), intestinal dysbiosis has also been found associated with this disease by facilitating adhesion of enteric pathogens. IBD includes Crohn's disease (CD) and ulcerative colitis (UC). Gut microbiota has been implicated to have roles in IBD pathogenesis, and the usage of antibiotics may help in reducing inflammation or prevented the same in murine models of disease and in patients.

20.3.1.3.2 Chronic Kidney Disease (CKD)

Intestinal dysbiosis is most often accompanied by defective intestinal barrier function, which promotes the production of bacterial by-products that are rapidly absorbed and retains in intestinal lumen. When increased absorption is coupled with reduced clearance of these substances by the kidneys, levels of gut derived toxins rise in circulation and may potentiate vascular calcification, atherosclerosis and adverse cardiovascular functioning, which are the clinical conditions in the later stage of CKD. Numerous epidemiologic studies have shown a relationship between gut-derived vascular toxins and cardiovascular events in patient with CKD (Valdes et al. 2018).

20.3.1.3.3 Gut Microbiome–Associated Cardiovascular Disease (CVD)

Metabolic origin of traditional CVD risk factors such as obesity, dyslipidaemia and insulin resistance indicates close linkage between the gut microbiome and CVD. Recent studies have implicated that atherosclerosis has generated the largest amount of data in association with gut microbiome. According to a hypothesis, gut microbiota metabolites elicit inflammatory cascade by translocating into bloodstream and promote atherosclerosis. Oral bacteria causing oral cavities have also been found in atherosclerosis plaque (Clifford and Hoffman 2015). In many CVDs, heart failure is considered as end stage with a higher rate of morbidity and mortality (Jin et al. 2019). Low intestinal perfusion and disruption of intestinal barrier have been reported as reasons for reduced cardiac output and blood redistribution. The translocation of microbiota and endotoxins into blood circulation leads to an aggravated systemic inflammation that further can lead to increased chances of heart failure (Peng et al. 2018).

Various studies suggested the direct and indirect link between gut microbiota and the development of hypertension (Jin et al. 2019). It is a complex clinical condition and can be influenced by number of factors. It is considered as modifiable risk factor for CVD. In some studies, *Prevotella, Faecalibacterium, Klebsiella, Clostridium* and *Streptococcus* are found in abundance in hypertensive patients, indicating a close association of host microflora to such a clinical condition. Other studies demonstrated that reduction in gut microbial diversity, short-chain fatty acid (SCFA)-producing bacteria and increase in sympathetic drive to the gut and lactic acid producing bacteria, respectively, have direct role in blood pressure regulation. Thus, it can be concluded that blood pressure is closely linked to diversity, richness and evenness of microbiome living in the gut and the improved gut microbiota may be a target for future therapies for hypertension (Peng et al. 2018).

20.3.1.3.4 Human Nervous System Association with Gut Microbiota

The gut-brain axis plays a major role in central nervous system (CNS) and intestinal and immune system functioning, as mentioned earlier. Gut-brain axis is the bidirectional biochemical signalling between GI and CNS activities, integrating efferent and afferent neural, endocrine, nutrient and immunological signals, providing gut microbiota and its derived metabolites a route to access the brain (Bull and Plummer 2014; Joscelyn and Kasper 2014). Wang and Kasper illustrated that this bidirectional communication system enables brain to command GI functions such as metabolism, peristalsis, mucin production and other immunological functions.

Gut microbiota is also found to have impact on hypothalamic-pituitary-adrenal (HPA) axis, thus playing a role in body's stress response (Gareau et al. 2008; Teitelbaum et al. 2008). The gut microbiota has also been shown to synthesise neurotransmitters and neuromodulators. Various neurotransmitters produced by bacterial species are presented in Table 20.2. Those released neurotransmitter then stimulates epithelial cells to synthesise modulators within the enteric nervous system (ENS) or directly acting on primary afferent axons (McVey Neufeld et al. 2013). Moreover, it has been shown that ENS plays a major role in fundamental gastrointestinal physiological functions such as motility, fluid secretions and blood flow. Furthermore, numerous studies have revealed the correlation between microbiome, microbiota derived products, antibiotics, prebiotics and probiotics and CNS (Wang and Kasper 2014).

Anxiety and stress, characteristic mood disorders, associated with nervous, endocrinal and immunological system have also been shown to have an association with the gut microbiota. Stressors such as chemical, biological or any environmental

Table 20.2 Various neurotransmitters synthesised by different bacterial species

Bacterial species	Neurotransmitter
Lactobacillus spp.	Gamma-amino butyric acid (GABA), acetylcholine
Bifidobacterium spp.	Gamma-amino butyric acid (GABA)
Escherichia spp.	Norepinephrine (NE), 5-hydroxy tryptamine-serotonin (5-HT)
Bacillus spp.	Norepinephrine (NE), dopamine (DA)
Streptococcus spp.	5-HT
Enterococcus spp.	5-HT

stimuli can act as an active component to trigger the anxiety and stress response which ultimately activates the hypothalamic pituitary adrenal (HPA) axis. Intestinal dysbiosis and gut pathogens have thereby been shown to cause stress and anxiety. Animal models have represented to ameliorate these disorders by using probiotic formulations (Messaoudi et al. 2011). Various output and input of role of gut microbiota is depicted in Fig. 20.2.

20.4 Microbiome as a Paradigm for Future Nutritional and Medicinal Strategies

The analysis of microbiome of human body shows both the pathogenic as well as beneficial microbial network (Ozturk et al. 2017). Customizable medicine based on an individual's microbiome is an excellent approach for therapeutic choices based either on exercises or/and medications which considers the genetic makeup of an individual, their health condition and quality of life (Hasani-Ranjbar and Larijani 2017). Human genomes are about 99.9% indistinguishable to one another; the generally steady human gene pool does not completely clarify all the phenotypic varieties amongst people. On the other hand, the bacterial biological system, living in each human body, that contributes multiple times larger number of qualities than the human genome, could be significantly unique in relation to an individual. Therefore, the human microbiome can be regarded as the true cause of numerous responses that can influence the structure and plenitude of microorganism (Hasani-Ranjbar and Larijani 2017) in the human body.

Recently various studies have shown that intestinal microbiota is fundamentally associated with various therapeutics as in cardiovascular diseases, cancer, etc. These include narrow spectrum antibiotics along with probiotics, prebiotics and synbiotics, faecal microbiota transplantation, nutritional modulators, immune modulators and phage therapy.

20.4.1 Effects of Antibiotic Abuse on Microbiota

In the USA and Europe, jointly, in the year 2015, because of the onset of antibiotic resistance amongst pathogens, around 50,000 deaths were witnessed, which was

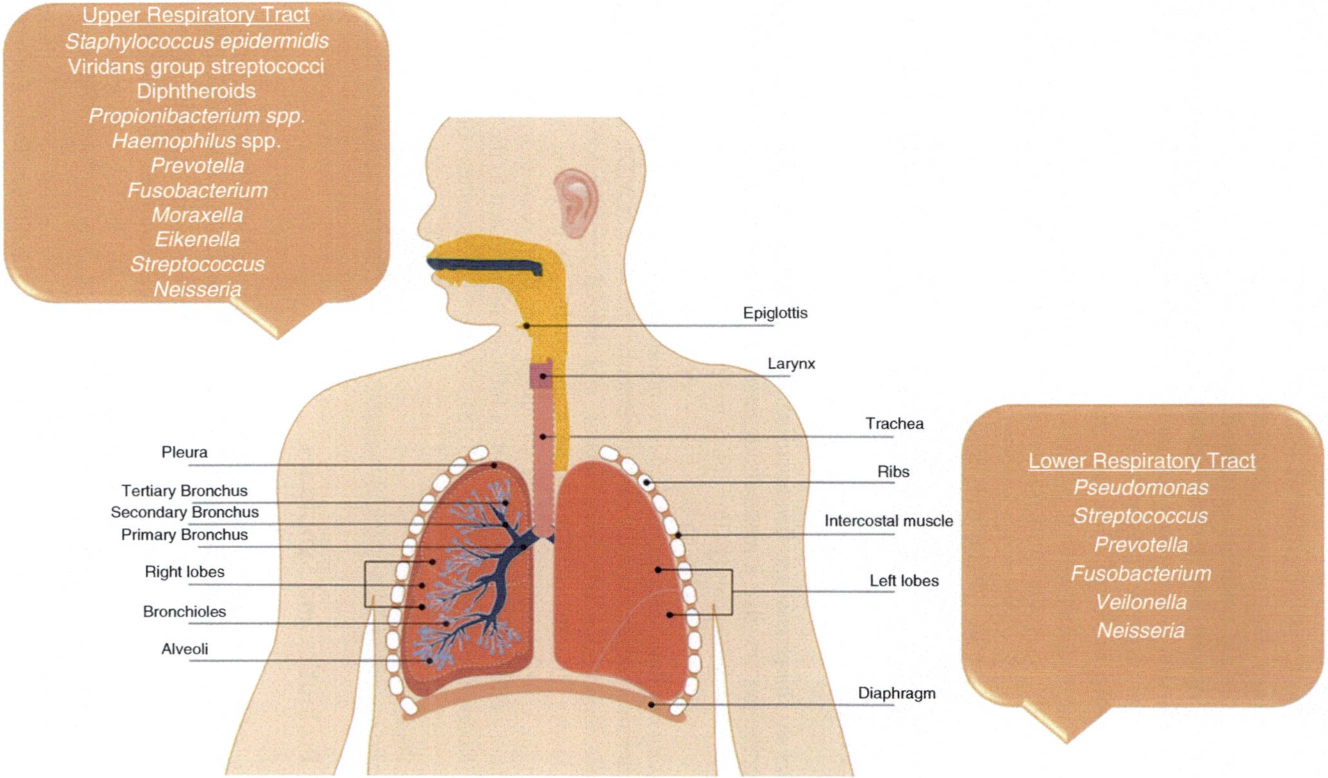

Fig. 20.2 Structures of human respiratory tract, the upper respiratory tract and the lower respiratory tract along with their prevalent microbiota

projected to increase to a score of around ten million deaths per year worldwide by 2050 (Langdon et al. 2016).

In addition to the development of resistance, the usage of antibiotics has been reported to disrupt the ecology of the human microbiome. Whereby, a dysbiotic microbiome loses the capability to perform vital functions such as nutrient supply, production of vitamin, and defence against the pathogens; thereby it leads to an eventual impairment of the metabolic, immunological and developmental system of the host. For an instance, drug-induced modification of the gut microflora can influence a group of Foxp3$^+$Treg cells that control demyelination in exploratory autoimmune encephalomyelitis (EAE) (Ochoa-Repáraz et al. 2010). Another study reflects the antigens produced by *Bacteroides fragilis* (capsule polysaccharide) can protect against EAE (CNS demyelination) and further in human multiple sclerosis (Ochoa-Repáraz et al. 2011). Similarly, the effects of common antimicrobial treatments on the gut microbiota have been illustrated in Table 20.3.

20.4.2 Probiotics

Probiotics are characterised as live microorganisms which are not part of the human host microbiome yet give a medical advantage to the host when administered or directed in sufficient quantity. Probiotics have been extensively studied in recent years as it imparts various health benefits by the metabolites it produces in the relief from certain intestinal disorders as well as controlling EAE. Clinical trials have been able to demonstrate that certain cardio metabolic disorders (CMD), such as type 2 diabetes mellitus (T2D), dyslipidaemia and arterial hypertension, as well as chronic kidney diseases (CKD) can be managed by the ingestion of probiotics (Neto et al. 2018). *Saccharomyces boulardii* has been shown to exert anti-inflammatory effect which helps to control inflammation related to the dysbiosis in lumen (Rodríguez-Nogales et al. 2018). Various metabolites produced by microorganisms present in the host and its associated health benefits are discussed in Table 20.4.

20.4.3 Prebiotics and Synbiotics

Prebiotics are those food components that cannot be digested by human body but can be selectively digested by the members of probiotics, and thereby serves as a food fibre for probiotics. Recent studies have demonstrated that the use of prebiotics can result in enhancing the ecological performance of the gut microbiota, thus promoting a much more beneficial community (Vandeputte et al. 2017). This conceptualizes that the human microbiota can be enhanced, stabilised and shifted by feeding with certain specific prebiotics such as carbohydrates. However, characterisation of the relationship between the prebiotic and probiotic is still a challenge. The point when the idea of synbiotic was first presented, two setups were proposed: first, where the prebiotic and probiotic segments were independent, each being in charge of a

Table 20.3 Effects of commonly used antimicrobials on gut microbiota

Antimicrobial class	Antimicrobial agent	Effects on faecal microbiota count			Reference
		Decrease in no. of microbiota	Increase in no. of microbiota	Constant no. of microbiota	
Penicillin	Piperacillin, tazobactum	*Bifidobacteria, Eubacteria, Lactobacilli, Enterobacter cloacae*	*Klebsiella pneumoniae*	Enterococci clostridia, Bacteroides	Morjaria et al. (2019) and Bhalodi et al. (2019)
	Ampicillin		*Enterobacter cloacae, Klebsiella pneumonia*		Kamal et al. (2019)
Cephalosporins	Cefuroxime	*Enterobacter cloacae*		*Klebsiella pneumoniae*	Kamal et al. (2019)
	Cefotaxime	Firmicutes, Actinobacteria, Bacteroidetes		Proteobacteria	Burdet et al. (2019)
	Ceftriaxone	Firmicutes, Actinobacteria, Bacteroidetes		Proteobacteria	Burdet et al. (2019)
Carbapenems	Meropenem		Actinobacteria Proteobacteria		Ye et al. (2019)
Fluoroquinolones	Ciprofloxacin			*Klebsiella pneumonia Enterobacter cloacae*	Kamal et al. (2019)
Aminoglycosides	Gentamicin		*Enterobacter cloacae*	*Klebsiella pneumoniae*	Kamal et al. (2019)

Table 20.4 Metabolites-producing microbiota and its associated health benefits

Microorganisms	Metabolites	Health benefits	References
Bacteroides fragilis	Polysaccharide A	Controls EAE and IBD by promoting Foxp3$^+$Treg quantity	Ochoa-Repáraz et al. (2010)
Lactobacilli Bifidobacteria	Lactic acid	Anti-inflammatory, promotes IL-10$^+$Foxp3$^+$Treg	Takata et al. (2011) and Kwon et al. (2013)
Escherichia coli	Vitamin B	Prevents vitamin K-deficiency bleeding, lower risk of type 2 diabetes	Tursunov et al. (2018), and Díaz-Rizzolo et al. (2019)
Bifidobacterium bifidum, Bifidobacterium longum, Bifidobacterium breve,	Complex vitamin B	Maintains the immune system, prevents cardiovascular diseases, chronic kidney diseases and improves nervous system	Yoshii et al. (2019), Sivamaruthi et al. (2019), Kobayashi et al. (2019), and Jena et al. (2018)
Bacteroides	Propionate and butyrate	Relieves from intestinal problems, relieves from tuberculosis	De Paepe et al. (2018) and Maji et al. (2018)
Propionibacteria	Propionate	Protects colon mucosa cells and prevents cancer, Prevent diarrhoea	Casanova et al. (2018) and Gaucher et al. (2019)
Veillonella, Bacteroides, Coprococcus, Lactobacillus, Ruminococcus	Acetate, lactate and propionate	Ecological performance of intestine is increased	El Hage et al. (2019)
Streptococcus thermophilus	Folic acid	Cures intestinal mucositis, has anti-cancer activity, promotes gut-bone signalling	Levit et al. (2018), Tarrah et al. (2018), and Schepper et al. (2017)

specific impact or medical advantage, and second, a symbiotic segment, where the probiotic was explicitly structured with a prebiotic substrate that would synergistically support the intensity, survival or metabolic movement of a related probiotic strain in the ecology of gastrointestinal system (Krumbeck et al. 2016).

As of late, two novel methodologies have been proposed for creating such synergistic synbiotics, both being dependent on their ecological function and wellness. First, the in vivo strategy which depends upon the determination and isolation of probiotic strain that would increase in number when a certain population of participants are administered with the specific prebiotic component (Krumbeck et al. 2015). Second, called as a multi-taxon insertion, depending on sequencing, based on the identification of genes, would examine the fitness of probiotic strain in relation to the type of prebiotic administered by the usage of libraries of transposon mutants (Wu et al. 2015). A list of prebiotics with their probiotic components are given in Table 20.5.

Table 20.5 List of prebiotic and its specific probiotic components

Prebiotic	Probiotic microorganisms	References
Inulin-type fructans	*Anaerostipes, Bilophila and Bifidobacterium*	Vandeputte et al. (2017)
Galacto-oligosaccharides	*Bifidobacterium*	Canfora et al. (2017)
Resistant Starch	*Bifidobacterium,* Proteobacteria	Alfa et al. (2018)
Xylooligosaccharide	*Blautia hydrogenotrophica; Bifidobacterium*	Long et al. (2019) and Carlson et al. (2017)
Whole fibre and pure inulin	Collinsella	Carlson et al. (2017)
Fructooligosaccharides	Lactobacillaceae; Bifidobacterium	Chen et al. (2017)

20.4.4 Faecal Microbiota Transplantation

Sometimes the mere usage of antibiotics is insufficient to treat few diseases and rather requires urgent alternatives to manage the severity of the clinical condition. This has led to the introduction of transplantation of the faecal microbiota. The process involves the separation and delivery of the faecal microbiota from stool of healthy donor to the gastrointestinal tract of the receiver patient, thereby enabling an efficient cure by normalising the microbiota composition. Recent investigations have explained the mechanism behind the faecal microbiota transplantation (FMT), which has been used to treat *Clostridium difficile* infection (CDI) for the restoration of the gut microbiome to gain the ability to inhibit *Clostridium difficile* indirectly by competing for nutrients. The faecal microbiota also prevents the colonisation by unwanted microorganism by the activation of immune system as well as by direct release of certain antimicrobial components and other metabolites that helps in the inhibition of vegetative and as well as the sporulated disease-causing organism (Khoruts and Sadowsky 2016). Table 20.6 shows the list of diseases and infections which can be treated by using FMT.

20.4.5 Nutritional Modulators

Intake of dietary medications for a long period of time impacts the structure and function of microbiome of the human body. The change in the availability of nutritional content of the diet causes corresponding alteration of the human microbiome. Recent findings have suggested that depending upon the intake of dietary, the microbiome richness diversifies. The lower is the genetic richness of microbiome; the lower will be the immune status of the person, relating to abnormal metabolic function and poor anti-inflammatory activity (Cotillard et al. 2013; Le Chatelier et al. 2013). The diet patterns may even contribute in managing different diseases (Table 20.7) by way of microbiome.

Table 20.6 Diseases and infections treated by FMT

Disease/infection	Mechanism	References
Clostridium difficile infection (CDI)	Competition for nutrients; direct suppression by antimicrobial peptides; bile-acid-mediated inhibition of spore germination and vegetative growth; activation of immune-mediated colonisation resistance	Khoruts and Sadowsky (2016)
Ulcerative colitis (UC)	Increasing the production of short-chain fatty acids, (butyrate); inhibitingTh1 differentiation, activity of T cells, leukocyte adhesion; production of inflammatory factors	Shen et al. (2018)
Irritable Bowel Syndrome (IBS)	Visceral hypersensitivity; altered barrier function, gastrointestinal motility and the gut-brain axis	Grover et al. (2014)
Obesity	Decrease adiposity; Alter metabolic phenotype by increased bacterial diversity	Marotz and Zarrinpar (2016)

Table 20.7 List of diseases that may be treated by dietary interventions

Diet pattern	Disease/infection	References
High fibre diet	Obesity	Menni et al. (2017)
	Ulcerative colitis	Silveira et al. (2017)
Fasting mimicking diet (FMD)	Multiple sclerosis and autoimmunity	Choi et al. (2016)
Calorie restriction	EAE	Piccio et al. (2008)
Sodium chloride restriction	Multiple sclerosis	Hernandez et al. (2015)
Western style diet	Anxiety	Ohland et al. (2013)
Poly unsaturated fatty acids	Depression	Gilbert et al. (2013)
Oestrogen	Breast cancer	Chen and Madak-Erdogan (2016)
Oleic acid	Ulcerative colitis	Fernández et al. (2020)
Polyphenols	Obesity	Henning et al. (2018)

20.4.6 Immune Modulators

Microbiome-based approaches involving antibiotics, probiotics, prebiotics and synbiotics, faecal microbiota transplantation and nutritional modulators correlate directly with the alteration of immune status of an individual focusing on the innate immunity. Thereby, the human microbiome is being affected for benefits with the change in immune status. Until now, there are inadequate information in this method; however, treatment of inflammatory diseases using steroids is in abundance. Intestinal innate and acquired immunity as well as systemic acquired immunity involves various mechanisms for the control of gut microbiome. Some of them include change in barrier function, expression of leptin, molecule β, human leukocyte antigen (HLA) class I and class II loci, activation of toll-like receptors, natural killer cells, CD4+ cells and Foxp3+ as well as the production of antimicrobial peptides and α-defensins (Ticinesi et al. 2019) for the proper functioning of the gut microbiome.

20.4.7 Phage Therapy

Phage therapy involves the introduction of explicit bacteriophages that targets a microorganism which in turn has the ability to generate a beneficial microbiome shift. However, a limitation to this strategy is the simultaneous resistance offered by the microorganism in play which is yet to be proven. Till date, none of the phage therapy approaches have been established as an FDA-approved drug. Recently, scientists are working hard to perform phage based killing of microbes, but it is much more complicated. CRISPR (clustered regularly interspaced short palindromic repeats) is one of the approaches to advance the limitations (LeMieux 2019). It is a tool derived from prokaryotic immune system empowered to study and modify organisms with ease and efficiency. The system helps to modify the gut genome of gut microorganisms and bacteriophages. This engineered CRISPR-Cas system ultimately can control gene expression and modulate production of metabolite and protein presenting a new approach for the development of drugs that can target the microbiome (Table 20.8).

20.5 Conclusion

Although these interventions are used as medicinal and nutritional strategies and have been clinically experimented for a successful result, more focus and deliberate attempts are being made for the establishment of such casualty in order to define the functional metabolic change. These therapies are yet not so widely utilised due to the requirement of huge amount of money and time. Hence, advancement in this field targeting the relation of human microbiome with health and diseases, identification of the composition and activity of microbiome as well as linking it, and most importantly using technical resources such as bioinformatics to incorporate and fill up the loopholes ascertaining to the models is quite difficult. Future research based on these directions is a key to solve the problems and hence enlightening with knowledge about the human microbiome and their influence on the human health.

Table 20.8 Microbiome therapy involving CRISPR-Cas system

Therapy model	Microbiome therapy	References
Additive therapy	Probiotics (bacteria)	Hidalgo-Cantabrana et al. (2017)
	Probiotics (yeast)	Liu et al. (2016)
Subtractive therapy	Lytic phages	Hwang et al. (2018) and Yosef et al. (2015)
	Antimicrobials	Park et al. (2017)
Modulatory therapy	Temperate phages	Park et al. (2017) and Yosef et al. (2015)
	CRISPRi gene regulation	Berlec et al. (2018)

References

Alfa MJ, Strang D, Tappia PS, Graham M, Van Domselaar G, Forbes JD, Laminman V, Olson N, DeGagne P, Bray D, Murray BL (2018) A randomized trial to determine the impact of a digestion resistant starch composition on the gut microbiome in older and mid-age adults. Clin Nutr 37(3):797–807

Aragón IM, Herrera-Imbroda B, Queipo-Ortuño MI, Castillo E, Del Moral JS, Gómez-Millán J, Yucel G, Lara MF (2018) The urinary tract microbiome in health and disease. Eur Urol Focus 4 (1):128–138

Bäckhed F, Fraser CM, Ringel Y, Sanders ME, Sartor RB, Sherman PM, Versalovic J, Young V, Finlay BB (2012) Defining a healthy human gut microbiome: current concepts, future directions, and clinical applications. Cell Host Microbe 12(5):611–622

Belkaid Y, Hand TW (2014) Role of the microbiota in immunity and inflammation. Cell 157 (1):121–141

Berer K, Krishnamoorthy G (2014) Microbial view of central nervous system autoimmunity. FEBS Lett 588(22):4207–4213

Berlec A, Škrlec K, Kocjan J, Olenic M, Štrukelj B (2018) Single plasmid systems for inducible dual protein expression and for CRISPR-Cas9/CRISPRi gene regulation in lactic acid bacterium Lactococcus lactis. Sci Rep 8(1):1009

Bhalodi AA, van Engelen TS, Virk HS, Wiersinga WJ (2019) Impact of antimicrobial therapy on the gut microbiome. J Antimicrob Chemother 74(Supp 1):i6–i15

Breitbart M, Hewson I, Felts B, Mahaffy JM, Nulton J, Salamon P, Rohwer F (2003) Metagenomic analyses of an uncultured viral community from human feces. J Bacteriol 185(20):6220–6223

Bull MJ, Plummer NT (2014) Part 1: the human gut microbiome in health and disease. Integr Med Clin J 13(6):17

Burdet C, Grall N, Linard M, Bridier-Nahmias A, Benhayoun M, Bourabha K, Magnan M, Clermont O, d'Humières C, Tenaillon O, Denamur E (2019) Ceftriaxone and cefotaxime have similar effects on the intestinal microbiota in human volunteers treated by standard-dose regimens. Antimicrob Agents Chemother 63(6):e02244–e02218

Byrd AL, Belkaid Y, Segre JA (2018) The human skin microbiome. Nat Rev Microbiol *16*(3):143

Canfora EE, van der Bee CM, Hermes GD, Goossens GH, Jocken JW, Holst JJ, van Eijk HM, Venema K, Smidt H, Zoetendal EG, Dejong CH (2017) Supplementation of diet with galacto-oligosaccharides increases bifidobacteria, but not insulin sensitivity, in obese prediabetic individuals. Gastroenterology 153(1):87–97

Carlson J, Erickson J, Hess J, Gould T, Slavin J (2017) Prebiotic dietary fiber and gut health: comparing the in vitro fermentations of beta-glucan, inulin and Xylooligosaccharide. Nutrients 9(12):1361

Casanova MR, Azevedo-Silva J, Rodrigues LR, Preto A (2018) Colorectal cancer cells increase the production of short chain fatty acids by Propionibacterium freudenreichii impacting on cancer cells survival. Front Nutr 5

Chen KL, Madak-Erdogan Z (2016) Estrogen and microbiota crosstalk: should we pay attention? Trends in Endocrinol Metab 27(11):752–755

Chen D, Yang X, Yang J, Lai G, Yong T, Tang X, Shuai O, Zhou G, Xie Y, Wu Q (2017) Prebiotic effect of fructooligosaccharides from Morinda officinalis on Alzheimer's disease in rodent models by targeting the microbiota-gut-brain axis. Front Aging Neurosci 9:403

Chial H (2008) DNA sequencing technologies key to the Human Genome Project. Nature Education 1(1):219

Choi IY, Piccio L, Childress P, Bollman B, Ghosh A, Brandhorst S, Suarez J, Michalsen A, Cross AH, Morgan TE, Wei M (2016) A diet mimicking fasting promotes regeneration and reduces autoimmunity and multiple sclerosis symptoms. Cell Rep 15(10):2136–2146

Clifford A, Hoffman GS (2015) Evidence for a vascular microbiome and its role in vessel health and disease. Curr Opin Rheumatol 27(4):397–405

Collado MC, Gueimonde M, Salminen S (2010) Probiotics in adhesion of pathogens: mechanisms of action. In: Bioactive foods in promoting health. Academic, Chennai, pp 353–370

Costello EK, Lauber CL, Hamady M, Fierer N, Gordon JI, Knight R (2009) Bacterial community variation in human body habitats across space and time. Science 326(5960):1694–1697

Cotillard A, Kennedy SP, Kong LC, Prifti E, Pons N, Le Chatelier E, Almeida M, Quinquis B, Levenez F, Galleron N, Gougis S (2013) Dietary intervention impact on gut microbial gene richness. Nature 500(7464):585

De Paepe K, Verspreet J, Verbeke K, Raes J, Courtin CM, Van de Wiele T (2018) Introducing insoluble wheat bran as a gut microbiota niche in an in vitro dynamic gut model stimulates propionate and butyrate production and induces colon region specific shifts in the luminal and mucosal microbial community. Environ Microbiol 9:3406–3426

De Vadder F, Grasset E, Holm LM, Karsenty G, Macpherson AJ, Olofsson LE, Bäckhed F (2018) Gut microbiota regulates maturation of the adult enteric nervous system via enteric serotonin networks. Proc Natl Acad Sci 115(25):6458–6463

Díaz-Rizzolo DA, Kostov B, López-Siles M, Serra A, Colungo C, González-de-Paz L, Martinez-Medina M, Sisó-Almirall A, Gomis R (2019) Healthy dietary pattern and their corresponding gut microbiota profile are linked to a lower risk of type 2 diabetes, independent of the presence of obesity. Clin Nutr

Dickson RP, Erb-Downward JR, Martinez FJ, Huffnagle GB (2016) The microbiome and the respiratory tract. Annu Rev Physiol 78:481–504

Eckburg PB, Bik EM, Bernstein CN, Purdom E, Dethlefsen L, Sargent M, Gill SR, Nelson KE, Relman DA (2005) Diversity of the human intestinal microbial flora. Science 308 (5728):1635–1638

El Hage R, Hernandez-Sanabria E, Calatayud Arroyo M, Props R, Van de Wiele T (2019) Propionate-producing consortium restores antibiotic-induced dysbiosis in a dynamic in vitro model of the human intestinal microbial ecosystem. Front Microbiol 10:1206

Elahi S, Ertelt JM, Kinder JM, Jiang TT, Zhang X, Xin L, Chaturvedi V, Strong BS, Qualls JE, Steinbrecher KA, Kalfa TA (2013) Immunosuppressive CD71+ erythroid cells compromise neonatal host defence against infection. Nature 504(7478):158

Faner R, Sibila O, Agustí A, Bernasconi E, Chalmers JD, Huffnagle GB, Manichanh C, Molyneaux PL, Paredes R, Brocal VP, Ponomarenko J, Sethi S, Dorca J, Monsó E (2017) The microbiome in respiratory medicine: current challenges and future perspectives. Eur Respir J 49(4):1–12

Fernández J, García de la Fuente V, García MTF, Gómez Sánchez J, Redondo BI, Villar J, Lombó F (2020) A diet based on cured acorn-fed ham with oleic acid content promotes anti-inflammatory gut microbiota and prevents ulcerative colitis in an animal mode. Lipids Health Dis 19(28):1–19

Gao L, Xu T, Huang G, Jiang S, Gu Y, Chen F (2018) Oral microbiomes: more and more importance in oral cavity and whole body. Protein Cell 9(5):488–500

Gareau MG, Silva MA, Perdue MH (2008) Pathophysiological mechanisms of stress-induced Intestina damage. Curr Mol Med 8(4):274–281

Gaucher F, Bonnassie S, Rabah H, Marchand P, Blanc P, Jeantet R, Jan G (2019) Adaptation of beneficial Propionibacteria, lactobacilli, and Bifidobacteria improves tolerance toward technological and digestive stresses. Front Microbiol 10

Gilbert K, Arseneault-Bréard J, Monaco FF, Beaudoin A, Bah TM, Tompkins TA, Godbout R, Rousseau G (2013) Attenuation of post-myocardial infarction depression in rats by n-3 fatty acids or probiotics starting after the onset of reperfusion. Br J Nutr 109(1):50–56

Gill SR, Pop M, DeBoy RT, Eckburg PB, Turnbaugh PJ, Samuel BS, Gordon JI, Relman DA, Fraser-Liggett CM, Nelson KE (2006) Metagenomic analysis of the human distal gut microbiome. Science 312(5778):1355–1359

Grice EA, Segre JA (2011) The skin microbiome. Nat Rev Microbiol 9(4):244

Grover M, Camilleri M, Smith K, Linden DR, Farrugia G (2014) On the fiftieth anniversary Postinfectious irritable bowel syndrome: mechanisms related to pathogens. Neurogastroenterol Motil 26(2):156–167

Hasani-Ranjbar S, Larijani B (2017) Human microbiome as an approach to personalized medicine. Altern Ther Health Med 23(6):8–9

Henning SM, Yang J, Hsu M, Lee RP, Grojean EM, Ly A, Tseng CH, Heber D, Li Z (2018) Decaffeinated green and black tea polyphenols decrease weight gain and alter microbiome populations and function in diet-induced obese mice. Eur J Nutr 57(8):2759–2769

Hernandez AL, Kitz A, Wu C, Lowther DE, Rodriguez DM, Vudattu N, Deng S, Herold KC, Kuchroo VK, Kleinewietfeld M, Hafler DA (2015) Sodium chloride inhibits the suppressive function of FOXP3+ regulatory T cells. J Clin Invest 125(11):4212–4222

Hidalgo-Cantabrana C, O'Flaherty S, Barrangou R (2017) CRISPR-based engineering of next-generation lactic acid bacteria. Curr Opin Microbiol 37:79–87

Hwang IY, Lee HL, Huang JG, Lim YY, Yew WS, Lee YS, Chang MW (2018) Engineering microbes for targeted strikes against human pathogens. Cell Mol Life Sci 75(15):2719–2733

Jena PK, Sheng L, Nagar N, Wu C, Barile D, Mills DA, Wan YJ (2018) The effect of synbiotics Bifidobacterium infantis and milk oligosaccharides on shaping gut microbiota community structure and NASH treatment. Data Brief 19:1025–1029

Jin M, Qian Z, Yin J, Xu W, Zhou X (2019) The role of intestinal microbiota in cardiovascular disease. J Cell Mol Med 23(4):2343–2350

Joscelyn J, Kasper LH (2014) Digesting the emerging role for the gut microbiome in central nervous system demyelination. Mult Scler J 20(12):1553–1559

Kamal SS, Hyldig N, Krych Ł, Greisen G, Krogfelt KA, Zachariassen G, Nielsen DS (2019) Impact of early exposure to cefuroxime on the composition of the gut microbiota in infants following cesarean delivery. J Pediatr 210:99

Khoruts A, Sadowsky MJ (2016) Understanding the mechanisms of faecal microbiota transplantation. Nat Rev Gastroenterol Hepatol 13(9):508

Kobayashi Y, Kuhara T, Oki M, Xiao JZ (2019) Effects of Bifidobacterium breve A1 on the cognitive function of older adults with memory complaints: a randomised, double-blind, placebo-controlled trial. Benefic Microbes 10(5):511–520

Krumbeck JA, Maldonado-Gomez MX, Martínez I, Frese SA, Burkey TE, Rasineni K, Ramer-Tait AE, Harris EN, Hutkins RW, Walter J (2015) In vivo selection to identify bacterial strains with enhanced ecological performance in synbiotic applications. Appl Environ Microbiol 81 (7):2455–2465

Krumbeck JA, Maldonado-Gomez MX, Ramer-Tait AE, Hutkins RW (2016) Prebiotics and synbiotics: dietary strategies for improving gut health. Curr Opin Gastroenterol 32(2):110–119

Kwon HK, Kim GC, Kim Y, Hwang W, Jash A, Sahoo A, Kim JE, Nam JH, Im SH (2013) Amelioration of experimental autoimmune encephalomyelitis by probiotic mixture is mediated by a shift in T helper cell immune response. Clin Immunol 146(3):217–227

Langdon A, Crook N, Dantas G (2016) The effects of antibiotics on the microbiome throughout development and alternative approaches for therapeutic modulation. Genome Med 8(1):39

LaTuga MS, Ellis JC, Cotton CM, Goldberg RN, Wynn JL, Jackson RB, Seed PC (2011) Beyond bacteria: a study of the enteric microbial consortium in extremely low birth weight infants. PLoS One 6(12):e27858

Le Chatelier E, Nielsen T, Qin J, Prifti E, Hildebrand F, Falony G, Almeida M, Arumugam M, Batto JM, Kennedy S, Leonard P (2013) Richness of human gut microbiome correlates with metabolic markers. Nature 500(7464):541

LeMieux J (2019) Phage therapy: turning the tables on Bacteria: when engineered to incorporate CRISPR components, phages may overwhelm bacterial defenses or transform bacterial functions. Genet Eng Biotechnol News 39(3):20–22

Levit R, Savoy de Giori G, de Moreno de LeBlanc A, LeBlanc JG (2018) Folate-producing lactic acid bacteria reduce inflammation in mice with induced intestinal mucositis. J Appl Microbiol 125(5):1494–1501

Liu JJ, Kong II, Zhang GC, Jayakody LN, Kim H, Xia PF, Kwak S, Sung BH, Sohn JH, Walukiewicz HE, Rao CV (2016) Metabolic engineering of probiotic Saccharomyces boulardii. Appl Environ Microbiol 82(8):2280–2287

Long J, Yang J, Henning SM, Woo SL, Hsu M, Chan B, Heber D, Li Z (2019) Xylooligosaccharide supplementation decreases visceral fat accumulation and modulates cecum microbiome in mice. J Funct Foods 52:138–146

Maji A, Misra R, Dhakan DB, Gupta V, Mahato NK, Saxena R, Mittal P, Thukral N, Sharma E, Singh A, Virmani R (2018) Gut microbiome contributes to impairment of immunity in pulmonary tuberculosis patients by alteration of butyrate and propionate producers. Environ Microbiol 20(1):402–419

Man WH, de Steenhuijsen Piters WA, Bogaert D (2017) The microbiota of the respiratory tract: gatekeeper to respiratory health. Nat Rev Microbiol 15(5):259

Marotz CA, Zarrinpar A (2016) Focus: microbiome: treating obesity and metabolic syndrome with fecal microbiota transplantation. Yale J Biol Med 89(3):383

McVey Neufeld KA, Mao YK, Bienenstock J, Foster JA, Kunze WA (2013) The microbiome is essential for normal gut intrinsic primary afferent neuron excitability in the mouse. Neurogastroenterol Motil 25(2):183–e88

Menni C, Jackson MA, Pallister T, Steves CJ, Spector TD, Valdes AM (2017) Gut microbiome diversity and high-fibre intake are related to lower long-term weight gain. Int J Obes 41(7):1099

Messaoudi M, Lalonde R, Violle N, Javelot H, Desor D, Nejdi A, Bisson JF, Rougeot C, Pichelin M, Cazaubiel M, Cazaubiel JM (2011) Assessment of psychotropic-like properties of a probiotic formulation (Lactobacillus helveticus R0052 and Bifidobacterium longum R0175) in rats and human subjects. Br J Nutr 105(5):755–764

Moffatt MF, Cookson WO (2017) The lung microbiome in health and disease. Clin Med 17 (6):525–529

Morjaria S, Schluter J, Taylor BP, Littmann ER, Carter RA, Fontana E, Peled JU, van den Brink MR, Xavier JB, Taur Y (2019) Antibiotic-induced shifts in fecal microbiota density and composition during hematopoietic stem cell transplantation. bioRxiv 1:606533

Morris A, Beck JM, Schloss PD, Campbell TB, Crothers K, Curtis JL, Flores SC, Fontenot AP, Ghedin E, Huang L, Jablonski K (2013) Comparison of the respiratory microbiome in healthy nonsmokers and smokers. Am J Respir Crit Care Med 187(10):1067–1075

Nayfach S, Pollard KS (2015) Average genome size estimation improves comparative metagenomics and sheds light on the functional ecology of the human microbiome. Genome Biol 6(1):51

Neto MP, de Souza AJ, da Silva LD, de Oliveira SR, de Lima Guimaraes KS, de Oliveira Y, de Souza EL, Magnani M, Vidal H, de Brito Alves JL (2018) Gut microbiota and probiotics intervention: a potential therapeutic target for management of cardiometabolic disorders and chronic kidney disease? Pharmacol Res 130:152–163

Obermeier B, Daneman R, Ransohoff RM (2013) Development, maintenance and disruption of the blood-brain barrier. Nat Med 19(12):1584

Ochoa-Repáraz J, Mielcarz DW, Ditrio LE, Burroughs AR, Begum-Haque S, Dasgupta S, Kasper DL, Kasper LH (2010) Central nervous system demyelinating disease protection by the human commensal Bacteroides fragilis depends on polysaccharide A expression. J Immunol 185 (7):4101–4108

Ochoa-Repáraz J, Mielcarz DW, Begum-Haque S, Kasper LH (2011) Gut, bugs, and brain: role of commensal bacteria in the control of central nervous system disease. Ann Neurol 69(2):240–247

Ohland CL, Kish L, Bell H, Thiesen A, Hotte N, Pankiv E, Madsen KL (2013) Effects of Lactobacillus helveticus on murine behavior are dependent on diet and genotype and correlate with alterations in the gut microbiome. Psychoneuroendocrinology 38(9):1738–1747

Ozturk AB, Turturice BA, Perkins DL, Finn PW (2017) The potential for emerging microbiome-mediated therapeutics in asthma. Curr Allergy Asthma Rep 17(9):62

Park JY, Moon BY, Park JW, Thornton JA, Park YH, Seo KS (2017) Genetic engineering of a temperate phage-based delivery system for CRISPR/Cas9 antimicrobials against Staphylococcus aureus. Sci Rep 7:44929

Peng J, Xiao X, Hu M, Zhang X (2018) Interaction between gut microbiome and cardiovascular disease. Life Sci 214:153

Piccio L, Stark JL, Cross AH (2008) Chronic calorie restriction attenuates experimental autoimmune encephalomyelitis. J Leukoc Biol 84(4):940–948
Pothoulakis C (2009) Anti-inflammatory mechanisms of action of Saccharomyces boulardii. Aliment Pharmacol Ther 30(8):826–833
Ransohoff RM, Engelhardt B (2012) The anatomical and cellular basis of immune surveillance in the central nervous system. Nat Rev Immunol 12(9):623
Rodríguez-Nogales A, Algieri F, Garrido-Mesa J, Vezza T, Utrilla MP, Chueca N, García F, Rodríguez-Cabezas ME, Gálvez J (2018) Intestinal anti-inflammatory effect of the probiotic Saccharomyces boulardii in DSS-induced colitis in mice: impact on microRNAs expression and gut microbiota composition. J Nutr Biochem 61:129–139
Ross AA, Doxey AC, Neufeld JD (2017) The skin microbiome of cohabiting couples. MSystems 2 (4):e00043–e00017
Sarkar S, Das B, Banerjee SK (2018) Insights into the human gut microbiome and cardiovascular diseases. J Pract Cardiovasc Sci 4(1):10
Schepper JD, Irwin R, Kang J, Dagenais K, Lemon T, Shinouskis A, Parameswaran N, McCabe LR (2017) Probiotics in gut-bone signaling. In: Understanding the gut-bone signaling axis. Springer, Cham, pp 225–247
Segal LN, Alekseyenko AV, Clemente JC, Kulkarni R, Wu B, Chen H, Berger KI, Goldring RM, Rom WN, Blaser MJ, Weiden MD (2013) Enrichment of lung microbiome with supraglottic taxa is associated with increased pulmonary inflammation. Microbiome 1:19
Shen ZH, Zhu CX, Quan YS, Yang ZY, Wu S, Luo WW, Tan B, Wang XY (2018) Relationship between intestinal microbiota and ulcerative colitis: mechanisms and clinical application of probiotics and fecal microbiota transplantation. World J Gastroenterol 24(1):5
Silveira AL, Ferreira AV, de Oliveira MC, Rachid MA, da Cunha Sousa LF, dos Santos MF, Gomes-Santos AC, Vieira AT, Teixeira MM (2017) Preventive rather than therapeutic treatment with high fiber diet attenuates clinical and inflammatory markers of acute and chronic DSS-induced colitis in mice. Eur J Nutr 56(1):179–191
Sivamaruthi BS, Kesika P, Suganthy N, Chaiyasut C (2019) A review on role of microbiome in obesity and Antiobesity properties of probiotic supplements. BioMed Res Int 2019:1
Soares RL (2014) Irritable bowel syndrome: a clinical review. World J Gastroenterol *20*(34):12144
Sobol CV (2014) How microbiome impact on the cardiovascular system. J Clin Trial Cardiol 1:1411
Swidsinski A, Loening-Baucke V, Lochs H, Hale LP (2005) Spatial organization of bacterial flora in normal and inflamed intestine: a fluorescence in situ hybridization study in mice. World J Gastroenterol 11(8):1131
Takata K, Kinoshita M, Okuno T, Moriya M, Kohda T, Honorat JA, Sugimoto T, Kumanogoh A, Kayama H, Takeda K, Sakoda S (2011) The lactic acid bacterium Pediococcus acidilactici suppresses autoimmune encephalomyelitis by inducing IL-10-producing regulatory T cells. PLoS One 6(11):e27644
Tang WW, Hazen SL (2017) The gut microbiome and its role in cardiovascular diseases. Circulation 135(11):1008–1010
Tarrah A, De Castilhos J, Rossi RC, Duarte VD, Ziegler D, Corich V, Giacomini A (2018) In vitro probiotic potential and anti-cancer activity of newly isolated folate-producing Streptococcus thermophilus strains. Front Microbiol 9:2214
Teitelbaum AA, Gareau MG, Jury J, Yang PC, Perdue MH (2008) Chronic peripheral administration of corticotropin-releasing factor causes colonic barrier dysfunction similar to psychological stress. Am J Physiol Gastroint Liver Physiol 295(3):G452–G459
Thaiss CA, Zmora N, Levy M, Elinav E (2016) The microbiome and innate immunity. Nature 535 (7610):65–74
Ticinesi A, Lauretani F, Tana C, Nouvenne A, Ridolo E, Meschi T (2019) Exercise and immune system as modulators of intestinal microbiome: implications for the gut-muscle axis hypothesis. Exerc Immunol Rev 25

Tomayko E, Pillsbury L, Pray L (2013) The human microbiome, diet, and health: workshop summary. National Academies Press, Washington, DC

Turnbough L, Wilson L (2007) " Take your medicine": nonadherence issues in patients with ulcerative colitis. Gastroenterol Nurs 30(3):212–217

Tursunov D, Yoshida Y, Yrysov K, Sabirov D, Alimova K, Yamamoto E, Reyer JA, Hamajima N (2018) Estimated costs for treatment and prophylaxis of newborn vitamin K deficiency bleeding in Tashkent, Uzbekistan. Nagoya J Med Sci 80(1):11

Valdes AM, Walter J, Segal E, Spector TD (2018) Role of the gut microbiota in nutrition and health. BMJ 361:k2179

Vandeputte D, Falony G, Vieira-Silva S, Wang J, Sailer M, Theis S, Verbeke K, Raes J (2017) Prebiotic inulin-type fructans induce specific changes in the human gut microbiota. Gut 66 (11):1968–1974

Wang Y, Kasper LH (2014) The role of microbiome in central nervous system disorders. Brain Behav Immun 38:1–2

Wang B, Yao M, Lv L, Ling Z, Li L (2017) The human microbiota in health and disease. Engineering 3(1):71–82

Wu M, McNulty NP, Rodionov DA, Khoroshkin MS, Griffin NW, Cheng J, Latreille P, Kerstetter RA, Terrapon N, Henrissat B, Osterman AL (2015) Genetic determinants of in vivo fitness and diet responsiveness in multiple human gut Bacteroides. Science 350(6256):5992

Yatera K, Noguchi S, Mukae H (2018) The microbiome in the lower respiratory tract. Respir Investig 56(6):432–439

Ye L, Chan EW, Chen S (2019) Selective and suppressive effects of antibiotics on donor and recipient bacterial strains in gut microbiota determine transmission efficiency of bla NDM-1-bearing plasmids. J Antimicrob Chemother 74(7):1867–1875

Yosef I, Manor M, Kiro R, Qimron U (2015) Temperate and lytic bacteriophages programmed to sensitize and kill antibiotic-resistant bacteria. Proc Natl Acad Sci 112(23):7267–7272

Yoshii K, Hosomi K, Sawane K, Kunisawa J (2019) Metabolism of dietary and microbial vitamin B family in the regulation of host immunity. Front Nutr 6:48

PART VIII

Microbes in Forensic Science

An Introduction to Microbial Forensics

21

Richa Aeri, Jasbir Arora, Priyanka Kapoor, and Arun Dev Sharma

Abstract

An incredible life is present beyond the scope of the naked eye. The one present underneath the lens of a microscope is the life of "microorganisms." Microorganisms are present everywhere in the environment. Human body serves as a host to a wide variety of microbes including bacteria and viruses. Though most of the microorganisms are harmless to humans, they can cause diseases in humans as well as in animals and plants. Microorganisms play a vital role in food and dairy industries; production of enzymes, amino acids, vitamins, antibiotics, etc.; genetic engineering; biotechnology; and so on. With advancement in science and technology, the antisocial elements are manufacturing bioweapons by using the microorganisms and the toxins produced by them. Microbial forensics refers to deployment of scientific principles to analyze microbial evidence. Technologies like massive parallel sequencing (MPS) and next-generation sequencing (NSG) help scientists to understand the role of microorganisms in the origin of biocrimes, cause and time of death, sexual assault, homicide, agricultural contamination, and medical malpractice. The microbial forensic investigation results in linking the causative agent with a specific group or an individual by following special procedures.

Keywords

Forensic microbiology · Microbial biodiversity · Biocrimes · Bioterrorism · Postmortem interval · Sexual assault · Forensic investigation

R. Aeri (✉) · A. D. Sharma
Department of Biotechnology, Lyallpur Khalsa College, Jalandhar, Punjab, India

J. Arora
Department of Anthropology, Panjab University, Chandigarh, Punjab, India

P. Kapoor
Department of Forensic Science and toxicology, Chandigarh University, Gharuan, Punjab, India

S. G. Sharma et al. (eds.), *Microbial Diversity, Interventions and Scope*,
https://doi.org/10.1007/978-981-15-4099-8_21

21.1 Introduction

The term forensic science broadly refers to the science pertaining to the administration of law and justice. Forensic science basically incorporates and utilizes the techniques and principles of basic sciences for the purpose of administration of law. Forensic science works in conjunction with the experts from many subjects like physics, chemistry, biology, zoology, botany, microbiology, etc. Thus, it is an interdisciplinary or a multidisciplinary field, where, depending upon the need of the case, scientists of different expertise assist the court of law. Almost every forensic science laboratory in the world has well-established disciplines like forensic ballistics, physics, chemistry, toxicology, biology, entomology, etc., but microbial forensics is comparatively new and a young field for the forensic laboratories and scientists as well.

Microorganisms are ubiquitous organisms and are attentive observers to our world. Microorganisms are present everywhere, including the air we breathe and the food we eat. Human population is host to a wide variety of microbes, including bacteria and viruses that are present on our skin, in our alimentary and respiratory tract (mouth and nose), conjunctiva, and so on. Specifically, the human skin is a habitat of countless numbers of eukaryotic microbes including fungi and protists and represents a diverse and dynamic ecosystem which has evolved to colonize almost every available habitat on the human body as well as within the integumentary system (Roth and James 1988; Grice et al. 2009). The presence of a large number of microorganisms on human skin enables the analysis of forensically interesting body sites (Costello et al. 2009; Sender et al. 2016).

According to reports, microbes are present in the ratio of 4:3 to the cells of human body and weigh about 0.2 kg of a 70-kg human body (Rana and Manhas 2018). The number of microbial fauna varies from <10 to $>10^7$ microorganisms per square cm (Bojar and Holland 2002). In addition to the microbial fauna found in the environment, forensic scientists have been exploring the microbes present in and on the human body. Apart from being present in high numbers, the microbial communities associated with humans are highly diverse at various body sites (Costello et al. 2009). This is because of the different thriving conditions present on the human body. For instance, the skin is dry, moist, or sebaceous at different sites, thereby causing different kinds of microbes to thrive on these different microenvironments (Grice et al. 2009). Hence, the interaction of microbes with humans is inevitable.

Though most of the microorganisms are harmless to us, however, their activities can produce diseases in humans, as well as in animals and plants. They need a host mechanism to survive and proliferate. Diseases are caused due to this adaptive strategy of microbes for replication and survival within the host (Cummings and Relman 2002). However, diseases may also be induced by a forced interaction between microbes and a host or by altering or manipulating the genome of the microbe. This interaction may be induced as a result of the acts of bioterrorism (Cummings and Relman 2002) which involves malafide human intention.

In the forensic scenario as well, microorganisms play a vital role and have a variety of forensic applications, the foremost being the use of microbes as biological weapons in biocrimes or bioterrorism. However, due to advancements in recent technology, specifically the massive parallel sequencing (MPS) and the next-generation sequencing (NSG), the scope of microbial forensics has been expanded, and microbial analysis can be used to understand the role of microorganisms in the process of decomposition, thereby aiding in estimation of postmortem interval. Additionally, microbes are present in the soil as forensic biomarkers; helps in personal identification from body fluids like saliva; assist in understanding the epidemiology of a disease, thereby aiding in linking a victim and a suspect through microbial diseases (Oliveira and Amorim 2018).

21.2 Microbes as Biological Agents in Biocrimes

Biocrime or bioterrorism refers to the threat raised due to use of microorganisms and their toxins and related products to commit acts of crime in order to terrorize people. Such acts of crime can be harmful to humans as well as animals and plants and can lead to the outbreak and spread of various microbial infections and diseases which may ultimately lead to an epidemic.

The Centers for Disease Control and Prevention (CDC) defines biocrimes or bioterrorism as the crimes that involve the deliberate usage of living organisms or the toxins produced by them to harm animals, plants, humans, or environment so as to disrupt the social stability and cause panic in the society (Jansen et al. 2014). The National Center for Biotechnology Information (NCBI) defines a biocrime as "an assault crime, except, instead of a gun or knife, the weapon used is a pathogen or a toxin" (Schutzer et al. 2005). Although commonly termed "bioterrorism," the purposes of such attacks may not necessarily be intended and confined to terrorize the established government structures but can also be influenced by religious, political, or ecological beliefs (Carus 2001).

The traditional terrorists use violence as a means to an end, while bioterrorists view biological agents as specialized tools to cause mass destruction by spreading diseases.

On the basis of risk capability of the microbes, the public health agency like CDC of the United States (2002) classified bioweapons into three categories: categories A, B, and C. Category A includes the agents responsible for high mortality, easy dissemination, and a high potential to cause major health concerns, while categories B and C are moderate and easy to disperse, respectively. Category A includes organisms like *Francisella tularensis*, *Bacillus anthracis*, *Yersinia pestis*, and hemorrhagic Ebola virus and botulinum neurotoxins, while category B includes *Brucella* species, abrin, ricin, *Clostridium perfringens* E. toxin, and *Burkholderia mallei*. Hantavirus and Nipah viruses are examples of microbes in category C (Smart 1997; Grundmann 2014; Jansen et al. 2014; Kaur et al. 2014; Janik et al. 2019).

The concept of biocrime is not new. It can be traced well throughout history but was not scientifically advanced (Bhatia et al. 2016). Biocrimes can be witnessed from the time of Bible. The Bible documents that God demanded the freedom of Israelite slaves by the Pharaoh. However, when the Pharaoh did not agree, God sent ten deadly plagues upon Egypt. One of them was Shkhim (the sixth plague) which is a skin disease causing boils on the whole body. It is believed that anthrax was a cause of these boils, and it spreads when Moses sprinkles soot in the air (Mickley 2010).

An incident considered to be the most plausible was the use of bioweapon in 1346 by the Mongols. The Tartar army besieged the city of Caffa in the fourteenth century. The soldiers of the Tartar army started to die due to plague, and the survivors threw the infected bodies over the wall into the city in order to spread the disease among the residents, which caused an epidemic resulting in the residents leaving the place and the troops taking over the city of Caffa (Carus 2017).

Another example of bioweapon which could be quoted here is of "ricin." Ricin is a plant toxin that is extracted from castor plant *Ricinus communis*. The CDC has classified it as a category B pathogen that is easy to disseminate and has low mortality. Many cases have been reported in the past where conviction has been made for the possession or misuse of ricin. In a well-known case, a Bulgarian exile, Georgi Markov who was a novelist and a play writer, had published anticommunist views. On September 7, 1978, while standing at the bus stop, he felt a stab in his right thigh. It was found that a man had injected a ricin pallet in his thigh using an umbrella-like weapon (Crompton and Gall 1980). Subsequently, he suffered from gastroenteritis and fever and reportedly died within 3 days of the occurrence of the event. Autopsy report concluded that the death was caused by ricin (Duncan and Smith 2012). The bioterrorist attacks have been a matter of concern in the present century since 2001. In 2001, following the attack on World Trade Center, an unknown person sent bacterium *B. anthracis* spores through letters to news and congressional offices in the United States. This agent is the causative agent of the disease anthrax. In this attack, 5 people died and 17 others were infected due to inhalation of anthrax spores. The Federal Bureau of Investigation (FBI) investigated the case, and Bruce Ivins was found to be the perpetrator of this attack. But, unfortunately, he committed suicide before he could face the trial in the court. This attack actually threw spotlight on the field of microbial forensics.

The bioweapons or microorganisms used for the commission of biocrimes may be natural, wild-type strains or genetically engineered microbes (Arora et al. 2002). The use of microbes as agents of bioterrorism is becoming a choice due to the ease of production and high toxicity (Rossodivita et al. 2019). Moreover, these are cost-effective weapons, need no expertise, and are less sophisticated but are more lethal and powerful (Budowle 2003; Moghaddasi et al. 2018). Further, most of the pathogens are endemically found and can be disseminated with ease. Furthermore, most of the organisms are quite stable and are transmittable (e.g., smallpox, anthrax, HIV, etc.), thereby causing potential threat with just a few initial outbreaks. Microbial forensics seeks to answer the basic questions including identification of the threat agent and linking it to a probable source, if possible. Due to the potential risks involved, the forensic scientists need to develop methods for identification and attribution of the microbes used in biocrimes. However, this becomes challenging as there are millions of species of microbes with a unique microbiome of their own. This makes it even more complicated to identify and individualize the causative microbe involved in any acts of biocrimes (Taylor et al. 2001).

The results of biowarfare often lead to the death of the host organism or to the toxicity of water or soil (Green et al. 2019; Janik et al. 2019). Apart from humans, the next potential threat is borne by agriculture and food, as both are exceptionally tempting targets of attack due to economic, social, as well as political impact they have (Rogers et al. 1999; Wilson et al. 2001; Budowle et al. 2005). The agents causing foodborne diseases can also be used as agents of bioterrorism, as no sophisticated equipment is required to spread the microbes involved; rather, the perpetrator generally needs a readily available, infested agent.

Microbial forensics, as the name suggests, refers to the deploying of scientific principles and techniques to analyze microbial evidence. Budowle (2003) defines microbial forensics as "a scientific discipline dedicated to analyze evidence from an act of bioterrorism, biocrime or inadvertent microorganism or toxin release for attribution purposes." Microbial evidence is the evidence produced from any act of bioterrorism, biocrime, or an inadvertent release of toxin or microorganism (Budowle et al. 2003).

The misuse of biological agents poses severe threat to the society as well as the general health and well-being of the individuals. As said by Murch (2003), "Science and technology are used to serve as independent 'witnesses' in criminal or civil matters, intelligence, and policy." Microbial forensics, hence, helps in the attribution of an offense, based upon the analysis of biological and microbial evidence. The disease-causing microbes and pathogens and the toxins released by them are identified and distinguished by the law enforcement agencies in collaboration with scientists from various fields like microbiology, genetics, public health, and so on.

21.3 Role of Microorganisms in the Decomposition Process

Decomposition is a natural process and refers to the breakdown of dead organic matter to smaller particles, thereby resulting in release of carbon dioxide, water, minerals, and energy into the atmosphere. It is a complex process in which both biotic and abiotic factors come into play and is aided by two different groups of organisms, namely, the scavengers and the microorganisms, which are considered as true decomposers (Saraswat et al. 2008).

The biotic factors include but are definitely not confined to the stature, size, and weight of the cadaver and presence of intrinsic and extrinsic bacteria, other microbes, insects, and scavengers. The abiotic factors include climatic conditions and humidity (Evans 1963; Vass 2001; Carter et al. 2008; Janaway et al. 2009; Pechal et al. 2013).

When an animal dies, a series of events take place, beginning from the shutdown of its immune system leading to a change in the internal temperature of the body and subsequently the initiation of the growth of microorganisms (Weiss et al. 2016). A dead body, at this stage, can in fact be compared to an ecosystem hosting a variety of bacteria, insects, and fungi (Hyde et al. 2013).

The increase in the internal temperature causes the skin of the cadaver to rupture, thereby allowing the interchange of air, microbes, and body fluids. A carcass releases large amounts of nitrogen into the environment, mostly in the form of ammonia, along with other elements like carbon and phosphorus. Decomposition is influenced by various taphonomic processes, which are in turn influenced by the geographical area where the cadaver has been discovered (Goff 1993).

Initially, the bacteria begin with the digestion of the intestine, thereby causing decomposition of the body from inside to the outside (Janaway et al. 2009). Autolysis of the cells of the cadaver also begins, and the internal bacteria undergo anaerobic respiration to release putrefying gases like hydrogen sulfide, methane, cadaverine, and putrescine (Vass 2001). However, if the cadaver is frozen immediately following the death, then the decomposition occurs from outside to inside, after the body is brought to a temperature above 0 °C. Due to the accumulation of gases in the internal organs, the body bloats. Putrefaction begins during the bloat stage, and the cadaver looks inflated like a balloon. The cadaver becomes a site of attraction as well as infestation for the adult blowflies (Calliphoridae).

Next to bloating is the decay stage, which is marked by the rupture of the outer layer of the skin resultantly leading to deflation of the body as the gases present in the abdomen escape and the decay initiates. The strong odor associated with decomposition is evident, and the cadaver at this stage becomes the breeding ground for large masses of Diptera larvae. Most of the Calliphoridae present generally complete their development and leave the decomposing remains. By the end of the decay stage, only the skin and the cartilages are visible (Catts and Haskell 1990; Hewadikaram and Goff 1991), and this marks the beginning of the post-decay stage, where Diptera are no longer the principal organisms in the cadaver. In dry and arid xerophytic and mesophytic conditions, Diptera generally gets replaced by the adult Coleoptera, who feeds on the cadaver, thereby removing the remaining flesh and cartilage so as to expose the clean bones. The post-decay stage is followed by the skeletal stage, wherein a number of soil-dwelling microorganisms can be seen for an initial period.

In order to determine the postmortem interval (PMI), the stages of decomposition form an important parameter. Several studies have reported the stages of decomposition as a function of temperature to find out the time intervals for decomposition. Megyesi et al. (2005) devised a scoring system to evaluate decomposition as a function of time by using bloating and purging of the head and trunk to mark the end of early decomposition. However, it has been reported that bloat and purge are quite proficient to evaluate all the stages of decomposition (Hyde et al. 2013).

Though bloat has been profusely used as a parameter to estimate time since death, limited literature is accounted for the action of internal microorganisms (Melvin et al. 1984; Hopkins et al. 2000; Howard et al. 2010). There are, however, many studies to show the role of microorganisms in decomposition in general (Melvin et al. 1984; Kakizaki et al. 2008; Meyers and Foran 2008; Stokes et al. 2009; Howard et al. 2010; Lenz and Foran 2010; Dickson et al. 2011; Butzbach et al. 2013; Schoenen and Schoenen 2013).

The geographical conditions like climate and topography also affect the process of decomposition. In dry and arid regions, with strong air current, the cadaver tends to mummify (Wescott 2018), while in warm and humid regions, the decomposition is quick (Powers et al. 2009; Forbes et al. 2014). Hence, the plentitude and variety and dissemination of the microorganisms in the cadaver are highly significant parameters in the determination of the postmortem interval.

21.4 Soil Microbes as Forensic Biomarkers

The study of soil is known as "pedology." Soil is a complex substance which contains minerals as well as decaying organic matter. This organic matter supports microbial life, the presence of which causes significant changes in the soil profile. Soil analysis is potent for linking a person, animal, or plant to a specific place. Off lately, most of the forensic studies on soil were based upon the chemical as well as the mineral profile of the soil (Tibbett and Carter 2009; Ruffell 2010). This required large amounts of soil sample and lacked a good mapped soil database (Zala 2007). However, due to extremely diverse microbial communities present in the soil, the focus is now shifting to the microbes for identification purposes.

The soil microbes thrive well in a specific pH, the alteration to which can affect the microbial density in a particular soil type. Generally, with an increase in the pH, the microbial biomass decreases (Santiago-Rodriguez TM 2016). Changes in pH may further lead to alterations in the cellular processes like DNA replication and transcription, thereby resulting in specific changes which may allow the specific population of microbes to survive in varied pHs (Eilers et al. 2012).

Biological factors including anthropogenic activities like agriculture and discharge of feces and sewage also influence the soil microbial diversity (Brookes and Mcgrath 1984; Elliott 1986; Stockdale and Brookes 2006). There are other external sources of alteration such as decomposition of a cadaver which can alter the soil composition. When soil comes in contact with a decomposing cadaver, the microbes in the immediately surrounding areas are likely to get highly affected. As a result of decomposition, significant quantity of decomposing organic matter from the cadaver may leak into the soil (Vass et al. 2004), thereby causing changes in the soil profile. These changes may be quantified and may assist in the crime scene reconstruction by relating the cadaver to the site of its presence by way of the analysis of soil microbes.

Most of the studies relating to soil microbiology have been undertaken using 16S ribosomal RNA gene sequencing using terminal restriction fragment length polymorphism analysis (T-RFLP) (Quaak and Kuiper 2011) or amplicon length heterogeneity-polymerase chain reaction (ALH-PCR) (Moreno et al. 2006). The rRNA gene is only about 1.5 kilobases long, due to which it is easy to sequence and is cost-effective (Gunn 2009). The study of soil microbes is advantageous as a very small sample size is required owing to the plentitude of soil microbes. Further, the techniques involved are comparatively cheap and automated (Hill et al. 2007; Sensabaugh 2009).

21.5 Personal Identification from Microbes Present in Saliva

Saliva is a fluid secreted by the salivary glands found in the mouth of humans and other animals. In addition to the salivary fluid, it contains exfoliated epithelial cells of the mouth as well as the microorganisms. Saliva is of utmost importance to forensic scientists as it is easy to sample and the technique used for collection is less invasive than sampling blood or urine.

Saliva may often be encountered in sexual offenses along with bite marks or lip prints and can aid in personal identification of the culprit. The methods of identification mainly depend upon analysis of human DNA. Nevertheless, the human DNA may be degraded or masked and may be present in less quantity. The bacterial DNA, therefore, is of immense utility as it is comparatively better resistant to degradation than human DNA (Leake et al. 2016). The differentiation between identical twins may also be possible by using bacterial DNA (Stahringer et al. 2012). Almost 99% of the environmental bacteria cannot be cultured in the lab (Handelsman 2004). However, with the advent of modern technological aids like the next-generation sequencing (NGS), the bacteria can be employed for a variety of forensic purposes. Studies have reported that it is possible to recover live bacteria from bite sites if they are undisturbed for 24 h (Borgula et al. 2003). It has further been reported that microbes present on the skin can be potent to link an individual to the objects touched by them (Fierer et al. 2010).

A lot of studies have shown the composition of the bacterial communities present in the saliva (Costello et al. 2009; Lazarevic et al. 2009; Zaura et al. 2009; Lazarevic et al. 2010; Caporaso et al. 2011), but the extent up to which it can differentiate between two individuals needs to be explored. A lot of research is required to establish the reliability of bacterial microbiome in personal identification.

21.6 Role of Microbes in Reconstruction of Crime Scene

Crime scene reconstruction is the process that helps the investigators to interpret the evidences found at the crime scene. Law enforcement agencies play a crucial role in assessing and investigating the evidences, wherein microbial forensics comes as an aid to analyze and characterize the microbial evidence for crime scene reconstruction.

21.6.1 Role in Epidemiology

The World Health Organization (WHO) defines epidemiology as the study of the distribution and determinants of health-related states or events (including disease) and the application of this study to control the diseases and other health problems. Epidemiology studies the occurrence of disease as well as factors responsible for the occurrence of disease in populations. Biocrimes also come under the purview of the general principles of epidemiologic investigation involving a biological agent

(Flowers et al. 2002). In case of an outbreak of a disease, certain important points have to be taken into consideration, namely, determination of the occurrence of an outbreak, identification of the population at risk, determination of the method of dispersal and dissemination of disease, and last but not the least, identification and characterization of the causal agent (Morse and Budowle 2006). In order to link the donor and recipient of an infectious disease, the associated genetic marker, exhibiting the following important criteria, has to be assessed:

(a) Sufficient variation must be exhibited by the genetic markers so as to exclude the unlinked individuals.
(b) Such variation must be easily identifiable.
(c) The genetic variation should neither arise at a very slow or a very fast pace so as to compromise one's ability to link the infected people together (Gunn 2009).

The use of epidemiologic methods for investigating the suspicious health problems/diseases or evidences relating to intentional acts or criminal behavior is called forensic epidemiology. In fact, forensic science and epidemiology must be treated as an integrated discipline, as many of the methods and procedures followed in epidemiology are similar to the ones used in microbial forensics (Goodman et al. 2003). The investigation under microbial forensics differs from epidemiological investigations on the point that the former are scrutinized and evaluated by the scientists as well as the law enforcement agencies while the latter is evaluated by scientists alone.

21.6.2 Cause and Manner of Death

When a person dies, one of the main objectives is to determine the cause and manner of death. Microbes can be of immense utility in some cases to determine whether an individual died due to a natural cause, such as disease, or a biological attack (Gunn and Sarah 2012). If the cause of death remains unknown till the time the individual was alive, then postmortem analysis along with serological and/or genomic analysis may help to deduce the underlying cause of death. If the disease-causing agent is identified prior to death, the analysis confirming the agent to be the cause of death may still be desired post death. Knowledge about the cause of death is helpful in treating other patients having similar ailment; and it may also help in preventing the dissemination of disease by either warning the public or quarantining a suspicious contaminated agent/source.

The role of microbes to determine the cause of death can be understood by quoting the example of SIDS (sudden infant death syndrome), which is one of the leading causes of post-neonatal infant mortality in the world. It has been suggested that microorganisms play a crucial role in many of the cases of SIDS (Gleeson and Cripps 2004). Studies have reported that viral infections caused by human herpesvirus-6, Epstein–Barr virus, and cytomegalovirus were present in infants that died from SIDS (Alvarez-Lafuente et al. 2008). SIDS has also been reported to be associated with bacteria like *Staphylococcus aureus*, *Escherichia coli*, *Streptococcus* spp., and *Haemophilus influenzae* (Blackwell 2004). The cause of death was inferred by relating the presence of specific microbes in the deceased individuals.

Attribution of microbial infection as a cause of death must be done with utmost care because numerous bacteria are otherwise present in the human body in a sterile state. However, these sterile bacteria may start taking advantage of the lack of a healthy immune system in case of death of the host (Janaway et al. 2009).

21.6.3 Drowning Cases

In cases of drowning, microbial communities present in the water may provide useful evidence about the cause of death as well as the place of drowning. The diatoms are one of the most common microorganisms which are indicative of drowning as the cause of death (Timperman 1972). Diatoms are unicellular, microscopic phytoplanktons containing silica cell walls. The silica cell wall is resistant to degradation from acid, enzymes, and temperature, due to which they are present in the drowned bodies (Lin et al. 2014; Rana and Manhas 2018). They have been found in lungs, blood, bone marrow, and other internal organs (Ludes et al. 1999). Diatoms are exclusively present in cases where the person drowned in natural water body like sea, lake, or river. The water bodies containing treated water like a swimming pool generally lack the diatoms due to the process of water treatment (Lin et al. 2014).

21.6.4 Sexual Assault Cases

Cases involving sexual assault are another category where microbial forensics helps link the victim to the suspect. In cases of sexual assault, a victim can generally furnish a statement about the occurrence of the crime. However, in certain instances, the victim cannot speak for himself, as in cases of child sexual abuse. In such cases, microbial forensic analysis is of utmost significance as it may help link the victim to the culprit by detecting the sexually transmitted pathogens. This may be done by using various methods like nucleic acid amplification tests (NAATs) (Jaureguy et al. 2016) and cultures to analyze the body fluids collected by way of swabs. The bed sheets, blankets, clothing, hair, as well as fibers may also be additionally analyzed. A number of pathogens are reported to cause sexually transmitted diseases (STDs), including *Neisseria gonorrhoeae*, *Chlamydia trachomatis*, and *Trichomonas vaginalis*, to name a few. STDs can be transmitted during sexual assault and can be of forensic significance in cases involving transmission of an infectious disease as a weapon of biocrime.

21.6.5 Human Immunodeficiency Virus (HIV)

HIV is another example of a natural disease-causing pathogen that has been reported as a means of commission of a biocrime (Ou et al. 1988; Ou et al. 1992; Robbins et al. 2003). Though viruses can infect all life forms, they cannot replicate without the host mechanism. A basic virus is comprised of a nucleic acid genome which is present inside a protein coat. Viruses may ether be composed of DNA or RNA. HIV is an RNA virus, which evolves expeditiously (Carrillo and Rock 2005) and may exhibit high rates of mutation (Drake et al. 1998; Preston et al. 1988; Sala and Wain-Hobson 2000; Svarovskaia et al. 2004). Therefore, there is a high probability that two HIV samples having common origin differ at a number of nucleotides within the genome. Hence, it is impossible to obtain an exact match between the virus molecular profile from an alleged donor and recipient.

Investigations in HIV cases involve both gathering the factual information referred to as epidemiological investigation as well as the genetic and/or serologic data. Further, phylogenetic analysis comes as an aid in supporting or contradicting the relationships of samples that have an alleged common origin in comparison to those which do not.

21.7 Collection of Microbial Evidence

Evidence collection is of paramount importance for any forensic investigation, including microbial forensic investigation. Degradation or contamination of evidence may prove detrimental for the investigation and in turn for the whole process of attribution (Budowle et al. 2003). Investigations pertaining to microbial forensics are potentially similar to other forensic investigations, wherein similar steps of crime scene investigation, evidence collection and preservation, analysis of evidence, and maintaining a chain of custody are to be followed. Microbiological evidence may include viable samples of the microbes, their toxins, nucleic acids, specimens from victims, laboratory equipment, dissemination devices and their contents, environmental samples, contaminated clothing, or trace evidence specific to the process that produced the particular biological agent. Therefore, the method of collection should be sensitive and reliable.

The biocrimes are generally reported by the first responders including health-care providers, police, paramedics, firefighters, or even the general public. The first responders are followed by the crime scene investigators. For most of the cases involving microorganisms, a plan describing the detailed strategy must be devised prior to the initial sampling. This is so because the microbes involved in that very particular case may be hazardous and fatal for the crime scene investigators. The investigators must wear protective equipment so as to protect them from the hazards related to the microbial evidences (CDC 2002).

The samples to be collected must include every material found on the scene which must be well labeled indicating the time and site of collection. The name of the person who has collected the sample should also be mentioned.

Evidence collection may be done following three general approaches, the foremost being collection of the whole item and transporting it to a safer place for further sampling and analysis. In case the whole item is too large to be collected and transported, a part of the item can be sampled and preserved by using techniques like vacuuming or filtration. For trace materials, swabbing or wiping the contaminated surfaces with suitable sample collection devices is suggested (Smith 2011). The sample collection may be done with the help of either dry or wet swabs, wipes, vacuums, filters, aspirating – needles, and so on (Buttner et al. 2001; Buttner et al. 2005). However, the sample collection methods as well as the devices have certain limitations, the primary being the lack of proper validation of the methods in use. Further, it cannot be said with certainty as to which method would be the most effective method for investigation of that particular evidence.

The general methodology for processing the microbial samples is outlined as follows:

- Sample collection
- Sample transportation
- Sample extraction
- Sample analysis (Junkins et al. 2017)

It is obvious that the procedure for each stage will depend upon the organisms to be collected. Extraction techniques will further determine what analyses can be done on each sample. In some cases, the quantity of the sample available may be limited. Moreover, the data on the techniques which could be efficient may or may not be available to the investigator. Further, owing to the diversity of microbes, there is a great deal of uncertainty associated with the analysis. A technique might work well for one organism but at the same time may be deleterious for another (Schutzer et al. 2005). Another critical point to take into account is that the procedures used for sample collection must be developed so as to preserve the traditional forensic evidences like hair, documents, fingerprints, DNA, etc., wherever possible. Minimal disturbance of evidence is appreciated in a forensic setup; hence, the methods of collection must preferably be noninvasive like sterile swabbing methods. Generally, microbes are present in larger numbers in the gastrointestinal (GI) tract and other body openings like buccal cavity and the anal orifice; therefore, collection from these regions is comparatively easier. It has been reported that in general, 0.25–0.50 g of microbial GI material per sample is sufficient for DNA isolation. It has further been suggested to collect replicate samples, if possible (Pechal et al. 2017).

There are many other points of significance which must be followed while on the field, the prime being that the investigator must be aware of as well as careful of the other animals present on the field that may either feed upon the cadaver or may even be harmful to the investigator. One must take precautions on the field so as to protect the cadaver as well as oneself from the necrophagous organisms present on the field. This may generally be achieved by using protective cage made of sturdy material covered by a layer of strong mesh so as to protect the cadaver from scavengers. The large mesh may be further lined by a fine net if required, to prevent the small insects from entering.

In the study of microbes for estimating PMI, insects that feed on the decomposing body as well as those on and under the soil surrounding the cadaver may be of interest (Pechal et al. 2014; Weiss et al. 2016). While collecting soil microbes associated with the decomposing cadaver, the placement of the cadaver on the soil is significant. It is so because the soil present under and around the cadaver is a mixture of microbes originally present in the soil and those which have leached from the decomposing cadaver (Carter et al. 2007). This in turn influences the diversity and density of soil microbes; hence, it is appropriate to sample soil from different depths and document properly the layer of soil which has been sampled (Ranjard et al. 2001; Kakirde et al. 2010). Further, the quantity of the soil sample must be adjusted so as to house the desired amount of microbiome. Soil samples collected within 5 cm from the surface generally requires 5–10 g of soil, while the soil present on the subsurface may require a larger volume so as to be sufficiently rich in biomass for DNA/RNA extraction. Separate sterile devices like tubes or pipes must be used for collecting different samples (Kakirde et al. 2010).

It is advisable to process the samples as soon as possible (within 2 h) so as to prevent any material changes like oxygenation in them. However, in practical field scenario, this may not be accomplished with ease (Rochelle et al. 1994). Therefore, it is advisable to protect the samples from extreme changes in temperature, pH, and ionic strength of a solution, if applicable (Wilfinger et al. 1997). Use of airtight containers must be done cautiously depending upon the type of microbes to be stored. For instance, storing the aerobic samples in airtight containers could affect the microbial composition as it will restrict the exchange of air and gasses. Conversely, storing the anaerobic communities in excess of air might contaminate the sample (Rochelle et al. 1994). When immediate transfer and processing of samples are not possible, it is advisable to have immediate, sample storage facility on the field to facilitate short-term sample storage. This may include innovations like ice packs, coolers, etc. Storage at −20 °C is recommended. Though freezing is used as an aid to preserve the sample in a majority of studies, however, some studies suggest that freezing changes the structure of microbial community (Kakirde et al. 2010; Carroll et al. 2012). Once frozen, the samples must not be thawed unless and until extraction has to be done. Repeated cycles of freezing and thawing must be avoided so as to avoid fragmentation, which may in turn affect characterization of the microbiome (Männistö et al. 2009; Cardona et al. 2012). It is needless to say that a checklist as well as a proper chain of custody must be maintained prior and subsequent to sample collection.

It is needless to say that the samples collected must be transported to the lab as soon as possible on ice or in liquid nitrogen and must be stored at −20 or − 80 °C till the time they are processed.

21.8 Analysis of Microbial Evidence

Classical technique for the identification of microbes involves the biochemical testing of the microorganism in pure culture in different growth media (Maccallum and Hastings 1899). After a microorganism is isolated and obtained in pure culture, it can be analyzed phenotypically, genotypically, or both. Phenotype refers to the set of observable characteristics like the color, colony morphology, growth pattern on agar and in broth, susceptibility to antibiotics, etc., of an individual which results from the interaction of its genotype with the environment. Genotype refers to the specific genetic combination which leads to the phenotype (Slonczewski and Foster 2013). Some microbes can even be identified by the biochemical processes performed by them, e.g., the ability to create the enzyme "catalase," which helps in the breaking down of hydrogen peroxide to oxygen and water (Slonczewski and Foster 2013).

However, though the classical methods are a valuable source for the identification of microorganisms, there are certain limitations of this method. It has been reported that every microorganism cannot be cultured in the lab. In fact, only 1–2% of the microorganisms are reported to be culturable (Amann et al. 1995; Wilson et al. 1997). Further, most of the organisms are fastidious and have specific environmental and nutritional requirements. Furthermore, the organism to be cultivated must be live, and the methods of commercial identification that are not optimized can therefore lead to misidentification. In view of the above limitations, it is very difficult, if not impossible, to obtain a forensically significant culture.

21.8.1 Extraction of DNA for Analysis

The method used for extraction of DNA tends to have a significant impact on the interpretation of results of microbial structure and function (Henderson et al. 2013; Wagner Mackenzie et al. 2015). The method of DNA extraction must be such so as to enable the efficient recovery of both the prokaryotic and the eukaryotic DNA. The microbial DNA can be extracted from the environment either directly, whereby the entire sample is lysed during cell lysis, or indirectly, where the microbial cells first get separated and subsequently undergo lysis during the cell lysis (Ogram et al. 1987; Parachin et al. 2010; Delmont et al. 2011). As both the methods mentioned here yield similar results (Delmont et al. 2011), the choice of method would depend upon the experimental requirement and setup. For example, if the aim of the experiment is to construct large-insert clone libraries, the indirect method would be a better choice as the direct method would lead to the shearing of the genomic DNA. On the other hand, if the aim involves deploying the DNA for 16 s rDNA or shotgun sequencing where small sample sizes are sufficient, the direct method would be better as it is less strenuous and quicker and yields high DNA content. However, the direct method has a disadvantage that it cannot remove PCR inhibitors from some environmental samples efficiently. Due to the development of commercial DNA purification kits, which can remove common PCR inhibitors, the direct method for DNA extraction has become quite popular in recent years.

The quantity as well as the quality of the extracted DNA may be evaluated using a number of different techniques. The purity of DNA is assessed by using UV spectroscopy; the presence of inhibitors is evaluated using a PCR inhibition assay, while the intactness is evaluated using gel electrophoresis. The quantification of DNA is mostly done by using quantitative polymerase chain reaction (qPCR) (Budowle et al. 2005).

Many real-time PCR assays are highly specific as well as sensitive and thereby shorten the time required for analysis when compared to the conventional PCR protocols, cultivation, as well as biochemical methods of identification. The next-generation sequencing which targets the nucleic acids is recently being deployed as a popular method to study genetic material from mixed microbial populations for analysis of microbial communities. The ability of the technique to identify the species up to the strain level from both live and dead microorganisms which are nonculturable has been very advantageous for the field of microbiology (Woese et al. 2000). Common sequencing techniques include 16S or 18S ribosomal RNA gene sequencing; next-generation sequencing (NGS), namely, pyrosequencing; and pulsed field gel electrophoresis (Gu et al. 2015). Techniques involving non-DNA-based analysis such as studying microbial chemistry and matrix using mass spectrometry can also be utilized (Seng et al. 2013).

21.9 Conclusion

Microorganisms are ubiquitous diverse organisms, harbored by the humans majorly on their skin as well as on the gastrointestinal and respiratory tract. The technological advances in microbial profiling are very significant for forensic purposes as they help in crime scene reconstruction by linking people, animals, plants, and objects to one another and/or to the scene of crime. The analysis and evaluation of the microbial profiles could be of immense use in estimating the postmortem interval (PMI). The soil profile of a crime scene may help to trace the location of the cadaver by studying the microbes present in the soil. Microorganisms have the potential to solve the cases related to origin of biocrimes, sexual assaults, homicides, agricultural contaminations, and medical malpractices. Microbial forensics quantifies molecular variations to find out the origin and the route of transmission of a particular microbial strain.

In order to be effectively applicable in the court of law, the techniques involved in the analysis of microbes must be standardized and validated. There is an utmost need of national and international collaborations to create a database on bioterrorism for the development of bioterrorism information system that can help to prevent the threat of bioattack. To combat the menace of biocrimes rapidly and effectively, there is a need to allocate funds for the establishment of hi-tech microbial forensic laboratories for research and development. Therefore, though forensic microbiology has vast applications and utility in criminal investigations and administration of law, however, a great deal of research in the subject is required so that its potential can be fully explored.

References

Álvarez-Lafuente R, Aguilera B, Suárez-Mier MP et al (2008) Detection of human herpesvirus-6, Epstein-Barr virus and cytomegalovirus in formalin-fixed tissues from sudden infant death: a study with quantitative real-time PCR. Forensic Sci Int 178(2–3):106–111

Amann R, Ludwig W, Schleifer KH (1995) Phylogenetic identification and in situ detection of individual microbial cells without cultivation. Microbiol Rev 59(1):143–169

Arora DR, Gautam V, Arora B (2002) Biological warfare: bioterrorism. Indian J Med Microbiol 20 (1):6–11

Bhatia M, Mishra B, Thakur A et al (2016) Concept of forensic microbiology and its applications. SMU Med J 3(1):275–294

Blackwell C (2004) Infection, inflammation and SIDS. FEMS Pathogens Dis 42(1):1–2

Bojar RA, Holland KT (2002) The human cutaneous microflora and factors controlling colonisation. World J Microbiol Biotechnol 18(9):889–903

Borgula LM, Robinson FG, Rahimi M, Chew KEK, Birchmeier KR, Owens SG, Kieser JA, Tompkins GR (2003) Recovery of oral bacteria from experimental bite marks. J Forensic Odontostomato 21:23–30

Brookes PC, McGrath SP (1984) Effect of metal toxicity on the size of the soil microbial biomass. J Soil Sci 35(2):341–346

Budowle B (2003) Defining a new forensic discipline: microbial forensics. Profiles in DNA 6 (1):7–10

Budowle B, Schutzer SE, Einseln A et al (2003) Building microbial forensics as a response to bioterrorism. Science 301(5641):1852–1853

Budowle B, Johnson MD, Fraser CM et al (2005) Genetic analysis and attribution of microbial forensics evidence. Crit Rev Microbiol 31(4):233–254

Buttner MP, Cruz-Perez P, Stetzenbach LD (2001) Enhanced detection of surface-associated bacteria in indoor environments by quantitative PCR. Appl Environ Microbiol 67(6):2564–2570

Buttner MP, Cruz P, Stetzenbach LD et al (2005) Evaluation of the biological sampling kit (BiSKit) for large-area surface sampling. Appl Environ Microbiol 70(12):7040–7045

Butzbach DM, Stockham PC, Kobus HJ et al (2013) Bacterial degradation of risperidone and paliperidone in decomposing blood. J Forensic Sci 58(1):90–100

Caporaso JG, Lauber CL, Costello EK et al (2011) Moving pictures of the human microbiome. Genome Biol 12(5):R50

Cardona OD, van Aalst MK, Birkmann J, Fordham M, McGregor G, Mechler R (2012) Determinants of risk: exposure and vulnerability. Cambridge University Press, Cambridge

Carrillo C, Rock DL (2005) Molecular epidemiology and forensics of RNA viruses. In: Breeze R, Budowle B, Schutzer S (eds) Microbial forensics. Academic Press, Amsterdam

Carter DO, Yellowlees D, Tibbett M (2007) Cadaver decomposition in terrestrial ecosystems. Naturwissenschaften 94(1):12–24

Carter DO, Yellowlees D, Tibbett M (2008) Temperature affects microbial decomposition of cadavers (Rattus rattus) in contrasting soils. Appl Soil Ecol 40(1):129–137

Carroll IM, Ringel-Kulka T, Siddle JP et al (2012) Characterization of the fecal microbiota using high-throughput sequencing reveals a stable microbial community during storage. PLoS One 7 (10):e46953

Carus WS (2001) Bioterrorism and biocrimes: the illicit use of biological agents since 1900. National Defense University, Washington, DC

Carus WS (2017) A short history of biological warfare: from from pre-history to the 21st century. National Defense University Press, Washington, DC

Catts EP, Haskell NH (1990) Entomology and death: a procedural guide. Joyce's Print Shop. Inc., Clemson, p 182

Centers for Disease Control and Prevention (CDC) (2002) Occupational health guidelines for remediation workers at Bacillus anthracis-contaminated sites–United States, 2001–2002. MMWR. Morbidity and mortality weekly report. 51(35):786

Costello EK, Lauber CL, Hamady M et al (2009) Bacterial community variation in human body habitats across space and time. Science 326(5960):1694–1697
Crompton R, Gall D (1980) Georgi Markov—death in a pellet. Med Leg J 48(2):51–62
Cummings CA, Relman DA (2002). Microbial Forensics-"Cross-Examining Pathogens". Science's Compass 296:1976–77
Delmont TO, Robe P, Clark I, Simonet P, Vogel TM (2011) Metagenomic comparison of direct and indirect soil DNA extraction approaches. J Microbiol Methods 86(3):397–400
Dickson GC, Poulter RT, Maas EW et al (2011) Marine bacterial succession as a potential indicator of postmortem submersion interval. Forensic Sci Int 209(1–3):1–10
Drake JW, Charlesworth B, Charlesworth D (1998) Rates of spontaneous mutation. Genetics 148 (4):1667–1686
Duncan RVI, Smith LA (2012) Of beans and beads: ricin and abrin in bioterrorism and biocrime. J Bioterr Biodef S2:002. https://doi.org/10.4172/2157-2526.S2-002
Eilers KG, Debenport S, Anderson S et al (2012) Digging deeper to find unique microbial communities: the strong effect of depth on the structure of bacterial and archaeal communities in soil. Soil Biol Biochem 50:58–65
Elliott ET (1986) Aggregate structure and carbon, nitrogen, and phosphorus in native and cultivated soils 1. Soil Sci Soc Am J 50(3):627–633
Evans WE (1963) Adipocere formation in a relatively dry environment. Med Sci Law 3:145–153
Fierer N, Lauber CL, Zhou N et al (2010) Forensic identification using skin bacterial communities. Proc Natl Acad Sci U S A 107(14):6477–6481
Flowers LK, Mothershead JL, Blackwell TH (2002) Bioterrorism preparedness. II: the community and emergency medical services systems. Emerg Med Clin North Am 20(2):457–476
Forbes SL, Perrault KA, Stefanuto PH et al (2014) Comparison of the decomposition VOC profile during winter and summer in a moist, mid-latitude (Cfb) climate. PLoS One 9(11):e113681
Gleeson M, Cripps A (2004) Development of mucosal immunity in the first year of life and relationship to sudden infant death syndrome. FEMS Immunol Med Microbiol 42(1):21–33
Goff ML (1993) Estimation of postmortem interval using arthropod development and successional patterns. Forensic Sci Rev 5:81–94
Goodman RA, Munson JW, Dammers K et al (2003) Forensic epidemiology: law at the intersection of public health and criminal investigations. J Law Med Ethics 31(4):684–700
Green MS, LeDuc J, Cohen D et al (2019) Confronting the threat of bioterrorism: realities, challenges, and defensive strategies. Lancet Infect Dis 19(1):e2–e13
Grice EA, Kong HH, Conlan S et al (2009) Topographical and temporal diversity of the human skin microbiome. Science 324(5931):1190–1192
Grundmann O (2014) The current state of bioterrorist attack surveillance and preparedness in the US. Risk Manage Healthc Policy 7:177–187
Gu Y, Mao X, Zha L, Hou Y, Yun L (2015) Development of a candidate method for forensic microbial genotyping using multiplex pyrosequencing combined with a universal biotinylated primer. Forensic Sci Int 246:e1–e6
Gunn A (2009) Essential forensic biology, 2nd edn. John Wiley & Sons
Gunn A, Sarah JP (2012) Microbes as forensic indicators. Trop Biomed 29(3):311–330
Handelsman J (2004) Metagenomics: application of genomics to uncultured microorganisms. Microbiol Mol Biol Rev 68(4):669–685
Henderson G, Cox F, Kittelmann S et al (2013) Effect of DNA extraction methods and sampling techniques on the apparent structure of cow and sheep rumen microbial communities. PLoS One 8(9):e74787
Hewadikaram KA, Goff ML (1991) Effect of carcass size on rate of decomposition and arthropod succession patterns. Am J Forensic Med Pathol 12(3):235–240
Hill PW, Marshall C, Williams CG, Blum H, Harmens H, Jones DL, Farrar JF (2007) The fate of photosynthetically-fixed carbon in grassland as modified by elevated CO and sward management. New Phytol 173(4):766–777

Hopkins DW, Wiltshire PE, Turner BD (2000) Microbial characteristics of soils from graves: an investigation at the interface of soil microbiology and forensic science. Appl Soil Ecol 14 (3):283–288

Howard GT, Duos B, Watson-Horzelski EJ (2010) Characterization of the soil microbial community associated with the decomposition of a swine carcass. Int Biodeterior Biodegradation 64:300–304

Hyde ER, Haarmann DP, Lynne AM et al (2013) The living dead: bacterial community structure of a cadaver at the onset and end of the bloat stage of decomposition. PloS one 8(10):e77733

Janaway RC, Percival SL, Wilson AS (2009) Decomposition of human remains. In: Microbiology and aging. Humana Press, Totowa, pp 313–334

Janik E, Ceremuga M, Saluk-Bijak J et al (2019) Biological toxins as the potential tools for bioterrorism. Int J Mol Sci 20(5):1181

Jansen HJ, Breeveld FJ, Stijnis C (2014) Biological warfare, bioterrorism, and biocrime. Clin Microbiol Infect 20(6):488–496

Jaureguy F, Chariot P, Vessieres A, Picard B (2016) Prevalence of chlamydia trachomatis and Neisseria gonorrhoeae infections detected by real-time PCR among individuals reporting sexual assaults in the Paris, France area. Forensic Sci Int 266:130–133

Junkins EN, Embriette RH, David OC (2017) Culture and long-term storage of microorganisms for forensic science. In: Carter DO, Tomberlin JK, Benbow ME, Metcalf JL (eds). John Wiley & Sons, New York

Kakirde K, Parsley L, Liles M (2010) Size does matter: application-driven approaches for soil metagenomics. Soil Biol Biochem 42(11):1911–1923

Kakizaki E, Takahama K, Seo Y et al (2008) Marine bacteria comprise a possible indicator of drowning in seawater. Forensic Sci Int 176(2–3):236–247

Kaur M, Gupte S, Aggarwal P et al (2014) Methods in microbial forensics. J Punjab Acad Forensic Med Toxicol 14(1):51–54

Lazarevic V, Whiteson K, Huse S et al (2009) Metagenomic study of the oral microbiota by Illumina high-throughput sequencing. J Microbiol Methods 79(3):266–271

Lazarevic V, Whiteson K, Hernandez D et al (2010) Study of inter-and intra-individual variations in the salivary microbiota. BMC Genomics 11(1):523

Leake SL, Pagni M, Falquet L (2016) The salivary microbiome for differentiating individuals: proof of principle. Microbes Infect 18(6):399–405

Lenz EJ, Foran DR (2010) Bacterial profiling of soil using genus-specific markers and multidimensional scaling. J Forensic Sci 55(6):1437–1442

Lin CY, Yen WC, Hsieh HM et al (2014) Diatomological investigation in sphenoid sinus fluid and lung tissue from cases of suspected drowning. Forensic Sci Int 244:111–1115

Ludes B, Coste M, North N, Doray S, Tracqui A, Kintz P (1999) Diatom analysis in victim's tissues as an indicator of the site of drowning. Int J Legal Med 112(3):163–166

Maccallum WG, Hastings TW (1899) A case of acute endocarditis caused by micrococcus zymogenes (Nov. Spec.), with a description of the microorganism. J Exp Med 4(5–6):521–534

Männistö MK, Tiirola M, Häggblom MM (2009) Effect of freeze-thaw cycles on bacterial communities of Arctic tundra soil. Microb Ecol 58(3):621–631

Megyesi MS, Nawrocki SP, Haskell NH (2005) Using accumulated degree-days to estimate the postmortem interval from decomposed human remains. J Forensic Sci 50(3):1–9

Melvin JR, Cronholm LS, Simson LR et al (1984) Bacterial transmigration as an indicator of time of death. J Forensic Sci 29(2):412–417

Meyers MS, Foran DR (2008) Spatial and temporal influences on bacterial profiling of forensic soil samples. J Forensic Sci 53(3):652–660

Mickley A (2010) Preventing the sixth plague: microbial forensics in the war against terrorism. Honors College Theses 1:91

Moghaddasi H, Arani LS, Zarghi A (2018) Features of bioterrorism information system. J Bioterror Biodefense 9:1–6

Moreno LI, Mills DK, Entry J, Sautter RT, Mathee K (2006) Microbial metagenome profiling using amplicon length heterogeneity-polymerase chain reaction proves more effective than elemental analysis in discriminating soil specimens. J Forensic Sci 51(6):1315–1322

Morse SA, Budowle B (2006) Microbial forensics: application to bioterrorism preparedness and response. Infect Dis Clin 20(2):455–473

Murch RS (2003) Microbial forensics: building a national capacity to investigate bioterrorism. Biosecur Bioterr Biodef Strategy Pract Sci 1(2):117–122

Ogram A, Sayler GS, Barkay T (1987) The extraction and purification of microbial DNA from sediments. J Microbiol Methods 7(2–3):57–66

Oliveira M, Amorim A (2018) Microbial forensics: new breakthroughs and future prospects. Appl Microbiol Biotechnol 102(24):10377–10391

Ou CY, Kwok S, Mitchell SW et al (1988) DNA amplification for direct detection of HIV-1 in DNA of peripheral blood mononuclear cells. Science 239(4837):295–297

Ou CY, Ciesielski CA, Myers G et al (1992) Molecular epidemiology of HIV transmission in a dental practice. Science 256(5060):1165–1171

Parachin NS, Schelin J, Norling B, Rådström P, Gorwa-Grauslund MF (2010) Flotation as a tool for indirect DNA extraction from soil. Appl Microbiol Biotechnol 87(5):1927–1933

Pechal JL, Crippen TL, Tarone AM et al (2013) Microbial community functional change during vertebrate carrion decomposition. PLoS One 8(11):e79035

Pechal JL, Crippen TL, Benbow ME et al (2014) The potential use of bacterial community succession in forensics as described by high throughput metagenomic sequencing. Int J Legal Med 128(1):193–205

Pechal JL, Schmidt CJ, Jordan HR, Benbow ME (2017) Frozen: thawing and its effect on the postmortem microbiome in two pediatric cases. J Forensic Sci 62(5):1399–1405

Powers JS, Montgomery RA, Adair EC et al (2009) Decomposition in tropical forests: a pan-tropical study of the effects of litter type, litter placement and mesofaunal exclusion across a precipitation gradient. J Ecol 97(4):801–811

Preston BD, Poiesz BJ, Loeb LA (1988) Fidelity of HIV-1 reverse transcriptase. Science 242 (4882):1168–1171

Quaak FCA, Kuiper I (2011). Statistical data analysis of bacterial t-RFLP profiles in forensic soil comparisons. Forensic Sci Int 210:96–101

Rana AK, Manhas S (2018) The future of forensic biology. J Biomed 3:13–18

Ranjard L, Poly F, Lata JC et al (2001) Characterization of bacterial and fungal soil communities by automated ribosomal intergenic spacer analysis fingerprints: biological and methodological variability. Appl Environ Microbiol 67(10):4479–4487

Robbins KE, Lemey P, Pybus OG et al (2003) US human immunodeficiency virus type 1 epidemic: date of origin, population history, and characterization of early strains. J Virol 77 (11):6359–6366

Rochelle P, Cragg B, Fry J et al (1994) Effect of sample handling on estimation of bacterial diversity in marine sediments by 16S rRNA gene sequence analysis. FEMS Microbiol Ecol 15 (1–2):215–225

Rogers P, Whitby S, Dando M (1999) Biological warfare against crops. Sci Am 280(6):70–75

Rossodivita A, Visconti A, Saporito T et al (2019) Bioterrorism: toxins as potential biological weapons-an emerging global health threat. Int J Infect Dis 79:55

Roth RR, James WD (1988) Microbial ecology of the skin. Annu Rev Microbiol 42(1):441–464

Ruffell A (2010) Forensic pedology, forensic geology, forensic geoscience, geoforensics and soil forensics. Forensic Sci Int 202(1–3):9–12

Sala M, Wain-Hobson S (2000) Are RNA viruses adapting or merely changing? J Mol Evol 51 (1):12–20

Santiago-Rodriguez TM, Cano RJ (2016) Soil microbial forensics. Microbiol Spectrum. 4(4). https://doi.org/10.1128/microbiolspec.EMF-0007-2015

Saraswat PK, Nirwan PS, Saraswat S et al (2008) Biodegradation of dead bodies including human cadavers and their safe disposal with reference to mortuary practice. J Indian Acad Forensic Med 30:273–280

Schoenen D, Schoenen H (2013) Adipocere formation—the result of insufficient microbial degradation. Forensic Sci Int 226(1–3):301e1–301e6

Schutzer SE, Budowle B, Atlas RM (2005) Biocrimes, microbial forensics, and the physician. PLoS Med 2(12):e337

Sender R, Fuchs S, Milo R (2016) Revised estimates for the number of human and bacteria cells in the body. PLoS Biol 14(8):e1002533

Seng P, Abat C, Rolain JM et al (2013) Identification of rare pathogenic bacteria in a clinical microbiology laboratory: impact of matrix-assisted laser desorption ionization–time of flight mass spectrometry. J Clin Microbiol 51(7):2182–2194

Sensabaugh G (2009) Microbial community profiling for the characterisation of soil evidence: forensic considerations. In: Ritz K, Dawson LA, Miller D (eds) Criminal and environmental soil forensics. Springer, London, p 49e60

Slonczewski JL, Foster JW (2013) Microbiology: an evolving science. W.W. Norton and Company, New York

Smart JK (1997) History of chemical and biological warfare: an American perspective. In: Medical aspects of chemical and biological warfare. Office of the Surgeon General, Washington, DC, pp 9–86

Smith JA (2011) Collection and preservation of microbial forensic samples. In: Microbial Forensics. Academic, London, pp 379–392

Stahringer SS, Clemente JC, Corley RP (2012) Nurture trumps nature in a longitudinal survey of salivary bacterial communities in twins from early adolescence to early adulthood. Genome Res 22(11):2146–2152

Stockdale EA, Brookes PC (2006) Detection and quantification of the soil microbial biomass–impacts on the management of agricultural soils. J Agric Sci 144(4):285–302

Stokes KL, Forbes SL, Tibbett M (2009) Freezing skeletal muscle tissue does not affect its decomposition in soil: evidence from temporal changes in tissue mass, microbial activity and soil chemistry based on excised samples. Forensic Sci Int 183(1–3):6–13

Svarovskaia ES, Xu H, Mbisa JL et al (2004) Human apolipoprotein B mRNA-editing enzyme-catalytic polypeptide-like 3G (APOBEC3G) is incorporated into HIV-1 virions through interactions with viral and nonviral RNAs. J Biol Chem 279(34):35822–35828

Taylor LH, Latham SM, Woolhouse ME (2001) Risk factors for human disease emergence. Philos Trans R Soc Lond Ser B Biol Sci 356(1411):983–989

Tibbett M, Carter DO (2009) Research in forensic taphonomy: a soil-based perspective. In: Criminal and environmental soil forensics. Springer, Dordrecht, pp 317–331

Timperman J (1972) The diagnosis of drowning. A review. Forensic Sci 1(4):397–409

Vass AA (2001) Beyond the grave-understanding human decomposition. Microbiol Today 28:190–193

Vass AA, Smith RR, Thompson CV (2004) Decompositional odor analysis database. J Forensic Sci 49(4):1–10

Wagner Mackenzie B, Waite DW, Taylor MW (2015) Evaluating variation in human gut microbiota profiles due to DNA extraction method and inter-subject differences. Front Microbiol 6:130

Weiss S, Carter DO, Metcalf JL et al (2016) Carcass mass has little influence on the structure of gravesoil microbial communities. Int J Legal Med 130(1):253–263

Wescott DJ (2018) Recent advances in forensic anthropology: decomposition research. Forensic Sci Res 3(4):278–293

Wilfinger WW, Mackey K, Chomczynski P (1997) Effect of pH and ionic strength on the spectrophotometric assessment of nucleic acid purity. BioTechniques 22(3):474–481

Wilson MJ, Weightman AJ, Wade WG (1997) Applications of molecular ecology in the characterization of uncultured microorganisms associated with human disease. Rev Med Microbiol 8 (2):91–103

Wilson TM, Gregg DA, King DJ et al (2001) Agroterrorism, biological crimes, and biowarfare targeting animal agriculture: the clinical, pathologic, diagnostic, and epidemiologic features of some important animal diseases. Clin Lab Med 21(3):549–592

Woese CR, Olsen GJ, Ibba M, Söll D (2000) Aminoacyl-tRNA synthetases, the genetic code, and the evolutionary process. Microbiol Mol Biol Rev 64(1):202–236

Zala K (2007) Dirty science: soil forensics digs into new techniques. Science 318(5849):386–387

Zaura E, Keijser BJ, Huse SM et al (2009) Defining the healthy "core microbiome" of oral microbial communities. BMC Microbiol 9(1):259

Microbial Forensics: A New Boon to Legal Justice

22

Jaskaran Singh, Shikha, and Ekampreet Kaur

Abstract

Forensic microbiology is a thriving new discipline combining both microbiology and forensic sciences. The growing field deals with profound effect of microorganisms weapons which identify and prioritize bioterrorism. Bioterrorism uses microorganisms as weapons which has been known to exist since centuries. The major components of a successful microbial forensics investigation are detection, identification, characterization, attribution and interpretation of weapon pathogen. Advance technologies in this field no doubt develop better resources and treatment for the microbial diseases which affect humans.

Keywords

Bioterrorism · Pathogen · Attribution · Forensic

22.1 Introduction

A microorganism or microbe is a microscopic organism which may exist in single celled form or in colonies (Schmedes et al. 2016). There are different varieties of microorganisms in our environment. These microbes are both boon and bane for us and for our surroundings.

Because of their universal nature, they can act as a commendable forensic indicator for investigation. Microbial forensics has been defined as "a scientific

J. Singh (✉) · Shikha
Department of Forensic Science, School of Bioengineering and Biosciences, Lovely Professional University, Phagwara, Punjab, India
e-mail: jaskaran.22220@lpu.co.in

E. Kaur
Department of Forensic Science, Punjabi University, Patiala, Punjab, India

S. G. Sharma et al. (eds.), *Microbial Diversity, Interventions and Scope*,
https://doi.org/10.1007/978-981-15-4099-8_22

discipline dedicated to analyzing evidence from a bioterrorism act, biocrime, or inadvertent microorganism/toxin release for attribution purposes"(Budowle et al. 2003).

Microbial forensic investigation includes characterisation and detection of both biological and non-biological evidence. Biological agents consist of protists, toxins, bacteria, fungi and viruses. Non-biological evidence, such as growth media, additives, and delivery devices, which can be useful in microbial forensics, potentially provides investigative leads and helps infer methods of manufacture and dissemination (Budowle et al. 2008).

In this study, proper classification of specific organism is considered beside its origin and impacts. Indeed our human body contains botches of bacteria inside it; consequently they can act as a proficient tool in forensic investigation. For instance, if the sample collected from the crime scene which could be a whitish liquid, the microbe will tell almost the specific fluid by the assistance of microbiome inside the human body.

The microorganisms are amassed inside the body on or inside various body fluids, tissues, skin or glands like lungs, placenta, ovarian follicles, seminal fluid, uterus and gastrointestinal tracts. This aggregate of microorganisms inside the human body is known as human microbiota. Microbes which reside on the human body are typically excluded from this elucidation. The human microbiome interprets precisely the collective genomes of resident microbes. Many microbes reside inside the human body. The conventional estimate is that the average human body is dwelled by 10 times as many non-human cells as human cells, but the recent estimate had dropped that value to 3:1 (Sender et al. 2016).

On the basis of their interaction with the host or human body, the microorganisms are classified into:

1. Commensal (they coexist without harming the host/human being)
2. Mutualistic (both host and parasite are benefitted)
3. Non-pathogenic (parasite can harm human hosts through their secretions or metabolites such as trimethylamine)

There are certain microbes which are favourable for the human body. These include:

22.2 Bifidobacteria

A diverse microbial community has evolved to adapt and survive in the human GIT and is commonly referred to as the gut microbiota (Thursby and Juge 2017). It helps in converting complex food into simpler form and helps in absorbing nutrients for providing energy to the body. The demerit of this bacteria is that it liberates gas. It helps the body to get rid of constipation and diarrhoea.

22.2.1 *Escherichia coli*

These are present inside the human intestine. This bacteria is useful in digestion and helps in providing nutrition to the body. Moreover, it fights with harmful bacteria inside our body and inhibit their growth.

Hence, their location can be easily determined from this human microbiota. We can easily analyse the type of sample by knowing the origin of the microbe. The most important thing is that these microbes are present universally. Therefore, they play a major role in contamination of the sample. In forensic scenario, we have no hold over the purity of the sample. If the sample is contaminated by bacteria, pathogens, fungi, etc., *Staphylococcus* and *Bacillus cereus* are the microorganisms that are most often recovered from contaminated blood. Forensic scientists can also analyse the time period, the progressive changes that occurred simultaneously by time since death decomposition analysis. The progression can typically be divided into a number of distinct stages:

1. Fresh
2. Bloated
3. Decay
4. Post-decay
5. Dry skeleton

Each stage is additionally associated with a rough time period during which it is likely to occur, subject to the components that can modify these time periods. Each of these stages are also associated with the arrival of distinctive species of microorganisms and insects. DNA sequencing is an everlasting technique used in the forensic scenario for the purpose of research. Moreover, this technique has various roles and applications which were not known earlier. Hence, DNA sequencing is the newer methodology in microbial analysis. A much better understanding of the environment, how it changes over time, and how it interacts and changes the ecology of its more extensive environment might have vital applications in forensic science. It could, for example, lead to new, more accurate ways of estimating time of death and of finding bodies that have been hidden (Pattnaik and Jana 2005).

In case of transmissions of diseases especially in sexual assault case, these microbes are utilitarian for forensic investigation as they tell about successive changes through their life cycle. The population of microbes increases when they act on any sample and their colonies may form. Variations are analysed in the sample and in the microbes and hence identification is done. Microbiology along with forensic science focuses on the reliability to describe the nature, origin and classification of these microorganisms along with their stimuli to different external and internal factors. If these microbes are present in a trace amount, forensic analysis can be achieved to yield desired results.

22.3 Bioterrorism

Bioterrorism is an illicit, menacing use of microbes or their toxins to cause death and destruction to humans, animals or plants (Murch 2001). Many of these microbes or bio-agents are isolated from the environment and are manipulated genetically to make them more hazardous and pathogenic so that they can cause destruction at a very high scale when used as a bioweapon.

The microbial agents used for this purpose include:

- *Bacillus anthracis*
- *Brucella abortus*
- *Brucella suis*
- *Clostridium*
- *Tularaemia*
- *Filoviruses and arena virus*

These are supplied/sprayed over large geographical areas and are transferred into the human body by inhaling, contact or via the gastrointestinal tract. In some cases these pathogens are mixed with the water bodies of geographical area which is to be destroyed or are mixed in the crops. The crop and water samples are collected by the investigators with great care and are sent to the laboratory for analysis. In the laboratory samples are analysed and interpreted by using certain tools and techniques, and hence the result is made. The techniques used most commonly are enlisted below.

These microorganisms are divided into three different categories:

Category A:	Easily disseminated
Example:	*Bacillus anthracis, Yersinia pestis, Variola major*
Category B:	Moderately easy to disseminate
Example:	*Brucella* spp., *Salmonella* sp., *Shigella* spp.
Category C:	Engineered for mass dissemination

22.4 Bioterrorism and Forensic Science

Forensic science plays a major role in bioterrorism. Microbial forensics deals with study related to classical microbiology, microbial genomics, phylogenetics and informatics. The samples collected from the attack site are analysed by various scientific tools and techniques. This includes dealing with knowledge of development and preparation of microorganisms, the arrangement of toxins, the different approaches for weaponisation and dispersal of biothreat agents, and the use of synthetic biology (Tims et al. 2010).

The techniques equipped for the analysis include:

DNA typing, genomic polymorphism, genetic mapping, massively parallel sequencing.

Microbial forensic investigators focus mainly on three important steps:

Identification: Identification includes identifying the microorganism used in the attack. Depending upon the destruction caused, the pathogen can be analysed by manifesting the signs and symptoms on the particular population (human/animals/plants) in the affected geographical area. It is the main step because if the pathogen is identified the causes, origin and study become quick easy.

Characterization: It deals with the successive events that took place while destruction. This includes whether the attack was intentional or unintentional. It is a very useful step for determining the cause of death.

Attributions: In the third step – the source – tracing is done, i.e. the origin of the pathogen is identified. It helps in locating the perpetrator and weapon strain pathogen.

The tasks involved in identification and attribution include:

Identifying and collecting sample
Handling and preservation of the collected sample
Choosing appropriate analytical methods
Casework analysis
Interpretation of results
Quality assurance and validation

Forensic science works by initially collecting the samples from attack site. Further the recognition of the attack is done; it is based upon parameters like intentional, unintentional, covert, overt attacks. The next step is diagnosing the diseases caused to particular inhabitants of attack area. This is followed by analysis of the specimens and their identification to claim out the origin of the pathogen and culprit. Last but not the least, validation is done in which the quality assurance and quality control is checked. Because microbial forensics focuses on tracking and establishing the link between microorganisms and individuals along with the locations, different strategies are implemented, depending on the nature of the attack and the type(s) of evidence collected (Budowle et al. 2010). In an overt attack, for instance, the evidence collected such as the package, the weapon and associated materials (hairs, fibres, fingerprints) can be analysed. In a covert attack, the evidence gathered from the crime scene may be limited to medical histories, diagnoses and isolates taken from victims.

Forensic science is the science of identification and comparison. The comparison of a genetic profile from a reference sample with that of an evidentiary sample can have three possible general outcomes: "match" or "inclusion", "exclusion", or "inconclusive." Because of lack of database, genetic testing and unknown diversity, there is a lot of inaccuracy, but on the flip side, the advancement in the forensic microbial tools triggers certain amount of accuracy which is very apprehensive (Budowle et al. 2005).

22.4.1 Techniques/Methods Used in Forensic Investigation: Microarray Analysis

This is useful in genomic study of pathogens and other microbes involved in bioweapon.

It involves complex and statistical tests.

Sample preparation →Hybrids→Washing→
Image acquisition→
Data analysis

Microarray is a pattern of SSDNA probes which are not allowed to move and are stuck on a chip or slide.

Hybridisation is used to detect specific DNA or RNA in a sample.

22.4.2 DNA Fingerprinting

In this technique, the DNA is extracted from the pathogen cell. Cutting of DNA is done by enzymes, then the DNA fragments are separated and transferred onto a paper. Radioactive probe is added an end X-ray film is set Electron Microscopy (Budowle et al. 2008).

22.4.3 Serological Assays

Serological assay technique is used for diagnosing infectious diseases. Some pathogens are highly infectious and are difficult to cultivate, so this technique is used to determine the diagnostic accuracy of specific tests.

22.4.4 PCR

PCR (polymerase chain reaction) is used as an effective tool in DNA fingerprinting. This technique is useful in individualization. For example, any minute DNA sample collected from the crime scene/attack site can be compared with other suspected samples (Hampton-Marcell et al. 2017). PCR-based DNA fingerprinting is also useful in solving maternal and paternal disputes. Quantitative polymerase chain reaction is used for detection of nucleic acids in many biological fields.

22.4.5 Subtractive Cloning

This technique is also termed as subtractive hybridization. This technology works in the way that it removes dsDNA formed by hybridization between standard and reference sample. It is used for identification of strain-specific DNA sequences in a variety of microbes like bacteria.

22.4.6 Advanced Techniques in Forensic Science

22.4.6.1 Next-generation sequencing/high-throughput sequencing/massively parallel sequencing (MPS)

This technique has overcome the problem of identification of unknown pathogens and microbes. It helps in detecting less amount/minute microbes in complex mixture sample.

MPS is useful in the characterisation of microbe and checks its abundance and nature (whether degraded or intact). It rapidly diagnoses and oversees the infections using culture-independent methods and hence trails the disease outbreaks.

MPS provides tactics for human microbiome and is useful in human identification, body fluid characterisation and time-since death decomposition analysis (Budowle and Chakroborty 2004).

22.4.7 Microbiome Profiling

Microbiome profiling is useful in identifying a person or lifestyle characteristic according to the studies. Humans shed 30 million bacterial cells into their vicinity every hour, e.g. bacterial community found on our fingerprints which could be traced on keyboard in cases where computer has been used and, hence, a person can be identified according to bacterial residue with the help of microbiome profiling.

22.4.8 Metagenome Classification

This technique is used for the extraction of essential information of the organism through the traces left by it in a given environmental sample. It helps in depicting origin and generating composition profile along with diversity of the microbe. It leads to fast and accurate classification after sequencing entire samples and it allows a database to be built without restriction.

22.5 Case Studies in Support of Microbial Forensics

- ***Salmonella enterica***

Salmonella enterica serovar Typhimurium is a Gram-negative, rod-shaped, facultative anaerobic pathogen. This pathogen infects both humans and animals. Infection due to *Salmonella* occurs due to contaminated food and water, which leads it to intestinal epithelium and triggers gastrointestinal diseases. When this bacteria first enters the human body, it initially propagates inside the intestinal tract and spreads throughout the peripheral lymphatic system, such as the bone marrow and causes typhoid fever.

Case Study

The **Rajneeshee bioterror attack (1984)** was the food-poisoning attack of 751 individuals in Dallas, Oregon. Salad bars contaminated with *Salmonella* were used for the attack at 10 local restaurants by Rajneeshee followers (later known as Osho) led by Ma Anand Sheela. Among them, 45 individuals were hospitalised but none died. The incident is considered to be the biggest bioterrorism attack in the United States until now (Lightfoot et al. 2001).

- **Scrub Typhus**

It is a mite-borne disease and is also well known as bush typhus. It is mainly caused by bacteria *Orientia tsutsugamushi* (Rickettsia tsutsugamushi) and through infected larval mites or chiggers which belong to family Trombiculidae and genus and subgenus *Leptotrombidium.* Symptoms of scrub typhus will begin after 10 days of being bitten by the bacteria like body chills, muscle pain, rashes, fever, chills, headache, etc.

Case Study

Scrub typhus disease was diagnosed in the troops of Assam and West Bengal during World War II. During the Indo-Pakistan War of 1965, the same symptoms appeared in Jammu-Sialkot sector troops. A similar scrub typhus outbreak in north-eastern India came under suspicion and India's defense and typhus outfits were alerted to this outbreak of pneumonic plague (Singh 2004).

- **Diatoms**

Diatoms are unicellular algae found in freshwater or sea water. They exhibit great role in investigation of drowning cases in the field of forensic sciences. Through diatoms forensic scientists can establish ante-mortem and post-mortem drowning. When a person gets submerged into water, the diatom will enter the lungs through water and then carried to distinct parts of the body through the circulation process. But when the person is already dead, the water will enter the lungs but there will be no circulation. By examining the body tissue for diatom, we can establish the ante-mortem or post-mortem drowning.

Case Study

A dead body of a male aged 30–40 years was found in a highly decomposed condition near Shimla bypass, Himachal Pradesh, India, in the month of November. The person was identified as Lal Pani Nullah (canyon). The cause of death could not be ascertained in autopsy. The femur bone was sent to the laboratory with water sample of canyon for the detection of diatom. After examination, diatoms *Cymbella sp., Aulacoseira sp. and Gyrosigma sp.* were detected in the water sample. However, diatoms could not be detected in femur bone, which signifies that it was post-mortem drowning (Kaushik et al. 2017).

22.6 Forensic Application

Microbial forensics elucidate the new and growing field of microbial forensic that is the science that will help in providing justice to victims and apprehending criminals and terrorists who use biological material to cause harm and destruction. Microbe forensic is a very vast field in forensic scenario. It is very helpful in identification and attribution of the pathogens used as a bioweapon. Forensic scientists use different tools and select appropriate analysis methods and tools according to the sample. Databases of these microbe genomes are burgeoning promptly and prove to be commendable for forensic analysis. Forensic microbiology works as a base for interpreting the result. It uses certain tools and techniques to explore the identity and origin and to initialize the investigation.

- **Helpful in Solving Crimes**

Microorganisms because of their universal nature are very helpful in solving crime cases. They are part of decay process; therefore, by analysing the extent of decomposition and type of microns, one can tell about the age of dead body and reason of the death. In case of drowning, the microbes inside the body will give detailed information regarding death (during post-mortem).

Using genomic analysis of the pathogen and identifying its origin in most cases can be solved i.e. bioterrorism attack by *Bacillus anthacris* in USA, 2001 was solved by genomic analysis. The 2001 anthrax attacks are also called Amerithrax. It took place in Washington, DC. This attack killed five people; it has become obvious that biocrimes can only be solved when genomic information is used to identify the source of an organism.

- **Identification of Source**

In forensic microbiology certain methodologies are there which can be used to detect and trace back the spread of microorganisms in the context to the crime. By forensic analysis of microbes, the perpetrator pathogen and type of events can be analysed. It can tell whether the cause was intentional or unintentional or whether it was covert or overt.

- **Scrutinizing the Human History by Studying Ancient Microbial DNA**

Study of ancient microbes and their DNA is known as paleomicrobiology. Paleomicrobiology is the study of microbes which are associated with primeval material. This field of science is hauled from various other branches like microbiology, anthropology, history, palaeontology and archaeology. It tells about the disease and infections that our ancestors had suffered from their dietary habits were also analysed through this method. So, this study tells about the evolution and lifestyle of ancient pathogens.

– **Tracking Culprits in Public Health Crimes Along with Their Sources**

If pathogens are mixed with either food, crops, water bodies, by monitoring their quality and the interaction between the host body and pathogen is analysed. This analysis is done by microarray analysis (DNA) sequencing qPCR. qPCR is used to detect, characterise and quantify nucleic acids for numerous applications. It involves fluorescent labelling of the data which quantifies the amplified DNA molecule.

References

Budowle B, Chakroborty R (2004) Genetic consideration for interpreting molecular microbial forensic evidence. Int Congr Ser 1261:56–58

Budowle B, Schutzer SE, Einseln A, Kelley LC, Walsh AC, Smith JAL, Marrone BL, Robertson J, Campos J (2003) Building microbial forensics as a response to bioterrorism. Science 301:1852–1853

Budowle B, Johnson MD, Fraser CM et al (2005) Genetic analysis and attribution of microbial forensics evidence. Crit Rev Microbiol 31(4):233–254

Budowle B, Schutzer SE, Morse SA, Martinez KF, Chakraborty R, Marrone BL, Messenger SL, Murch RS, Jackson PJ, Williamson P, Harmon R, Velsko SP (2008) Criteria for validation of methods in microbial forensics. Appl Environ Microbiol 74:5599–5607

Budowle B, Schutzer S, Breeze R, Keim P, Morse S (eds) (2010) Microbial forensics. Academic, Burlington, 744 p

Hampton-Marcell J a T, Lopez JV, Gilbert A (2017) The human microbiome: an emerging tool in forensics. Microb Biotechnol 10:228–230

Kaushik N, Pal KS, Sharma A, Thakur G (2017) Role of diatoms in diagnosis of death due to drowning: case studies. Int J Med Toxicol Forensic Med 7(1):59–65

Lightfoot N, Wale M, Spencer R, Nicoll A (2001) Appropriate responses to bioterrorist threats. BMJ 323(7318):877–878

Murch RS (2001) Forensic perspective on bioterrorism and the proliferation of biological weapons. In: Layne SP, Beugelsdijk TJ, Patel CKN (eds) Firepower in lab automation in the fight against infectious diseases and bioterrorism. Joseph Henry Press, Washington, DC

Pattnaik P, Jana A (2005) Microbial forensics: application in bioterrorism. J Environ Forensics 6 (2):197–204

Schmedes S, Sajantila A, Budowle B (2016) Expansion of microbial forensics. J Clin Microbiol 54 (8):1964–1974

Sender R, Fuchs S, Milo R (2016) Revised estimates for the number of human and bacteria cells in the body. PLoS Biol 14:1–14

Singh P (2004) Scrub typhus, a case report: military and regional significance. Med J Armed Forces India 60(1):89–90

Thursby E, Juge N (2017) Introduction to the human gut microbiota. Biochem J 474 (11):1823–1836

Tims S, van Wamel W, Endtz HP, van Belkum A, Kayser M (2010) Microbial DNA fingerprinting of human fingerprints: dynamic colonization of fingertip microflora challenges human host inferences for forensic purposes. Int J Legal Med 124:477–481

Part IX

Microbes in Space

Microbes Supporting Life Off-Planet

23

Shiwani Guleria Sharma and Mohit Sharma

Abstract

Space microbiology pays attention toward possible microbial interactions in outer space or in distant shell of earth. Environmental conditions of outermost shell and space craft and their effect on microbiome need to be explored for basic understanding and for possible applied aspects. Various exposure facilities are developed for such studies. This chapter provides an overview about main aspects of space microbiologist, locations for research, their environment conditions, sample processing tools, and future aspects of space microbiology.

Keywords

Space · International Space Station · Next-generation sequencing · Sequencing tools · PCR

Current information and findings of space microbiology related to future space exploration programs need to be critically reviewed. Emphasis has not been given on recent research that are of relevance to microorganisms in space. There are various adverse conditions for microbes in space which include radiations, vacuum, microgravity, unfavorable temperature, and hostile environment. Microbes need to protect them from these adverse conditions in order to survive in space. For space microbiologists, environmental conditions of space and their effect on microbes are the main aspects to understand (Horneck et al. 2010). Following are main aspects need to cover:

S. G. Sharma (✉)
Department of Microbiology, School of Bioengineering and Biosciences, Lovely Professional University, Phagwara, Punjab, India

M. Sharma
Molecular Genetics laboratory, Dayanand Medical College & Hospital, Ludhiana, Punjab, India

S. G. Sharma et al. (eds.), *Microbial Diversity, Interventions and Scope*,
https://doi.org/10.1007/978-981-15-4099-8_23

- Understanding of relationship between space environment and basic biological mechanisms of microbes
- Effect of gravity on extracellular, subcellular cellular levels
- Biological effects of the radiation field in space
- Understanding of factors of upper boundary of Earth's biosphere on biological survival
- Study of microorganisms in bioregenerative life support systems
- Control and characterization of spacecraft microflora
- Monitoring of microflora associated with microbial crew
- Study of pathogenic microflora in spacecraft and other health-related concerns

During the 1970s, microbial ecology equipment device was used to expose microorganisms to space conditions during translunar *Apollo* trips (Taylor 1974). Biostack experiments were also used to take microbes outside the Earth's magnetic shield (Taylor 1974). Mostly locations used for study of microorganisms in space were:

- *International Space Station* (ISS) (Bagnioli et al. 2007)
- Earth-orbiting robotic spacecraft (Innocenti and Mesland 1995)
- Human-tended spacecraft (Horneck et al. 1984)
- Space stations (Rettberg et al. 2002)

The International Space Station (ISS) is present 400 km above the Earth's surface, so they are the largest space platform for space research in low Earth orbit. The main concern about ISS is their environmental condition. ISS is a closed system and has lots of extreme condition which oppose growth of microbes. However, there are extremophiles which can survive extreme environmental conditions. According to the National Aeronautics and Space Administration's (NASA) maintenance protocol, culture-based methods are used to keep check on microbial population in air, surfaces and water samples of ISS (Checinska et al. 2015; NASA 2005). But there are limitations of culture-based methods. Limited population of microbes can be cultivated using standard culture-based protocol. It is not possible to have complete information about diversity of ISS (Pace 2009). Thus, it is better to use molecular-based techniques for diversity analysis of ISS flora with more accuracy and broad range. These can identify and quantify both culturable and unculturable organisms. The mostly used molecular methods are quantitative polymerase chain reaction (qPCR) and targeted amplicon sequencing, Illumina-based 16S rRNA sequencing and internal transcribed spacer (ITS) region. Main concern while using these methods is sample collecting tools. Reliable and compact samples and sample collecting instruments are required for study of space microbial flora. But for these molecular methods, reliable and compact samples are required. However, there is lack of such simple and reliable sample-collecting instruments on ISS. Selective molecular sequencing techniques have been used in ISS such as Sanger sequencing for microbial characterization. Various space flight requirements such as diagnosis of disease, gene expression, population metagenomics, and genetic mutations are potentially addressed by sequencing technologies. Thus, other sequencing techniques such as next-generation sequencing (NGS) can be implemented for improvement in the microbial characterization. Use of next-generation sequencing (NGS) in closed system has just started, although it is already in use in different microbial fields such as ecology (Checinska Sielaff et al. 2019; Yergeau et al. 2012; La Duc et al. 2012).

MinION™ DNA sequencer (Oxford Nanopore Technologies, Oxford, UK) is commercially available sequencer for space flight. In this device, nucleic acid molecule passes through nanopore present in membrane and result in current change. This change in current is measured in order to sequence DNA and RNA. The current change is due to DNA or RNA sequence that is passing the pore at a given time. MinION has already tested for remote locations on the Earth (Johnson et al. 2017; Edwards et al. 2017; Hoenen et al. 2016; Quick et al. 2016) and space environment. Its success reports are available for Earth's remote areas. It has also been tested on International Space Station (ISS) which was Orbiting 400 km above the Earth, and its speed was 28,000 km/h. However, MinION faced challenges in case of operation aboard the ISS. Challenges include disruption of flow cell membrane due to launch effects, air bubble formation in sample, difficulty in removal of air bubble, damage to nanopore or flow cell membrane due to air bubbles and blockage of nanopore. According to report of Sarah et al. (2017) performed on aboard the ISS, genomic DNA has been extracted from a bacterium (*Escherichia coli*), a virus (*Enterobacteria phage lambda*), and model mammalian organism, the mouse (*Mus musculus*), after nanopore DNA sequencing experiments. Simulated analyses of in-flight data using an automated metagenomic pipeline and de novo assembly algorithms demonstrate the feasibility of on-board, real-time analysis and genomic assembly for future sequencing applications in space (Sarah et al. 2017).

After discussion about importance of sequencing and molecular sequencing tools in space microbiology for diversity analysis, another essential aspect is sample collection and processing technology (Vaishampayan et al. 2013). There is lack of technically rigorous methods for sample collection and culturing of many microorganisms in space. Due to this unavailability of specific methods, it is still difficult to study human-associated microbial populations in the ISS environment. This is the main challenge to overcome in space microbiology (Checinska et al. 2019).

In viable microbial community analysis, use of the propidium monoazide (PMA) reagent enhances accuracy. Before DNA extraction protocol, PMA addition eliminates compromised membrane cells. Total microbial population, i.e., cells with disrupted membrane, dead cells, intact cells, viable cells, and free DNA information, is obtained in case of PMA-untreated samples. PMA functions as follows to differentiate between intact and compromised cells:

- It is of high molecular weight.
- Not able to penetrate in intact cell.
- Can penetrate cell with disrupted/ compromised membrane.
- Can bind to free DNA.

– Can bind to DNA inside cells of disrupted membrane.

Because of abovementioned properties, PMA binds to DNA of dead cells and makes them unavailable for amplification (Nocker et al. 2006, 2007; Lin et al. 2011; Vaishampayan et al. 2013; Jager et al. 2018)

The National Aeronautics and Space Administration (NASA) is working on the closed system of ISS for exploring microbial diversity and environment effect on microbiome. NASA engineers are also using Earth-based clean rooms with controlled temperature, humidity and circulation for periodic quality assurance. Audit of required and certified cleanliness levels of ISS and Earth-based rooms is regularly done by NASA quality assurance department. The environment of ISS is with zero gravity and high carbon dioxide levels and space radiations. On the other hand, Earth-based clean rooms are constantly replenished with fresh air. Major difference between environment of ISS and earth-based clean room is human traffic. In earth-based clean room, human traffic is at least 50+ people in a given working day; however, in the ISS, only six astronauts are allowed at a single time (Checinska et al. 2015).

These space research studies are based on possible commercial applications of microbes. By understanding the effects of space environment on biological responses, these can be applied on applied research. Applied aspects that can be elucidated by space flight research can be production of pharmaceutical products, commercially important secondary metabolite production, development of vaccine, detection of microbial virulence, and drug resistance in space. Humans are already exploring the surface of Mars and Moon which increases the possibility of microbial space research. However, issues related to human health are always maintained priority. Space craft, a closed system, creates environment for contamination because of the presence of consumables and waste products in closed cabin. Therefore, analysis of space microbiological experiments should account interaction between biological and physical phenomena associated with environment both within and external to space habitat.

References

Bagnioli P, Sabbatini M, Horneck G (2007) Astrobiology experiments in low Earth orbit: facilities, instrumentation, and results. In: Horneck G, Rettberg P (eds) Complete course in astrobiology. Wiley-VCH, New York, pp 273–320

Checinska A, Probst AJ, Vaishampayan P, White JR, Kumar D, Stepanov VG, Fox GE, Nilsson HR, Pierson DL, Perry J et al (2015) Microbiomes of the dust particles collected from the International Space Station and Spacecraft Assembly Facilities. Microbiome 3:50–68

Checinska Sielaff A, Urbaniak C, Malli Mohan GB, Stepanov VG, Tran Q, Wood JM, Minich J, McDonald D, Mayer T et al (2019) Characterization of the total and viable bacterial and fungal communities associated with the International Space Station surfaces. Microbiome 7:50

Edwards A et al (2017) Deep sequencing: intra-terrestrial metagenomics illustrates the potential of off-grid nanopore DNA sequencing. BioRxiv 133413

Hoenen T et al (2016) Nanopore sequencing as a rapidly deployable Ebola outbreak tool. Emerg Infect Dis 22:331

Horneck G, Bucker H, Reitz G, Requardt H, Dose K, Martens KD, Mennigmann HD, Weber P (1984) Microorganisms in the space environment. Science 225:226–228
Horneck G, Klaus DM, Mancinelli RL (2010) Space microbiology. Microbiol Mol Biol Rev 74 (1):121–156
Innocenti L and Mesland D A M (ed.). 1995. EURECA scientific results. Advances in space research 16(8) Elsevier, Bedford
Jager T, Alexander J, Kirchen S, Dotsch A, Wieland A, Hiller C, Schwartz T (2018) Live-dead discrimination analysis, qPCR assessment for opportunistic pathogens, and population analysis at ozone wastewater treatment plants. Environ Pollut 232:571–579
Johnson SS, Zaikova E, Goerlitz DS, Bai Y, Tighe SW (2017) Real-time DNA sequencing in the antarctic dry valleys using the Oxford nanopore sequencer. J Biomol Tech 28(1):2–7
La Duc MT, Vaishampayan P, Nilsson HR, Torok T, Venkateswaran K (2012) Pyrosequencing-derived bacterial, archaeal, and fungal diversity of spacecraft hardware destined for Mars. Appl Environ Microbiol 78(16):5912–5922
Lin WT, Luo JF, Guo Y (2011) Comparison and characterization of microbial communities in sulfide-rich wastewater with and without propidium monoazide treatment. Curr Microbiol 62 (2):374–381
NASA (2005) MR050L microbial analysis of ISS surfaces using the surface sampler kit (SSK) Edited by JSC28913 MRIDM. NASA, Houston
Nocker A, Cheung CY, Camper AK (2006) Comparison of propidium monoazide with ethidium monoazide for differentiation of live vs. dead bacteria by selective removal of DNA from dead cells. J Microbiol Methods 67(2):310–320
Nocker A, Sossa-Fernandez P, Burr MD, Camper AK (2007) Use of propidium monoazide for live/dead distinction in microbial ecology. Appl Environ Microbiol 73(16):5111–5117
Pace NR (2009) Mapping the tree of life: progress and prospects. Microbiol Mol Biol Rev 73 (4):565–576
Quick J et al (2016) Real-time, portable genome sequencing for Ebola surveillance. Nature 530:228–232
Rettberg P, Eschweiler U, Strauch K, Reitz G, Horneck G, Wanke H, Brack A, Barbier B (2002) Survival of microorganisms in space protected by meteorite material: results of the experiment EXOBIOLOGIE of the PERSEUS mission. Adv Space Res 30:1539–1545
Sarah LC-W, Chiu CY, John KK, Stahl SE, Rubins KH, McIntyre ABR et al (2017) Nanopore DNA sequencing and genome assembly on the International Space Station. Sci Rep 7:18022
Taylor G (1974) Space microbiology. Annu Rev Microbiol 28:121–137
Vaishampayan P, Probst AJ, La Duc MT, Bargoma E, Benardini JN, Andersen GL et al (2013) New perspectives on viable microbial communities in low-biomass cleanroom environments. ISME J 7(2):312–324
Yergeau E, Lawrence JR, Sanschagrin S, Waiser MJ, Korber DR, Greer CW (2012) Next-generation sequencing of microbial ocmmunities in the Athabasca River and its tributaries in relation to oil sands mining activities. Appl Environ Microbiol 78(21):7626–7637

The manufacturer's authorised representative in the EU is Springer
Nature Customer Service Centre GmbH, Europaplatz 3, 69115 Heidelberg,
Germany. If you have any concerns regarding our products, please
contact ProductSafety@springernature.com

Printed and bound by CPI Group (UK) Ltd, Croydon, CR0 4YY
30/09/2026
02225317-0014